Springer Proceedings in Mathematics & Statistics

Volume 43

For further volumes:
http://www.springer.com/series/10533

Springer Proceedings in Mathematics & Statistics

This book series features volumes composed of select contributions from workshops and conferences in all areas of current research in mathematics and statistics, including OR and optimization. In addition to an overall evaluation of the interest, scientific quality, and timeliness of each proposal at the hands of the publisher, individual contributions are all refereed to the high quality standards of leading journals in the field. Thus, this series provides the research community with well-edited, authoritative reports on developments in the most exciting areas of mathematical and statistical research today.

Jonathan M. Borwein • Igor Shparlinski
Wadim Zudilin

Editors

Number Theory
and Related Fields

In Memory of Alf van der Poorten

 Springer

Editors
Jonathan M. Borwein
Centre for Computer Assisted Research
    Mathematics and its Applications
School of Mathematical & Physical Sciences
University of Newcastle
Callaghan, NSW, Australia

Igor Shparlinski
Department of Computing
Macquarie University
Sydney, NSW, Australia

Wadim Zudilin
Centre for Computer Assisted Research
    Mathematics and its Applications
School of Mathematical & Physical Sciences
University of Newcastle
Callaghan, NSW, Australia

ISSN 2194-1009   ISSN 2194-1017 (electronic)
ISBN 978-1-4939-0217-0   ISBN 978-1-4614-6642-0 (eBook)
DOI 10.1007/978-1-4614-6642-0
Springer New York Heidelberg Dordrecht London

Alf van der Poorten

Preface

The International Number Theory Conference in Memory of Alf van der Poorten was held in Newcastle, NSW, Australia, 12–16 March 2012; see http://carma.newcastle.edu.au/alfcon/. This conference was followed by a less formal *International Workshop* at Macquarie University, Sydney, NSW, Australia, 19–20 March 2012.

The purpose of these meetings was to commemorate the research and influence of Alf van der Poorten on number theory and generally on mathematics. It also presented an exciting opportunity to:

- Promote number theoretic research and graduate study in Australia, and initiate collaborative research in number theory within Australia
- Attract several distinguished international visitors and strengthen the existing and establish new collaborative international links between Australian and international researchers
- Give a comprehensive account of recent achievements in theoretic and computational number theory and its applications to cryptography and theoretical computer science

The meeting in Newcastle attracted a large number of excellent international participants (about half of 75 participants) from essentially all parts of the world, and fully served its purpose. Many of the participants attended the later workshop at Macquarie University. Both were noteworthy for the quality of the science and of the companionship.

The volume presents a collection of carefully refereed papers in honour of Alf van der Poorten that for the most part originated with talks given at both meetings. In addition, it starts with a detailed academic appreciation of Alf's life and work written by his close friend and former colleague, David Hunt.

Both meetings were generously supported by the Australian Mathematical Society, Australian Mathematical Science Institute, the Centre for Computer Assisted Research Mathematics and its Applications (CARMA), University of Newcastle and Macquarie University.

Finally, we would like to thank all those who made the two meetings and this volume possible—funders, organizers, participants, speakers, authors, referees, and publishers.

Callaghan, NSW, Australia   Jonathan M. Borwein
North Ryde, NSW, Australia   Igor Sparlinski
Callaghan, NSW, Australia   Wadim Zudilin

Contents

Life and Mathematics of Alfred Jacobus van der Poorten (1942–2010)

David Hunt

Abstract Alf van der Poorten was born in the Netherlands in 1942 and educated in Sydney after his family's move to Australia in 1951. He was based in Sydney for the rest of his life but travelled overseas for professional reasons several times a year from 1975 onwards. Alf was famous for his research in number theory and for his extensive contributions to the mathematics profession both in Australia and overseas.

The scientific work of Alf van der Poorten ranges widely across mathematics, but nearly all of his research was in pure mathematics. Most of his work is contained in his published books and papers, and the bibliography at the end of this memoir contains 219 items. A majority of these publications are discussed in this 51-page biographical memoir.

1 Introduction

Alf van der Poorten was born in Amsterdam in 1942 and educated in Sydney after his family's move to Australia in 1951. He was based in Sydney for the rest of his life but travelled overseas for professional reasons several times a year from 1975 onwards. Alf was famous for his research in number theory and his contributions to the mathematics profession both in Australia and overseas. His honours include an honourary doctorate from Université Bordeaux, the Szekeres Medal of the Australian Mathematical Society and Membership of the Order of Australia (AM).

D. Hunt (✉)
School of Mathematics and Statistics, UNSW, Sydney NSW 2052, Australia
e-mail: d.hunt@unsw.edu.au

J.M. Borwein et al. (eds.), *Number Theory and Related Fields: In Memory of Alf van der Poorten*, Springer Proceedings in Mathematics & Statistics 43,
DOI 10.1007/978-1-4614-6642-0_1, © Springer Science+Business Media New York 2013

2 Early Life and Education

Alf's parents were David van der Poorten and Marianne van der Poorten–Stokvis. Marianne Stokvis was born in Amsterdam on 2 April 1914. Both her parents came from large Jewish families with connections in the diamond trade. In 1929 Marianne left school and worked as an international telephonist for Western Union. Her language skills and excellent memory for numbers were highly regarded. David van der Poorten was born on 5 December 1906 and was 6 years old when his father died. David, the youngest child, was sent to a Jewish orphanage in Amsterdam in August 1913. Here David received a good education and religious instruction. He was treated well by the governors of the orphanage and was encouraged to continue his schooling after the usual age of 16. David named his son Alfred Jacobus after two of the governors. David studied medicine at the University of Amsterdam, qualified as a doctor and set up private practice. In 1939 David and Marianne were married.

Alf was the second of the three van der Poorten children. He was born in Amsterdam on 16th May 1942 at a time when the Netherlands was under Nazi occupation. In September 1943, Marianne and David went into hiding in Amsterdam and found foster parents for their two small children, Alf and his older sister Malieke. Consequently Alf spent the next 2 years as "Fritsje", the youngest child of the Teerink family in Amersfoort. Alf's parents were captured but were among the few who returned from the concentration camps. The family was reunited in late 1945, David re-established his medical practice, and a daughter, Rose, was born in 1947.

Alf's parents decided to move to Australia, away from the dangers of Europe. David flew ahead to restart his university studies, and the rest of the family migrated to Sydney aboard the SS Himalaya and arrived in April 1951. David, as a foreign-trained doctor, had to return to university for 3 years, while Marianne took in boarders to assist with family expenses. Alf and his sister started at local primary schools, and the family made every effort to speak English and establish themselves in Sydney. The family settled in Bellevue Hill where David became a successful general practitioner and Alf enjoyed attending a challenging "opportunity" school at Woollahra.

As a "refo" or the more politically correct "new Australian", Alf was quick to learn English and within 2 years had won a place at the prestigious and selective Sydney Boys High School. This was a New South Wales version of the English grammar school. Alf modestly claimed to have "survived the ministrations of Sydney Boys High School". In fact he excelled, and was ranked among the top few students in New South Wales in the Leaving Certificate Examination. After a year in a youth leadership programme in Israel, Alf accepted a cadetship in mathematics at the University of New South Wales (UNSW) in 1961. UNSW is Sydney's "second" university in terms of date of foundation and is situated some 7 km south of the city.

Alf was a student and then an academic at UNSW for 18 years. Altogether he collected four degrees from the university. He first graduated in 1965 from a

Bachelor of Science degree with Honours in Pure Mathematics. He next completed his doctorate in 1968. Alf avoided the compulsory general studies subjects of the science degree by pursuing a concurrent major sequence in Philosophy, which he later converted to a Bachelor of Arts with Honours. Lest his education be totally impractical, he then proceeded to complete a Masters in Business Administration.

During the 1960s Alf was extremely active politically. He was President of the UNSW Students' Union Council from 1964 to 1965 and President of the UNSW University Union from 1965 to 1967. He served on the Council of the University from 1967 to 1973. He was also National President of the Jewish youth movement Betar, President of the Australian Union of Jewish Students and President of the NSW Zionist Youth Council. In 1966 he received the Australian Youth Citizenship Award "for his attainments in community service, academic achievement and youth leadership". Alf joined the Board of Directors of the University Co-operative Bookshop Ltd., a large chain of campus bookshops, in 1965 and remained a Director until 1982, serving as Chairman of the Board from 1979.

3   Academic Career at the University of New South Wales, 1969–1978

Alf was interested in all parts of mathematics, and as an undergraduate at UNSW, he was particularly influenced by John Blatt and George Szekeres. He began research with George Szekeres on multi-dimensional continued fraction algorithms. However, encouraged by George, Alf also made frequent trips to Canberra to talk to Kurt Mahler, who was then a Professor in the Department of Mathematics, Research School of Physical Sciences, at the Australian National University. At that time, Mahler had returned to his love of transcendental number theory and, particularly, the methods inspired by Hermite's original proof of the transcendence of e. Mahler became Alf's doctoral supervisor, and under his guidance, Alf began working on a form of multi-dimensional continued fraction algorithm for systems of functions, which Mahler had discovered many years earlier (this was Mahler's theory of "perfect systems of functions") but still had not published at the time. In his doctoral thesis, Alf showed that these ideas of Mahler could be used to give completely new proofs of a series of well-known theorems of the Hungarian mathematician Paul Turán. His results were published in[5, 6, 8–10]. Interestingly, similar work was done quite independently, and essentially at the same time, by Rob Tijdeman in his doctoral thesis in Holland, and they subsequently did some joint work on these questions.

It is no surprise that Alf's second paper was a note on "Transcendental entire functions mapping every algebraic number field into itself". As his career developed, Alf became particularly interested in diverse aspects of number theory and related subject areas. Alf obtained his first permanent academic job in 1969 when he was appointed Lecturer in Pure Mathematics at UNSW. Since George Szekeres and Max Kelly had no interest in administrative details, Alf had become de facto Head of Department by the time that I first met him in late 1970. Alf was promoted to Senior Lecturer in 1972 and to Associate Professor in 1976.

In 1972, John Loxton joined the department and over the next decade Alf and John wrote 21 joint research papers—all of them are serious contributions to number theory. A number of these papers were contributions to Baker's method of linear forms in the logarithms of algebraic numbers and some extended Mahler's method for determining transcendence and algebraic independence. Other papers related to the growth of recurrence sequences and algebraic functions satisfying functional equations, and there was a sequence of papers investigating arithmetic properties of certain functions in several variables. They also wrote about regular sequences and automata. The collaboration of Alf van der Poorten and John Loxton was clearly the most important in both their mathematical careers.

In 1974–1975, Alf spent a year overseas with six months in each of Leiden and Cambridge (UK). He took full advantage of this time outside Australia to rapidly build his network of mathematical collaborators. In 1978, a second study leave mainly spent in Kingston, Ontario, as well as in Bordeaux was pivotal in Alf's career development, and it was at this time that he was offered a chair in mathematics at Macquarie University which is situated some 15 km northwest of the city of Sydney.

4   Professor of Mathematics at Macquarie University, 1979–2002

At the time of his appointment to Macquarie University, Alf had already published over forty research papers out of a career total of almost 220 listed in the bibliography at the end of this memoir. Any thought that Alf would rest on his laurels was immediately dismissed. Once he made the move to Macquarie, both Alf's scholarship and his research output rapidly increased with a constant stream of overseas visitors, mainly courtesy of the Australian Research Council. MathSciNet lists at least sixty-five of Alf's co-authors, with the most frequent collaborators being John Loxton, Michel Mendès France (MMF) and Hugh Williams. The list also includes Enrico Bombieri, Richard Brent, John Coates, Paula Cohen, Bernhard Dwork, Kutso Inkeri, Sidney Morris, Bernhard Neumann, Jeffrey Shallit, Igor Shparlinski, Robert Tijdeman and Michel Waldschmidt.

Not only did Alf's academic career prosper at Macquarie, but he became heavily involved in the administration of the university. He served as Head of the School of Mathematics, Physics, Computing and Electronics from 1980 to 1987 and again from 1991 to 1996. As Head of School (equivalent to being a Dean), he had a long-lasting influence on the quality of all the appointments made to the school. As well as building up mathematics, he helped to create a world-class laser research group. In the periods he was not Head of School, he served on the Academic Senate of the University and in that capacity was an ex-officio member of the University Council as well as a member of the Senior Executive of the University. Alf retired from Macquarie University in 2002.

Alf always attempted to prove that "mathematics is beautiful, elegant and fun, a language as worthwhile learning as any other", as he remarked to a reporter in 1979. In part this was done by giving vivid and interesting colloquium talks on five continents. In addition, he put real effort into expository writing. One example of this was in 1978 when "A proof that Euler missed ... Apéry's proof of the irrationality of $\zeta(3)$",[46], appeared in the first volume of the Mathematical Intelligencer. Roger Apéry was a Professor of mathematics and mechanics at the University of Caen not known for his research in number theory. Apéry's seminar at Marseille–Luminy on this surprising result was very terse and was not believed by some in the audience. Henri Cohen sorted out most of the details and Alf published a full proof[46]. I heard him give a very successful seminar on this material at Rutgers University in late 1978. This talk was given at over 20 departments in the United States and Canada during the fall of that year. (It was claimed that his seminar was at the time the best ever given at the University of Waterloo.) As a second example, his fine book "Notes on Fermat's Last Theorem"[141] was published by Wiley-Interscience in 1996. Alf claimed that reading it "required little more than one year of university mathematics and an interest in formulas". This was not quite true but nevertheless the book was awarded an American Publishers' Award for Excellence in Mathematics.

The UNSW began as a technological university in 1950 before developing into a large comprehensive university by about 1980. As a young academic in the 1960s Alf cut his teeth teaching groups of 200 or more engineering students first- and second-year courses in Algebra and in Calculus. He was probably at his best when teaching honours students, many of whom went on to gain higher degrees in both Mathematics and other disciplines, as acknowledged below.

William McCallum, University Distinguished Professor in Mathematics, University of Arizona:

"As a first year student at UNSW in 1974 Alf's command and presence as a teacher convinced me I wanted to be a mathematician. I remember vividly those lectures in the autumn of 1974, watching the steady flow of elegant mathematical arguments flow from Alf's pen onto the overhead projector in his characteristic crisp calligraphy. It was Alf who showed me that mathematics could be beautiful and deep—and suave and sophisticated. Later in my undergraduate career, Alf passed on to me some of his other passions—transcendence theory, Catalan's equation and science fiction."

Frank Calegari, Associate Professor in Mathematics, Northwestern University:

"I first encountered Alf's unique style when I was a member of the Australian International Mathematical Olympiad (IMO) team and we came across his wonderfully entertaining paper on Apéry's proof of the irrationality of $\zeta(3)$. I assumed from his name that he was Dutch, and was then pleasantly surprised when I met Alf at the IMO team send-off reception and to find that he lived and worked in Sydney! He then and there told me what the p-adic numbers were and immediately offered me a job at any time in the future! I took him up on his offer at the end of my first year at Melbourne University, and spent six weeks in his office annex learning about elliptic curves, the Riemann–Roch theorem, the Weil conjectures, and, of course a lot of great stuff about recurrence relations (Skolem–Mahler–Lech) and continued fractions. I have always appreciated the time he spent talking to me—it was clear that he cared a great deal about young Australian mathematicians."

Neil Ormerod, Professor of Theology, The Australian Catholic University:

"Though I moved out of mathematics into another discipline I have very fond memories of Alf as both a lecturer, a research supervisor while John Loxton was on leave and as a bridge player of dubious skill. I recall his advice to me not to mix my studies with student politics—do what I say, not what I did! His lectures were always challenging, very well-organised, entertaining and supplemented by interesting extension notes. As Bill McCallum, whom I remember well, said Alf left an indelible impression on all students he taught."

William Hart, Research Fellow, University of Warwick:

"As a former doctoral student of his, I was perhaps too naive to appreciate all of the character that was Alf. But I do nevertheless remember him fondly for his highly entertaining talks and for promoting the idea that one must always 'sneak up' on a mathematical problem and to not scare it away with too much machinery."

After Alf moved to Macquarie in 1979, his administrative commitments increased as did his overseas travel for conferences and for longer visits. Consequently, most of his teaching was within the advanced undergraduate program where he taught special units which allowed him to wander far and wide mathematically.

Alf supervised nine successful higher-degree students listed below. In the case of Michael Hirschhorn, George Andrews was his principal supervisor.

- David P. A. Mooney, MSc, 1977, UNSW, "On the Diophantine equation $X^m + Y^m = Z^m$"
- Andrew M. Adams, MSc, 1978, UNSW, "Applications of the Gelfond–Baker method to elementary number theory"
- Michael D. Hirschhorn, PhD, 1979, UNSW, "Developments in the theory of partitions"
- Deryn Griffiths, PhD, 1992, Macquarie University, "Power series expansions of algebraic functions"
- Takao Komatsu, PhD, 1994, Macquarie University, "Results on fractional parts of linear functions of n and applications to Beatty sequences"
- Drew S. Vandeth, PhD, 2000, Macquarie University, "Mahler functions"
- Xuan C. Tran, PhD, 2000, Macquarie University, "Periodic continued fractions in function fields"
- Roger D. Patterson, PhD, 2004, Macquarie University, "Real quadratic fields with large class number"
- William B. Hart, PhD, 2005, Macquarie University, "Evaluation of the Dedekind Eta function"

Although not as peripatetic as Paul Erdös, Alf did generate United Airlines frequent flyer points on a very regular basis. Clearly he was welcome everywhere and these visits nearly always led to research publications and typically invitations from Alf for colleagues to visit Macquarie. It is neither possible nor sensible to try to list the 100 or so conferences that Alf attended over the years, but below is an attempt to list his longer visiting appointments and his host:

Rijksuniversiteit Leiden, 1974, Rob Tijdeman
University of Cambridge, 1975, John Coates
Queen's University, Kingston, Ontario, 1978, Paulo Ribenboim
Université Bordeaux I, 1981, MMF
Technische Hogeschool Delft, 1982, Mike Keane
Université Bordeaux I, 1982–1983, MMF
Institute for Advanced Study, Princeton, 1986, Enrico Bombieri
Rijksuniversiteit Leiden, 1986, Rob Tijdeman
Université Bordeaux I, 1987, MMF
Mathematics Sciences Research Institute, Berkeley, 1989, Irving Kaplansky
Rutgers University, 1989, Hendrik Iwanevic
Institute for Advanced Study, Princeton, 1989, Enrico Bombieri
Technische Universiteit, Delft, 1990, Mike Keane
Université Bordeaux I, 1991, MMF
Université Aix–Marseille, 1991, Gérard Rauzy
Basel, 1995, David Masser
Lille, 2001, Paula Cohen–Tretkoff
Simon Fraser University, Vancouver, 2003, Jon Borwein
Dalhousie University, Halifax, 2004, Jon Borwein
University of East Anglia, Norwich, 2004, Graham Everest
Università Roma III, 2004, Francesco Pappalardi
Brown University, Providence, 2005, Joseph Silverman
Dalhousie University, Halifax, 2007, Jon Borwein

5 Contributions to the Mathematics Profession

Alf made a number of significant contributions to the mathematics profession both within Australia and overseas. He was the founding Editor of the Australian Mathematical Society Gazette, served on the Council of the Australian Mathematical Society for many years and was President from 1996 to 1998. As President, he worked hard to reduce any remaining tension between pure and applied mathematicians in Australia. During his tenure as President the Council agreed to underwrite the fifth International Congress on Industrial and Applied Mathematics, ICIAM 2003, the first large mathematics congress to be held in Australia. This was an important and visionary undertaking for the Society.

In 1994–1995, Alf chaired a Working Party, with Noel Barton as Executive Officer, on behalf of the National Committee for Mathematics to report on "Mathematical Sciences Research and Advanced Mathematical Services in Australia"[148]. In 1998 he was a member of the Canadian National Science and Engineering Research Council (NSERC) site visit committee to evaluate Canada's three national Mathematical Sciences Research Institutes, and in 2003 he was a member of the Association of Universities in the Netherlands (VSNU) review committee for academic research in mathematics. In 1998, Alf joined the new Committee on Electronic Information Communication (CEIC) of the International Mathematical Union (IMU) and was reappointed in 2002 and 2006; he was also one of Australia's three delegates to the quadrennial Assembly of the IMU at Dresden (1998), at Beijing (2002) and at Santiago de Compostela (2006). Three of Alf's publications[191, 192, 195] analyze "Peer Refereeing". Publication[192] also includes contributions from Steven Krantz and Greg Kuperberg. A number of his observations are linked to changes in technology and reflect his work on the IMU committee. However, Alf emphasizes the duty of the Editor to maintain standards.

Alf's contribution to mathematics was recognised by the award of a Doctorate Honoris Causa by the Université Bordeaux I in 1998, for which he was nominated by MMF. In his acceptance speech Alf noted that he had visited the department so many times that he considered it to be his third university. In recognition of both his own mathematics and his services to mathematics in Australia, Alf, jointly with Ian Sloan, was awarded the inaugural George Szekeres Medal of the Australian Mathematical Society in 2002. He was also appointed a Member of the Order of Australia (AM) in the Australia Day Honours List 2004, for "service to mathematical research and education, particularly in the field of number theory".

Away from mathematics most of Alf's interests were rather sedentary. His writing of mathematics seemed to occur during the ad-breaks in televised football. He watched all the main codes of football as well as cricket, baseball and golf. Above all, he supported the St George Rugby League football team from the early 1950s onwards. When mathematical inspiration was hard to come by, he read science fiction and mysteries. Alf claimed never to have thrown a book away, and he thus owned some five thousand science fiction books and several thousand mysteries; but he was "not a collector, just a keeper". Since he also owned a couple of thousand mathematics books, almost no part of his family home was without bookshelves.

Alf is survived by his mother Marianne, his wife Joy, his children Kate and David, and four grandchildren Elizabeth, James, Gabrielle and Ellie.

6   Scientific Work

The scientific work of Alf van der Poorten ranges widely across mathematics, but nearly all his research was in pure mathematics. Most of his work is contained in his published books and papers, and the bibliography at the end of this memoir contains 219 items. In addition, he made many other contributions through his editorial work, his visits to overseas mathematics departments, his lectures, his reviews, his training of research students and his participation in and organisation of conferences and his hosting of visitors to Australia.

Alf's published research is primarily in number theory but touches on many areas of pure mathematics. Within algebra publications in field theory, linear algebra, associative rings, power series and topological groups have appeared. Within analysis there are many papers on functions of a complex variable and functional equations. In addition, Alf wrote extensively on automata and recreational mathematics. His published work includes a number of monographs, reports and conference proceedings. Finally, as well as a number of formal book reviews, he wrote 100s of reviews of papers both for Mathematical Reviews/MathSciNet and Zentralblatt für Mathematik. A few of Alf's papers were incorrect and in addition there were a number where a corrigendum was later published. I believe that all of the erroneous proofs were later corrected by other authors, sometimes years down the track. I have attempted to acknowledge each of these instances at the appropriate place in this memoir.

The items in the bibliography are listed in chronological order mainly using the date given by Mathematical Reviews/MathSciNet. No distinction is made between major research papers and short notes, of which there are quite a number. Formal book reviews are included but not, of course, individual Mathematical Reviews. There are a very small number of publications which are not either mathematical or related to the mathematics profession.

A large number of Alf's publications are expository. In some cases, such as "Notes on Fermat's Last Theorem"[141], this is clear from the title, but in other cases it is not. The reviews of these expository papers are often very complimentary and talk about Alf's clear but individualistic style of writing which often included humourous anecdotes.

The organisation of Alf's papers is based on the American Mathematical Society Mathematical Offprint Scheme. We begin with eight sections on Number Theory. As Rose [13] said:

"Number theory deals mainly with properties of integers and rational numbers; it is not an organised theory in the usual sense but a vast collection of individual topics and results, some with coherent sub-theories and a long list of unsolved problems."

Those who knew Alf well can appreciate why he worked primarily in number theory.

A   Continued Fractions

At least 20 of Alf's papers are about or heavily use continued fractions.

A.1   Expository

A 38-page paper[80] by Alf is part of the London Mathematical Society (LMS) Lecture Note Series No. 109 on Diophantine Analysis. In this Alf uses the 2×2 matrix representation of the simple continued fraction process to prove a number of the well-known results on continued fractions. It also includes a proof of "A necessary and sufficient condition that an odd prime p be a sum of two squares is that $p \equiv 1 \pmod 4$" using continued fractions. (This proof is originally due to

H. J. Smith.) The article continues with the relationship between the continued fraction expansion of $\sqrt{D}$ and the solutions of the Diophantine equation $x^2 - Dy^2 = t$. Explicit calculations of solutions are included.

A.2 Continued Fractions of Quadratic Integers

It is well-known that the continued fraction process can be described using 2×2 matrices of the form $\begin{pmatrix} a & 1 \\ 1 & 0 \end{pmatrix}$.

In a sequence of papers,[110, 132, 158–160, 167, 194, 197], published from 1991 onwards, together with various co-authors, Alf proves many results about the continued fraction expansion of numbers from quadratic fields, often using little more than the 2×2 matrix decomposition. Several of the papers simply give new proofs of results by various authors. Others contain new results as well as extending known results.

The most cited of these papers is[160] which is a joint paper with Hugh Williams. In this paper they construct an infinite family of positive integers D such that the period length of the continued fraction expansion of the fundamental unit of $Q(\sqrt{D})$ is independent of D. This result is new.

The most unusual paper is[194] which is a joint paper with Alf's doctoral student Roger Patterson. Here, with apologies to Louis Armstrong, one can learn about jeepers, creepers, sleepers, beepers, kreepers and leapers. This work arose from a letter from Irving Kaplansky to Richard Mollin, Hugh Williams and Kenneth Williams dated 23 November 1998. In this letter Kaplansky [6] reported on his investigation of the length of the period of the continued fraction expansion of the square root $\sqrt{D_n}$ of families of discriminants of the form

$$D_n = ax^{2n} + bx^n + c.$$

In his thesis Patterson produces more numerical evidence and various interesting examples including evidence for the conjecture that no jeepers exist.

Publication[197] gives new results on all quadratic integers for which the palindromic part of the continued fraction of the square root is given as $a_1, a_2, \ldots, a_{r-1}$. On the other hand,[167] gives a new proof of a result due to Edward Burger on when two irrational numbers belong to the same field.

A.3 Continued Fractions of Formal Power Series and Folded Continued Fractions

The familiar continued fraction algorithm, usually applied to real numbers, can also be applied to formal Laurent series in X^{-1}. The partial quotients are polynomials in X. Such continued fractions can be rewritten using the Mendès-France folding lemma.

Continued fractions of formal power series can be used to obtain continued fraction expansions of classes of numbers. The series used are often lacunary (i.e., relatively sparse) and the numbers obtained are often transcendental (think of Liouville and $\Sigma_n/10^{n!}$).

In a sequence of nine papers,[115, 119, 120, 124, 143, 144, 155, 163, 185], together with various joint authors, these ideas and applications are developed.

As an example, in[115], jointly authored with Jeffrey Shallit, Alf exhibits uncountable many numbers whose continued fraction expansions have only 1 and 2 as partial quotients. It is no surprise that all these numbers are transcendental.

As a second example, in[120], the same authors explain the very unexpected example

$$2^{-1} + 2^{-2} + 2^{-3} + 2^{-5} + \cdots + 2^{-F_k} + \cdots$$

$$= [0, 1, 10, 6, 1, 6, 2, 14, 4, 124, 2, 1, 2, 2039, 1, 9, 1, 1, 1, 262111, \ldots]$$

where F_k are the Fibonacci numbers $1, 2, 3, 5, \ldots$. This is achieved by studying the associated formal power series $\Sigma_k X^{-F_k}$. The explanation involves both the Fibonacci and the Lucas numbers.

As a third example, of a different type of result, in[185], Alf provides an unusual explanation for the well-known symmetry in the expansion of a quadratic irrational integer using the folding lemma.

A.4   Continued Fractions of p-Adic Numbers

Given Alf's interest in both continued fractions and p-adic analysis, it is not surprising that in[119] Alf writes on continued fractions of p-adic numbers. There are, perhaps unfortunately, several possible ways of expanding a p-adic number as a continued fraction. The most studied is due to Schneider. The analogy with the standard continued fraction soon breaks down. For example, periodicity implies α is quadratic, but there are quadratic p-adic numbers that have non-periodic Schneider continued fractions. Publication[119] is a survey of work of Becker, Tilborghs and de Weger.

A.5   Elliptic and Hyperelliptic Function Fields and Algebraic Numbers

As mentioned in Sect. A.3 a power series

$$F(X) = \sum_{h=-m}^{\infty} f_h X^{-h}, \quad f_h \in \mathbb{F}$$

where $\mathbb{F}$ is a field, has a continued fraction expansion.

Set $F_0 = F$ and let the partial quotient a_0 to be the polynomial part of F. The complete quotient $F_1 = 1/(F_0 - a_0)$. Iteratively define a_i to be the polynomial part of F_i and $F_{i+1} = 1/(F_i - a_i)$. Thus, $F = (a_0, a_1, a_2, \ldots)$. If D is a polynomial of even degree whose leading coefficient is a square, then $\sqrt{D}$ has such a power series expansion.

In[187] the authors (Alf and Xuan Tran) investigate quasi-elliptic integrals. These are integrals of the form

$$\int \frac{f(x)\,dx}{\sqrt{D(x)}} = \log\left(p(x) + q(x)\sqrt{D(x)}\right)$$

which one would expect to be inverse elliptic functions. They classify all the polynomials D of degree 4 over the rationals that have an f and an integral, as above. In these cases $\sqrt{D(x)}$ has a periodic continued fraction. More details of this are given in[168]. A related paper is[176] in which the rate of increase in height of the partial quotients and convergents of continued fraction expansions of square roots of generic polynomials is studied. Striking contrasting numerical examples are given.

In[193] Alf considers the elliptic surface $Y^2 = (X^2 + u)^2 + 4v(X + w)$, where $v = u + w^2$. He explicitly expands Y as a continued fraction of an element of $\mathbb{Q}(u, v, w)((X^{-1}))$. He relates the period of the continued fraction with the order of the torsion point. He shows that the first few partial quotients permit one to recover all families of elliptic surfaces with rational coefficients and rational torsion point at infinity.

Finally I should mention[134] in which, with Bombieri, they study computational problems connected with the calculation of the continued fractions of algebraic numbers. As so often occurs, the difficulty relates to locating one root of the polynomial.

B   Arithmetic Algebraic Geometry

B.1   *Elliptic Curves*

Elliptic curves have been studied for at least 200 years, in part because they are the curves of degree 3, and the curves of degree 2 had been classified in the seventeenth century, or perhaps 2,000 years earlier. Elliptic curves can be classified by their conductor, a positive integer, and the smallest possible conductor is 11.

In the late 1970s the Taniyama–Shimura–Weil conjecture that all elliptic curves defined over $\mathbb{Q}$ of a given conductor N are parametrised by modular functions for the subgroup $\Gamma_0(N)$ of the modular group was not settled. For $N = 11$ the question was settled by Alf, with three joint authors, in[43, 49]. In[49] it was stated that "the general question seems to be shrouded in mystery and quite inaccessible at present". In these papers it was shown that there are, up to isomorphism, exactly three elliptic curves of conductor 11. Since they were known to be isogenous, this work confirmed the Taniyama–Shimura–Weil conjecture for curves of conductor 11. The technique used in[49] is to reduce the question to solving a Diophantine equation of Thue–Mahler type. Recently discovered sharp inequalities for linear forms in the logarithms of algebraic numbers enabled the authors to bound solutions of the equations. The gap between H (a function of the parameters in an equation derived from the Diophantine equation) equal to 20 and 11^{15} was dealt with using the primitive multiprecision software available at that time. In 2012 the situation has changed in many ways. First and foremost Andrew Wiles has proved the full Taniyama–Shimura–Weil conjecture. Secondly, using modern software, curves of a given conductor have been classified in thousands of cases.

Publication[190] is a much cited mathematical survey on "Recurrence Sequences" by Graham Everest, Alf van der Poorten, Igor Shparlinski and Thomas Ward. In chapter 1 elliptic divisibility sequences and Somos-k sequences are introduced, and the whole of chapter 10 is devoted to elliptic divisibility sequences. In[200] Alf analyses the data yielded by the general expansion of the square root of the general monic quartic polynomial. He notes that each line of the expansion corresponds to addition of the divisor at infinity, and obtains "elliptic sequences" satisfying Somos relations at the same time deriving several new results on such sequences. In[205] Alf and Christine Swart show that the definition of Somos sequences via the recurrence is internally coherent without involving elliptic function theory. In addition, it is proved that every Somos-4 sequence is also a Somos-5,6,7,... sequence.

B.2   Curves of Higher Degree

Between 2005 and 2008 Alf published five papers on hyperelliptic curves[194, 201, 203, 207, 209].

In[194], jointly with Roger Patterson, Alf links continued fractions of quadratic integers to hyperelliptic curves with torsion divisor of relatively high order.

In[201] the ideas described in Sect. B.1 are extended to relate certain Somos-6 recursive sequences to arithmetic progressions on Jacobians of hyperelliptic genus 2 curves. This is achieved by analysing the continued fraction expansion of the square root of a general monic sextic polynomial.

In[203] Alf and Francesco Pappalardi considered a square-free polynomial $D(x)$ over a field K of characteristic 0. They discuss various conditions under which the ring $K[x, \sqrt{D(x)}]$ has non-constant units.

In[207] Alf further discusses the correspondence between Somos sequences, continued fractions of quadratic irrational functions and sequences of divisors on elliptic curves, particularly of genus 1 and 2. Finally, in[209] Alf continues the theme of the continued fraction expansion of the square root of monic polynomials of even degree. In the quartic and sextic cases, he observes explicitly that the parameters that appear in the continued fraction expansion generate sequences of Somos type.

C Irrationality and Transcendence

C.1 Irrationality

The study of rational approximations to algebraic numbers goes back to Liouville who showed that if α is an algebraic number of degree $n > 1$, then for all $\beta > 0$ and $\delta > 0$, the inequality $|\alpha - \frac{x}{y}| < \beta/y^{n+\delta}$ has only finitely many integer solutions x and y. This has the well-known corollary that $\Sigma_{n=1}^{\infty} 1/2^{n!}$ is transcendental, the first such result ever proved.

In an often cited paper[147] Alf and Enrico Bombieri derive a number of effective measures of irrationality for cubic extensions of the rationals. The paper uses results of Bombieri and Mueller and combines techniques of Thue and Siegel with Dyson's lemma on nonvanishing of two-variable auxiliary polynomials.

The genesis of[46] is discussed in the section on Alf's time at Macquarie University. The mathematics is also very interesting and is well summarised in I. John Zucker's Mathematics Review [MR0547748]:

"It has long been known that $\zeta(s) = \Sigma_1^{\infty} n^{-s}, s > 1$ may be written as $(-1)^{k-1} ((2\pi)^{2k}/2(2k)!)B_{2k}$ when $s = 2k$ is an even positive integer, where B_{2k} denotes the $2k$th Bernoulli number. Since these are rational; $\zeta(2k) = R\pi^{2k}$ with R rational, thus $\zeta(2k)$ is irrational. However, up to now nothing has been shown about the arithmetic nature of $\zeta(2k + 1)$. This entertaining report discusses a proof of the irrationality of $\zeta(3)$ due to R. Apéry. This is based on the fact that, given a sequence of rational numbers p_n/q_n such that $\beta \neq p_n/q_n$, if $|\beta - p_n/q_n| > q_n^{-1-\delta}$ and $\delta > 0$ for $n = 1, 2, \ldots$, then β is irrational. It is shown that for p_n, q_n satisfying the recurrence relation (1) $n^3 u_n + (n-1)^3 u_{n-2} = (34n^3 - 51n^2 + 27n - 5)u_{n-1}$, $n \geq 2$ with $p_0 = 0, p_1 = 6; q_0 = 1, q_1 = 5$ then $|\zeta(3) - p_n/q_n| > q_n^{-(\theta+\varepsilon)}$ with $\theta = 13.417820$, and hence $\zeta(3)$ is irrational. A similar result is obtained for $\zeta(2) = \pi^2/6$, which was already known to be irrational. Several intriguing formulas for $\zeta(3)$ and $\zeta(2)$ are given, for example, $\zeta(3) = \frac{5}{2}\Sigma_1^{\infty}(-1)^{n-1}/(n^3\binom{2n}{n})$. {One formula is incorrectly given: it should read $\zeta(3) = \frac{5}{4}\text{Li}_3(\omega^{-2}) + (\pi^2/6)\log\omega - \frac{5}{6}\log^3\omega$, where $\omega = (1 + \sqrt{5})/2$ and $\text{Li}_3(x) = \Sigma_1^{\infty} x^3/n^3$.} Although generalisations for proving $\zeta(2k + 1)$ irrational with $k > 1$ do not appear to be forthcoming, it seems possible to examine Dirichlet L-functions of small argument by this approach. Thus Catalan's constant $G = 1 - 1/3^2 + 1/5^2 - 1/7^2 + \cdots = L_{-4}(2)$ and $1 - 1/2^2 + 1/4^2 - 1/5^2 + 1/7^2 - 1/8^2 + \cdots = L_{-3}(2)$ appear in formulas similar to those for $\zeta(3)$. Thus $\Sigma_1^{\infty} 1/(n^3\binom{2n}{n}) = (\sqrt{3}/2)\pi L^{-3}(2) - \frac{4}{3}\zeta(3)$ and $\Sigma_1^{\infty} 2^n/(n^3\binom{2n}{n}) = (\pi^2/8)\log 2 - \frac{35}{16}\zeta(3) + \pi G$. A crucial step would seem to require a recurrence relation similar to (1)."

Publication[46] has been cited at least 76 times over the last 30 years, and serious attempts have been made to prove that both $\zeta(5)$ and $\zeta(7)$ are irrational. In 2000, Tanguy Rivoal, [15], proved that there are infinitely many irrational numbers among the numbers $\zeta(2n + 1)$ with $n \geq 1$, and in 2001, Wadim Zudilin, [23], proved that at least one of the four numbers $\zeta(5), \zeta(7), \zeta(9)$ and $\zeta(11)$ is irrational. Note also

Wadim Zudilin's survey paper, [24]. There are a few further results in this direction, but they are far from the expected result that all numbers $\zeta(2n+1)$ with $n \geq 1$ are transcendental and, even more, that they are algebraically independent over the field $\mathbb{Q}(\pi)$.

In[52] Alf gives an exposition of some remarkable identities involving values of polylogarithms $L_k(z) = \Sigma_1^\infty z^n n^{-k}$. In some cases new proofs are given. In particular, a simple derivation is given for $\zeta(2) = 3\Sigma_1^\infty n^{-2} \binom{2n}{n}^{-1}$ and its analogues for $\zeta(3)$ and $\zeta(4)$. These ideas are used to (slightly) recast Apéry's proof. Publication[53] covers much of the same material as[46, 52] but gives some extra motivation and extra details of some of the proofs.

C.2   General Theory of Transcendence

Alf published 14 papers[7, 17, 28, 36, 41, 42, 44, 45, 57, 59, 60, 87, 92, 94], which I have included under this heading. Seven of them are joint with John Loxton, and all were published before 1990. Since Alf's doctoral supervisor was Kurt Mahler, a number of the papers make heavy use of ideas from Mahler's papers.

In a very early paper[7], Alf considers integrals of rational functions along contours Γ in the complex plane. (The contour Γ must be closed, or have algebraic or infinite end points.) The integral

$$I = \int \frac{P(z)}{Q(z)} \, dz$$

is considered where $P(z)$ and $Q(z)$ are polynomials. I is shown to be algebraic if and only if a simple expression involving the zeros of $Q(z)$ and the residues of the function is zero. As a striking simple corollary, if $\deg(P(z)) < \deg(Q(z))$ and the zeros of $Q(z)$ are distinct, then $\int \frac{P(z)}{Q(z)} dz$ is either transcendental or zero. This included a then recent example of Alan Baker

$$\int_0^1 \frac{dx}{1+x^3} = \frac{1}{3}\left(\log 2 + \frac{\pi}{\sqrt{3}}\right) \quad \text{is transcendental.}$$

In 1929–1930 Kurt Mahler published a series of three papers, [8–10], on the transcendence of values at algebraic points of functions satisfying a functional equation. There is various evidence that Mahler's ideas had been largely forgotten until about 1970. However, by 1975, John Loxton and Alf had significantly generalised the method, and using modern techniques such as Baker's linear forms in logarithms results had constructed many new examples of transcendental numbers. Perhaps the best place for a reader to start is[36], a survey article describing the results on both transcendence and algebraic independence obtained by the authors using developments of Mahler's methods.

For the non-expert[34] gives a number of examples of functions in terms of factorials, others in terms of powers of 2 and others in terms of Fibonacci numbers which yield families of algebraically independent numbers. Infinite products which yield transcendental numbers are also discussed.

In[57] Alf and MMF consider numbers that come from paper folding. The mechanical procedure of paper folding generates an uncountable family of infinite sequences of "fold patterns". The authors calculate the associated Fourier series and show that the sequences are almost periodic and hence deterministic. Also they show that the paper-folding numbers defined by the sequences are all transcendental.

Finally in[92] John Loxton and Alf claim to show that the decimal expansions of algebraic irrational numbers cannot be generated by a finite automaton. The result appears as a corollary of a theorem on algebraic independence. This result can be thought of as being part way towards the natural conjectures that these expansions should be random, or even normal. As has been pointed out by several colleagues, not only is the proof of one of the main results of[92] fallacious but Alf was unable to correct his proof. Alf fully acknowledged this in a talk he gave in August 1997 at Penn State University. Fortunately Boris Adamczewski, Yann Begeaud and Florian Luca, [1, 2], correctly proved that an algebraic irrational is not automatic. As I understand it, John Loxton and Alf only used a version of one of Mahler's methods in transcendence theory, whereas in [1] the transcendence theorem is a consequence of Schlickewei's p-adic extension of Schmidt's subspace theorem.

C.3   Linear Forms in Logarithms

Between 1975 and 1987 Alf published ten papers,[25, 26, 33, 34, 38–40, 54, 66, 82], with various co-authors in which he improved the bounds involved in Baker's inequality for linear forms in logarithms of algebraic numbers. Between 1966 and 1968 Alan Baker published four papers in Mathematika in which he developed the theory. Many number theorists worked to improve the results including Baker. A large number of applications were found including the striking result of Rob Tijdeman that the only solution to Catalan's Diophantine equation $a^b - c^d = 1$ is $3^2 - 2^3 = 1$. Alf's paper[7] discussed above is another straightforward application. An excellent exposition of the theory is in Alan Baker's "Transcendental Number Theory" [3].

If $\alpha_1, \alpha_2, \ldots, \alpha_n$ are non-zero elements of a number field K of degree D over the rationals, then such numbers are said to be multiplicatively dependent if there are rational integers $b_1, b_2, \ldots, b_n$, not all zero, such that

$$\alpha_1^{b_1} \alpha_2^{b_2} \ldots \alpha_n^{b_n} = 1.$$

Provided care is taken with branches of the logarithm function one obtains D linear equations

$$b_1 \log(\sigma \alpha_1) + b_2 \log(\sigma \alpha_2) + \cdots + b_n \log(\sigma \alpha_n) = 0$$

where σ runs through the D embeddings of K into the complex numbers. So multiplicative dependence in number fields is linked to linear forms in logarithms of algebraic numbers.

In[33, 66] and the corrigendum[39], Alf and John Loxton investigate this situation and show, inter alia, that the $b_1, b_2, \ldots, b_n$ can be chosen to have absolute values relatively small in terms of n, D and the sizes of $\alpha_1, \alpha_2, \ldots, \alpha_n$. These results can be used when applying Baker's method.

In[40] Alf claimed to have proved a sequence of four theorems about "Linear forms in logarithms in the p-adic case". These were essentially p-adic analogues of inequalities proved by Baker in the characteristic zero situation. Unfortunately, these results were seriously flawed and were eventually corrected by Kunrui Yu in an impressive sequence of three papers [19–21]. Perhaps a good starting point is Yu's survey paper, [22].

Finally, I mention[182] in which Alf and Paula Cohen investigate the best results one can expect using the Thue–Siegel method as developed by Bombieri in his equivariant approach to effective irrationality measures of roots of high order of algebraic numbers, in the non-archimedean setting.

D Diophantine Approximation

D.1 Approximations to Algebraic Numbers

As mentioned in Sect. C.1 the question of approximating algebraic numbers by rationals was initiated by Liouville in 1844. He showed that:

For any algebraic number α with degree $n > 1$ there exists $c = c(\alpha) > 0$ such that $|\alpha - p/q| > c/q^n$ for all rationals p/q ($q > 0$).

In the first half of the twentieth century, considerable progress was made by Thue, Siegel, Dyson, Gel'fond and Schneider culminating in Roth's result in 1955

$$\left| \alpha - \frac{p}{q} \right| > \frac{c}{q^\kappa} \qquad \text{for any } \kappa > 2.$$

Some of these results were effective in the sense that c was determined. In Roth's theorem the c was simply proved to exist, that is, the result was ineffective. An excellent introduction to this topic is in Chap. 7 of Alan Baker's book, Transcendental Number Theory [3].

Alf made a number of contributions to this theory including[33, 39, 46, 66, 147, 182] discussed in Sects. C.1 and C.3. Other contributions are discussed in this section and the next.

In[91] and the corrigendum[98] Enrico Bombieri and Alf employ a Dyson-type lemma to obtain a new and sharp formulation of Roth's theorem on the approximation of algebraic numbers by algebraic numbers. The authors then obtain a refinement of a result of Davenport and Roth on the number of exceptions to Roth's inequality and a sharpening of the Cugiani–Mahler theorem. In[93] Alf makes some conjectures closely related to the results of[91].

Publication[106] is entitled "Dyson's Lemma Without Tears". "Without tears" means only zeros not infinities are counted which is claimed to make the argument more elementary. Dyson's lemma depends on constructing a polynomial $P \in \mathbb{C}[x, y]$ with many mixed partial derivatives zero at the k points with algebraic coordinates. The points where these conditions are satisfied lie within Newton triangles related to the parameters of the curve. The authors prove a "remarkable generalisation" of Dyson's inequality where the areas of the Newton triangles are essentially replaced by the areas of the actual Newton polygons of the curve $P = 0$ at the k points.

In[134] Alf and Enrico Bombieri discuss the computational problems connected with the calculation of terms of the continued fraction of algebraic numbers. They develop explicit formulas for finding the "next" partial quotient without having to calculate the complete quotient. In the case of the polynomial $X^3 - 2$ this yields explicit formulas for the continued fraction expansion of $\sqrt[3]{2}$, "contrary to received wisdom".

In[138] Alf together with Enrico Bombieri and David Hunt, building on[91, 106, 134] and other work of Bombieri was led to consider determinants in which the entries were products of binomial coefficients related to the shape of the Newton triangle. Calculation of a large number of examples using Richard Brent's multiprecision Fortran package led to many new conjectures about values of such determinants, all unknown at the time. To give the flavour a typical result:

$$\text{Let} \quad l!! = \prod_{k=0}^{l-1} k! \quad \text{then} \quad \Delta = \pm \left(\frac{(l!!)^3 (3l)!!}{((2l)!!)^3} \right)^{\binom{N_2+2}{3}},$$

a "powerful determinant" indeed. A later publication by Alf,[175] did shed some light on these interesting conjectures.

D.2   Continued Fractions of Formal Power Series

A recurring theme in many of Alf's papers is to calculate the continued fraction expansion of a formal power series with finite principal part in which the partial quotients are polynomials. These can then be specialised and used to calculate Diophantine approximations. At least [107, 113, 115, 120, 121, 144, 155, 163, 164, 168, 187, 193, 197, 200, 202, 203] fall into this category. A number of these papers are discussed elsewhere in this memoir, for example, in Sect. B, and some are discussed in the following paragraphs.

A typical example of such a paper is[107]. Let $E = \{0, 1, 4, 5, 16, 17, \ldots\}$ be the set of non-negative integers which are sums of distinct powers of 4. If $\alpha = 3 \sum_{h \in E} 10^{-h}$ then $\alpha^{-1} = 3 \sum_{h \in 2E} 10^{-h-1}$, an example due to Blanchard. The origin of this example is clearly in the factorisation into many factors of $10^{2^a} - 1$; but in this paper Alf and MMF consider the continued fraction expansion of the infinite product

$$f = \prod_{h=0}^{\infty} \left(1 + X^{-\lambda_n} \right) \quad \text{with} \quad \lambda_{n+1} > 2 \left(\lambda_0 + \cdots + \lambda_n \right)$$

which includes Blanchard's example with $\lambda_n = 4^n$ and $X = 10$.

In[113] the introduction begins with "Continued fractions of formal power series are not well understood and there is a dearth of instructive examples". This is true; however, there are dichotomies to be explained. With many, perhaps most, infinite products, the degrees (as polynomials) of the partial quotients are bounded, or even linear, while the coefficients grow rapidly or as Alf says "at a furious rate". In other cases "most" of the partial quotients are of low degree, but spikes of high degree occur.

In this paper Alf with Jean–Paul Allouche and MMF show that the partial quotients of $\Pi_{h=0}^{\infty}(1 + X^{-3^h})$ are all linear, whereas for $\Pi_{h=0}^{\infty}(1 + X^{-5^h})$ and for $\Pi_{h=0}^{\infty}(1 + X^{-7^h})$ two out of three of the partial quotients are linear. Publication[121] is in part a continuation of[113].

On the other hand, in[120], Alf and Jeffrey Shallit show that the continued fraction expansion of $\Sigma_{n=1}^{\infty} X^{F_n}$, where $\{F_n\}$ are the Fibonacci numbers, has partial quotients of unbounded degree but shows that they are degree ≤ 3 or $X^{F_n}, X^{L_{n+2}}$ or $X^{L_{n+2}} - X^{F_{n+1}}$, where $\{L_n\}$ are the Lucas numbers, $L_n = F_{n-1} + F_{n+1}$.

Publication[144] is closely linked to[115] and to a number of papers by Jeffrey Shallit, MMF and others. Let

$$g_\varepsilon(X) = \sum_{i \geq 0} \varepsilon_i X^{-2^i} \quad \text{and} \quad h_\varepsilon(X) = X g_\varepsilon(X)$$

where $\varepsilon_0 = 1$ and $\varepsilon_i = \pm 1$, $i \geq 1$. In an earlier paper by Jeffrey Shallit together with the results of[115], it is shown that the continued fraction for $g_\varepsilon(X)$ has partial quotients in the set $\{0, X-1, X+1, X, X+2, X-2\}$, whereas for $h_\varepsilon(X)$ they lie in the set $\{1, X, -X\}$. In[144] Alf with four co-authors proves the surprising result that the denominators of the convergents of both $g_\varepsilon(X)$ and $h_\varepsilon(X)$ are polynomials in $\mathbb{Z}[X]$ with coefficients $0, 1$ and -1. A number of further approximation results are also found. In[163] the results are generalised, simplified and refined.

Publication[155] is a well-written survey of the area. By 1999 the situation where the partial quotients in continued fractions are linear polynomials is termed "normality"; see[164]. In[164] reduction modulo p is considered.

E   Sequences

E.1   Recurrence Sequences

By far the most important publication by Alf in this area and possibly his most important publication is[190], the 2003 monograph "Recurrence Sequences", written by Alf together with Graham Everest, Igor Shparlinski and Thomas Ward, of 318 pages published by the American Mathematical Society. It is a remarkable tour de force demonstrating amazing scholarship. Perhaps the best way to do justice to the work is to reproduce here Yann Bugeaud's Mathematical Review [MR1990179]:

"Recurrence sequences appear almost everywhere in mathematics and computer science. Their study is, as well, plainly of intrinsic interest and has been a central part of number theory for many years. Surprisingly enough, there was no book in the literature entirely devoted to recurrence sequences: only some surveys, or chapters of books, dealt with the subject. With the book under review, the authors fill this gap in a remarkable way. Its content can be summarized as follows.

In the first eight chapters, general results concerning linear recurrence sequences are presented. The topics include various estimates for the number of solutions of equations, inequalities and congruences involving linear recurrence sequences. Also, there are estimates for exponential sums involving linear recurrence sequences as well as results on the behaviour of arithmetic functions on values of linear recurrence sequences. Apart from basic results from the theory of finite fields and from algebraic number theory, there are three important tools.

The first one is p-adic analysis, which is used in the proof of the Skolem–Mahler–Lech theorem. This theorem asserts that the set of zeros of a linear recurrence sequence $(u_n)_n$ (that is, the set of indices n for which $u_n = 0$) over a field of characteristic zero comprises a finite set together with a finite number of arithmetic progressions. Also, p-adic analysis is at the heart of the proof of the Hadamard quotient problem and may in some cases be applied to get very good estimates for the number of solutions of equations.

A second tool is Baker's theory of linear forms in the logarithms of algebraic numbers. It yields effective growth rate estimates (under some restrictions) and many arithmetical results: lower bounds for the greatest prime factor of the nth-term of a linear recurrence sequence $(u_n)_n$, finiteness of the number of perfect powers in $(u_n)_n$ (under some restrictions), etc.

A third tool is the Schmidt subspace theorem, and in particular its applications to sums of S-units. Specifically, linear recurrence sequences provide a special case of S-unit sums. As an example of a striking result obtained thanks to this theory, it has been proved by Hans Peter Schickewei and Wolfgang Schmidt, [17], that the number of zeros in any nondegenerate linear recurrence sequence of order n over a number field of degree d is at most equal to $(2n)^{35n^3} d^{6n^2}$.

In Chapters 9 to 14, a selection of applications are given, together with a study of some special sequences. Chapter 10 deals with elliptic divisibility sequences, an area with geometric and Diophantine methods coming to the fore. Other applications include graph theory, dynamical systems, pseudorandom number generators, computer science and coding theory.

Rather than giving complete proofs, the authors often prefer to point out the main arguments of the proofs, or to sketch them. But for every result not proved in their book, they give either a direct reference or a pointer to an easily available survey in which a proof can be found.

With its 1382 bibliographical references, this well-written book will be extremely useful for anyone interested in any of the many aspects of linear recurrence sequences."

In two closely related papers,[97, 102], Alf, together with Jean–Paul Bézivin and Attila Pethő, give a "full characterisation of divisibility sequences". Divisibility sequences are recurrence sequences (a_h), that is sequences satisfying linear homogeneous recurrence relations with constant coefficients, with the property that whenever $h|k$, then $a_h|a_k$. A well-known example is the sequence of Fibonacci numbers. Marshall Hall and Morgan Ward in the 1930s suggested that all divisibility sequences are, essentially, just termwise products of second-order recurrence sequences generalising the Fibonacci numbers. The proof of the characterisation uses the factorisation theory for exponential polynomials and a deep arithmetic result on the Hadamard quotient of rational functions, which Alf had earlier proved.

An interesting result due to Alf and Hans Peter Schickewei in[112] deals with the number of solutions to $a_h = 0$ where the sequence (a_h) satisfies a recurrence relation. The result is that this number $< C_2$ which is explicitly computable and depends on n, the order of the recurrence, the degree of the number field and the number of primes involved in the roots of the companion polynomial. The method is very classical and uses Strassman's theorem on the number of p-adic zeros of a power series. The interest is partly because the value of C_2 turns out to be much better than the bound obtained by the much deeper "subspace theorem" of Wolfgang Schmidt. (These comments come from the Mathematical Reviews by F. Beukers [MR1126359].)

Publication[136], written by Alf and Gerry Myerson, was published in the American Mathematical Monthly. It is written for a general mathematical audience and explains in a few pages the Skolem–Mahler–Lech theorem and its application to proving that if zeros in a recurrence sequence occur infinitely often, then it vanishes on an arithmetic progression with a common difference l that depends only on the roots of the companion polynomial. The paper introduces generalised power sums, exponential polynomials and p-adic numbers and proves the S-M-L theorem using ideas from complex analysis. It is indeed very readable.

E.2 Taylor Series and Rational Functions

In a sequence of three papers, two joint with Robert Rumely, Alf deals with results and conjectures related to recurrences, exponential polynomials and the Taylor coefficients of rational functions.

In[71] a rational function $r(x)/s(x) = \Sigma_{h \geq 0} f_h x^h$ defined over a field of characteristic zero with $\deg r < \deg s$ is considered. The results include growth conditions on $|f_h|$ and the fact that at most finitely many values appear more than once in (f_h). In addition, a result for discovering rational functions and a result about when a Hadamard quotient is a rational function are proved.

In[84] Alf establishes the Pisot kth root conjecture. Suppose $\Sigma_{h=0}^{\infty} b_h x^h$ is the Taylor expansion of a rational function $f(x)$. Suppose each b_h is a kth power of an element a_h from a finitely generated extension of $\mathbb{Q}$. The conjecture is that $\Sigma_{h=0}^{\infty} a_h x^h$ is also a rational function. The conjecture is established subject to two reasonable and probably essential conditions. In[85] similar results related to finitely generated subgroups of $F^\times$ are established.

F   Diophantine Equations

F.1   Quadratic Diophantine Equations

If D is a square-free positive integer, then there is a well-known connection between the continued fraction expansion of $\sqrt{D}$ and solutions to Diophantine equations of the form $x^2 - Dy^2 = t$, in particular for small values of t such as 1, -1 and -3. In papers[80, 129, 140, 166, 217] these issues are discussed.

As discussed earlier in this memoir,[80] is a presentation of the simple continued fraction algorithm using the 2×2 matrix decomposition. In this paper $X^2 - DY^2 = 1$, $X^2 - DY^2 = -1$, $X^2 - DY^2 = 2$ and $X^2 - DY^2 = -2$ are considered, and the solutions are shown to lie part way along the continued fraction expansion of $\sqrt{D}$.

Publication[129], a joint paper with Richard Mollin and Hugh Williams, is more substantial. The paper considers the Diophantine equation $X^2 - DY^2 = -3$. From the introduction:

"Our finding halfway to a solution of $X^2 - DY^2 = -3$ includes a necessary and sufficient condition for that equation to be solvable at all, and then explains how to find the solutions doing only half the expected amount of work. It will become clear that our ideas should enable one to make useful remarks about the equation with general k, including the somewhat mysterious k larger than $\sqrt{D}$.

We have alluded to the fact that if $X^2 - DY^2 = -1$ then halfway to a solution we find consecutive complete quotients $(\sqrt{D} + P_h)/Q_h$ and $(\sqrt{D} + P_{h+1})/Q_{h+1}$, with the fact that $Q_h = Q_{h+1}$ signalling halfway. Moreover, the well-known symmetry of the cycle says that the second half of the expansion is exactly the reverse of the first half. Similarly, halfway to a solution of $X^2 - DY^2 = 1$ is signalled by successive complete quotients with $P_h = P_{h+1}$.

We find analogous signals halfway to a solution of $X^2 - DY^2 = -3$ and prove that the expansion has *twisted* symmetry. Namely, its second half is essentially the reverse of its first half disguised by having been multiplied by 3."

The theory is developed in terms of ideals of the form $\langle Q, \sqrt{D} + P \rangle$. Equivalence of ideals, reduced ideals, ambiguity and composition of ideals are discussed. Examples treated include $D = 1729$, $D = 1891$ and $D = 5719$.

In[140] the relationship between solutions to $X^2 - DY^2 = d$ and solutions to $P^2 - dQ^2 = D$ is discussed. The work relied on[129]. Publication[166] makes some more progress on norm-form equations.

F.2   Higher Degree Equations Including Fermat's Equation

Although some of the papers dealt with in other sections cover Diophantine equations, we discuss here papers whose principal purpose is to discuss such equations. Most of these relate to Fermat's last theorem (FLT).

It has been said on many occasions, mainly by publisher's editors, that each equation in a book halves the number of sales. Alf's monograph,[141], "Notes on Fermat's Last Theorem", published by Wiley-Interscience as part of the Canadian Mathematical Society Series of Monographs and Advanced Texts sold many thousands of copies. Perhaps if he had removed all the equations, it would have sold more copies than any of Dan Brown's novels or even the bible. Notwithstanding the equations the book did win the Association of American Publishers' Professional/Scholarly Award for Excellence in Mathematics, 1996.

The publishers blurb includes:

"This book displays the unique talents of author Alf van der Poorten in mathematical exposition for mathematicians. Here, mathematics' most famous question and the ideas underlying its recent solution are presented in a way that appeals to the imagination and leads the reader through related areas of number theory. The first book to focus on FLT since Andrew Wiles presented his celebrated proof, *Notes on Fermat's Last Theorem* surveys 350 years of mathematical history in an amusing and intriguing collection of tidbits, anecdotes, footnotes, exercises, references, illustrations and more."

The publishers also note that:

"This book offers the first serious treatment of FLT since Wiles' proof. It is based on a series of lectures given by the author to celebrate Wiles' achievements, with each chapter explaining a separate area of number theory as it pertains to FLT. Together, they provide a concise history of the theorem as well as a brief discussion of Wiles' proof and implications. Requiring little more than one year of university mathematics and some interest in formulas, this overview provides many useful tips and cites numerous references for those who desire more mathematical detail".

It is unfortunate that the American Mathematical Society in its Mathematical Reviews of the book simply replicated the publisher's description and did not find someone who could have analysed the monograph and given the critical appraisal which it certainly deserved. On the other hand, Serge Lang was upset with just one sentence "This book offers the first serious treatment of FLT since Wiles' proof". In my opinion, in context this is unexceptional although on its own, it is over the top. In addition, Lang objected to the discussion in the book of a conjecture widely known as the "Taniyama–Shimura–Weil conjecture". Alf spends some effort discussing this naming including the analogy with Pell's equation. This was not enough for Lang who accuses the Canadian Mathematical Society of "Hype and false promotion" and Alf of "Lack of scholarship". See Lang [7]. Lang's conclusion is bizarre:

"I object to van der Poorten's misrepresentation of history and to his lack of scholarship in the matter (of the naming). I also object to van der Poorten's book having received the scientific endorsement of the Canadian Mathematical Society under the circumstances. I think the judgment of the CMS in sponsoring the publication of van der Poorten's book deserves scrutiny and questioning."

Alf[152] chose to use humour in his right of reply letter which was either unfortunate or perhaps all that Lang deserved.

In[128] in the Australian Mathematical Society Gazette, Alf publishes a light-hearted and entertaining article, in part extracted from[141]. The article includes additional historical notes and comments from Wiles and Frey on the recent history of FLT. Alf was always prepared to publicise mathematics, and one of the most unusual was[169] which is several thousand words in Italian about FLT and related matters in the "Enciclopedia Italiana". Alf also wrote a book review[178] on "The Fermat Diary" by C. J. Mozzochi which puts "faces and personalities" to the names associated with the proof—Gerhard Frey, Ken Ribet, Andrew Wiles and Richard Taylor. "This eyewitness account and wonderful collection of photographs capture the marvel and unfolding drama of this great mathematical and human story.".

Two papers are linked to Kustaa Inkeri. In[51] Alf and Kustaa Inkeri prove three interesting results on special cases of FLT giving bounds on solutions with effectively computable constants using results from Baker's method. In[174] Alf wrote an interesting expository paper on Inkeri's mathematics and other related problems in number theory. (Incidently those special cases were not resolved until Wiles solved FLT.)

Finally, in[162], Alf with Béla Brindza, Ákos Pintér and Michel Waldschmidt found explicit bounds on solutions to Thue's equation using strong results which complement those of Bombieri and Schmidt.

F.3 Other Topics Including S-Units, Catalan's Equation and Exponential Equations

Publication[34], written by Alf with Tarlok Shorey, Rob Tijdeman and Andrzej Schinzel, is titled "Applications of the Gel'fond–Baker Method to Diophantine Equations". It is widely cited. The paper begins with a valuable survey of the applications of approximation methods to Diophantine equations. Some new results are obtained, using the latest version of Baker's inequality for linear forms in the logarithms of algebraic numbers and its analogue in the p-adic domain. Thus, improvements are obtained on what is known concerning the greatest prime factor of a binary form and the greatest prime factor of the expression $ax^m + by^n$.

In[32] Alf studies a p-adic version of Catalan's equation:

$$x^u - y^v = (p_1^{\omega_1} \ldots p_s^{\omega_s})^{\mathrm{lcm}(u,v)}.$$

He shows that if $p_1, \ldots, p_s$ are distinct primes, then there is an effectively computable constant $C = C(p_1, \ldots, p_s)$, such that all solutions $(x, y, u, v, \omega_1, \ldots, \omega_s) \in \mathbb{Z}^{s+4}$ (with $x, y, u, v \geq 2$, $u + v \geq 5$ and $(x, y) = 1$) of the equation are bounded by C. An analogous result for algebraic number fields is also obtained.

In[109] Alf and Hans Peter Schickewei study "additive relations in fields" which are generalised Diophantine equations. The study goes back at least to a 1933 paper of Mahler. The context is that K is an algebraic number field, S is a finite set of prime ideals in K, Γ the group of S-units in K, and $a_1, \ldots, a_n \in K^\times$, $n \geq 2$. It is proved that the S-unit equation $\sum_{i=1}^{n} a_i x_i = 1$ has only finitely many solutions $x_1, \ldots, x_n$ in Γ such that $\sum_{i \in I} a_i x_i \neq 0$ for each non-empty proper subset I of $\{1, 2, \ldots, n\}$.

The results are extended to the situation where K is an arbitrary field of characteristic zero and Γ is an arbitrary finitely generated multiplicative subgroup of $K^\times$. This highly cited paper has led to a number of developments in the area.

In[117] Alf and Igor Shparlinski study the number of zeros of exponential polynomials. They apply Strassman's theorem to p-adic power series satisfying linear differential equations with polynomial coefficients. The approach leads to an estimation of the number of integer zeros of polynomials on a given interval and thence to an investigation of the number of p-adic small values of a function on such an interval.

G   Algebraic Number Theory

G.1   Polynomials

We discuss three papers which are primarily about polynomials.

In[114] and the corrigendum[126] Alf and Bernard Dwork study power series which satisfy a polynomial equation in two variables. Using recent powerful methods the authors improve "The Eisenstein constant", so called because the work goes back to a theorem of Eisenstein. To be more precise, suppose $y = \sum a_j x^j$ is a power series with coefficients algebraic over $\mathbb{Q}$. Suppose x, y is a solution of $f(x,y) = 0$, $f(x,y) \in \mathbb{Z}[x,y]$ with coefficients bounded by H and of degree n in y and degree m in x. Eisenstein's theorem is that there exists integers $A, B \geq 1$, such that $AB^j a_j \in \mathbb{Z}$ for all j. Much research by various authors has gone into improving estimates for B. In this paper the method of proof is based upon the p-adic theory of differential equations. The main application is an estimate involving Runge's theorem asserting that for a polynomial $f(x,y) \in \mathbb{Z}[x,y]$, irreducible in $\mathbb{Z}[x,y]$, but reducible in $\mathbb{Q}[[x]][y]$ or $\mathbb{Q}[[y]][x]$, the equation $f(x,y) = 0$ has only a finite number of solutions in $\mathbb{Z}^2$. The results of this article have application in estimating that number of solutions. Publication[114] is widely cited.

In[131] Alf builds on work of Gourin, Ritt and Schnizel to study irreducible polynomials in n variables $x_1, \ldots, x_n$ over algebraically closed fields of characteristic zero which become reducible when some variables are replaced by powers of themselves. The main result is a specific upper bound for the number of factors of such a polynomial, except in some trivial cases. The bound is independent of how large these replacing powers may be.

In a very straightforward paper in the American Mathematical Monthly,[180], Alf and Gerhard Woeginger investigate expressions such as $1.2.3.4 + 1 = 5^2$, $2.3.4.5 + 1 = 11^2$ and $3.4.5.6 + 1 = 19^2$. They consider

$$P_{n,c}(x) = x(x+1)(x+2)\dots(x+n-1) + c$$

and determine all values of n and c such that $P_{n,c}(x)$ is the square of a polynomial. The only values are $n = 2$, $c = 1/4$ and $n = 4$, $c = 1$. Other results are obtained.

G.2   Quadratic Extensions

There are a number of Alf's other publications related to quadratic extensions of the integers, for example, in Sect. A.2.

Publication[132], by Alf and Richard Mollin, is entitled "A Note on Symmetry and Ambiguity". They present the relationship between quadratic irrationals whose continued fraction expansion has symmetric period and ambiguous ideal cycles in real quadratic fields. The Mathematical Reviews by Andrew Rockett [MR1322789] is unusually clear.

"Let δ be a real quadratic integer with algebraic conjugate $\overline{\delta}$, trace $t = \delta + \overline{\delta}$ and norm $n = \delta\overline{\delta}$. Given a real quadratic irrational $\gamma \in \mathbb{Q}(\delta)$, without loss of generality we may suppose that there are rational integers P, Q such that $\gamma = (\delta + P)/Q$ with $Q | \mathrm{norm}(\delta + P)$. Then the $\mathbb{Z}$-module $\langle Q, \delta + P \rangle$ is an ideal in the order $\mathbb{Z}[\delta]$ of the quadratic field $\mathbb{Q}(\delta)$. Using uniform notations and proceeding from first principles, the authors describe the correspondence between numbers $(\delta + P)/Q$ whose regular continued fractions have symmetric periodic blocks, ideals $\langle Q, \delta + P \rangle$ which are ambiguous since they are equal to their conjugates, and quadratic forms

$$Qx^2 - (t + 2P)xy + ((n + tP + P^2)/Q)y^2$$

which are ambiguous because they are improperly equivalent to themselves, as well as the connections via Pell's equation between numbers of norm -1 and decompositions as sums of squares. Given the long history of these ideas and the different terminologies in the literature, this article provides a welcome overview and includes many useful incidental remarks."

Another essentially expository paper[127] is entitled "Explicit formulas for units in certain quadratic number fields". From the abstract, there is a class of quadratic number fields for which it is possible to find an explicit continued fraction expansion of a generator and hence find an explicit formula for the fundamental unit. One therefore displays a family of quadratic fields with relatively large regulator.' In this derivation Alf uses rather complicated products of 2×2 matrices to model the continued fraction expansions, nevertheless yielding an easier and more transparent construction and verification of the fundamental units than in earlier works on the same material.

Another expository paper is[167] by Alf and Edward Burger in which they give a condition for two real numbers with purely periodic continued fraction expansions to belong to the same quadratic field.

Unusually for Alf[173] and the Corrigenda[188] is a computer verification of a conjecture for a range of primes. The paper is joint with Herman te Riele and Hugh Williams and deals with the Ankeny–Artin–Chowla conjecture (AACC). Suppose $p \equiv 1 \pmod 4$ and let t, u be rational integers such that $(t + u\sqrt{p})/2$ is the fundamental unit of the real quadratic field $\mathbb{Q}(\sqrt{p})$. The AACC asserts that p will not divide u. This is equivalent to p will not divide the $(p-1)/2$th Bernoulli number. A third formulation is that if (t, u) is the smallest solution of $t^2 - pu^2 = \pm 4$, then $p \nmid u$. The conjecture was known to hold for primes $p \leq 10^9$ and in this paper the calculations are pushed through to 2×10^{11} (see Corrigenda). The new method is based on some clever new ideas. If $\varepsilon = (t + u\sqrt{p})/2$ and $\varepsilon^k = (X_k + Y_k\sqrt{p})/2$ with $X_k, Y_k \in \mathbb{Z}$, one easily proves that $2^{k-1} Y_k \equiv kt^{k-1} u \pmod p$. Therefore, if k is not divisible by p, $p|u$ if and only if $p|Y_k$. The main idea of the new algorithm is to find a k for which ε^k is easy to compute. The natural candidate for k is the class number h of $K = \mathbb{Q}(\sqrt{p})$, using the formula $2h\log\varepsilon = \sqrt{p}\, L(1, \chi_p)$ where $L(1, \chi_p)$ is the Dirichlet L-function of K. Shank's infrastructure method is used to compute an accurate value of $h\log\varepsilon$. Altogether the work is quite a tour de force.

In[211] Alf, Roger Patterson and Hugh Williams characterise a generalised Shanks sequence. They consider generalisations of Shanks' sequence of quadratic fields $\mathbb{Q}(\sqrt{S_n})$ where $S_n = (2^n + 1)^2 + 2^{n+2}$. Quadratic fields of this type are of interest because it is possible to explicitly determine the fundamental unit. If a sequence of quadratic fields given by $D_n = A^2 x^{2n} + Bx^n + C^2$ satisfies certain conditions (notably that the regulator is of $O(n^2)$), then the exact form such a sequence must take is determined.

H Other Contributions to Number Theory

H.1 Elementary Number Theory

A number of Alf's publication should be listed under this heading including at least[4, 50, 75, 180, 198, 199, 206]. Some have been discussed elsewhere. A paper such as[206] appears to be surprising except for the insight it gives into the thinking of John Horton Conway.

A very early paper[4] of Alf and Charles Cox looks at the sequence of primes $\{p_i\}$ defined by $p_1 = 2$ and p_{n+1} is the highest prime factor of $1 + p_1 p_2 \cdots p_n$. They show that most primes between 5 and 53 do not occur in the sequence $\{p_i\}$ and ask (1) whether the set of primes not in $\{p_i\}$ is infinite and (2) whether the set $\{p_i\}$ is monotonic increasing. Perhaps no one has followed up this paper in the last 44 years.

The study of pseudoprimes, strong pseudoprimes, Carmichael numbers and the like has a long history. A composite number N is said to be a strong pseudoprime for the base C if with $n - 1 = 2^s d$, $(2, d) = 1$ either $C^d \equiv 1$ or $C^{2^r} \equiv 1 \pmod{N}$, some r, $0 \leq r < s$. In[50] Alf and Andrzej Rotkiewicz showed every arithmetic progression $ax + b$, $x = 0, 1, 2, \ldots$, $(a, b) = 1$ contains an infinite number of odd (composite) strong pseudoprimes for each base $C \geq 2$. This is probably a rather counter-intuitive result.

Publication[75] is a filler in Volume 7 of The Mathematical Intelligencer entitled "The longest prime". The number 1979339339 is prime as are 19, 197, 1979, ..., and this is the longest such prime in base 10 starting with a 1. Alf suggests that the longest prime in base b is not clearly of bounded length and that proving this may be a hard problem.

H.2   Radix Representation

In[76] Alf, Kurt Mahler and Derrick Lehmer study integers with all digits 0 or 1. To be precise let $g \geq 2$ be an integer, and L be the language of all non-negative integers which in base g employ only the digits 0 or 1. For $a = 0, 1, \ldots, k - 1$, let $L(a)$ be the subset of those $h \in L$ satisfying the congruence $h \equiv a \pmod{k}$. Then $L(a)$ is either infinite or empty. Various numerical examples and a simple criterion to effectively distinguish the two cases are given.

In[86] Alf and John Loxton show that every odd integer is a quotient of elements in the set L of integers whose expansion to the base 4 does not contain a 2. The proof uses the subset S of integers whose expansion only contains 0 or 1 and a pigeon-hole principle argument.

H.3   Quadratic Forms and the Composition of Binary
Quadratic Forms

The study of binary quadratic forms goes back at least to Gauss, who introduced both equivalence of forms and composition of forms of a given discriminant. With various authors Alf has made significant contributions both to the theory and to exposition of the ideas.

In[177] Alf begins with the "cute" observation due to Hendrik Lenstra, that $12^2 + 33^2 = 1233$ which satisfies the Diophantine equation $a^2 + b^2 = 10^n a + b$, which is a particular case of the more general Diophantine equation (*) $x^2 + y^2 = n$. Several methods exist to solve (*) such as those due to Hermite–Serret and Cornacchia and require prime factorisation of n. In this paper Alf relates (*) to the problem of representing a positive integer by a positive definite binary quadratic form using the method of reducing a symmetric matrix. The paper is intended to provide a nice and easily comprehended story to tell students, and it certainly succeeds.

Publication[132] could have been discussed here, but it is described in detail in Sect. G.2.

Three of Alf's papers,[161, 161, 184], deal with Dedekind's eta function. The first two are joint with Kenneth Williams and third is joint with Robin Chapman, although it can be thought of as a continuation of[161]. Let $\eta(z)$ be the Dedekind modular form of weight 1/2, given by the infinite product

$$\eta(z) = e^{\frac{\pi i z}{12}} \prod_{m=1}^{\infty} (1 - e^{2\pi i m z}), \quad \text{Im } z > 0.$$

The important paper[161] explicitly determines $|\eta(z)|$ for certain quadratic irrationalities z in the upper half plane. Let F be an imaginary quadratic field. Let d denote the discriminant of F. Thus, d is a negative integer, with $d \equiv 0$ or $1 \pmod 4$, $F = \mathbb{Q}(\sqrt{d})$, and the largest positive integer m such that $m^2 | d$ and $d/m^2 \equiv 0$ or $1 \pmod 4$ is $m = 1$. Let a, b and c be integers satisfying $d = b^2 - 4ac$, $a > 0$, $\gcd(a,b,c) = 1$. Such integers exist since we may take $a = 1, b = 0, c = -d/4$ if $d \equiv 0 \pmod 4$ and $a = 1, b = 1, c = (1-d)/4$ if $d \equiv 1 \pmod 4$. The main result of this 50-page paper is the explicit determination of $|\eta((b+\sqrt{d})/2a|$, which is too complicated to explain here.

The proof in part depends on analytical number theory estimates of functions related to $P_k(n)$, the number of proper representations of a positive integer n by the class k of the (form) class group $H(d)$ of discriminant d.

In order to illustrate the fact that the formulas can be used, numerical details are provided for the cases $d = -15$ and $d = -31$. For example,

$$\left| \eta\left(\frac{1+\sqrt{-15}}{2}\right) \right| \quad \text{and} \quad \left| \eta\left(\frac{1+\sqrt{-15}}{4}\right) \right|$$

are given in terms of powers of π, values of the Γ-function and $\frac{1+\sqrt{5}}{2}$.

It is further shown that the Chowla–Selberg formula follows from the main theorem (of this paper) and finally that the Chowla–Selberg formula for genera is also a consequence of the main theorem.

The corrigendum[161] is unusually long at 15 pages, but the correction is needed since a wrong version of an estimate of a function closely related to $\pi(x)$ was used in the original paper. The corrigendum looks as though it could have come straight out of Hardy and Wright.

In[184] Alf and Robin Chapman give a succinct and reorganised presentation of the argument in[161] leading to the Chowla–Selberg formula. The key to the substantial improvements is the recognition of the quantities $f(K,L)$ critical to the argument in[161] as characters of generalised ideal class groups. (The reviewer for Mathematical Reviews/MathSciNet was very impressed by this work.)

Finally in this section there are two papers of Alf's written about NUCOMP,[186, 189]. NUCOMP is an algorithm introduced by Shanks [16] in 1989 for computing the reduced composite of two positive definite binary quadratic forms of discriminant Δ. Publication[189] contains a detailed explanation of the Shanks–Atkin NUCOMP with composition and reduction carried out "simultaneously" for imaginary and real quadratic fields. Extensive testing in both the numerical and function field cases (by Michael Jacobson, reported elsewhere) confirms that NUCOMP as described here is in fact efficient for composition both of indefinite and of definite forms once the parameters are large enough to compensate for NUCOMP's extra overhead. An example with $D = 10209$ is detailed which takes 27 steps to the mid-point of the period of the continued fraction of $\omega = (1 + \sqrt{10209})/2$.

The second paper[186], joint with Michael Jacobson, reports on computational aspects of both NUCOMP and NUDUPL.

H.4   Probabilistic Theory

Publication[101], published jointly with Hans Peter Schickewei, is entitled "A Diophantine problem in harmonic analysis". In 1989 at Luminy, Jean Bourgain conjectured that:

Suppose α is an algebraic number of degree d. Assume that, for each $n \in \mathbb{Z}^+$, $\deg \alpha^n = d$. Then there exists a constant $c = c(\alpha)$ such that: given any subspace W of dimension $\leq d - 1$ of the d-dimensional $\mathbb{Q}$-vector space $\mathbb{Q}(\alpha)$, the set $\{n \in \mathbb{Z}^+, \alpha^n \in W\}$ contains fewer than $c(\alpha)$ elements.

Bourgain needed this result in his proof of the Riesz–Raikov theorem for algebraic numbers. The main theorem proved in this paper depends on recent results on the number of solutions of S-unit equations and thus relies on quantitative versions of the Schmidt subspace theorem in Diophantine approximation.

From the introduction to[96] by Alf and Ross Moore:

"This paper was to be a description of the work of Mendès France and various of his collaborators on how one may felicitously and instructively attach thermodynamic quantities to plane curves. However, the focus of the paper changed as we became interested in questions arising from the preparation of the pictures that were to illustrate the ideas to be presented. We have described and reformulated the work of Berry and Goldberg on renormalisation of certain curves containing fantastic curls and twists.

The curves are 'exponential curves'. That is, given a sequence of real numbers (θ_n), the curve is a broken line whose Nth vertex in the complex plane is $S_N = \sum_{k=0}^{N-1} \exp(2i\pi\theta_k)$. In the paper[96] the authors explore in detail the case $\theta_k = \tau k^2$, $\tau \in \mathbb{R}$. The novelty in the paper resides in the fine quality of the graphs showing very explicitly the renormalisation principle. Higher order Gaussian sums of other sequences such as $\theta_n = (\log n)^d$ where $1 \leq d \leq 6$ and $\theta_n = n^{3/2}$ are also treated."

I Algebra

I.1 Field Theory and Polynomials

A very large number of Alf's publications could have been included here but for various reasons are discussed elsewhere. In addition to the research papers discussed below, we note[48], a translation from Dutch of Hendrik Lenstra's three papers on Euclidean number fields published in Volume 2 of the Mathematics Intelligencer.

In an early application of Schmidt's subspace theorem,[72], Alf investigates additive relations in number fields. For the main theorem let K be a field of characteristic zero and H a finitely generated subgroup of $K^\times$. Then for each $m = 1, 2, \ldots$, there are at most finitely many nontrivially distinct relations $U_1 + U_2 + \cdots + U_m = 1$ with the U_i in H. A more general result about algebraic numbers is proved with the main theorem as a corollary. Both[65, 109] can perhaps be thought of as continuations of[72]. Publication[72] is a very important contribution to field theory since it was perhaps the first to investigate a generalisation of Siegel's unit equation. Nowadays this consequence of Schmidt's subspace theorem plays a central role in Diophantine geometry.

In[165], Alf, Marek Karpinski and Igor Shparlinski obtain new algorithms for testing whether a multivariate polynomial over a p-adic field given by a black box is identical to zero. Zero testing of polynomials in residue rings is also discussed. The results complement known results on zero testing of polynomials over the integers, the rationals and finite fields.

In[218], Alf and Ákos Pintér give a lower bound on the number of simple and distinct zeros of elements of a function field over an algebraically closed field, defined by linear recurrence sequences. An example shows that the result is rather sharp in the case of binary sequences.

I.2 p-Adic Fields

With MMF in[77], Alf studies automata and p-adic numbers in order to settle questions formulated in analogy with the Gel'fond–Schneider theorem. The following is V. Drobot's Mathematical Review [MR0844590].

"For a prime p let $\mathbb{F}_p$ be the field with p elements. As usual, let $\mathbb{F}_p(X)$ denote the field of rational functions over $\mathbb{F}$; $\mathbb{F}_p((X))$ is the field of all formal Laurent series, $\mathbb{F}_p[[X]]$ denotes the subring of all formal Taylor series and $\mathbb{Z}_p$ is the ring of p-adic integers. Notions of transcendent, algebraic and irrational elements in this context are defined in the usual way. If $\lambda = \sum_0^\infty \lambda_n p^n$, $0 \le \lambda_n < p$, is in $\mathbb{Z}_p$ and f is in $\mathbb{F}_p[[X]]$, the authors define $f^\lambda = \prod_0^\infty (f(X^{p^n}))^{\lambda_n}$. Two conjectures are then formulated, in analogy with the Gel'fond–Schneider theorem. Conjecture 1: If f is an algebraic element of $\mathbb{F}_p[[X]]$ with $f_0 = 1$ and λ is an algebraic irrational element of $\mathbb{Z}_p$, then f^λ is a transcendental element of $\mathbb{F}_p[[X]]$. Conjecture 2 has to do with a special case of the above with $p = 2$ and $f = 1 + X$ and deals with complexity of the digits of λ. It is formulated in terms of a finite two-automata. At the end of the paper the authors announce that they have proved Conjecture 1, requiring, in fact, only that λ be irrational."

Another of Alf's survey papers is[122] which is entitled "Power series representing algebraic functions". In the paper Alf surveys recent work on algebraic power series, citing numerous results, estimates and connections, together with some challenging questions. The exposition of the relationships among finite automata, algebraic series and Hadamard products is particularly well done. A number of interesting examples illustrate problems in lifting and reducing algebraic series between characteristics zero and p.

I.3   Linear Algebra and Determinants

Alf wrote just five papers on linear algebra.

In[2] he pointed out that if one takes the usual ten vector space axioms and leaves out just two of the additive group axioms—the existence of a zero and of negatives—then there are structures which satisfy the eight axioms but are not vector spaces. Although this is no surprise at all, the note was motivated by an error in a Dutch textbook on Analytic Geometry and Linear Algebra.

Rather more interesting is[116] by Alf and Ross Talent in which they consider complex inner product spaces. To be precise they take V a vector space over a subfield K of the complex numbers. Length on V is defined as a map to the non-negative reals satisfying $\|\alpha\mathbf{v}\| = |\alpha|.\|\mathbf{v}\|$, $\alpha \in K$. It is proved that length, together with the standard parallelogram law $\|\mathbf{u} + \mathbf{v}\|^2 + \|\mathbf{u} - \mathbf{v}\|^2 = 2\|\mathbf{u}\|^2 + 2\|\mathbf{v}\|^2$, induces a (unique) inner product on V.

One of Alf's very early survey papers is[23] on "Some determinants that should be better known". He evaluates some determinants which occur in research on transcendental numbers and related topics and which do not appear in the classical works of T. Muir. The topics covered are Wronskian determinants, determinants of systems of approximating polynomials, generalised Vandermonde determinants and generalised Kronecker products related to periodic functions.

The remaining two papers[138, 175] are discussed in Sect. D.1 on approximations to algebraic numbers although a number of new families of determinants are calculated in these papers.

I.4   Associative Rings and Power Series Rings

A number of Alf's papers are probably best categorised as being about ring theory although they are always motivated by serious questions from number theory. The exception is[130] in which Hanna Neumann's proof that $x^3 = x$ for all x in R implies R is commutative is presented.

In[70] Alf considers Hadamard operations on rational functions. He proves a result on the Hadamard quotient of two rational functions. This theorem was previously announced by Yves Pourchet, but he had not published complete proofs. The context is to let K be a field of characteristic zero and R be a finitely generated subring of K. Let $\sum c(k)X^k$ and $\sum b(k)X^k$ be the Taylor series at the origin of two rational functions on K such that there exists a sequence $(a'(k))$, elements of R, satisfying $a'(k)b(k) = c(k)$ for all k; then there exists a rational function $\sum a(k)X^k$ such that $a(k)b(k) = c(k)$, for all k. The proof relies essentially on the theory of p-adic analytic functions and is a very nice application of that theory. In[68], dedicated to Mahler on his 80th birthday, Alf outlines the proof of the Hadamard quotient theorem.

Publication[88] is a well-cited paper by Alf. It is clearly an extension of[70]. The main result is that the Hadamard quotient of two rational functions is rational if and only if its Taylor coefficients are all elements of a ring finitely generated over $\mathbb{Z}$. More of the mathematical details are given in[88]. To establish rationality, a criterion based on the vanishing of high-order Hankel–Kronecker determinants is used. Robert Rumely decided to fill in all the details of the proof in [14]. This work of Alf's on the Hadamard quotient problem is considered to be a major contribution to the theory of rational functions.

Publication[97] is entitled "Some facts that should be better known, especially about rational functions". This 32-page paper is a very highly cited survey article, and we reproduce here the Mathematical Review written by Robert Rumely [MR1123092].

"This paper is a survey of results centered about when a formal power series $f(x) = \sum a_n x^n$ represents a rational function. If $f(x)$ is rational, then the coefficients a_n belong to a finitely generated ring over $\mathbb{Z}$, and the a_n satisfy a finite linear recurrence, which means that they are the values of a generalised power sum. But what are sufficient conditions for rationality? One is led to interpolation of generalized power sums (over $\mathbb{C}$ and over the p-adics) by exponential polynomials, and to Ritt's factorization theory for them. One is led to the specialization and lifting of rational functions, to transcendence estimates for the growth and vanishing of the coefficients, to linear equations in S-units. Some of the main results are as follows. The Skolem–Mahler–Lech theorem: If a rational function $\sum a_n x^n$ has infinitely many zero Taylor coefficients, then the indices n for which $a_n = 0$ (with finitely many exceptions) lie in a finite number of arithmetic progressions. The author's Hadamard quotient theorem: If $\sum a_n x^n$ and $\sum b_n x^n$ are rational (with the $b_n \neq 0$), and if the quotients a_n/b_n belong to a finitely generated ring, then $\sum (a_n/b_n)x^n$ is rational. Hankel's criterion for rationality based on the vanishing of the Hankel determinants, the Pólya–Carlson–Bertrandias–Dwork theorem from capacity theory, the Bézivin–Robba theory of 'Pólya operators' and many other results are discussed. The paper closes with a new application of this machinery, a solution to an old problem of M. Ward, the classification of all 'divisibility sequences', sequences $\{a_n\}$ such that $a_n | a_m$ whenever $n | m$. The canonical example of a divisibility sequence is, of course, the Fibonacci sequence."

Publication[103] is interesting in that it is not a proof of a theorem but simply an argument using divergent power series, illustrating a possible proof. The "result" is a conjecture, due to Pisot and Schutzenberger, that $\sum a_h x^h$ is the Taylor expansion of a rational function provided that the a_h all belong to a field finitely generated over $\mathbb{Q}$ and there is a non-zero polynomial f so that $\sum f(a_h) X^h$ represents a rational function. The suggested proof resembles Alf's earlier proof of the Hadamard quotient theorem.

Publication[151], joint with Graham Everest, is Alf's only paper on group rings although the target is rings of exponential polynomials. Let G be a divisible torsion-free ordered abelian group, and let U be a unique factorisation domain, and let $R = UG$. An element of R factors uniquely (up to units) as a product of irreducibles of U, a product of polynomials $P_i(\phi_i)$, where the ϕ_i are not rational multiples of one another, and a product of irreducibles of R. This result generalises a factorisation theorem of J. F. Ritt for exponential polynomials.

I.5 Group Theory

Alf published just two short papers,[13, 14] on topological groups jointly with Sidney Morris. One paper investigates extremely disconnected topological groups. The other paper studies closed subgroups of products of the reals.

J Analysis

J.1 Functions of a Complex Variable

A very large number of Alf's early papers are in complex analysis not in number theory. The influence of Mahler, Szekeres and Turán can be clearly seen. The papers include[3, 5–10, 12, 15, 16, 18–21, 27, 37, 45, 47].

Some of these are simple results but others require considerable technical facility, which Alf presumably largely learnt from Kurt Mahler.

In[3] Alf constructs a transcendental entire function which maps every algebraic number field into itself. The example also has the property that all its derivatives also assume algebraic values at every point in every algebraic number field. Publication[7] has been discussed in Sect. C.2.

In his 1953 book [18], "Eine neue Methode in der Analysis und deren Anwendungden", Paul Turán proved a number of theorems giving lower bounds for sums of powers. Turán's second main theorem is:

Let $\alpha_1, \alpha_2, \ldots, \alpha_m$ be complex numbers and $a_1, a_2, \ldots, a_m$ complex constants not all zero, and assume $|\alpha_1| \geq |\alpha_2| \geq \ldots \geq |\alpha_m|$ then

$$\max_{n+1 \leq \mu \leq n+m} \left| \sum_{k=1}^{m} a_k \alpha_k^{\mu} \right| \geq \left(\frac{m}{24e^2(n+2m)} \right)^m \min_l |a_1 + \cdots + a_l|$$

Over the years the result has been improved in various ways. In[5] Alf generalises the theorem to exponential sums with polynomial coefficients. Alf also obtained generalisations of Turán's first main theorem and showed that in various ways his result is best possible.

In[6] Alf went further in generalising the second main theorem by replacing exponential sums with polynomial coefficients to functions of the form $\sum p_{kq}(z)(1 - z)^{\alpha_k} \log^{q-1}(1 - z)$ subsuming identities which Mahler has shown to contain transcendence results on the exponential and logarithmic functions and Diophantine results of the Thue–Siegel–Roth type. In[10] Alf significantly improves Turán's first main theorem in other directions. In[8] Alf significantly improves the theorem of Dancs and Turán on the number of zeros that an exponential polynomial has in a square of a given side length in the complex plane.

The interesting question of polynomial approximation to vectors of functions is discussed by Alf in[9]. (At its heart is the Taylor polynomial approximations to a function.) Alf discusses significant contributions to the theory by Mahler and Coates in terms of perfect systems and recasts the material by obtaining interesting explicit identities, based on complex integration, related to such systems. These include, as special cases, identities obtained earlier by Mahler, Coates and others. A further contribution to the theory of perfect systems is in[12]. This involves functions of the form $z^{\alpha} \exp(\beta z)$.

In[15] Alf and Rob Tijdeman write "On common zeros of exponential polynomials". A conjecture of Shapiro is that, in the ring of exponential polynomials $\sum_k a_k e^{\alpha_k z}$, with $\{a_k\}$ and $\{\alpha_k\}$ complex, if two elements have infinitely many zeros in common, then they possess a common factor with infinitely many zeros. Using primarily algebraic arguments, and the Skolem–Mahler–Lech theorem they establish the conjecture in an important special case.

In[16] Alf studies the distribution of zeros of exponential polynomials $F(z) = \sum_{j=1}^{m} p_j(z) \exp(\alpha_j z)$ where $\alpha_1, \ldots, \alpha_m$ are distinct numbers, and $p_1(z), \ldots, p_m(z)$ are complex polynomials. Moreno [11] gave precise information as to the location of the strips containing the zeros in the special case of $F(z) = \sum_{j=1}^{m} \exp(\alpha_j z)$. Alf shows how the ideas of Moreno can be applied in the general case. In[18] Alf with Rob Tijdeman and Marc Voorhoeve further study the class of exponential polynomials. They find estimates for the number of zeros on the real line by relating it to the number of zeros of certain Wronskian determinants. Also, in[18], they obtain upper bounds for the number of zeros of an exponential polynomial in an arbitrary disc in the complex plane. The proof is heavily analytic.

Publication[37] is a 20-page survey article "On the number of zeros of functions". The object of the paper is to give a complete description of a technique that leads to estimates for the number of zeros of certain classes of functions in discs of a given radius and centre in the complex plane. Alf principally considers the case of exponential polynomials. The principal motive for this formulation resides in the application to the theory of transcendental numbers. The main result of this paper shows that an exponential polynomial cannot be small at too many points. Complete proofs are given throughout.

The Padé table is a method of generating rational functions which approximate a given function. In[47] Alf and John Loxton describe Latin and German polynomial systems introduced by Mahler. They show how these schemes fit into an even more general theory of algebraic approximation of functions. They give some examples where the construction can be explicitly given.

J.2   p-Adic Analysis

There are two papers on "Zeros of p-adic exponential polynomials" published 12 years apart. The first,[22], is by Alf and the second,[83], is joint with Robert Rumely. In[22] Alf uses a classical result of Strassman to obtain an estimate for the number of zeros in the p-adic numbers of p-adic exponential polynomials. The method improves on estimates obtained by Tarlok Shorey and by Michel Waldschmidt by using p-adic analogues of the method applicable in the complex case. In[83] the first upper bounds for the total number of zeros in the disc of convergence are given. The proof of the main result requires a thorough analysis of the location of the final corner of the Newton polygon of the exponential polynomial.

In[100] Alf, Vichian Laohakosol and John Loxton study "Integer-valued p-adic functions". A classical theorem of Polya states that among all entire transcendental functions taking integer values at the non-negative integers, the one with the least rate of growth is 2^z. The authors investigate Polya's theorem in the p-adic setting where very different problems arise since the functions are only defined locally.

J.3   Functional Equations

A number of papers in other sections of this memoir include considerations of functional equations. In this section we discuss those papers whose prime concern is functional equations.

In 1930, Mahler, [12], proved the following lemma as a step towards finding conditions under which certain functions take transcendental values:

If $f(z_1, \ldots, z_n)$ is an algebraic function of the n complex variables $z_1, \ldots, z_n$ and satisfies the functional equation

$$f(z_1, \ldots, z_n)^r = f(z_1^r, \ldots, z_n^r)$$

where $r > 1$ is an integer, then

$$f(z_1, \ldots, z_n) = \eta z_1^{e_1} \cdots z_n^{e_n}$$

where $e_1, \ldots, e_n$ are rational numbers and η is an $(r-1)$th root of unity.

In[24], Alf and John Loxton show that algebraic functions satisfying a functional equation generalising the one above must be of a very special shape. Conversely, they obtain a purely algebraic proof of the transcendence of solutions not of that shape. Publication[35] is in some ways a development of[24]. It is entitled "A class of hypertranscendental functions". A hypertranscendental function is a function such as Riemann's zeta function which does not satisfy any algebraic differential equation. The set up is that $T = (t_{ij})$ is an $n \times n$ matrix with non-negative integer entries, (for technical reasons T is nonsingular and 1 is not an eigenvalue of T^k for $k = 1, 2, \ldots$). Define $T : \mathbb{C}^n \to \mathbb{C}^n$ by $w = Tz$, $z = (z_1, \ldots, z_n)$ and $w = (w_1, \ldots, w_n)$ with

$$w_i = \prod_{j=1}^{n} z_j^{t_{ij}} \quad (1 \leq i \leq n).$$

The functional equations considered in[35] are of the form $f(Tz) = af(z) + b(z)$, where a is a non-zero constant and $b(z)$ is a rational function of z. The main theorem shows that the $p+q$ functions $f_1(z), \ldots, f_p(z), g_1(z), \ldots, g_q(z)$ are algebraically independent given that they satisfy $p+q$ functional equations of the form $f(Tz) = af(z) + b(z)$ plus several additional conditions. The paper is motivated by several papers of Mahler.

Publications[41, 42] are closely related to[24, 35]. In[41] Alf and John Loxton investigate when functions studied in[24] take transcendental values at algebraic points. In[42] families of functions as studied in[35] are further considered and conditions are obtained under which the values of these functions at algebraic points are algebraically independent.

In[44], the third in this series of papers by Alf and John Loxton, they obtain a general transcendence theorem for the solutions of much more complicated functional equations depending in part on an infinite sequence of integer matrices. There are several applications to extensions of known theorems, but a particularly striking consequence is that, for any irrational number ω, the function $\sum_{h=1}^{\infty} [h\omega] z^h$ takes transcendental values at all algebraic points α with $0 < |\alpha| < 1$.

Publication[45] by Alf and John Loxton is entitled "Algebraic independence of the Fredholm series". They consider functions analytic in a neighbourhood of the origin in the complex plane that satisfy a functional equation $g(z) = ag(z^r) + b(z)$, where $r > 1$ is an integer, a is a non-zero complex number and $b(z)$ is a polynomial. They first give necessary and sufficient conditions for families of such functions to be algebraically independent. They then give necessary and sufficient conditions for values of a single such function or a family of such functions at algebraic points to be algebraically independent. One example of such a function is $f(z) = \sum_{h=0}^{\infty} z^{2^h}$ which satisfies the functional equation $f(z) = f(z^2) + z$. This function is known as the Fredholm series. f has a number of interesting properties.

J.4   Other Topics in Analysis Including Padé Approximation

Publication[29] is entitled "Hermite interpolation and p-adic exponential polynomials". By employing a precise form of the Hermite interpolation formula, Alf obtains a best possible bound for the number of zeros of p-adic exponential polynomials. A second result is the best possible bound on the coefficients, if the exponential polynomial is small at sufficiently many points. He also obtains an extrapolation lemma which is more efficient than is extrapolation with the aid of the Schnirelman integral. In[108] Alf generalises the approximation theory described in Mahler's paper "Perfect systems" to linked simultaneous approximations and proves the existence of nonsingular approximation and of transfer matrices by generalising Coates' normality zig-zag theorem. This paper was published in the Kurt Mahler memorial volume of the Journal of the Australian Mathematical Society.

Finally there is[181] entitled "Life on the Edge" and published as a note in the MAA Monthly. The note begins by deriving the series:

$$\sum_{n=1}^{\infty} \frac{\cos n\theta}{n} = -\log\left(2\sin\frac{\theta}{2}\right) \text{ and } \sum_{n=1}^{\infty} \frac{\sin n\theta}{n} = \frac{\pi}{2} - \frac{\theta}{2}$$

which are true for $0 < \theta < 2\pi$. He then considers $\theta \to 0$ and $\theta \to 2\pi$. Then by integrating from 0 to π several times, he obtains $\zeta(2) = \pi^2/6$, $\zeta(4) = \pi^4/90$, etc. It contains very instructive examples for a beginning student.

K   Automata

K.1   Paper Folding, Automata and Transcendence

Probably the most important mathematical collaboration Alf developed in the second half of his career is that with MMF. As has been said elsewhere in this memoir, Alf thought of Université de Bordeaux I as his third university. In this section all of Alf's papers on automata and their applications to questions in number theory are collected although many could have been discussed elsewhere.

Three survey papers on "Folds",[62–64], written jointly with Michel Dekking and MMF appeared in Volume 4 of the Mathematical Intelligencer in 1982. Perhaps the best way to describe this survey is to include the very clear Mathematical Reviews of William Beyer [MR0685559]:

"These three papers are an exposition of the following subjects and their connections: paper-folding sequences, finite automata which 'recognize' paperfolding sequences, substitution automata (also called uniform tag machines), a theorem of Cobham connecting a g-automaton with a substitution automaton, the Thue–Morse sequence, generalized Rudin–Shapiro sequences, the Fredholm sequence, functional equations for these sequences, Mahler functions and Mahler systems of functional equations, entropy and dimension of infinite plane curves, continued fractions, algebraic and transcendental numbers, and finally dragon curves. Despite this potpourri, the work is unified by the simple concept of the paperfolding sequence.

The authors state that very little of the material is original. Much of the material is of recent origin and adequate references are given. The level of the exposition is high and its beauty reflects the care and effort of the authors. A sequence of statements called 'observations' replaces the traditional sequence of theorems. However, there seems to be no difference between a theorem and an 'observation'. Full proofs are often left to the reader, but adequate background is usually given to permit this. Some 'observations' are merely quoted from the literature. One especially interesting (to the reviewer) observation: 'If $a_0.a_1a_2a_3\ldots$ is algebraic, but not rational, then the sequence of digits (a_h) is not generated by a finite automaton. In this sense, the digits of the expansion of an algebraic number are "random".' Despite the efforts at clarity, readers unfamiliar with the subject will have to work hard to read these papers in detail. There are places where the reviewer wishes more had been said. The papers are pleasantly sprinkled with literary quotations."

Alf's first paper on paper folding[57] was joint with MMF. A paper-folding sequence (p.f.s.) is a sequence $f = (f_n) \in \{-1, 1\}^{\mathbb{N}}$ generated by folding a sheet of paper in the negative or positive direction. All p.f.s.'s are obtained from an arbitrary number α, $0 < \alpha < 1$, that is, from an arbitrary binary "decimal" $0.a_0a_1a_2\ldots(a_i \in \{0, 1\})$ in the following way:

$$f_{2^k} = 2a_k - 1 \ (k \geq 0), \quad f_{2^k(2l+1)} = (-1)^l f_{2^k}.$$

It is proved that a p.f.s. is almost periodic (in the sense of Besicovitch). With the operator $T : (g_n) \to (g_{2n})$ on $\{-1, 1\}^{\mathbb{N}}$ a sequence $f^{(k)}$ of p.f.s.'s can be defined by $f^{(0)} = f$ (a given p.f.s.), $f^{(k)} = T f^{(k-1)}$. The generating function

$$F_k(X) = \sum_{n \geq 0} f_n^{(k)} X^n \qquad (k \geq 0)$$

of the sequence $f^{(k)}$ satisfies the following system of functional equations

$$F_k(X^2) = F_{k-1}(X) - f_1^{(k-1)} X/(X^2 + 1) \qquad (k \geq 1).$$

Such systems were studied by Alf and John Loxton in[44], and with techniques from there it is proved that $\sum_{n\geq 0} f_n \alpha^n$ is transcendental if α $(0 < |\alpha| < 1)$ is algebraic and (f_n) is a p.f.s.

One of the originators of the links between automata and numbers expressed in a fixed base is Alan Cobham; see [4, 5]. In[58] Alf reports on Tag Machines or Substitutional Automata with deletion number one and how they correspond to systems of functional equations involving transformations originally studied by Mahler in the 1930s. This study also links to algebraic functions over finite fields. This note serves as an appendix to the paper of MMF in the same proceedings.

In a long and complicated paper[60] Alf and John Loxton collect facts about sequences $\alpha = \{\alpha_n\}$ with entries belonging to a finite alphabet. They point out that Cobham and MMF and his collaborators C. Christol, J. Kamae and G. Rauzy have shown that the following constructions all lead to essentially the same class of sequences.

1. The sequence α is generated by a finite automaton. The finite automaton operates by accepting the string of digits of n in some fixed base as input and producing the output α_n.
2. The sequence α is a fixed point of a uniform substitution. A substitution is described by giving a rule which replaces each symbol of the alphabet by a finite string of symbols of the alphabet. It is uniform if these strings all have the same length.
3. The function $f(z) = \sum \alpha_n z^n$ is algebraic over $\mathbb{F}_q(z)$, the field of rational functions over the finite field $\mathbb{F}_q$ of q elements. (The alphabet of symbols is a subset of $\mathbb{F}_q$.)
4. The functions $f_1(z) = \sum \alpha_n z^n, f_2(z), \dots, f_m(z)$ satisfy a system of functional equations of the shape $f_i(z) = \sum_{j=1}^m a_{ij}(z) f_j(z^q)$ in which the coefficients $a_{ij}(z)$ are rational functions.

These constructions yield the class of *regular sequences*. The main result of[60] establishes the algebraic independence of the values at algebraic points of functions as described in 4. In particular, subject to some technical requirements on the system in 4, they show that the number $f_1(1/b) = \sum_{n=0}^{\infty} \alpha_n/b^n$ is either rational or transcendental.

Publication[78] is a short note by Alf and MMF entitled "Automata and the arithmetic of formal power series". An elegant proof using finite automata is given of the following theorem. Let F be a finite field of characteristic p and suppose $f = \sum_{n\geq 0} f_n X^n$ is a nonconstant formal power series with $f_0 = 1$. Let A be a p-adic integer and define

$$f^A = \sum_{n\geq 0} \binom{A}{n}(f-1)^n.$$

If both f and f^A are algebraic over $F(X)$, then A is rational.

An excellent review of the many results pertaining to the algebraic properties of power series in one or several variables over an integral domain, usually $\mathbb{F}_p$ or $\mathbb{Z}_p$, is[104], a 20-page paper entitled "Rational functions, diagonals, automata and arithmetic" by Alf and Leonard Lipshitz. Publication[122] is an addendum to[104].

Another paper which treats folded sequences is[163] discussed in Sect. A.2. Chapter 14.1 of Alf's book[190], discussed in Sect. E.1, gives a survey of "Finite automata and power series" in the broad. Finally, although not on automata, the paper[165] by Alf, Marek Karpinski and Igor Shparlinski discusses "Zero testing of p-adic and modular polynomials". They obtain new algorithms for whether a multivariate polynomial over a p-adic field given by a black box is identically zero. They also consider zero testing of polynomials in residue rings. Their results complement known results on zero testing of polynomials over the integers, the rationals and finite fields.

L   Recreational Mathematics, Obituaries and Biographies

L.1   Recreational Mathematics

Papers[69, 81, 198, 212, 217] belong here. Perhaps[75, 206] which are included in Sect. H.1 could have been discussed here as well.

Publication[69], "A dozen years is but a day", is another insight into the mind of John Conway. The note explains a method of mentally calculating the day of the week—"The doomsday algorithm" as Conway calls it. For example, 13 July 1945 is a Friday. Provided it is regularly practised, it is quite straightforward to remember and apply.

In[81], "A bridge problem", Alf constructs a bridge deal in which (against best defence) declarer can make a grand slam in each of the four suit contracts but cannot make even a small slam in a no-trump contract. This answers a question of S. M. Ulam.

Publication[198], is couched as a discussion between Alf and an amateur. The sieve of Eratosthenes, prime pairs, prime n-tuples and the $\pi(x+y)$-conjecture are discussed.

Publication[212] is a long delayed report on an investigation carried out by two undergraduates from the University of Manitoba, Kurt Thomsen and Mark Wiebe, during a visit to Macquarie University. They were asked to generalise the "Self-similar sum of squares" $12^2 + 33^2 = 1233$. They solved the problem of constructing all examples of self-similar sums of squares. The argument, once found, is very straightforward.

In[217], Alf and Raymond Breu, again motivated by Lenstra's $12^2 + 33^2 = 1233$, study quadratic forms $f(x,y) = ux^2 + vxy + wy^2$ with u,v,w integers and representations of the form $f(a,b) = 10^k a + b$ with a,b integers $< 10^k$. A method of searching for solutions and sequences of solutions subject to mildly technical divisibility conditions and some relatively simple Diophantine equations are described. Many concrete examples are given. The methods are completely elementary and, in particular, do not use the theory of quadratic forms.

L.2   Biographies and Obituaries

It was quite to be expected for Alf to honour Kurt Mahler with a number of obituaries[89, 105, 111, 125]. Publication[89] was published in the Gazette of the Australian Mathematical Society and included brief biographies written by Bernhard Neumann and by Alf—Alf's was more informal. Publication[105] was a brief obituary published by the Australian Academy of Science. Publication[111] was a full 38-page obituary published by the Journal of the Australian Mathematical Society. As an obituary it is a little unusual since as well as 13 pages on Mahler's life and mathematics and a full list of Mahler's publications it also includes 15 pages written by Mahler in 1971 entitled "Fifty years as a mathematician". Finally in[125] Alf and John Coates give a somewhat fuller analysis of Mahler's mathematical works.

Alf's lifelong interest in science fiction, noted elsewhere, led to[211], a brief biography of the science fiction writer Arthur Bertram Chandler.

M   Books, Reports and Conference Proceedings

There are 12 of Alf's publications,[56, 79, 123, 133, 139, 141, 148, 172, 190, 196, 215, 219], which naturally fit under this heading.

Publication[56] is Alf's short rather cryptic foreword to Leonard Lewin's impressive book "Polylogarithms and Associated Functions".

Publication[123] is an English translation of Vladimir Sprindžuk's book on Classical Diophantine Equations published in 1993. This volume had a difficult genesis since the author started it in the USSR in 1982 and Alf with the help of Ross Talent, Deryn Griffiths and Chris Pinner finished it in 1993. A serious attempt has been made to retain the idiosyncratic style of the author even where it was difficult to determine the author's intentions. It is a very interesting book.

Publication[148], Mathematical Sciences: Adding to Australia, is mentioned on p. 15.

Publications[79, 133, 139, 196, 215] are conference proceedings coedited by Alf with five different colleagues. All of them are of the highest quality and reflect Alf's commitment to proper refereeing and his insistence on rigorous editorial work. Taken together they represent a huge amount of work.

Publications[141, 172, 190, 219] are Alf's three major monographs. Publication[141] and the Japanese translation[172] are his "Notes on Fermat's Last Theorem" discussed in Sect. F.2. Publication[190] is his monograph "Recurrence Sequences" with three joint authors discussed in Sect. E.1. Alf's final publication,[219], on continued fractions is still to appear, but given the strength of his joint authors, I trust it will be of the same high quality as Alf's other monographs.

Acknowledgements Firstly, I acknowledge Jon Borwein who 18 months ago encouraged me to write this monograph which has allowed me to appreciate the whole of Alf van der Poorten's published work. I hope that I have done justice to his memory and to his mathematics.

Secondly, I thank Joy, Kate, David and Marianne van der Poorten who contributed most of the first few pages. In particular, I thank Joy for reading the whole manuscript and also for providing many of the details about Alf's life.

Thirdly, I thank John Coates, Jeffrey Shallit, MMF and Michel Waldschmidt for the very substantial improvements that I was able to make to the memoir using their suggestions.

References

1. B. Adamczewski, Y. Bugeaud, F. Luca, Sur la complexité des nombres algébriques (On the complexity of algebraic numbers). C.R. Math. Acad. Sci. Paris **339**(1), 11–14 (2004)
2. B. Adamczewski, Y. Bugeaud, On the independence of expansions of algebraic numbers in an integer base. Bull. Lond. Math. Soc. **39**(2), 283–289 (2007)
3. A. Baker, *Transcendental Number Theory* (Cambridge University Press, London, 1975)
4. A. Cobham, Uniform tag sequences. Math. Systems Theory **6**, 164–172 (1972). See also: On the Hartmanis-Stearns problem for a class of tag machines, Technical report RC2178, IBM Research Centre, Yorktown Heights, New York, 1968
5. A. Cobham, On the base-dependence of sets of numbers recognisable by finite automata. Math. Systems Theory **3**, 186–192 (1969)
6. I. Kaplansky, letter to Richard Mollin, Kenneth Williams and Hugh Williams, 23 November 1998. Available on the web page for^{210}
7. S. Lang, Letter to the editor. Canad. Math. Soc. Notes **29**, 5–6 (1997)
8. K. Mahler, Arithmetische Eigenschaften der Lösungen einer Klasse von Functionalgleichungen. Math. Ann. **101**, 243–366 (1929)
9. K. Mahler, Über das Verschwinden von Potenzreihen mehrerer Veränderlichen in speziellen Punktfolgen. Math. Ann. **103**, 573–587 (1930)
10. K. Mahler, Arithmetische Eigenschaften einer Klasse transzendental-transzendenter Funktionen. Math. Z. **32**, 545–585 (1930)
11. C.J. Moreno, The zeros of exponential polynomials (1). Comp. Math. **26**, 69–78 (1973)
12. K. Mahler, Über das Verschwinden von Potenzreihen mehrerer Veränderlichen in speziellen Punktfolgen. Math. Ann. **103**, 573–587 (1930)
13. H.E. Rose, *A Course in Number Theory*, 2nd edn. (Oxford University Press, Oxford, 1994)
14. R. Rumely, Notes on van der Poorten's proof of the Hadamard quotient theorem: part I, part II. In *Séminaire de Théorie des Nombres*, Paris 1986–87. Progress in Mathematics, vol. 75 (Birhäuser, Boston, 1989), pp. 349–382, 383–409
15. T. Rivoal, La fonction zêta de Riemann prend une infinité de valeurs irrationnelles aux entiers impairs (There are infinitely many irrational values of the Riemann zeta function at odd integers). C.R. Acad. Sci. Paris Sér. I Math. **331**(4), 267–270 (2000).
16. D. Shanks, On Gauss and composition, I, II. In *Proc. NATO ASI on Number Theory and Applications*, ed. by R.A. Mollin (Kluwer Academic Press, 1989), pp. 163–179
17. H.P. Schlickewei, W.M. Schmidt, The number of solutions of polynomial-exponential equations. Compositio Math. **120**, 193–225 (2000)
18. P. Turán, *Eine neue Methode in der Analysis und deren Anwendungen* (Akadémiai Kiadó, Budapest 1953)
19. K. Yu, *p*-adic logarithmic forms and group varieties. I. J. Reine Angew. Math. **502**, 29–92 (1998)
20. K. Yu, *p*-adic logarithmic forms and group varieties. II. Acta Arith. **89**(4), 337–378 (1999)
21. K. Yu, *p*-adic logarithmic forms and group varieties. III. Forum Math. **19**(2), 187–280 (2007)

22. K. Yu, Report on p-adic logarithmic forms. In *A Panorama of Number Theory or the View from Baker's Garden (Zürich, 1999)* (Cambridge University Press, Cambridge, 2002), pp. 11–25
23. V.V. Zudilin, One of the numbers $\zeta(5)$, $\zeta(7)$, $\zeta(9)$, $\zeta(11)$ is irrational (Russian). Uspekhi Mat. Nauk **56**(4), 149–150 (2001); Translation in Russian Math. Surveys **56**(4), 774–776 (2001)
24. V.V. Zudilin, On the irrationality of the values of the Riemann zeta function (Russian). Izv. Ross. Akad. Nauk Ser. Mat. **66**(3), 49–102 (2002); Translation in Izv. Math. **66**(3), 489–542 (2002)

Publications of Alf van der Poorten

1. On Gödel's theorem. Cogito, J. UNSW Socratic Society **1**, 34–39 (1966)
2. A note on the independence of the axioms for a vector space. J. Austral. Math. Soc. **7**, 425–428 (1967)
3. Transcendental entire functions mapping every algebraic number field into itself. J. Austral. Math. Soc. **8**, 192–193 (1968)
4. On a sequence of prime numbers (with C.D. Cox). J. Austral. Math. Soc. **8**, 571–574 (1968)
5. Generalisations of Turán's main theorems on lower bounds for sums of powers. Bull. Austral. Math. Soc. **2**, 15–37 (1970)
6. A generalisation of Turán's main theorems to binomials and logarithms. Bull. Austral. Math. Soc. **2**, 183–195 (1970)
7. On the arithmetic nature of definite integrals of rational functions. Proc. Amer. Math. Soc. **29**, 451–456 (1971)
8. On a theorem of S. Dancs and P. Turán. Acta Math. Acad. Sci. Hungar. **22**, 359–364 (1971/1972)
9. Perfect approximation of functions. Bull. Austral. Math. Soc. **5**, 117–126 (1971)
10. Generalizing Turán's main theorems on lower bounds for sums of powers. Acta Math. Acad. Sci. Hungar. **24**, 93–96 (1973)
11. A note on recurrence sequences. J. Proc. Roy. Soc. New South Wales **106**, 115–117 (1973)
12. A note on simultaneous polynomial approximation of exponential functions (with J.H. Loxton). Bull. Austral. Math. Soc. **11**, 333–338 (1974)
13. Extremally disconnected topological groups (with S.A. Morris). J. Proc. Roy. Soc. New South Wales **107**, 114–115 (1974)
14. Closed subgroups of products of reals (with D.C. Hunt and S.A. Morris). Bull. Lond. Math. Soc. **7**, 124–128 (1975)
15. On common zeros of exponential polynomials (with R. Tijdeman). Enseignement Math. (2) **21**, 57–67 (1975)
16. A note on the zeros of exponential polynomials. Compositio Math. **31**, 109–113 (1975)
17. Propriétés arithmétiques et algébriques de fonctions satisfaisant une classe d'équations fonctionnelles. *Séminaire de Théorie des Nombres* (Univ. Bordeaux I, Talence, 1974–75), Exp. No.7, Lab. Théorie des Nombres, Centre Nat. Recherche Sci. (Talence, 1975), p. 13
18. On the number of zeros of certain functions (with M. Voorhoeve and R. Tijdeman). Nederl. Akad. Wetensch. Proc. Ser. A **78**, 407–416 (1975); Indag. Math. **37**, 407–416 (1975)
19. Wronskian determinants and the zeros of certain functions (with M. Voorhoeve). Nederl. Akad. Wetensch. Proc. Ser. A **78**; Indag. Math. **37**, 417–424 (1975)
20. Arithmetic implications of the distribution of integral zeros of exponential functions. *Séminaire Delange-Pisot-Poitou (Théorie des Nombres)* (16e année: 1974/75), Fasc. 1, Exp. No. 11, Secrétariat Mathématique (Paris 1975), p. 9
21. Zéros communs de plusieurs polynômes exponentielles. *Séminaire de Théorie des Nombres* (Univ. Bordeaux I, Talence, 1974–75), Exp. No. 8, Lab. Théorie des Nombres, Centre Nat. Recherche Sci. (Talence, 1975), p. 6
22. Zeros of p-adic exponential polynomials. Nederl. Akad. Wetensch. Ser. A **79**; Indag. Math. **38**, 46–49 (1976)

23. Some determinants that should be better known. J. Austral. Math. Soc. Ser. A **21**, 278–288 (1976)
24. On algebraic functions satisfying a class of functional equations (with J.H. Loxton). Aequationes Math. **14**, 413–420 (1976)
25. On Baker's inequality for linear forms in logarithms. Math. Proc. Camb. Phil. Soc. **80**, 233–248 (1976)
26. Computing the effectively computable bound in Baker's inequality for linear forms in logarithms (with J.H. Loxton). Bull. Austral. Math. Soc. **15**, 33–57 (1976)
27. On the distribution of zeros of exponential polynomials. Math. Slovaca **26**, 299–307 (1976)
28. On the transcendence and algebraic independence of certain somewhat amusing numbers. *Séminaire de Théorie des Nombres* (Univ. Bordeaux I, Talence, 1975–76), Exp. No. 14, Lab. Théorie des Nombres, Centre Nat. Recherche Sci. (Talence, 1976), p. 13
29. Hermite interpolation and *p*-adic exponential polynomials (to George Szekeres on his 65th birthday). J. Austral. Math. Soc. Ser. A **22**, 12–26 (1976)
30. More about π. Function (Monash University) **1**, 17–17 (1977)
31. On the growth of recurrence sequences (with J.H. Loxton). Math. Proc. Camb. Phil. Soc. **81**, 369–376 (1977)
32. Effectively computable bounds for the solutions of certain Diophantine equations. Acta Arith. **33**, 195–207 (1977)
33. Computing the lower bound for linear forms in logarithms. *Séminaire Delange-Pisot-Poitou (Théorie des nombres)*, 17e année: (1975/76), Fasc. 1, Exp. No. 17, Secrétariat Math. (Paris, 1977), p. 9
34. Applications of the Gel'fond-Baker method to Diophantine equations (with A. Schinzel, T.N. Shorey and R. Tijdeman). In *Transcendence Theory: Advances and Applications*, ed. by A. Baker, D.W. Masser. Proc. Conf., Univ. Cambridge, Cambridge (Academic, London, 1977), pp. 59–77.
35. A class of hypertranscendental functions (with J.H. Loxton). Aequationes Math. **16**, 93–106 (1977)
36. Transcendence and algebraic independence by a method of Mahler (with J.H. Loxton). In *Transcendence Theory: Advances and Applications*, ed. by A. Baker, D.W. Masser. Proc. Conf., Univ. Cambridge, Cambridge (Academic, London, 1977), pp. 211–226.
37. On the number of zeros of functions. Enseignement Math. (2) **23**, 19–38 (1977)
38. Multiplicative relations in number fields (with J.H. Loxton). Bull. Austral. Math. Soc. **16**, 83–98 (1977)
39. Corrigenda and addenda: Computing the effectively computable bound in Baker's inequality for linear forms in logarithms. Bull. Austral. Math. Soc. **15**, 33–57 (1976); Multiplicative relations in number fields. ibid. **16**, 83–98 (1977) (with J.H. Loxton). Bull. Austral. Math. Soc. **17**, 151–155 (1977)
40. Linear forms in logarithms in the *p*-adic case. In *Transcendence Theory: Advances and Applications*, ed. by A. Baker, D.W. Masser. Proc. Conf., Univ. Cambridge, Cambridge (Academic, London, 1977), pp. 29–57.
41. Arithmetic properties of certain functions in several variables (with J.H. Loxton). J. Number Theory **9**, 87–106 (1977)
42. Arithmetic properties of certain functions in several variables, II (with J.H. Loxton). J. Austral. Math. Soc. Ser. A **24**, 393–408 (1977)
43. The polynomial $x^3 + x^2 + x - 1$ and elliptic curves of conductor 11. *Séminaire Delange-Pisot-Poitou (Théorie des nombres)* (18e année: 1976/77), Fasc. 2, Exp. No. 17, Secrétariat Math. (Paris, 1977), p. 7
44. Arithmetic properties of certain functions in several variables, III (with J.H. Loxton). Bull. Austral. Math. Soc. **16**, 15–47 (1977)
45. Algebraic independence properties of the Fredholm series (with J.H. Loxton) (to Kurt Mahler on his 75th birthday). J. Austral. Math. Soc. Ser. A **26**, 31–45 (1978)
46. A proof that Euler missed ... Apéry's proof of the irrationality of $\zeta(3)$: An informal report. Math. Intelligencer **1**, 195–203 (1978/1979)

47. Multidimensional generalizations of the Padé table (with J.H. Loxton). Rocky Mountain J. Math. **9**, 385–393 (1979)
48. Euclidean number fields. 1,2,3 (by H.W. Lenstra) translated from the Dutch and revised by A.J. van der Poorten. Math. Intelligencer **2**, 6–15, 73–77, 99–103 (1979/1980)
49. Elliptic curves of conductor 11 (with M.K. Agrawal, J.H. Coates and D.C. Hunt). Math. Comp. **35**, 991–1002 (1980)
50. On strong pseudoprimes in arithmetic progressions (with A. Rotkiewicz). J. Austral. Math. Soc. Ser. A **29**, 316–321 (1980)
51. Some remarks on Fermat's conjecture (with K. Inkeri). Acta Arith. **36**, 107–111 (1980)
52. Some wonderful formulae ... footnotes to Apéry's proof of the irrationality of $\zeta(3)$. *Séminaire Delange-Pisot-Poitou, Théorie des nombres* 20e année: 1978/79, Fasc. 2, Exp. No. 29, Secrétariat Math. (Paris, 1980), p. 7
53. Some wonderful formulas ... an introduction to polylogarithms. In *Queen's Papers in Pure and Appl. Math., Proceedings of the Queen's Number Theory Conference*, vol. 54 (Kingston, Ont., 1979/1980), pp. 269–286
54. Linear forms in logarithms, effectivity and elliptic curves. Mém. Soc. Math. France, 2e Série, **2**, 71–71 (1980)
55. Identifying a rational function (with J.P. Glass, J.H. Loxton). C. R. Math. Rep. Acad. Sci. Canada **3**, 279–284 (1981)
56. Foreword to Leonard Lewin, in *Polylogarithms and Associated Functions* (North-Holland Publishing Co., New York, 1981), pp. xvii+359
57. Arithmetic and analytic properties of paper folding sequences (with M. Mendès France). Bull. Austral. Math. Soc. **24**, 123–131 (1981)
58. Substitution automata, functional equations and functions algebraic over a finite field, Papers in algebra, analysis and statistics (Hobart, 1981). Contemp. Math., vol. 9, Amer. Math. Soc., Providence, R.I., (1981), pp. 307–312
59. Bounds for solutions of systems of linear equations (with J.D. Vaaler). Bull. Austral. Math. Soc. **25**, 125–132 (1982)
60. Arithmetic properties of the solutions of a class of functional equations (with J.H. Loxton). J. Reine Angew. Math. **330**, 159–172 (1982)
61. Identification of rational functions: lost and regained. C. R. Math. Rep. Acad. Sci. Canada **4**, 309–314 (1982)
62. Folds (with M. Dekking, M. Mendès France). Math. Intelligencer **4**, 130–138 (1982)
63. Folds. II: Symmetry disturbed (with M. Dekking, M. Mendès France). Math. Intelligencer **4**, 173–181 (1982)
64. Folds. III: More morphisms (with M. Dekking, M. Mendès France). Math. Intelligencer **4**, 190–195 (1982)
65. The growth conditions for recurrence sequences (with Hans Peter Schickewei). In *Macquarie Math. Reports*, vol. 41 (Macquarie University, NSW 2109, Australia, 1982), p. 27
66. Multiplicative dependence in number fields (with J.H. Loxton). Acta Arith. **42**, 291–302 (1983)
67. Effective approximation of algebraic numbers, d'après Enrico Bombieri. In *Seminar on Number Theory* (Univ. Bordeaux I, Talence, 1983), Exp. No. 9 (Talence, 1982/1983), p. 9
68. p-adic methods in the study of Taylor coefficients of rational functions (to Kurt Mahler on his 80th birthday). Bull. Austral. Math. Soc. **29**, 109–117 (1984)
69. A dozen years is but a day. Austral. Math. Soc. Gazette **11**, 33–34 (1984)
70. Hadamard operations on rational functions. In *Study Group on Ultrametric Analysis*, 10th year, 1982/83, Exp. No. 4 (Inst. Henri Poincaré, Paris, 1984), p. 11
71. Some problems of recurrent interest. In *Topics in Classical Number Theory*, Colloq. Math. Soc. János Bolyai, Budapest 1981, vol. 34 (North Holland, Amsterdam, 1984), pp. 1265–1294
72. Additive relations in number fields. In *Seminar on Number Theory*, Paris 1982/83. Progr. Math., vol. 51 (Birkhäuser, Boston, MA 1984), pp. 259–266
73. Thue's method and curves with prescribed singularities. In *Macquarie Math. Reports*, vol. 48 (Macquarie University, NSW 2109, Australia, 1984), p. 14

74. Arithmetic properties of the solutions of a class of functional equations, II (with J.H. Loxton). In *Macquarie Math. Reports*, vol. 62 (Macquarie University, NSW 2109, Australia, 1985), p. 18

75. The longest prime. Math. Intelligencer **7**, 40–40 (1985)

76. Integers with digits 0 or 1 (with D.H. Lehmer, K. Mahler). Math. Comp. **46**, 683–689 (1986)

77. Automata and *p*-adic numbers (with M. Mendès France). In *Analytic and Elementary Number Theory*, (Marseille, 1983). *Publ. Math. Orsay*, vol. 86–1 (Univ. Paris XI, Orsay, 1986), pp. 114–118.

78. Automata and the arithmetic of formal power series (with M. Mendès France). Acta Arith. **46**, 211–214 (1986)

79. Diophantine analysis, Proceedings of the number theory section of the 1985 Australian Mathematical Society convention held at the University of New South Wales, Kensington, May 1985. Edited jointly with J.H. Loxton. *London Mathematical Society Lecture Note Series*, vol. 109 (Cambridge University Press, Cambridge, 1986), pp. viii+170

80. An introduction to continued fractions. In *Diophantine Analysis* (Kensington, 1985). London Math. Soc. Lecture Notes Ser. vol. 109, ed. by J.H. Loxton, A.J. van der Poorten (Cambridge University Press, Cambridge, 1986), pp. 99–138.

81. A bridge problem. Austral. Math. Soc. Gazette **13**, 92–92 (1986)

82. A lower bound for linear forms in the logarithms of algebraic numbers (with J.H. Loxton, M. Mignotte, M. Waldschmidt). C. R. Math. Rep. Acad. Sci. Canada **9**, 119–124 (1987)

83. Zeros of *p*-adic exponential polynomials. II. (with R.S. Rumely). J. London Math. Soc. (2) **36**, 1–15 (1987)

84. A note on the Hadamard *k*th root of a rational function (with R.S. Rumely). J. Austral. Math. Soc. Ser. A **43**, 314–327 (1987)

85. Remarks on generalised power sums (with R.S. Rumely). Bull. Austral. Math. Soc. **36**, 311–329 (1987)

86. An awful problem about integers in base four (with J.H. Loxton) (to Paul Erdős on his 75th birthday). Acta Arith. **49**, 193–203 (1987)

87. Remarks on automata, functional equations and transcendence. In *Séminaire de théorie des nombres* (Univ. Bordeaux I, Talence, 1986–1987), p. 11. Exp. No. 27, lab. de Théorie des Nombres, Centre Nat. Recherche Sci. (Talence, 1987)

88. Solution de la conjecture de Pisot sur le quotient de Hadamard de deux fractions rationelles (Solution of Pisot's conjecture on the Hadamard quotient of two rational functions). C. R. Acad. Sci. Paris Sér. I Math. **306**, 97–102 (1988)

89. Kurt Mahler: 1903–1988 (with B.H. Neumann). Austral. Math. Soc. Gaz. **15**, 25–27 (1988)

90. Remarks on the continued fractions of algebraic numbers. In *Groupe de Travail en Théorie Analytique et Élémentaire des Nombres*, 1986–87. Publ. Math. Orsay, vol. 88 (Univ. Paris XI, Orsay, 1988), pp. 89–90.

91. Some quantitative results related to Roth's theorem (with E. Bombieri). J. Austral. Math. Soc. Ser. A **45**, 233–248 (1988)

92. Arithmetic properties of automata: regular sequences (with J.H. Loxton). J. Reine Angew. Math. **392**, 57–69 (1988)

93. Remarks on Roth's theorem. In *Séminaire de Théorie des Nombres*, Paris 1986–87. Progr. Math., vol. 75 (Birkhäuser Boston, Boston, MA, 1988), pp. 443–452.

94. Indépendance algébrique de certaines séries formelles (Algebraic independence of certain power series), (with J.-P. Allouche, M. Mendès France). Bull. Soc. Math. France **116**, 449–454 (1988)

95. From geometry to Euler identities (with M. Mendès France). Theoret. Comput. Sci. **65**, 213–220 (1989)

96. On the thermodynamics of curves and other curlicues (with R.R. Moore). In *Miniconference on Geometry and Physics*, ed. by M.N. Barber, M.K. Murray. Proc. Centre Math. Anal. Austral. Nat. Univ., vol. 22 (Canberra, 1989), pp. 82–109.

97. Some facts that should be better known; especially about rational functions. In *Number Theory and Applications*, ed. by Richard A. Mollin. NATO – Adv. Sci. Inst. Ser. C Math. (Banff, AB, 1988) (Kluwer Acad. Publ., Dordrecht, 1989), pp. 497–528.

98. Corrigenda: Some quantitative results related to Roth's theorem (with E. Bombieri). J. Austral. Math. Soc. Ser. A **48**, 154–155 (1990)

99. Notes on continued fractions and recurrence sequences. In *Number Theory and Cryptography*, ed. by J.H. Loxton. London Math. Soc. Lecture Note Ser., vol. 154 (Cambridge Univ. Press, Cambridge, 1990), pp. 86–97.

100. Integer-valued p-adic functions (with V. Laohakosol, J.H. Loxton). In *Number Theory*, ed. by K. Győry, G. Halász, vol. II (Budapest, 1987). Colloq. Math. Soc. János Bolyai, vol. 51 (North-Holland, 1990), pp. 829–849

101. A Diophantine problem in harmonic analysis (with Hans Peter Schickewei). Math. Proc. Cambridge Phil. Soc. **108**, 417–420 (1990)

102. A full characterisation of divisibility sequences (with J.-P. Bézivin, A. Pethő). Amer. J. Math. **112**, 985–1001 (1990)

103. A divergent argument concerning Hadamard roots of rational functions. In *Analytic Number Theory*, ed. by B.C. Berndt, H.G. Diamond, H. Halberstam, A. Hilderbrand. Proceedings of a conference in honour of Paul T. Bateman (Allerton Park, IL, 1989). Progr. Math., vol. 85 (Birkhäuser, Boston, MA, 1990), pp. 413–427

104. Rational functions, diagonals, automata and arithmetic (with L. Lipshitz). In *Number Theory*, ed. by Richard A. Mollin (Banff, AB, 1988) (de Gruyter, Berlin, 1990), pp. 339–358

105. Kurt Mahler (Brief obituary), in *The Australian Academy of Science Year Book* (1990/1991), pp. 75–77.

106. Dyson's lemma without tears (with F. Beukers, R. Yager). Indag. Math. (N.S.) **2**, 1–8 (1991)

107. Some explicit continued fraction expansions (with M. Mendès France). Mathematika **38**, 1–9 (1991)

108. Generalised simultaneous approximation of functions (dedicated to the memory of Kurt Mahler). J. Austral. Math. Soc. Ser. A **51**, 50–61 (1991)

109. Additive relations in fields (with Hans Peter Schickewei). J. Austral. Math. Soc. Ser. A **51**, 154–170 (1991)

110. Fractions of the period of the continued fraction expansion of quadratic integers. Bull. Austral. Math. Soc. **44**, 155–169 (1991)

111. Obituary: Kurt Mahler, 1903–1988. J. Austral. Math. Soc. Ser. A **51**, 343–380 (1991)

112. Zeros of recurrence sequences (with Hans Peter Schickewei). Bull. Austral. Math. Soc. **44**, 215–223 (1991)

113. An infinite product with bounded partial quotients (with J.-P. Allouche, M. Mendès France). Acta Arith. **59**, 171–182 (1991)

114. The Eisenstein constant (with B.M. Dwork). Duke Math. J. **65**, 23–43 (1992)

115. Folded continued fractions (with J. Shallit). J. Number Theory **40**, 237–250 (1992)

116. A note on length and angle (with R.C. Talent). Nieuw Arch. Wisk. (4) **10**, 19–25 (1992).

117. On the number of zeros of exponential polynomials and related questions (with I.E. Shparlinski). Bull. Austral. Math. Soc. **46**, 401–412 (1992)

118. Recurrence sequences, continued fractions, and the HQT, prepared for *Proc. 23rd Annual Iranian Mathematics Conference*, ed. by S.M. Shahrtash (Bakhtaran, Iran, April 1992) Macquarie Number Theory Reports, No. 92–109 (August 1992)

119. Schneider's continued fraction. In *Number Theory with an Emphasis on the Markov Spectrum*, ed. by William Moran, Andrew D. Pollington (Provo, UT, 1991). Lecture Notes in Pure and Appl. Math., vol. 147 (Dekker, New York, 1993), pp. 271–281.

120. A specialised continued fraction (with J. Shallit). Canad. J. Math. **45**, 1067–1079 (1993)

121. Some infinite products with interesting continued fraction expansions (with C. Pinner, N. Saradha). J. Théor. Nombres Bordeaux **5**, 187–216 (1993)

122. Power series representing algebraic functions. In *Séminaire de Théorie des Nombres*, ed. by S. David (Paris, 1990–1991). Progr. Math., vol. 108 (Birkhäuser, Boston, MA, 1993), pp. 241–262.

123. Preparation of and foreword to English translation of *Classical Diophantine Equations* ed. by V.G. Sprindžuk. Lecture Notes in Mathematics, vol. 1559 (Springer, Berlin, 1993), pp. xii+228

124. Continued fractions of formal power series (dedicated to Paulo Ribenboim). In *Advances in Number Theory*, ed. by F. Gouvêa, N. Yui (Kingston, ON, 1991). Proceedings of the 3rd Conference of the CNTYA (Oxford Univ. Press, New York, 1993), pp. 453–466

125. Kurt Mahler 26 July 1903–26 February 1988 (with J.H. Coates). Biographical Memoirs of Fellows of the Royal Society (The Royal Society London) **39**, 265–279 (1994); Historical Records of Australian Science **9**, 369–385 (1993)

126. Corrigenda: The Eisenstein constant (with B. Dwork). Duke Math. J. **76**, 669–672 (1994)

127. Explicit formulas for units in certain quadratic number fields. In *Algorithmic Number Theory*, 1st Algorithmic Number Theory Symposium (Cornell University, Ithaca, NY, 1994). Lecture Notes in Comput. Sci., vol. 887 (Springer, Berlin, 1994), pp. 194–208

128. Remarks on Fermat's Last Theorem. Austral. Math. Soc. Gaz. **21**, 150–159 (1994)

129. Halfway to a solution of $X^2 - DY^2 = -3$ (with R.A. Mollin, H.C. Williams). J. Théor. Nombres Bordeaux **6**, 421–457 (1994)

130. Concerning commuting. Austral. Math. Soc. Gazette **21**, 68–68 (1994)

131. Factorisation in fractional powers. Acta Arith. **70**, 287–293 (1995)

132. A note on symmetry and ambiguity (with R. Mollin). Bull. Austral. Math. Soc. **51**, 215–233 (1995)

133. Computational algebra and number theory. Papers from the CANT2 Meeting held at Sydney University, Sydney, November 1992, ed. by W. Bosma, A. van der Poorten, Math. Appl., vol. 325 (Kluwer Acad. Publ., Dordrecht, 1995), pp. xiv+321

134. Continued fractions of algebraic numbers' (with E. Bombieri). In *Computational Algebra and Number Theory*, ed. by W. Bosma, A. van der Poorten (Sydney, 1992). Math. Appl., vol. 325 (Kluwer Acad. Publ., Dordrecht, 1995), pp. 137–152

135. The Lech-Mahler Theorem … with tears, Macquarie University Reports (1995), p. 7

136. Some problems concerning recurrence sequences (with G. Myerson). Amer. Math. Monthly **102**, 698–705 (1995)

137. On sequences of polynomials defined by certain recurrence relations (with I. Shparlinski). Acta Sci. Math. (Szeged) **61**, 77–103 (1995)

138. Determinants in the study of Thue's method and curves with prescribed singularities (with E. Bombieri and D. Hunt). Experiment. Math. **4**, 87–96 (1995)

139. Number-theoretic and algebraic methods in computer science. In *Proceedings of the International Conference (NTAMCS '93)* held in Moscow, June-July 1993, ed. by Alf J. van der Poorten, I.E. Shparlinskii, H.G. Zimmer (World Scientific Publishing Co., Inc, River Edge, NJ, 1995), pp. x+205

140. Explicit quadratic reciprocity. In *Number-Theoretic and Algebraic Methods in Computer Science*, Moscow, 1993 (World Sci. Publ., River Edge, NJ, 1995), pp. 175–180

141. Notes on Fermat's last theorem. In *Canadian Mathematical Society Series of Monographs and Advanced Texts* (Wiley Inc., New York 1996), pp. xviii+222

142. On linear recurrence sequences with polynomial coefficients (with I. Shparlinski). Glasgow Math. J. **38**, 147–155 (1996)

143. Continued fraction expansions of values of the exponential function and related fun with continued fractions. Nieuw Arch. Wisk. (4), **14**, 221–230 (1996)

144. Convergents of folded continued fractions (with J.-P. Allouche, A. Lubiw, M. Mendès France, J. Shallit). Acta Arith. **77**, 77–96 (1996)

145. A note on Hadamard roots of rational functions. Symposium on Diophantine Problems (dedicated to Wolfgang Schmidt on the occasion of his 60th birthday). Rocky Mountain J. Math. **26**, 1183–1197 (1996)

146. Substitution invariant Beatty sequences (with T. Komatsu). Japan J. Math. (N.S.) **22**, 349–354 (1996)

147. Effective measures of irrationality for cubic extensions of number fields (with E. Bombieri, J. Vaaler). Ann. Scuola Norm. Sup. Pisa Cl. Sci. (4), **23**, 211–248 (1996)

148. Chairman of Working Party appointed by the National Committee for Mathematics of the Australian Academy of Science. In *Mathematical Sciences: Adding to Australia* (Australian Research Council, Discipline Research Strategies) (Australian Government Publishing Service, Canberra, 1996), pp. *xxiii* + 120

149. A comparative study of algorithms for computing continued fractions of algebraic numbers (with R.P. Brent, H.J.J. te Riele). In *Algorithmic Number Theory*, ed. by Henri Cohen. Proc. 2nd International Symposium, ANTS-II, Talence, France, May 1996. Lecture Notes in Computer Science, vol. 1122 (Springer, Berlin, 1996), pp. 35–47.

150. Divisibility sequences and other Hadamard problems, abstract. In *International Conf. on Diophantine Analysis and its Applications*, in Honour of Acad. V.G. Sprindžuk (1936–1987), Minsk (Belarus), September 1996, 25–25

151. Factorisation in the ring of exponential polynomials (with G.R. Everest). Proc. Amer. Math. Soc. **125**, 1293–1298 (1997)

152. Pace Serge Lang (Letter to the Editor). Canad. Math. Soc. Notes **29**, 5–5 (1997)

153. Entries on Fermat's Last Theorem, #249; Skolem-Mahler-Lech Theorem, #453. In *Encyclopadia of Mathematics*, Editor-in-Chief, Prof. dr. M. Hazewinkel, supplement vol. I (Kluwer Academic Publishers, 1997)

154. A proof that Euler missed ... Apéry's proof of the irrationality of $\zeta(3)$; An informal report. The Mathematical Intelligencer **1**, 195–203 (1979). Reprinted in *Pi: A source book*, ed. by Lennart Berggren, Jonathan Borwein, Peter Borwein (Springer, 1997), pp. 439–447

155. Formal power series and their continued fraction expansion. In *Algorithmic Number Theory*, ed. by J. Buhler. Proc. 3rd International Symposium, ANTS-III, Portland, Oregon, 1998. Lecture Notes in Comput. Sci., vol. 1423 (Springer, Berlin, 1998), pp. 358–371

156. Speech in response on receiving a Dhc at Université Bordeaux I, 8 Jan 1998

157. Review of Kumiko Nishioka, *Mahler Functions and Transcendence*. Lecture Notes in Mathematics, vol. 1631 (Springer, 1996); Bull. Lond. Math. Soc. **30**, 663–664 (1998)

158. A note on Jacobi symbols and continued fractions (with P.G. Walsh). Amer. Math. Monthly, **106**, 52–56 (1999)

159. Beer and continued fractions with periodic periods. In *Number Theory* (Ottawa, ON, 1996). CRM Proc. Lecture Notes, vol. 19, Amer. Math. Soc., Providence, RI, (1999), pp. 309–314.

160. On certain continued fraction expansions of fixed period length (with H.C. Williams). Acta Arith. **89**, 23–35 (1999)

161. Values of the Dedekind eta function at quadratic irrationalities (with K.S. Williams). Canad. J. Math. **51**, 176–224 (1999); Corrigendum: ibid. **53**, 434–448 (2001)

162. On the distribution of solutions of Thue's equation (with B. Brindza, Á. Pintér, M. Waldschmidt). In *Number Theory in Progress*, Proc. Conf. in honour of Andrzej Schinzel on the occasion of his 60th birthday. Number Theory in Progress, ed. by K. Győry, H. Iwaniec, J. Urbanowicz, vol. 1 (de Gruyter, Berlin, 1999), pp. 35–46.

163. On lacunary formal power series and their continued fraction expansion (with M. Mendès France, J. Shallit). In *Number Theory in Progress* Proc. Conf. in honour of Andrzej Schinzel on the occasion of his 60th birthday. Number Theory in Progress, ed. by K. Győry, H. Iwaniec, J. Urbanowicz, vol. 1 (de Gruyter, Berlin, 1999), pp. 321–326.

164. Reduction of continued fractions of formal power series, in Continued fractions: from analytic number theory to constructive approximation, ed. by Bruce Berndt, Fritz Gesztesy, (Columbia, Missouri, May 1998). Contemp. Math., vol. 236, Amer. Math. Soc., Providence, RI, (1999), pp. 343–355

165. Zero testing of p-adic and modular polynomials (with M. Karpinski, I. Shparlinski). J. Theoret. Comp. Sci. **233**, 309–317 (2000)

166. Continued fractions, Jacobi symbols and quadratic Diophantine equations (with R. Mollin). Canad. Math. Bull. **43**, 218–225 (2000)

167. On periods of elements from real quadratic fields (with E. Burger). In *Constructive, Experimental, and Nonlinear Analysis*, Limoges, 1999. CMS Conf. Proc., vol. 27, Amer. Math. Soc., Providence, RI, (2000), pp. 35–43.

168. Quasi-elliptic integrals and periodic continued fractions (with X. Tran). Monatsh. Math. **131**, 155–169 (2000)

169. La dimostrazione dell'ultimo teorema di Fermat. In *Enciclopedia Italiana*, Appendice 2000, Roma 2000, pp. 341–344

170. Entry on Fermat-Goss-Denis Theorem. In *Encyclopaedia of Mathematics*, Editor-in-Chief, Prof. dr. M. Hazewinkel, supplement vol. 2 (Kluwer Academic Publishers, 2000), pp. 217–218

171. Voordeliger dan science fiction (Konrad Knopp, five Dover books on the theory of functions, in 1960 prices at US$1.35 per volume): Welk wiskundeboek heeft in uw leven op u de meeste undruk gemaakt? Nieuw Arch. Wisk. (5) **1**, 169–170 (2000)

172. Japanese translation by Itaru Yamaguchi of *Notes on Fermat's Last Theorem* (Morikita, Tokyo, 2000), p. 285

173. Computer verification of the Ankeny-Artin-Chowla conjecture for all primes less than 100 000 000 000 (with H. te Riele, H. Williams). Math. Comp., **70**, 1311–1328 (2001)

174. On number theory and Kustaa Inkeri. In *Number theory*, Proc. Turku Symposium on Number Theory in Memory of Kustaa Inkeri, Tutku, May 31–June 4, 1999, ed. by Matta Jutila, Tauno Metsänkylä (de Gruyter, Berlin, 2001), pp. 281–292

175. A powerful determinant. Experiment. Math. **10**, 307–320 (2001)

176. Non-periodic continued fractions in hyperelliptic function fields (Dedicated to George Szekeres on his 90th birthday). Bull. Austral. Math. Soc. **64**, 331–343 (2001)

177. The Hermite-Serret algorithm and $12^2 + 33^2$. In *Cryptography and Computational Number Theory* (CCNT'99), (National University of Singapore, 1999), ed. by K.-Y. Lam, I. Shparlinski, H. Wang, C. Xing (Birkhäuser, Basel, 2001), pp. 129–136

178. Review of C.J. Mozzochi, *The Fermat Diary*, American Mathematical Society, Providence, RI, 2000, pp. xii+196. In Austral. Math. Soc. Gazette **28**, 164–165 (2001)

179. Review of Andrzej Schinzel (with an appendix by Umberto Zannier), *Polynomials with Special Regard to Reducibility* (Cambridge University Press, Cambridge, 2000), pp. x+558. In Encyclopedia of mathematics and its applications 77. Nieuw Arch. Wisk. (5) **2**, 281–282 (2001)

180. Squares from products of consecutive integers (with G.J. Woeginger). Amer. Math. Monthly **109**, 459–462 (2002)

181. Life on the edge. Amer. Math. Monthly, **109**, 850–853 (2002)

182. Ideal constructions and irrationality measures of roots of algebraic numbers (with P.B. Cohen-Tretkoff). Illinois J. Math. **46**, 63–80 (2002)

183. Integer sequences and periodic points (with G. Everest, Y. Puri, T. Ward). J. Integer Seq. **5**, (2002) Article 02.2.3 (electronic), p.10

184. Binary quadratic forms and the eta function (with R. Chapman). In *Number Theory for the Millennium*, Proc. Millennial Conf. Number Theory, I (Urbana, IL, 2000) ed. by M. Bennett et al. (A K Peters, Natick, MA, 2002), pp. 215–227.

185. Symmetry and folding of continued fractions (dedicated to Michel Mendès France on his 65th birthday). J. Théor. Nombres Bordeaux **14**, 603–611 (2002)

186. Computational aspects of NUCOMP (with M. Jacobson Jr). In *Algorithmic Number Theory*, ed. by C. Fieker, D. Kohel. Proc. 5th International Symposium, ANTS-V, Sydney, NSW, Australia. Lecture Notes in Computer Science, vol. 369 (Springer, Berlin, 2002), pp. 120–133

187. Periodic continued fractions in elliptic function fields. (with X. C. Tran). In *Algorithmic Number Theory*, ed. by C. Fieker, D. Kohel. Proc. Fifth International Symposium, ANTS-V, Sydney, NSW, Australia. Lecture Notes in Computer Science, vol. 2369 (Springer, Berlin, 2002), pp. 390–404

188. Corrigenda and addition to: Computer verification of the Ankeny-Artin-Chowla conjecture for all primes less than 100 000 000 000 (with H. te Riele, H. Williams). Math. Comp. **72**(241), 521–523 (2003)

189. A note on NUCOMP. Math. Comp. **72**(244), 1935–1946 (2003) (electronic)

190. Recurrence sequences (with G. Everest, I. Shparlinski, T. Ward), *Mathematical Surveys and Monographs*, vol. 104, American Mathematical Society, Providence, RI, (2003), pp. xiv+318

191. Peer refereeing … will it be missed? In *Electronic Information and Communication in Mathematics*, ed. by Fengshan Bai, Bernd Wegner. Lecture Notes in Computer Science, vol. 2730 (Springer, Berlin, 2003), pp. 132–143

192. Three views of peer review (with Steven Krantz, Greg Kuperberg). Notice. Amer. Math. Soc. **50**, 678–682 (2003)

193. Periodic continued fractions and elliptic curves. In *High Primes and Misdemeanours*, lectures in honour of the 60th birthday of H. Williams, ed. by A. van der Poorten, A. Stein. Fields Inst. Commun., vol. 41, Amer. Math. Soc., Providence, RI, (2004), pp. 353–365

194. Jeepers, creepers, . . . (with R. Patterson). In *High Primes and Misdemeanours*, lectures in honour of the 60th birthday of H. Williams, ed. by A. van der Poorten, A. Stein. Fields Inst. Commun., vol. 41, Amer. Math. Soc., Providence, RI, (2004), pp. 305–316.

195. Advice to Referees . . . (with A. Stein). In *High Primes and Misdemeanours*, lectures in honour of the 60th birthday of H. Williams, ed. by A. van der Poorten, A. Stein. Fields Inst. Commun., vol. 41, Amer. Math. Soc., Providence, RI, (2004), pp. 391–392.

196. High primes and misdemeanours: lectures in honour of the 60th birthday of Hugh Cowie Williams. Selected papers from the *International Conference on Number Theory* held in Banff, AB, May 24–30, 2003, ed. by A. van der Poorten, A. Stein. Fields Inst. Commun., vol. 41, Amer. Math. Soc., Providence, RI, (2004), pp. xiv+318

197. Quadratic irrational integers with partly prescribed continued fraction expansion (in memory of Professor Dr Bela Brindza). Publ. Math. Debrecen **65**, 481–496 (2004)

198. On the twin prime conjecture (with Teur Ah-Mah). Aust. Math. Soc. Gaz. **31**, 36–39 (2004)

199. Squares from products of integers (with W. Banks). Aust. Math. Soc. Gaz. **31**, 40–42 (2004)

200. Elliptic curves and continued fractions. J. Integer Seq., **8**, (2005) Article 05.2.5 (electronic), p. 19

201. Curves of genus 2, continued fractions, and Somos sequences. J. Integer Seq., **8**, (2005) Article 05.3.4 (electronic), p. 9

202. Specialisation and reduction of continued fraction expansions of formal power series (to J.-L. Nicolas on his 60th birthday). Ramanujan J. **9**, 83–91 (2005)

203. Pseudo-elliptic integrals, units, and torsion (with F. Pappalardi). J. Austral. Math. Soc. **79**, 335–347 (2005)

204. Review of *automatic sequences*, by J.-P. Allouche, J. Shallit (Cambridge University Press, 2003). *Math. Comp.* **74**, 1039–1040 (2005)

205. Recurrence relations for elliptic sequences: every Somos 4 is a Somos k (with C. Swart). Bull. London Math. Soc. **38**, 546–554 (2006)

206. Exactly one hundred nontrivial composites. Austral. Math. Soc. Gaz. **33**, 326–327 (2006)

207. Determined sequences, continued fractions, and hyperelliptic curves. In *Algorithmic Number Theory*, Proc. Seventh International Symposium, ANST 7, TU Berlin July 2006. Lecture Notes in Computer Science, vol. 4076 (Springer, Berlin, 2006), pp. 393–405

208. Hyperelliptic curves, continued fractions, and Somos sequences. In *Proc. RIMS Conference on Analytic Number Theory and Related Areas*, RIMS August 2006 (Kyoto, October 18–22, 2004), pp. 98–107

209. Hyperelliptic curves, continued fractions, and Somos sequences. In *Dynamics and Stochastics*, IMS Lecture Notes Monogr. Ser., vol. 48 (Inst. Math. Statist., Beachwood, OH, 2006), pp. 212–224

210. Characterization of a generalized Shanks sequence (with R. Patterson, H. Williams). Pacific J. Math. **230**, 185–215 (2007)

211. Arthur Bertram Chandler (1912–1984), science fiction author. Austral. Diction. Biography **17**, 204–205 (2007)

212. A curious cubic identity and self-similar sums of squares (with K. Thomsen and M. Wiebe). Math. Intelligencer **29**, 39–41 (2007)

213. Continued fractions in function fields defined over an infinite field . . . with emphasis on the quadratic case. In *Proceedings of the Conference on Algorithmic Number Theory*, ed. by A.-M. Ernvall-Hytöten, M. Julita, J. Karhumäki, A. Lepistö, vol. 46 (Turku Centre for Computer Science General Publication, 2007), pp. 132–141

214. Review of *Introduction to Modern Number Theory*, by Yu. Manin, A. Panchishkin, 2nd edn. Aust. Math. Soc. Gaz. **34**, 47–48 (2007)

215. Algorithmic number theory, *Proceedings of the 8th International Symposium* (ANTS-VIII) held in Banff, AB, May 17–22, 2008, ed. by Alfred J. van der Poorten, Andreas Stein. Lecture Notes in Computer Science, vol. 5011 (Springer, Berlin, 2008), pp. x+455
216. Sequences of Jacobian varieties with torsion divisors of quadratic order (with R. Patterson, H. Williams). Funct. Approx. Comment. Math. **39**, 345–360 (2008)
217. Self-similar values of quadratic forms: a remark on pattern and duality (with R. Breu). J. Recreat. Math. **35**, 202–212 (2010)
218. A simple observation on simple zeros (with Á. Pintér). Arch. Math. (Basel) **95**, 355–361 (2010)
219. Neverending Fractions (with Jonathan M. Borwein, Jeffrey Shallit, Wadim Zudilin). Australian Mathematical Society Lecture Series (Cambridge University Press) (To appear)

Ramanujan–Sato-Like Series

Gert Almkvist and Jesús Guillera

Abstract Using the theory of Calabi–Yau differential equations we obtain all the parameters of Ramanujan–Sato-like series for $1/\pi^2$ as q-functions valid in the complex plane. Then we use these q-functions together with a conjecture to find new examples of series of non-hypergeometric type. To motivate our theory we begin with the simpler case of Ramanujan–Sato series for $1/\pi$.

Key words Ramanujan–Sato-like series • Examples of complex series for $1/\pi$ • Calabi–Yau differential equations • Mirror map • Yukawa coupling • Examples of non-hypergeometric series for $1/\pi^2$

1 Introduction

In his famous paper in 1914 S. Ramanujan published 17 formulas for $1/\pi$ [18], all of hypergeometric form

$$\sum_{n=0}^{\infty} \frac{(1/2)_n (s)_n (1-s)_n}{n!^3} (a+bn) z^n = \frac{1}{\pi}.$$

Here $(c)_n = c(c+1)(c+2)\cdots(c+n-1)$ is the Pochhammer symbol, $s = 1/2, 1/3$, $1/4$, or $1/6$ and z, b, a are algebraic numbers. The most impressive is

G. Almkvist (✉)
Institute of Algebraic Meditation, Fogdaröd 208, 24333 Höör, Sweden
e-mail: gert.almkvist@yahoo.se

J. Guillera
Avda. Cesreo Alierta 31, Zaragoza, Spain
e-mail: jguillera@gmail.com

J.M. Borwein et al. (eds.), *Number Theory and Related Fields: In Memory of Alf van der Poorten*, Springer Proceedings in Mathematics & Statistics 43,
DOI 10.1007/978-1-4614-6642-0_2, © Springer Science+Business Media New York 2013

$$\sum_{n=0}^{\infty} \frac{\left(\frac{1}{2}\right)_n \left(\frac{1}{4}\right)_n \left(\frac{3}{4}\right)_n}{(1)_n^3} \frac{1}{99^{4n}} (26390n + 1103) = \frac{9801\sqrt{2}}{4\pi}, \tag{1.1}$$

which gives eight decimal digits of π per term. All the 17 series were rigorously proved in 1987 by the Borwein brothers [9]. Independently, the Borwein [9] and the Chudnovsky brothers [12] studied and proved Ramanujan series of the form

$$\sum_{n=0}^{\infty} \frac{(6n)!}{(3n)!n!^3} (a + bn)z^n = \frac{1}{\pi}. \tag{1.2}$$

The value of z can be found in the following way: Let us take the Chudnovsky brothers series (of Ramanujan type with $s = 1/6$)

$$\sum_{n=0}^{\infty} \frac{(6n)!}{(3n)!n!^3} (10177 + 261702n) \frac{1}{(-5280^3)^n} = \frac{880^2\sqrt{330}}{\pi}.$$

The series

$$w_0 = \sum_{n=0}^{\infty} \frac{(6n)!}{(3n)!n!^3} z^n = \sum_{n=0}^{\infty} 12^{3n} \cdot \frac{\left(\frac{1}{2}\right)_n \left(\frac{1}{6}\right)_n \left(\frac{5}{6}\right)_n}{(1)_n^3} z^n$$

satisfies the differential equation

$$\left(\theta^3 - 24z(2\theta + 1)(6\theta + 1)(6\theta + 5)\right) w_0 = 0,$$

where $\theta = zd/dz$. A second solution is

$$w_1 = w_0 \ln z + 744z + 562932z^2 + 570443360z^3 + \cdots .$$

Define

$$q = \exp\left(\frac{w_1}{w_0}\right) = z + 744z^2 + 750420z^3 + 872769632z^4 + \cdots .$$

Then

$$J(q) = \frac{1}{z(q)} = \frac{1}{q} + 744 + 196884q + 21493760q^2 + \cdots$$

is the famous modular invariant and

$$J\left(-e^{-\pi\sqrt{67}}\right) = -5280^3.$$

A similar construction, getting a different $J := 1/z$, can be made starting with any third order differential equation which is the symmetric square of a second-order differential equation. This kind of series are called Ramanujan–Sato series for $1/\pi$ because Sato discovered the first example of this type, one involving the Apéry numbers [10].

Similarly, the first formulas for $1/\pi^2$, found by the second author, were of hypergeometric type, using a function

$$w_0 = \sum_{n=0}^{\infty} \frac{(1/2)_n (s_1)_n (1-s_1)_n (s_2)_n (1-s_2)_n}{n!^5} z^n,$$

where the 14 possible pairs (s_1, s_2) are given in [14] or [6] and w_0 satisfies a fifth-order differential equation

$$\left(\theta^5 - z(\theta + \frac{1}{2})(\theta + s_1)(\theta + 1 - s_1)(\theta + s_2)(\theta + 1 - s_2) \right) w_0 = 0.$$

This differential equation is of a very special type. It is a Calabi–Yau equation with a fourth-order pullback with solutions y_0, y_1, y_2, y_3, where

$$w_0 = y_0 (\theta y_1) - (\theta y_0) y_1.$$

This was used in [6, 14], where one new hypergeometric formula was found. Unfortunately, fifth-order Calabi–Yau differential equations are quite rare. The simplest non-hypergeometric cases are Hadamard products of second- and third-order equations (labeled $A * \alpha$, etc., in [5]). Seven formulas of this kind have been found [6], like, for example,

$$\sum_{n=0}^{\infty} \binom{2n}{n}^2 \sum_{i=0}^{n} \frac{(-1)^i 3^{n-3i} (3i)!}{i!^3} \binom{n}{3i} \binom{n+i}{i} \frac{(-1)^n}{36^n} (803n^2 + 416n + 68) = \frac{486}{\pi^2},$$

$$(1.3)$$

which involves the Almkvist–Zudilin numbers. Two of the others were proved by Zudilin [20]. In this paper we explore more complicated fifth-order equations, most of them found by the first author (#130 was found by Verrill).

To find q_0 in the $1/\pi$ case, we solve the equation $\alpha(q) = \alpha_0$, where α_0 is a rational and

$$\alpha(q) = \frac{\ln^2 |q|}{\pi^2}.$$

The real solutions are $q_0 = \pm e^{-\pi \sqrt{\alpha_0}}$. As there are many examples in the literature with q_0 real, in this paper we will show some series corresponding to $q_0 = e^{i \pi r_0} e^{-\pi \sqrt{\alpha_0}}$, where r_0 is a rational such that $e^{i \pi r_0}$ is complex. If we calculate $J_0 = J(q_0)$, then $z_0 = 1/J_0$. In the $1/\pi^2$ case we have two functions,

$$\alpha(q) = \frac{\frac{1}{6} \ln^3 |q| - T(q) - h\zeta(3)}{\pi^2 \ln |q|}, \quad \tau(q) = \frac{\frac{1}{2} \ln^2 |q| - (\theta_q T)(q)}{\pi^2} - \alpha(q), \quad (1.4)$$

where h is an invariant and $T(q)$ essentially is the Gromov–Witten potential in string theory. Solving the equation $\alpha(q) = \alpha_0$ numerically, where α_0 is rational, we get an approximation of q_0. Replacing q_0 in the second equation, we get τ_0. We conjecture that the corresponding series is of Ramanujan type for $1/\pi^2$ if, and only if, τ_0^2 is also rational. The success in finding the examples of this paper depends heavily on our experimental method to get the invariant h. It uses the critical value $z = z_c$, the radius of convergence for the power series w_0. From the conjecture $(dz/dq)(q_c) = 0$ we get an approximation of q_c and using the PSLQ algorithm to find an integer relation among the numbers

$$\frac{\ln^3 |q_c|}{6} - T(q_c), \qquad \pi^2 \ln |q_c|, \qquad \zeta(3),$$

we obtain simultaneously α_c and the invariant h. Replacing α_c and q_c in the second equation we get τ_c.

In the $1/\pi$ case, the algebraic but nonrational z_0 dominate the rational solutions (see the tables in [4]). In the case $1/\pi^2$ the only known series with a nonrational z_0 is

$$\sum_{n=0}^{\infty} \frac{\left(\frac{1}{2}\right)_n^3 \left(\frac{1}{3}\right)_n \left(\frac{2}{3}\right)_n}{(1)_n^5} \left(\frac{15\sqrt{5} - 33}{2}\right)^{3n} \times$$

$$\left[(1220/3 - 180\sqrt{5})n^2 + (303 - 135\sqrt{5})n + (56 - 25\sqrt{5})\right] = \frac{1}{\pi^2}, \quad (1.5)$$

which was discovered by the second author [15]. See also the corresponding mosaic supercongruences in [16].

We obtain the q-functions for all the parameters of general Ramanujan–Sato-like series for $1/\pi$ and $1/\pi^2$. Contrary to the series for $1/\pi$ in which everything can be proved rigorously by means of modular equations, in the case $1/\pi^2$, we can only evaluate the functions numerically and then guess the algebraic values when they exist. A modular-like theory which explains the rational and algebraic quantities observed is still not available [19]. For an excellent account of these topics, see [22].

2 Ramanujan–Sato-Type Series for $1/\pi$

Certain differential equations of order 3 are the symmetric square of a differential equation of second order. Suppose

$$\theta^3 w = e_2(z)\theta^2 w + e_1(z)\theta w + e_0(z)w, \qquad \theta = z\frac{d}{dz}, \qquad (2.1)$$

is the symmetric square of the second-order equation

$$\theta^2 y = c_1(z)\theta y + c_0(z)y, \qquad 3c_1(z) = e_2(z). \qquad (2.2)$$

We define the following function

$$P(z) = \exp \int \frac{-2c_1(z)}{z} dz,$$

with $P(0) = 1$, which plays an important role in the theory. In the examples of this paper $P(z)$ is a polynomial but we have also found examples for which $P(z)$ is a rational function.

The fundamental solutions w_0, w_1, w_2 of the third-order differential equation are connected to the fundamental solutions y_0, y_1 of the second-order equation by

$$w_0 = y_0^2, \quad w_1 = y_0 y_1, \quad w_2 = \frac{1}{2} y_1^2 \tag{2.3}$$

[7, Prop. 9]. We define the wronskians

$$W(w_i, w_j) = \begin{vmatrix} w_i & \theta w_i \\ w_j & \theta w_j \end{vmatrix}, \quad W(y_0, y_1) = \begin{vmatrix} y_0 & \theta y_0 \\ y_1 & \theta y_1 \end{vmatrix}.$$

Observe that this notation is not the same as in [1], where in the definition of $W(y_0, y_1)$ we have y_0' and y_1' instead of θy_0 and θy_1.

Theorem 2.1. *We have*

$$W(w_0, w_1) = \frac{y_0^2}{\sqrt{P}}, \quad W(w_0, w_2) = \frac{y_0 y_1}{\sqrt{P}}, \quad W(w_1, w_2) = \frac{y_1^2}{2\sqrt{P}}. \tag{2.4}$$

Proof. Using (2.3), we get

$$W(w_0, w_1) = y_0^2 W(y_0, y_1), \quad W(w_0, w_2) = y_0 y_1 W(y_0, y_1)$$

and

$$W(w_1, w_2) = \frac{1}{2} y_1^2 W(y_0, y_1).$$

If we denote with f the wronskian $W(y_0, y_1)$, then from (2.2), we see that $\theta f = c_1(z) f$. This implies $f = 1/\sqrt{P}$. $\qquad\square$

2.1 Series for $1/\pi$

Let $q = e^{i\pi r} e^{-\pi \tau}$. If the function

$$w_0(z) = \sum_{n=0}^{\infty} A_n z^n$$

satisfies a differential equation of order 3 as above, then we will find two functions $b(q)$ and $a(q)$ with good arithmetical properties, such that

$$\sum_{n=0}^{\infty} A_n \left(a(q) + b(q)n \right) z^n(q) = \frac{1}{\pi}.$$

The interesting cases are those with z, b, a algebraic. They are called Ramanujan–Sato-type series for $1/\pi$.

The usual q-parametrization is

$$q = \exp\left(\frac{y_1}{y_0}\right) = \exp\left(\frac{w_1}{w_0}\right),$$

and we can invert it to get z as a series of powers of q. The function $z(q)$ is the mirror map, and for this kind of differential equations, it has been proved that it is a modular function. We also define $J(q) := 1/z(q)$.

Theorem 2.2. *The functions* $\alpha(q)$, $b(q)$, $a(q)$ *such that*

$$\sum_{j=0}^{2} \left[(w_j)a + (\theta w_j)b \right] x^j = e^{i\pi rx} \left(\frac{1}{\pi} - \frac{\pi}{2}\alpha x^2 \right) \text{ truncated at } x^3 \qquad (2.5)$$

are given by

$$\alpha(q) = \tau^2(q), \quad b(q) = \tau(q)\sqrt{P(z)}, \quad a(q) = \frac{1}{\pi w_0}\left(1 + \frac{\ln|q|}{w_0} q \frac{dw_0}{dq} \right). \qquad (2.6)$$

In addition, if r and τ_0^2 are rational, then $z(q_0)$, $b(q_0)$, $a(q_0)$ are algebraic.

Proof. First, we see that $q = e^{i\pi r} e^{-\pi \tau}$ implies that

$$\tau(q) = -\frac{\ln|q|}{\pi}.$$

We can write (2.5) in the following equivalent form:

$$(w_0)a + (\theta w_0)b = \frac{1}{\pi},$$

$$(w_1)a + (\theta w_1)b = ir, \qquad (2.7)$$

$$(w_2)a + (\theta w_2)b = -\frac{\pi}{2}(\alpha + r^2).$$

In what follows, we will use the wronskians (2.4). As we want this system to be compatible, we have

$$\begin{vmatrix} w_0 & \theta w_0 & \frac{1}{\pi} \\ w_1 & \theta w_1 & ir \\ w_2 & \theta w_2 & -\frac{\pi}{2}(\alpha + r^2) \end{vmatrix} = 0. \tag{2.8}$$

Expanding along the last column, we get

$$\frac{1}{2\pi}\left(\frac{y_1}{y_0}\right)^2 - ir\left(\frac{y_1}{y_0}\right) - \frac{\pi}{2}(\alpha + r^2) = 0.$$

Hence

$$\frac{1}{\pi}\frac{\ln^2 q}{2} - ir\ln q - \frac{\pi}{2}(\alpha + r^2) = 0.$$

As $\ln q = \ln|q| + i\pi r$, we obtain the function $\alpha(q)$. To obtain b we apply Cramer's method to the system formed by the two first equations of (2.7). We get

$$b = \left(ir - \frac{1}{\pi}\frac{w_1}{w_0}\right)\sqrt{P(z)} = \left(\frac{i\pi r - \ln q}{\pi}\right)\sqrt{P(z)} = -\frac{\ln|q|}{\pi}\sqrt{P(z)}.$$

Then, replacing w_1 with $w_0 \ln q$ in the second equation of (2.7) and solving the system formed by the two first equations, we obtain the identity

$$w_0 = \frac{q}{z\sqrt{P(z)}}\frac{dz}{dq}.$$

Finally, using the two last formulas and the first equation of (2.7), we derive the formula for $a(q)$ in (2.6). From $b = \tau\sqrt{1-z}$ we see that b takes algebraic values when r and τ^2 are rational. By an analogue to the argument given in [14, Sect. 2.4], we see that the same happens to $a(q)$. □

2.2 Examples of Series for $1/\pi$

There are many examples in the literature (see [8] and the references in it), but until very recently, all of them were with $r = 0$ (series of positive terms) or with $r = 1$ (alternating series). The first example of a complex series was found and proved, with a hypergeometric transformation, by the second author and Wadim Zudilin in [17, Eq. 44]. Other complex series, proved by modular equations or hypergeometric transformations, are in [11], like, for example,

$$\sum_{n=0}^{\infty}\frac{(4n)!}{n!^4}\left(\frac{10+2\sqrt{-3}}{28\sqrt{3}}\right)^{4n}\left((320-55\sqrt{-3})n + (52-12\sqrt{-3})\right) = \frac{98\sqrt{3}}{\pi}.$$
$$\tag{2.9}$$

Tito Piezas (Ramanujan-type complex series available at Tito Piezas's web-site, personal communication)found numerically and then guessed the series

$$\sum_{n=0}^{\infty} \frac{(2n)!(3n)!}{n!^5} \frac{3+(17-i)n}{(2(7+i)(2+i)^4)^n} = \frac{33-6i}{4}\frac{1}{\pi},\qquad (2.10)$$

which involves only Gaussian rational numbers. It leads to taking $q = e^{\frac{2\pi i}{3}}e^{-\frac{4\sqrt{2}}{3}\pi}$, and of course, it is possible to prove it rigorously using modular equations. For our following examples, we have chosen the sequence of numbers

$$A_n = \binom{2n}{n}\sum_{k=0}^{n}\binom{n}{k}\binom{2k}{k}\binom{2n-2k}{n-k},$$

which is the Hadamard product $\binom{2n}{n} * (d)$ (see [7]). The differential equation is

$$\left(\theta^3 - 8z(2\theta+1)(3\theta^2+3\theta+1) + 128z^2(\theta+1)(2\theta+1)(2\theta+3)\right)w = 0.$$

The polynomial $P(z)$ is $P(z) = (1-16z)(1-32z)$ and

$$J(q) = \frac{1}{z(q)} = \frac{1}{q} + 16 + 52q + 834q^3 + 4760q^5 + 24703q^7 + \cdots.$$

For $q = ie^{-\pi\frac{\sqrt{13}}{2}}$, we find

$$\sum_{n=0}^{\infty} A_n \frac{(-1+6i)+(-9+33i)n}{(16+288i)^n} = \frac{52+91i}{\sqrt{13(1+18i)^3}} \cdot \frac{50}{\pi}.\qquad (2.11)$$

For $q = ie^{-\pi\frac{\sqrt{37}}{2}}$, we get

$$\sum_{n=0}^{\infty} A_n \frac{(11842+11741i)+112665(1+i)n}{(16-14112i)^n} = \left(\frac{37}{1+882i}\right)^{\frac{3}{2}} \cdot \frac{2\cdot5^3\cdot29^3}{\pi}.\qquad (2.12)$$

Taking $q = e^{i\frac{\pi}{4}}e^{-\pi\frac{\sqrt{15}}{4}}$, we find

$$\sum_{n=0}^{\infty} A_n \left[4(-3\sqrt{3}+13i)+3(29\sqrt{3}-44i)n\right]\left(\frac{2}{65-15\sqrt{3}i}\right)^n = \frac{14^2\sqrt{5}}{\pi}.\qquad (2.13)$$

We give a final example with the sequence of Domb's numbers [10], called (α) in [7]

$$A_n = \sum_{k=0}^{n}\binom{n}{k}^2\binom{2k}{k}\binom{2n-2k}{n-k}.$$

The differential equation is

$$\left(\theta^3 - 2z(2\theta + 1)(5\theta^2 + 5\theta + 2) + 64z^2(\theta + 1)^3 \right) w = 0.$$

We have $P(z) = (1 - 4z)(1 - 16z)$ and

$$J(q) = q^{-1} + 6 + 15q + 32q^2 + 87q^3 + 192q^4 + \cdots.$$

For $q = e^{\frac{i\pi}{3}} e^{-\frac{2\sqrt{2}}{3}\pi}$, we find

$$\sum_{n=0}^{\infty} A_n \frac{(1+i) + (4+2i)n}{(16 + 16i)^n} = \frac{6}{\pi}. \tag{2.14}$$

Taking the real and imaginary parts, we get

$$\sum_{n=0}^{\infty} A_n \frac{1 + 4n}{32^n} (\sqrt{2})^n \cos \frac{n\pi}{4} + \sum_{n=0}^{\infty} A_n \frac{1 + 2n}{32^n} (\sqrt{2})^n \sin \frac{n\pi}{4} = \frac{6}{\pi},$$

and

$$\sum_{n=0}^{\infty} A_n \frac{1 + 2n}{32^n} (\sqrt{2})^n \cos \frac{n\pi}{4} = \sum_{n=0}^{\infty} A_n \frac{1 + 4n}{32^n} (\sqrt{2})^n \sin \frac{n\pi}{4}.$$

The first author is preparing a collection of series for $1/\pi$ in [4]. Although we have guessed our examples from numerical approximations, the exact evaluations can be proved rigorously by using modular equations [11].

3 Ramanujan–Sato-Like Series for $1/\pi^2$

A Calabi–Yau differential equation is a fourth-order differential equation

$$\theta^4 y = c_3(z)\theta^3 y + c_2(z)\theta^2 y + c_1(z)\theta y + c_0(z)y, \qquad \theta = z\frac{d}{dz}, \tag{3.1}$$

where $c_i(z)$ are quotients of polynomials of z with rational coefficients, which satisfies several conditions [6]. There are two functions associated to these equations which play a very important role, namely, the mirror map and the Yukawa coupling. The mirror map $z(q)$ is defined as the functional inverse of

$$q = \exp(\frac{y_1}{y_0})$$

and the Yukawa coupling as

$$K(q) = \theta_q^2 \left(\frac{y_2}{y_0} \right), \qquad \theta_q = q \frac{d}{dq}.$$

We define $T(q)$ as the unique power series of q such that $T(0) = 0$ and

$$\theta_q^3 T(q) = 1 - K(q).$$

The function

$$\Phi = \frac{1}{2} \left(\frac{y_1}{y_0} \frac{y_2}{y_0} - \frac{y_3}{y_0} \right) = \frac{1}{6} \ln^3 q - T(q) \tag{3.2}$$

is well known in string theory and is called the Gromov–Witten potential (see [13, p. 28]).

3.1   Pullback

The solution of some differential equations of fifth order can be recovered from the solutions of a fourth-order Calabi–Yau differential equation. We say that they admit a pullback. If (3.1) is the ordinary pullback of the differential equation

$$\theta^5 w = e_4(z)\theta^4 w + e_3(z)\theta^3 w + e_2(z)\theta^2 w + e_1(z)\theta w + e_0(z)w, \tag{3.3}$$

then we know that w_0, w_1, w_2, w_3, w_4 can be recovered from the four fundamental solutions y_0, y_1, y_2, y_3 of (3.1) in the following way:

$$w_0 = \begin{vmatrix} y_0 & y_1 \\ \theta y_0 & \theta y_1 \end{vmatrix}, \quad w_1 = \begin{vmatrix} y_0 & y_2 \\ \theta y_0 & \theta y_2 \end{vmatrix}, \quad w_3 = \frac{1}{2} \begin{vmatrix} y_1 & y_3 \\ \theta y_1 & \theta y_3 \end{vmatrix}, \quad w_4 = \frac{1}{2} \begin{vmatrix} y_2 & y_3 \\ \theta y_2 & \theta y_3 \end{vmatrix},$$

$$\tag{3.4}$$

$$w_2 = \begin{vmatrix} y_0 & y_3 \\ \theta y_0 & \theta y_3 \end{vmatrix} = \begin{vmatrix} y_1 & y_2 \\ \theta y_1 & \theta y_2 \end{vmatrix}. \tag{3.5}$$

We define the following function

$$P(z) = \exp \int \frac{-2c_3(z)}{z} dz,$$

with $P(0) = 1$, which plays an important role in the theory. In the Yifan Yang's pullback the corresponding coefficient is $4c_3(z)$ instead of $c_3(z)$. In all the examples of this paper $P(z)$ is a polynomial.

We denote as $W(w_i, w_j, w_j)$ and $W(w_i, w_j)$ the following wronskians [1]:

$$W(w_i, w_j, w_k) = \begin{vmatrix} w_i & \theta w_i & \theta^2 w_i \\ w_j & \theta w_j & \theta^2 w_j \\ w_k & \theta w_k & \theta^2 w_k \end{vmatrix}, \qquad W(w_i, w_j) = \begin{vmatrix} w_i & \theta w_i \\ w_j & \theta w_j \end{vmatrix}. \tag{3.6}$$

Due to different definition and notation, f in [1] is $1/\sqrt[4]{P(z)}$ here and the powers of x (z here) do not appear now. We will need the following wronskians of order 3 (see [1]):

$$W(w_1, w_2, w_3) = \frac{1}{2} \frac{y_1 y_2 - y_0 y_3}{\sqrt{P}}, \quad W(w_0, w_2, w_3) = \frac{y_1^2}{\sqrt{P}}, \quad W(w_0, w_1, w_3) = \frac{y_0 y_1}{\sqrt{P}},$$

$$W(w_0, w_1, w_2) = \frac{y_0^2}{\sqrt{P}}, \quad W(w_0, w_1, w_4) = \frac{y_0 y_2}{\sqrt{P}}, \quad W(w_1, w_2, w_4) = \frac{y_2^2}{2\sqrt{P}},$$

and

$$W(w_0, w_2, w_4) = \frac{y_0 y_3 + y_1 y_2}{2\sqrt{P}}.$$

We will also need the following wronskians of order 2 (see [1]):

$$W(w_0, w_1) = \frac{y_0^2}{\sqrt[4]{P}}, \quad W(w_0, w_2) = \frac{y_0 y_1}{\sqrt[4]{P}}, \quad W(w_1, w_2) = \frac{y_0 y_2}{\sqrt[4]{P}}.$$

3.2 Series for $1/\pi^2$

Suppose that the function

$$w_0(z) = \sum_{n=0}^{\infty} A_n z^n, \tag{3.7}$$

is a solution of a fifth-order differential equation which has a pullback to a Calabi–Yau differential equation. We will determine functions $a(q)$, $b(q)$, $c(q)$ in terms of $\ln|q|$, $z(q)$ and $T(q)$, such that

$$\sum_{n=0}^{\infty} A_n z(q)^n (a(q) + b(q)n + c(q)n^2) = \frac{1}{\pi^2}.$$

The interesting cases are those for which z, c, b, a are algebraic numbers. We will call them Ramanujan–Sato-like series for $1/\pi^2$. In this paper we improve and generalize to the complex plane the theory developed in [6, 15]. We let $q = |q| e^{i\pi r}$ and consider an expansion of the form

$$\sum_{j=0}^{4} \left[(w_j)a + (\theta w_j)b + (\theta^2 w_j)c \right] x^j$$

$$= e^{i\pi rx} \left(\frac{1}{\pi^2} - \alpha x^2 + h \frac{\zeta(3)}{\pi^2} x^3 + \frac{\pi^2}{2} (\tau^2 - \alpha^2) x^4 \right) \text{ truncated at } x^5. \tag{3.8}$$

The number h is a rational constant associated to the differential operator D such that $Dw_0 = 0$. The motivation of this expansion is due to the fact that in the case of Ramanujan–Sato-like series for $1/\pi^2$ (z, c, b, a algebraic), we have experimentally observed that r, α, and τ^2 are rational while h is a rational constant (see the remark at the end of this section). We have the equivalent system

$$(w_0)a + (\theta w_0)b + (\theta^2 w_0)c = \frac{1}{\pi^2},$$

$$(w_1)a + (\theta w_1)b + (\theta^2 w_1)c = \frac{i}{\pi}r,$$

$$(w_2)a + (\theta w_2)b + (\theta^2 w_2)c = -\frac{r^2}{2} - \alpha, \qquad (3.9)$$

$$(w_3)a + (\theta w_3)b + (\theta^2 w_3)c = i\pi r\left(-\frac{r^2}{6} - \alpha\right) + h\frac{\zeta(3)}{\pi^2},$$

$$(w_4)a + (\theta w_4)b + (\theta^2 w_4)c = \pi^2\left(\frac{r^4}{24} + \frac{\tau^2 - \alpha^2}{2} + \frac{r^2}{2}\alpha\right) + \frac{i}{\pi}h\zeta(3)r.$$

This system allows us to develop the theory. In the next theorem we obtain α and τ as non-holomorphic functions of q.

Theorem 3.1. *We have*

$$\alpha(q) = \frac{\frac{1}{6}\ln^3|q| - T(q) - h\zeta(3)}{\pi^2\ln|q|}, \qquad (3.10)$$

and

$$\tau(q) = \frac{\frac{1}{2}\ln^2|q| - (\theta_q T)(q)}{\pi^2} - \alpha(q). \qquad (3.11)$$

Proof. In the proof we use the wronskians above. As we want the system (3.9) to be compatible, we have

$$\begin{vmatrix} w_0 & \theta w_0 & \theta^2 w_0 & p_0 \\ w_1 & \theta w_1 & \theta^2 w_1 & p_1 \\ w_2 & \theta w_2 & \theta^2 w_2 & p_2 \\ w_3 & \theta w_3 & \theta^2 w_3 & p_3 \end{vmatrix} = 0, \qquad (3.12)$$

where p_0, p_1, etc., stand for the independent terms. Expanding the determinant along the last column, we obtain

$$-p_0\left(\frac{y_1 y_2 - y_0 y_3}{2}\right) + p_1(y_1^2) - p_2(y_0 y_1) + p_3(y_0^2) = 0.$$

Then, dividing by y_0^2, we get

$$-p_0\frac{1}{2}\left(\frac{y_1}{y_0}\frac{y_2}{y_0} - \frac{y_3}{y_0}\right) + p_1\left(\frac{y_1}{y_0}\right)^2 - p_2\left(\frac{y_1}{y_0}\right) + p_3 = 0.$$

Hence

$$-p_0 \left(\frac{1}{6} \ln^3 q - T(q) \right) + p_1 \ln^2 q - p_2 \ln q + p_3 = 0.$$

Using $\ln q = \ln |q| + i\pi r$ and replacing p_0, p_1, p_2, and p_3 with there values in the system, we arrive at (3.10). Then, from the first, second, third, and fifth equations and using the the function $\alpha(q)$ obtained already, we derive (3.11). □

In the next theorem we obtain c, b, a as non-holomorphic functions of q.

Theorem 3.2.

$$c(q) = \tau(q) \sqrt[4]{P(z)}, \tag{3.13}$$

$$b(q) = \frac{z(q)}{\theta_q z(q)} \left(\frac{1}{\pi^2} \left(\theta_q^2 T(q) - \ln |q| \right) - \tau(q) \frac{\theta_q L(q)}{L(q)} \right) \sqrt[4]{P(z)}, \tag{3.14}$$

$$a(q) = \frac{1}{w_0(q)} \left(\frac{1}{\pi^2} - (\theta w_0) b(q) - (\theta^2 w_0) c(q) \right), \tag{3.15}$$

with

$$L(q) = \frac{y_0^2}{\sqrt[4]{P(z)}} = \frac{w_0(q)}{\sqrt[4]{P(z)}} \frac{\theta_q z(q)}{z(q)} = \frac{1}{\sqrt{P(q)} K(q)} \left(\frac{\theta_q z(q)}{z(q)} \right)^3, \tag{3.16}$$

where y_0 is the ordinary pullback.

Proof. Solving for c by Cramer's rule from the three first equations of (3.9), we get

$$\frac{c}{\sqrt[4]{P(z)}} = \frac{1}{\pi^2} \left(\frac{y_2}{y_0} \right) - \frac{i}{\pi} r \left(\frac{y_1}{y_0} \right) - \frac{r^2}{2} - \alpha.$$

Hence

$$\frac{c}{\sqrt[4]{P(z)}} = \frac{1}{\pi^2} \left(\frac{1}{2} \ln^2 q - \theta_q T \right) - \frac{i}{\pi} r \ln q - \frac{r^2}{2} - \alpha.$$

Replacing $\ln q$ with $\ln |q| + i\pi r$, we obtain (3.13). Then, solving for b from the two first equations of (3.9), we obtain

$$b = \frac{1}{\pi^2} \frac{w_0}{L} \left(i\pi r - \frac{w_1}{w_0} \right) - c(z) \frac{\theta_z L}{L}, \tag{3.17}$$

where $L = w_0(\theta w_1) - w_1(\theta w_0)$. But, as $q = \exp(y_1/y_0)$, we obtain

$$\theta_q \left(\frac{y_2}{y_0} \right) = \frac{q}{\theta_z q} \theta_z \left(\frac{y_2}{y_0} \right) = \frac{y_0 \theta y_2 - y_2 \theta y_0}{y_0 \theta y_1 - y_1 \theta y_0} = \frac{w_1}{w_0}.$$

Applying θ_q to the two extremes of it, we get

$$\theta_q \left(\frac{w_1}{w_0} \right) = \frac{w_0(\theta w_1) - w_1(\theta w_0)}{w_0^2} \frac{\theta_q z}{z} = K(q) = 1 - \theta_q^3 T(q),$$

which implies

$$\frac{w_1}{w_0} = \ln q - \theta_q^2 T(q) \tag{3.18}$$

and

$$w_0 \theta w_1 - w_1 \theta w_0 = \frac{w_0^2 K(q) z(q)}{\theta_q z(q)} = L(q). \tag{3.19}$$

But

$$L = w_0(\theta w_1) - w_1(\theta w_0) = \frac{y_0^2}{\sqrt[4]{P(z)}}. \tag{3.20}$$

In [1] we have the formula

$$y_0^2 = \left(\frac{\theta_q z(q)}{z(q)}\right)^3 \frac{1}{\sqrt[4]{P(q)} K(q)}. \tag{3.21}$$

From (3.19)–(3.21), we obtain

$$w_0 = \left(\frac{\theta_q z(q)}{z(q)}\right)^2 \frac{1}{\sqrt[4]{P(q)} K(q)}. \tag{3.22}$$

From the three last identities we arrive at (3.16). From (3.17), (3.18), and (3.22) we deduce (3.14). The proof of (3.15) is trivial from the first equation of (3.9). □

The relevant fact is that the functions $\alpha(q)$, $\tau(q)$, $c(q)$, $b(q)$, $a(q)$ have good arithmetical properties. This is stated in the following conjecture which is crucial to discover Ramanujan–Sato-like series for $1/\pi^2$:

Conjecture

Let $\alpha_0 = \alpha(q_0)$, $\tau_0 = \tau(q_0)$, $z_0 = z(q_0)$, $a_0 = a(q_0)$, etc. If two of the quantities α_0, τ_0^2, z_0, a_0, b_0, c_0 are algebraic, so are all the others. Even more, in that case, α_0 and τ_0^2 are rational.

Remark

As one has

$$\sum_{n=0}^{\infty} \frac{A_{n+x}}{A_x} z^{n+x} = w_0 + w_1 x + w_2 x^2 + w_3 x^3 + w_4 x^4 + O(x^5),$$

we can write (3.8) in the following way:

$$\frac{1}{A_x} \sum_{n=0}^{\infty} z^{n+x} A_{n+x} \left(a + b(n+x) + c(n+x)^2 \right)$$

$$= e^{i\pi r x} \left(\frac{1}{\pi^2} - \alpha x^2 + h \frac{\zeta(3)}{\pi^2} x^3 + \frac{\pi^2}{2} (\tau^2 - \alpha^2) x^4 \right) + \mathcal{O}(x^5).$$

The rational constant h appears (and can be defined) by the coefficient of x^3 in the expansion of A_x (analytic continuation of A_n) [6, Eq. 4]. In the hypergeometric cases we know how to extend A_n to A_x because the function Γ is the analytic continuation of the factorial. To determine h in the non-hypergeometric cases, we will not use this definition because it is not clear how to extend A_n to A_x in an analytic way. Instead, we will use the following conjecture:

Conjecture

The radius of convergence z_c of $w_0(z)$ is the smallest root of $P(z) = 0$ and

$$\frac{dz}{dq}(q_c) = 0.$$

In addition, α_c is rational. Hence there is a relation with integer coefficients among the numbers $\frac{1}{6} \ln^3 |q_c| - T(q_c)$, $\pi^2 \ln |q_c|$, and $\zeta(3)$, which we can discover with the PSLQ algorithm, and it determines the invariant h. This solution corresponds to the degenerated series $z = z_c$, $c(q_c) = b(q_c) = a(q_c) = 0$.

3.3 New Series for $1/\pi^2$

To discover Ramanujan-like series for $1/\pi^2$, we first obtain the mirror map, the Yukawa coupling, and the function $T(q)$. Solving the equation

$$\frac{dz(q)}{dq} = 0,$$

we get the value q_c which corresponds to z_c. Let $q = e^t e^{i\pi r}$, where $t < 0$ is real. If we choose a value of r, then we can write (3.10) in the form

$$\alpha(t) = \frac{\frac{1}{6} t^3 - T(q) - h\zeta(3)}{\pi^2 t}.$$

For $r = 0$ we get series of positive terms and for $r = 1$, we get alternating series. Solving numerically the equation $\alpha(t) = \alpha_0$, where α_0 is rational, we find an approximation of t_0 and hence also an approximation of q_0. Substituting this q_0 in (3.11), we get the value of τ_0. If τ_0^2 is also rational, then with the mirror map, we get the corresponding approximation of z_0. To discover the exact algebraic number z_0, we use the Maple function MinimalPolynomial which finds the minimal polynomial of a given degree, then we use the functions $c(q)$, $b(q)$, $a(q)$ to get the numerical values c_0, b_0, and a_0. To recognize the exact algebraic values of these parameters, we use minimal polynomial again. It is remarkable that in the "divergent" cases, we can compute c_0, b_0, a_0 with high precision by using formula (3.22) for w_0.

Big Table

In [5] there is a collection of many differential equations of Calabi–Yau type. We select some of those which are pullbacks of differential equations of fifth order. The ones not mentioned below gave no result. The symbol # stands as a reference of the equation in the Big Table. In [2] one can learn the art of finding Calabi–Yau differential equations. In Table 1, we show the invariants corresponding to the cases #60, #130, #189, #355, and #356. For all the cases cited above we have found examples of Ramanujan-like series for $1/\pi^2$. In Table 2 we show those examples, indicating the algebraic values of $\alpha - \alpha_c$, $z_c^{-1} \cdot z$, a, b, and c for which we have

$$\sum_{n=0}^{\infty} \tilde{A}_n \left(z_c^{-1} \cdot z \right)^n (a + bn + cn^2) = \frac{1}{\pi},$$

where $\widetilde{A_n} = A_n z_c^n$. If $|z_c^{-1} \cdot z| > 1$, then the series diverges, but we avoid the divergence considering the analytic continuation given by the parametrization with q. In all cases we have found congruences mod p^5 or mod p^3 (see [21]).

For #355 an explicit formula for A_n is not known but we can easily compute these numbers from the fifth-order differential equation $Dw = 0$, where D is the following operator:

$$\theta^5 - 2z(2\theta + 1)(43\theta^4 + 86\theta^3 + 77\theta^2 + 34\theta + 6)$$
$$+ 48z^2(\theta + 1)(2\theta + 1)(2\theta + 3)(6\theta + 5)(6\theta + 7).$$

Table 1 Table of invariants

$$
\#60 \qquad A_n = \sum_{k=0}^{n} \binom{n}{k}^2 \binom{2k}{k} \binom{2n-2k}{n-k} \binom{n+k}{n} \binom{2n-k}{n}
$$

$$
P(z) = (1-16z)^2(1-108z)^2, \quad z_c = \frac{1}{2^2 \cdot 3^3}, \quad \alpha_c = \frac{1}{3}, \quad \tau_c^2 = \frac{2}{23}, \quad h = \frac{50}{23}
$$

$$
\#130 \qquad A_n = \sum_{\substack{i+j+k+l \\ +m+s=n}} \left(\frac{n!}{i!\,j!\,k!\,l!\,m!\,s!} \right)^2
$$

$$
P(z) = (1-4z)^2(1-16z)^2(1-36z)^2, \quad z_c = \frac{1}{36}, \quad \alpha_c = \frac{1}{6}, \quad \tau_c^2 = \frac{2}{45}, \quad h = \frac{2}{3}
$$

$$
\#189 \qquad A_n = \binom{2n}{n} \sum_{j,k} \binom{n}{j}^2 \binom{n}{k}^2 \binom{j+k}{n}^2
$$

$$
P(z) = (1-4z)^2(1-256z)^2, \quad z_c = \frac{1}{256}, \quad \alpha_c = \frac{1}{2}, \quad \tau_c^2 = \frac{8}{21}, \quad h = \frac{30}{7}
$$

$$
\#355 \qquad \text{Explicit formula for } A_n \text{ not known}
$$

$$
P(z) = (1-64z)^2(1-108z)^2, \quad z_c = \frac{1}{108}, \quad \alpha_c = \frac{1}{3}, \quad \tau_c^2 = \frac{4}{33}, \quad h = \frac{30}{11}
$$

$$
\#356 \qquad A_0 = 1, \quad A_{n>0} = 2\binom{2n}{n} \sum_{k=0}^{[n/4]} \frac{n-2k}{3n-4k} \binom{n}{k}^2 \binom{2k}{k} \binom{2n-2k}{n-k} \binom{3n-4k}{2n}
$$

$$
P(z) = (1-108z)^2(1-128z)^2, \quad z_c = \frac{1}{128}, \quad \alpha_c = \frac{1}{3}, \quad \tau_c^2 = \frac{1}{10}, \quad h = \frac{14}{5}
$$

Complex Series for $1/\pi^2$

Another method to obtain series for $1/\pi^2$ is by applying suitable transformations to the already known series for $1/\pi^2$; see [3, 6, 20]. Although we can use this technique to obtain other real Ramanujan-like series for $1/\pi^2$, our interest here is to find examples of Ramanujan-like complex series for $1/\pi^2$. For that purpose we will use the following very general transformation:

$$
\sum_{n=0}^{\infty} A_n z^n = \frac{1}{1-z} \sum_{n=0}^{\infty} a_n \left[u \left(\frac{z}{1-z} \right)^m \right]^n, \qquad A_n = \sum_{k=0}^{n} u^k \binom{n}{mk} a_k,
$$

where $u = 1$ or $u = -1$ and m is a positive integer (check that both sides satisfy the same Calabi–Yau differential equation). For example, translating the hypergeometric series

$$
\sum_{n=0}^{\infty} \frac{(3n)!(4n)!}{n!^7} (252n^2 + 63n + 5)(-1)^n \left(\frac{1}{24} \right)^{4n} = \frac{48}{\pi^2}, \tag{3.23}
$$

taking $u = -1$ and $m = 4$, we find four series, one of them is the complex series

$$
\sum_{n=0}^{\infty} A_n \left(9072n^2 + (9072 - 756i)n + (2875 - 516i) \right) \left(\frac{1}{1-24i} \right)^n = \frac{27504 + 3454i}{\pi^2}, \tag{3.24}
$$

G. Almkvist and J. Guillera

Table 2 Table of examples

#	$\alpha_0 - \alpha_c$	$z_c^{-1} \cdot z_0$	a_0	b_0	c_0
60	$\dfrac{4}{23}$	$\dfrac{1}{2}$	$\dfrac{3}{3 \cdot 23}$	$\dfrac{20}{3 \cdot 23}$	$\dfrac{40}{3 \cdot 23}$
	$\dfrac{8}{23}$	$\dfrac{3^3}{5^3}$	$\dfrac{40}{5^2 \cdot 23}$	$\dfrac{282}{5^2 \cdot 23}$	$\dfrac{616}{5^2 \cdot 23}$
	$\dfrac{43}{46}$	$\dfrac{1}{48}$	$\dfrac{706}{2^5 \cdot 3^2 \cdot 23}$	$\dfrac{5895}{2^5 \cdot 3^2 \cdot 23}$	$\dfrac{16380}{2^5 \cdot 3^2 \cdot 23}$
	$\dfrac{3}{46}$	-2	$\dfrac{178}{2^5 \cdot 23}$	$\dfrac{719}{2^5 \cdot 23}$	$\dfrac{860}{2^5 \cdot 23}$
130	$\dfrac{1}{6}$	$-\dfrac{3^2}{4^2}$	$\dfrac{21}{96}$	$\dfrac{74}{96}$	$\dfrac{85}{96}$
	0	-4	$\dfrac{38}{54}$	$\dfrac{94}{54}$	$\dfrac{65}{54}$
189	$\dfrac{2}{21}$	$\dfrac{8^2}{9^2}$	$\dfrac{48}{3^5 \cdot 7}$	$\dfrac{328}{3^5 \cdot 7}$	$\dfrac{680}{3^5 \cdot 7}$
	$\dfrac{4}{7}$	$\dfrac{1}{3^2}$	$\dfrac{87}{2^4 \cdot 3^2 \cdot 7}$	$\dfrac{710}{2^4 \cdot 3^2 \cdot 7}$	$\dfrac{1840}{2^4 \cdot 3^2 \cdot 7}$
	$\dfrac{19}{42}$	$-\dfrac{2^4}{3^4}$	$\dfrac{843}{2^2 \cdot 3^5 \cdot 7}$	$\dfrac{5750}{2^2 \cdot 3^5 \cdot 7}$	$\dfrac{12610}{2^2 \cdot 3^5 \cdot 7}$
	$\dfrac{139}{42}$	$-\dfrac{2^4}{21^4}$	$\dfrac{1655799}{2^2 \cdot 3^5 \cdot 7^5}$	$\dfrac{24749870}{2^2 \cdot 3^5 \cdot 7^5}$	$\dfrac{122761930}{2^2 \cdot 3^5 \cdot 7^5}$
355	$\dfrac{1}{14}$	$\dfrac{4^2}{3^2}$	$\dfrac{51}{252}$	$\dfrac{254}{252}$	$\dfrac{370}{252}$
	$\dfrac{1}{11}$	$\dfrac{3}{4}$	$\dfrac{1}{2^2 \cdot 3 \cdot 11}$	$\dfrac{12}{2^2 \cdot 3 \cdot 11}$	$\dfrac{30}{2^2 \cdot 3 \cdot 11}$
	$\dfrac{5}{22}$	$-\dfrac{3^2}{4^2}$	$\dfrac{9}{2^2 \cdot 11}$	$\dfrac{42}{2^2 \cdot 11}$	$\dfrac{60}{2^2 \cdot 11}$
	$\dfrac{5}{11}$	$\dfrac{27}{196}$	$\dfrac{21}{2^2 \cdot 7 \cdot 11}$	$\dfrac{164}{2^2 \cdot 7 \cdot 11}$	$\dfrac{390}{2^2 \cdot 7 \cdot 11}$
	$\dfrac{13}{11}$	$\dfrac{1}{108}$	$\dfrac{3119}{2^2 \cdot 3^6 \cdot 11}$	$\dfrac{29860}{2^2 \cdot 3^6 \cdot 11}$	$\dfrac{93090}{2^2 \cdot 3^6 \cdot 11}$
	$\dfrac{1}{22}$	-3	$\dfrac{16}{3 \cdot 11}$	$\dfrac{60}{3 \cdot 11}$	$\dfrac{60}{3 \cdot 11}$
356	$\dfrac{1}{5}$	$\dfrac{1}{2}$	$\dfrac{2}{160}$	$\dfrac{27}{160}$	$\dfrac{74}{160}$
	1	$\dfrac{1}{50}$	$\dfrac{74}{800}$	$\dfrac{679}{800}$	$\dfrac{2002}{800}$
	$\dfrac{7}{10}$	$-\dfrac{1}{2^4}$	$\dfrac{158}{1280}$	$\dfrac{1113}{1280}$	$\dfrac{2618}{1280}$

where

$$A_n = \sum_{k=0}^{n} (-1)^k \binom{n}{4k} \frac{(3k)!(4k)!}{k!^7}.$$

Transformations preserve the value of the invariants h, α_c, and τ_c, and the series (3.24) has $2(\alpha - \alpha_c) = 3$ and $\tau = 3\sqrt{3}$ because it is a transformation of (3.23); see [6]. Looking at the transformation with $u = -1$ and $m = 4$, we see that the mirror maps z and z' corresponding to A_n and a_n, are related in the following way:

$$z = \frac{\sqrt[4]{z'}}{1 + \sqrt[4]{z'}}.$$

Write $z = z(q)$ and $z' = z'(q')$. Then, the first terms of $J(q)$ are

$$J(q) = \frac{1}{q} + 1 + 582q^3 + 277263q^7 + 167004122q^{11} + \cdots,$$

with $q = \sqrt[4]{q'}$. Writing, as usual, $q = e^{i\pi r}|q|$, we deduce that as the series (3.23) has $r = 1$, then the series (3.24) has $r = 1/4$.

4 Addendum

The method used in this paper, to find h, $\alpha_c = \alpha(q_c)$ and $\tau_c = \tau(q_c)$, is valid for those Calabi–Yau differential equations such that $K(q_c) = 0$, where q_c is a solution of $dz/dq = 0$. In these cases, we conjecture that $z(q_c)$ is the smallest root of $P(z)$ and that $a(q_c) = b(q_c) = c(q_c) = 0$. But from Theorem 3.2, we see that $b(q_c) = 0$ implies that $\tau_c = f(q_c)$, where

$$f(q) = \frac{1}{\pi^2} \left(\theta_q^2 T(q) - \ln|q| \right) \frac{L(q)}{\theta_q L(q)},$$

which allows us to obtain the critical value of τ. Then, replacing $q = q_c$ in (3.11), we can obtain α_c. Finally, replacing $q = q_c$ in (3.10), we obtain the value of h. As $q_c > 0$, the formula for h can be written in the form $h = h(q_c)$, where

$$h(q) = \frac{1}{\zeta(3)} \left(\Phi(q) - \ln(q)\, \theta_q \Phi(q) - \ln(q) \frac{L(q)}{\theta_q L(q)} \theta_q^2 \Phi(q) \right),$$

and $\Phi(q)$ is the Gromov–Witten potential (3.2). The advantage of this way of getting the invariants τ_c, α_c, and h is that we use explicit formulas instead of the PSLQ algorithm.

References

1. G. Almkvist, Calabi–Yau differential equations of degree 2 and 3 and Yifan Yang's pullback (2006). arXiv: math/0612215
2. G. Almkvist, The art of finding Calabi–Yau differential equations, in *Gems in Experimental Mathematics*, ed. by T. Amdeberhan, L.A. Medina, V.H. Moll. Contemp. Math., vol 517 (Amer. Math. Soc., 2010), Providence, RI, pp. 1–18. arXiv:0902.4786
3. G. Almkvist, Transformations of Jesus Guillera's formulas for $1/\pi^2$ (2009). arXiv:0911.4849
4. G. Almkvist, Some conjectured formulas for $1/\pi$ coming from polytopes, K3-surfaces and Moonshine (manuscript). arXiv 1211.6563
5. G. Almkvist, C. van Enckevort, D. van Straten, W. Zudilin, Tables of Calabi–Yau equations (2005). arXiv: math/0507430
6. G. Almkvist, J. Guillera, Ramanujan-like series and String theory. Exp. Math. **21**, 223–234 (2012). eprint arXiv:1009.5202
7. G. Almkvist, W. Zudilin, Differential equations, mirror maps and zeta values, in *Mirror Symmetry*, ed. by V.N. Yui, S.-T. Yau, J.D. Lewis. AMS/IP Studies in Advanced Mathematics, vol 38 (International Press/Amer. Math. Soc., 2007), Providence, RI, pp. 481–515. eprint math.NT/0402386
8. N.D. Baruah, B.C. Berndt, H.H. Chan, Ramanujan's series for $1/\pi$: a survey. Am. Math. Monthly **116**, 567–587 (2009). Available at Bruce Berndt's web-site
9. J.M. Borwein, P.B. Borwein, *Pi and the AGM: A Study in Analytic Number Theory and Computational Complexity*. Canadian Mathematical Society Series of Monographs and Advanced Texts (Wiley, New York, 1987)
10. H.H. Chan, S.H. Chan, Z. Liu, Domb's numbers and Ramanujan–Sato type series for $1/\pi$. Adv. Math. **186**, 396–410 (2004)
11. H.H. Chan, J. Wan, W. Zudilin, Complex series for $1/\pi$. Ramanujan J. **29**, 135–144 (2012)
12. D. Chudnovsky, G. Chudnovsky, Approximations and complex multiplication according to Ramanujan, in *Ramanujan Revisited: Proceedings of the Centenary Conference*, University of Illinois at Urbana-Champaign, ed. by G. Andrews, R. Askey, B. Berndt, K. Ramanathan, R. Rankin (Academic, Boston, 1987), pp. 375–472
13. D.A. Cox, S. Katz, *Mirror Symmetry and Algebraic Geometry* (AMS, Providence, 1999)
14. J. Guillera, A matrix form of Ramanujan-type series for $1/\pi$, in *Gems in Experimental Mathematics*, ed. by T. Amdeberhan, L.A. Medina, V.H. Moll. Contemp. Math., vol 517 (Amer. Math. Soc., 2010), Providence, RI, pp. 189–206. arXiv:0907.1547
15. J. Guillera, Collection of Ramanujan-like series for $1/\pi^2$. Unpublished manuscript available at J. Guillera's web site
16. J. Guillera, Mosaic supercongruences of Ramanujan-type. Experimental Math. **21**, 65–68 (2012)
17. J. Guillera, W. Zudilin, "Divergent" Ramanujan-type supercongruences. Proc. Am. Math. Soc. **140**(3), 765–777 (2012). eprint arXiv:1004.4337
18. S. Ramanujan, Modular equations and approximations to π. Q. J. Math. **45**, 350–372 (1914)
19. Y. Yang, W. Zudilin, On Sp_4 modularity of Picard–Fuchs differential equations for Calabi–Yau threefolds, (with an appendix by V. Pasol), in *Gems in Experimental Mathematics*, ed. by T. Amdeberhan, L.A. Medina, V.H. Moll. Contemp. Math., vol 517 (Amer. Math. Soc., 2010), Providence, RI, pp. 381–413. arXiv:0803.3322
20. W. Zudilin, Quadratic transformations and Guillera's formulae for $1/\pi^2$. Mat. Zametki **81**(3), 335–340 (2007). English transl., Math. Notes **81**(3), 297–301 (2007). arXiv:math/0509465v2
21. W. Zudilin, Ramanujan-type supercongruences. J. Number Theory **129**(8), 1848–1857 (2009). arXiv:0805.2788
22. W. Zudilin, Arithmetic hypergeometric series. Russ. Math. Surv. **66**(2), 369–420 (2011). Russian version in Uspekhi Mat. Nauk **66**(2), 163–216 (2011)

On the Sign of the Real Part of the Riemann Zeta Function

Juan Arias de Reyna, Richard P. Brent, and Jan van de Lune

In fond memory of Alfred Jacobus (Alf) van der Poorten 1942–2010

Abstract We consider the distribution of the argument of the Riemann zeta function on vertical lines with real part greater than 1/2, and in particular two densities related to the argument and to the real part of the zeta function on such lines. Using classical results of Bohr and Jessen, we obtain an explicit expression for the characteristic function associated with the argument. We give explicit expressions for the densities in terms of this characteristic function. Finally, we give a practical algorithm for evaluating these expressions to obtain accurate numerical values of the densities.

1 Introduction

Several authors, including Edwards [9, pg. 121], Gram [11, pg. 304], Hutchinson [13, pg. 58], and Milioto [24, §2], have observed that the real part $\Re\zeta(s)$ of the Riemann zeta function $\zeta(s)$ is "usually positive". This is plausible

J. Arias de Reyna
Facultad de Matemáticas, Universidad de Sevilla, Apdo. 1160, 41080 Sevilla, Spain
e-mail: arias@us.es

R.P. Brent (✉)
Mathematical Sciences Institute, Australian National University, Canberra, ACT 0200, Australia
e-mail: alf@rpbrent.com

J. van de Lune
Langebuorren 49, 9074 CH Hallum, The Netherlands
e-mail: j.vandelune@hccnet.nl

J.M. Borwein et al. (eds.), *Number Theory and Related Fields: In Memory of Alf van der Poorten*, Springer Proceedings in Mathematics & Statistics 43, DOI 10.1007/978-1-4614-6642-0_3, © Springer Science+Business Media New York 2013

because the Dirichlet series $\zeta(s) = 1 + 2^{-s} + 3^{-s} + \cdots$ starts with a positive term, and the other terms n^{-s} may have positive or negative real part. In this paper our aim is to make precise the statement that $\Re\zeta(s)$ is "usually positive" for $\sigma := \Re(s) > \frac{1}{2}$.

Kalpokas and Steuding [17], assuming the Riemann hypothesis, have given a sense in which the statement is also true on the critical line $\sigma = \frac{1}{2}$. They showed that the mean value of the set of real values of $\zeta(\frac{1}{2} + it)$ exists and is equal to 1.

We do not assume the Riemann hypothesis, and our results do not appear to imply anything about the existence or non-existence of zeros of $\zeta(s)$ for $\sigma > \frac{1}{2}$.

Our results depend on the classical results of Bohr and Jessen [4, 5] concerning the value distribution of $\zeta(s)$ in the half-plane $\sigma > \frac{1}{2}$. Since Bohr and Jessen, there have been many further results on the value distribution of various classes of L-functions. See, for example, Joyner [16], Lamzouri [19–21], Laurinčikas [22], Steuding [27], and Voronin [31]. However, for our purposes the results of Bohr and Jessen are sufficient.

After defining our notation, we summarise the relevant results of Bohr and Jessen in Sect. 2. The densities $d(\sigma)$ and $d_-(\sigma)$, defined in Sect. 3, can be expressed in terms of the characteristic function $\psi_\sigma(x)$ of a certain random variable $\Im S$ associated with $\arg\zeta(\sigma + it)$. We consider ψ_σ and a related function $I(b,x)$ in Sects. 4–7. In Theorem 1 we use the results of Bohr and Jessen to obtain an explicit expression for $\psi_\sigma(x)$. Theorem 2 relates $\log I(b,x)$ to certain polynomials $Q_n(x)$ which have nonnegative integer coefficients with interesting congruence properties, and Theorem 3 gives an asymptotic expansion of $I(b,x)$ which shows a connection between $I(b,x)$ and the Bessel function J_0. Theorem 4 shows that $\psi_\sigma(x)$ decays rapidly as $x \to \infty$.

The explicit expression for ψ_σ is an infinite product over the primes and converges rather slowly. In Sect. 8 we show how the convergence can be accelerated to give a practical algorithm for computing $\psi_\sigma(x)$ to high accuracy.

In Sect. 9 we show how $d(\sigma)$ and $d_-(\sigma)$ can be computed using $\psi_\sigma(x)$ and give the results of numerical computations in Sect. 10. Finally, in Sect. 11 we comment on how our results might be generalised.

Elliott [10] determined the characteristic function $\Psi_\sigma(x)$ of a limiting distribution associated with a certain sequence of L-functions. We note that Elliott's $\Psi_\sigma(x)$ is the same function as our $\psi_\sigma(x)$. For a possible explanation of this coincidence, using the concept of *analytic conductor*, we refer to [15, Ch. 5]. Here we merely note that Elliott's method of proof is quite different from our proof of Theorem 1 and applies only to sequences of L-functions $L(s,\chi)$ for which χ is a non-principal Dirichlet character.

Notation

$\mathbb{Z}$, $\mathbb{Q}$, $\mathbb{R}$, and $\mathbb{C}$ denote, respectively, the integers, rationals, reals, and complex numbers. The real part of $z \in \mathbb{C}$ is denoted by $\Re z$ and the imaginary part by $\Im z$.

When considering $\zeta(s)$, we always have $\sigma := \Re s$. Unless otherwise specified, $\sigma > \frac{1}{2}$ is fixed.

Consider the open set G equal to $\mathbb{C}$ with cuts along $(-\infty + i\gamma, \beta + i\gamma]$ for each zero or pole $\beta + i\gamma$ of $\zeta(s)$ with $\beta \geq \frac{1}{2}$. Since $\zeta(s)$ is holomorphic and does not vanish on G, we may define $\log \zeta(s)$ on G. We take the branch such that $\log \zeta(s)$ is real and positive on $(1, +\infty)$. On G we define $\arg \zeta(s)$ by

$$\log \zeta(s) = \log|\zeta(s)| + i \cdot \arg \zeta(s).$$

P is the set of primes, and $p \in P$ is a prime. When considering a fixed prime p, we often use the abbreviations $b := p^{\sigma}$ and $\beta := \arcsin(1/b)$.

$|B|$ or $\lambda(B)$ denotes the Lebesgue measure of a set $B \subset \mathbb{C}$ (or $B \subset \mathbb{R}$). A set $B \subset \mathbb{C}$ is said to be *Jordan measurable* if $\lambda(\partial B) = 0$, where ∂B is the boundary of B.[1]

${}_2F_1(a,b;c;z)$ denotes the hypergeometric function of Gauss; see [1, 8].

2 Classical Results of Bohr and Jessen

In [4, 5] Bohr and Jessen study several problems regarding the value distribution of the zeta function. In particular, for $\sigma > \frac{1}{2}$ and a given subset $B \subset \mathbb{C}$, they consider the limit

$$\lim_{T \to \infty} \frac{1}{2T} |\{t \in \mathbb{R} : |t| < T, \log \zeta(\sigma + it) \in B\}|.$$

They prove that the limit exists when B is a rectangle with sides parallel to the real and imaginary axes.

Bohr and Jessen also characterise the limit. In modern terminology, they prove [5, Erster Hauptsatz, pg. 3] the existence of a probability measure $\mathbb{P}_{\sigma}$, absolutely continuous with respect to Lebesgue measure, such that for any rectangle B as above the limit is equal to $\mathbb{P}_{\sigma}(B)$. Finally, they give a description of the measure $\mathbb{P}_{\sigma}$. To express it in modern language, consider the unit circle $\mathbb{T} = \{z \in \mathbb{C} : |z| = 1\}$ with the usual probability measure μ (i.e., $\frac{1}{2\pi} d\theta$ if we identify $\mathbb{T}$ with the interval $[0, 2\pi)$ in the usual way). Let P be the set of prime numbers. We may consider $\Omega := \mathbb{T}^P$ as a probability space with the product measure $\mathbb{P} = \mu^P$. Each point of Ω is a sequence $\omega = (z_p)_{p \in P}$, with each $z_p \in \mathbb{T}$. Thus, z_p may be considered as a random variable. The random variables z_p are independent and uniformly distributed on the unit circle.

Proposition 1. *Let $\sigma > \frac{1}{2}$ and for each prime number q, let z_q be the random variable defined on Ω such that $z_q(\omega) = z_q$ when $\omega = (z_p)_{p \in P}$. The sum of random variables*

---

[1] A bounded set B is Jordan measurable if and only if for each $\varepsilon > 0$, we can find two finite unions of rectangles with sides parallel to the real and imaginary axes, say S and T, such that $S \subseteq B \subset T$ and $\lambda(T \setminus S) < \varepsilon$ (see, e.g. Halmos [12]).

$$S := - \sum_{p \in P} \log(1 - p^{-\sigma} z_p) = \sum_{p \in P} \sum_{k=1}^{\infty} \frac{1}{k} p^{-k\sigma} z_p^k$$

converges almost everywhere, so S is a well-defined random variable.

Proof. The random variables $Y_p := -\log(1 - p^{-\sigma} z_p)$ are independent. The mean value of each Y_p is zero since

$$\mathbb{E}(Y_p) = \frac{1}{2\pi} \int_0^{2\pi} \sum_{k=1}^{\infty} \frac{1}{k} p^{-k\sigma} e^{ik\theta} \, d\theta = 0.$$

It can be shown in a similar way that $E(|Y_p|^2) \sim p^{-2\sigma}$. Thus, $\sum_p E(|Y_p|^2)$ converges. A classical result of probability theory [12, Thm. B, Ch. IX] proves the convergence almost everywhere of the series for S.   □

The measure $\mathbb{P}_\sigma$ of Bohr and Jessen is the distribution of the random variable S. For each Borel set $B \subset \mathbb{C}$, we have

$$\mathbb{P}_\sigma(B) = \mathbb{P}\{\omega \in \Omega : S(\omega) \in B\}.$$

The main result of Bohr and Jessen is that, for each rectangle R with sides parallel to the axes,

$$\mathbb{P}_\sigma(R) = \lim_{T \to \infty} \frac{1}{2T} |\{t \in \mathbb{R} : |t| < T, \log \zeta(\sigma + it) \in R\}| \tag{1}$$

and the limit exists. It is easy to deduce that (1) is also true for each Jordan-measurable subset $R \subset \mathbb{C}$ and for sets R of the form $\mathbb{R} \times B$, where B is a Jordan-measurable subset of $\mathbb{R}$.

3   Some Quantities Related to the Argument of the Zeta Function

Define a measure μ_σ on the Borel sets of $\mathbb{R}$ by $\mu_\sigma(B) := \mathbb{P}_\sigma(\mathbb{R} \times B)$. If we take a Jordan subset $B \subset \mathbb{R}$, the main result of Bohr and Jessen implies that

$$\mu_\sigma(B) = \lim_{T \to \infty} \frac{1}{2T} |\{t \in \mathbb{R} : |t| < T, \arg \zeta(\sigma + it) \in B\}|.$$

The measure μ_σ is the distribution function of the random variable $\Im S$. In fact

$$\mu_\sigma(B) = \mathbb{P}_\sigma(\mathbb{R} \times B) = \mathbb{P}\{\omega \in \Omega : S(\omega) \in \mathbb{R} \times B\} = \mathbb{P}\{\omega \in \Omega : \Im S(\omega) \in B\}.$$

We are interested in the functions $d(\sigma)$, $d_+(\sigma)$, and $d_-(\sigma)$ defined by

$$d(\sigma) := \lim_{T \to \infty} \frac{1}{2T} |\{t \in \mathbb{R} : |t| < T, |\arg \zeta(\sigma + it)| > \pi/2\}|,$$

$$d_+(\sigma) := \lim_{T \to \infty} \frac{1}{2T} |\{t \in \mathbb{R} : |t| < T, \Re\zeta(\sigma + it) > 0\}|,$$

$$d_-(\sigma) := \lim_{T \to \infty} \frac{1}{2T} |\{t \in \mathbb{R} : |t| < T, \Re\zeta(\sigma + it) < 0\}|.$$

Informally, $d_+(\sigma)$ is the probability that $\Re\zeta(\sigma + it)$ is positive; $d_-(\sigma)$ is the probability that $\Re\zeta(\sigma + it)$ is negative. We show in Sect. 10 that $d(\sigma)$ is usually a good approximation to $d_-(\sigma)$. Observe that $d(\sigma) = 1 - \mu_\sigma([-\pi/2, \pi/2])$, $d_+(\sigma) + d_-(\sigma) = 1$, and $d_+(\sigma) = \sum_{k \in \mathbb{Z}} \mu_\sigma(2k\pi - \pi/2, 2k\pi + \pi/2)$.

4   The Characteristic Function ψ_σ

Recall that the *characteristic function* $\psi(x)$ of a random variable Y is defined by the Fourier transform $\psi(x) := E[\exp(ixY)]$. We omit a factor 2π in the exponent to agree with the statistical literature.

Proposition 2. *The characteristic function of the random variable $\Im S$ is given by*

$$\psi_\sigma(x) = \prod_p I(p^\sigma, x), \tag{2}$$

where writing $b := p^\sigma$, $I(b, x)$ is defined by

$$I(b, x) := \frac{1}{2\pi} \int_0^{2\pi} \exp\left(-ix \arg(1 - b^{-1}e^{i\theta})\right) d\theta. \tag{3}$$

Proof. By definition

$$\psi_\sigma(x) = \int_\Omega \exp(ix\Im S(\omega)) \, d\omega = \int_\Omega \prod_p \exp\left(-ix \arg(1 - p^{-\sigma}z_p)\right) d\omega.$$

By independence the integral of the product is the product of the integrals, so

$$\psi_\sigma(x) = \prod_p \int_\Omega \exp\left(-ix \arg(1 - p^{-\sigma}z_p)\right) d\omega.$$

Each random variable z_p is distributed as $e^{i\theta}$ on the unit circle, so

$$\psi_\sigma(x) = \prod_p \frac{1}{2\pi} \int_0^{2\pi} \exp\left(-ix \arg(1 - p^{-\sigma}e^{i\theta})\right) d\theta = \prod_p I(p^\sigma, x). \qquad \square$$

5   The Function $I(b,x)$

In this section we study the function $I(b,x)$ defined by (3). It is easy to see from (3) that $I(b,x)$ is an even function of x. Hence, from (2), the same is true for $\psi_\sigma(x)$.

Proposition 3. *Let $b > 1$ and $\beta = \arcsin(b^{-1})$. Then*

$$
I(b,x) = \frac{1}{\pi} \int_0^\pi \cos\left(x \arctan \frac{\sin t}{b - \cos t} \right) dt
$$

$$
= \frac{2b}{\pi} \int_0^\beta \frac{\cos(xt)\cos t}{\sqrt{1 - b^2 \sin^2 t}} \, dt = \frac{2}{\pi} \int_0^1 \cos\left(x \arcsin \frac{t}{b} \right) \frac{dt}{\sqrt{1 - t^2}}.
$$

Proof. By elementary trigonometry we find

$$
\arg(1 - b^{-1} e^{it}) = -\arctan \frac{\sin t}{b - \cos t}. \tag{4}
$$

Substituting in (3) gives

$$
I(b,x) = \frac{1}{2\pi} \int_0^{2\pi} \exp\left(ix \arctan \frac{\sin t}{b - \cos t} \right) dt = \frac{1}{\pi} \int_0^\pi \cos\left(x \arctan \frac{\sin t}{b - \cos t} \right) dt.
$$

To obtain the second representation, note that $\arctan(\sin t/(b - \cos t))$ is increasing on the interval $[0, \gamma]$ and decreasing on $[\gamma, \pi]$, where $\gamma = \arccos b^{-1}$. We split the integral on $[0, \pi]$ into integrals on $[0, \gamma]$ and $[\gamma, \pi]$. In each of the resulting integrals we change variables, putting $u := \arctan(\sin t/(b - \cos t))$. Then

$$
t = \arccos\left(b \sin^2 u \pm \cos u \sqrt{1 - b^2 \sin^2 u} \right),
$$

where the sign is "+" on the first interval and "−" on the second interval. After some simplification, the second representation follows. The third representation follows by the change of variables $t \mapsto \arcsin(t/b)$.   □

Lemma 1. *For $|t| < 1$ and all $x \in \mathbb{C}$,*

$$
\cos(2x \arcsin t) = {}_2F_1(-x, x; \tfrac{1}{2}; t^2) = 1 + \sum_{n=1}^\infty \frac{(2t)^{2n}}{(2n)!} \prod_{j=0}^{n-1} (j^2 - x^2). \tag{5}
$$

Proof. In [1, Eqn. 15.1.17] (also [8, Eqn. 15.4.12]) we find the identity

$$
\cos(2az) = {}_2F_1(-a, a; \tfrac{1}{2}; \sin^2 z).
$$

Replacing a by x and z by $\arcsin t$, we get the first half of (5). The second half follows from the definition of the hypergeometric function.   □

Remark 1. An independent proof uses the fact that $f(t) := \cos(2x\arcsin t)$ satisfies the differential equation $(1-t^2)f''(t) - tf'(t) + 4x^2 f(t) = 0$, where primes denote differentiation with respect to t.

Remark 2. When $x \in \mathbb{Z}$, the series (5) reduces to a polynomial.

Proposition 4. *For $b > 1$ we have*

$$I(b,2x) = {}_2F_1(-x,x;1;b^{-2}) = 1 + \sum_{n=1}^{\infty} \frac{1}{b^{2n}n!^2} \prod_{j=0}^{n-1}(j^2 - x^2).$$

Proof. From Proposition 3, we have

$$I(b,2x) = \frac{2}{\pi} \int_0^1 \cos\left(2x\arcsin\frac{t}{b}\right) \frac{dt}{\sqrt{1-t^2}}.$$

The expression of $I(b,2x)$ as a sum follows from Lemma 1, using a well-known integral for the beta function $B(n+\frac{1}{2},\frac{1}{2})$:

$$\frac{2}{\pi} \int_0^1 \frac{t^{2n}\,dt}{\sqrt{1-t^2}} = \frac{1}{\pi} B\left(n+\tfrac{1}{2},\tfrac{1}{2}\right) = \frac{(2n)!}{n!^2 2^{2n}}.$$

The identification of $I(b,2x)$ as ${}_2F_1(-x,x;1;b^{-2})$ then follows from the definition of the hypergeometric function ${}_2F_1$. □

Corollary 1. *If $x \in \mathbb{Z}$, $b^2 \in \mathbb{Q}$ and $b > 1$, then $I(b,2x) \in \mathbb{Q}$.*

Proof. Since $I(b,2x)$ is even, we can assume that $x \geq 0$. Applying Euler's transformation [1, (15.3.4)] to the hypergeometric representation of Proposition 4, we obtain $I(b,2x) = (1-b^{-2})^x {}_2F_1(-x,1-x;1;1/(1-b^2))$, but the series for ${}_2F_1(-x,1-x;1;z)$ terminates, so is rational for $z \in \mathbb{Q}$. □

We can now prove our first main result, which gives an explicit expression for the characteristic function ψ_σ defined in Sects. 2–4.

Theorem 1. *For $\sigma > \frac{1}{2}$, the characteristic function ψ_σ of Proposition 2 is the entire function given by the convergent infinite product:*

$$\psi_\sigma(2x) = \prod_p \left(1 + \sum_{n=1}^{\infty} \frac{1}{n!^2} \prod_{j=0}^{n-1}(j^2 - x^2) \cdot \frac{1}{p^{2n\sigma}}\right). \tag{6}$$

Proof. The identity (6) follows from Propositions 2 and 4. Since $\sum p^{-2\sigma}$ converges, the infinite product (6) converges for all $x \in \mathbb{C}$. □

6 The Function $\log I(b,x)$

The explicit formula for ψ_σ given by Theorem 1 is not suitable for numerical computation because the infinite product over primes converges too slowly. In Sect. 8 we show how this difficulty can be overcome. First we need to consider the function $\log I(b,x)$.

Theorem 2. *Suppose that* $b > \max(1,|x|)$. *There exist even polynomials* $Q_n(x)$ *of degree* $2n$ *with* $Q_n(0) = 0$ *and nonnegative integer coefficients* $q_{n,k}$ *such that*

$$\log I(b,2x) = -\sum_{n=1}^{\infty} \frac{Q_n(x)}{n!^2} \frac{1}{b^{2n}} = -\sum_{n=1}^{\infty} \sum_{k=1}^{n} \frac{q_{n,k}}{n!^2} \frac{x^{2k}}{b^{2n}}. \tag{7}$$

The polynomials $Q_n(x)$ *are determined by the recurrence*

$$Q_1(x) = x^2, \quad Q_{n+1}(x) = (n!)^2 x^2 + \sum_{j=0}^{n-1} \binom{n}{j} \binom{n}{j+1} Q_{j+1}(x) Q_{n-j}(x). \tag{8}$$

Also, the polynomials $Q_n(x)$ *satisfy*

$$|Q_n(x)| \leq n!(n-1)! \max(1,|x|)^{2n}. \tag{9}$$

Proof. By Proposition 4 there exist even polynomials P_n with $P_n(0) = 0$, such that

$$I(b,2x) = 1 + \sum_{n=1}^{\infty} \frac{P_n(x)}{n!^2} \frac{1}{b^{2n}}.$$

It follows that

$$\log I(b,2x) = \sum_{k=1}^{\infty} \frac{(-1)^{k+1}}{k} \left(\sum_{n=1}^{\infty} \frac{P_n(x)}{n!^2} \frac{1}{b^{2n}} \right)^k.$$

It is clear that expanding the powers gives a series of the desired form (7).

To prove the recurrence for the Q_n, we temporarily consider x as fixed and define $f(y) := I(y^{-1/2}, 2x)$. Then, by (7),

$$\log f(y) = -\sum_{n=1}^{\infty} \frac{Q_n}{n!^2} y^n. \tag{10}$$

By Proposition 4 we have $f(y) = {}_2F_1(x, -x; 1; y)$, so $f(y)$ satisfies the hypergeometric differential equation

$$y(1-y)f'' + (1-y)f' + x^2 f = 0,$$

where primes denote differentiation with respect to y. Define $g(y) := f'(y)/f(y)$. Then it may be verified[2] that $g(y)$ satisfies the Riccati equation

$$y(g' + g^2) + g + \frac{x^2}{1-y} = 0. \tag{11}$$

Let $g(y) = \sum_{n=0}^{\infty} g_n y^n$, where the g_n are polynomials in x, e.g. $g_0 = -x^2$. Equating coefficients in (11), we get the recurrence

$$g_n = -\left(\frac{1}{n+1}\right)\left(x^2 + \sum_{j=0}^{n-1} g_j g_{n-1-j}\right), \quad \text{for } n \geq 0. \tag{12}$$

Now, from (10) and the definitions of f and g, we have

$$g(y) = \frac{f'(y)}{f(y)} = \frac{d}{dy}\log f(y) = -\frac{d}{dy}\sum_{n=1}^{\infty} \frac{Q_n(x)}{n!^2}y^n,$$

so we see that

$$g_n = -\frac{Q_{n+1}}{n!(n+1)!}. \tag{13}$$

Substituting (13) in (12) and simplifying, we obtain the recurrence (8). $\qquad\square$

From the recurrence (8) it is clear that $Q_n(x)$ is an even polynomial of degree $2n$, such that $Q_n(0) = 0$. Writing $Q_n(x) = \sum_{k=1}^{n} q_{n,k}x^{2k}$, we see from the recurrence (8) that the coefficients $q_{n,k}$ are nonnegative integers.

In view of (13), the inequality (9) is equivalent to $|g_n(x)| \leq \max(1, |x|)^{2n+2}$, which may be proved by induction on n, using the recurrence (12).

Finally, in view of (9), the series in (7) converge for $b > \max(1, |x|)$. $\qquad\square$

Corollary 2. If $b > 1$, then $I(b, 2x)$ is nonzero in the disc $|x| < b$.

Proof. This follows from the convergence of the series for $\log I(b, 2x)$. $\qquad\square$

Proposition 5. The numbers $q_{n,k}$ are determined by $q_{n,1} = (n-1)!^2$ for $n \geq 1$ and

$$q_{n+1,k} = \sum_{j=0}^{n-1}\binom{n}{j}\binom{n}{j+1}\sum_{r=\mu}^{\nu} q_{j+1,r}q_{n-j,k-r} \tag{14}$$

for $2 \leq k \leq n+1$, where $\mu = \max(1, k-n+j)$ and $\nu = \min(j+1, k-1)$. Also, $q_{n,k}$ is a positive integer for each $n \geq 1$ and $1 \leq k \leq n$.

---

[2]Usually a Riccati equation is reduced to a second-order linear differential equation; see, for example, Ince [14, §2.15]. We apply the standard argument in the reverse direction.

Proof. The recurrence is obtained by equating coefficients of x^{2k} in (8). Positivity of the $q_{n,k}$ for $1 \leq k \leq n$ follows. $\square$

Remark 3. We may consider the sum over r in (14) to be over all $r \in \mathbb{Z}$ if we define $q_{n,k} = 0$ for $k < 1$ and $k > n$. The given values μ and ν correspond to the nonzero terms of the resulting sum.

Corollary 3. *We have* $\sum_{k=1}^{n} q_{n,k} = n! (n-1)!$.

Proof. This is easily obtained if we substitute $x = 1$ in the recurrence (8). $\square$

Corollary 4. *We have*

$$q_{n,n} = 2^{2n} n! (n-1)! \sum_{k=1}^{\infty} \frac{1}{j_{0,k}^{2n}} \tag{15}$$

where $(j_{0,k})$ *is the sequence of positive zeros of the Bessel function* $J_0(z)$.

Proof. Define $q_n := q_{n,n}$. With $k = n+1$, the recurrence (14) gives, for $n \geq 1$,

$$q_{n+1} = \sum_{j=0}^{n-1} \binom{n}{j} \binom{n}{j+1} q_{j+1} q_{n-j} = \sum_{j=1}^{n} \binom{n}{j} \binom{n}{j-1} q_j q_{n-j+1}.$$

This recurrence appears in Carlitz [7, Eqn. (4)], where it is shown that the solution satisfies (15). $\square$

Remark 4. The sequence (q_n) is A002190 in Sloane's On-Line Encyclopedia of Integer Sequences (OEIS), where the generating function $-\log(J_0(2\sqrt{x}))$ is given. The numbers q_n enjoy remarkable congruence properties. In fact, (15) is analogous to Euler's identity $|B_{2n}| = 2(2n)! \sum_{k=1}^{\infty} (2\pi k)^{-2n}$, and the numbers q_n are analogous to Bernoulli numbers. We refer to Carlitz [7] for further discussion.

Remark 5. There are other recurrences giving the polynomials Q_n and the numbers $q_{n,k}$ (Table 1). We omit discussion of them here due to space limitations.

Table 1 The coefficients $q_{n,k}$

$n \setminus k$	1	2	3	4	5	6	7
1	1						
2	1	1					
3	4	4	4				
4	36	33	42	33			
5	576	480	648	720	456		
6	14400	10960	14900	18780	17900	9460	
7	518400	362880	487200	648240	730800	606480	274800

7 Bounds and Asymptotic Expansions

Since $I(b,x)$ is an even function of x, there is no loss of generality in assuming that $x \geq 0$ when giving bounds or asymptotic results for $I(b,x)$. This simplifies the statement of the results. Similarly remarks apply to $\psi_\sigma(x)$, which is also an even function.

Consider the first representation of $I(b,x)$ in Proposition 3. If b is large, then

$$\arctan\left(\frac{\sin\theta}{b - \cos\theta}\right) = \frac{\sin\theta}{b} + \mathcal{O}\left(b^{-2}\right).$$

However, it is well-known [32, §2.2] that the Bessel function $J_0(x)$ has an integral representation

$$J_0(x) = \frac{1}{\pi} \int_0^\pi \cos(x\sin\theta)\, d\theta. \tag{16}$$

Thus, we expect $I(b,x)$ to be approximated in some sense by $J_0(x/b)$. A more detailed analysis confirms this (see Proposition 7 and Corollary 9). The connection with Bessel functions makes Corollary 4 less surprising than it first appears.

Proposition 6. *For all $b > 1$ and $x \in \mathbb{R}$, we have $|I(b,x)| \leq 1$.*

Proof. This follows from the final integral in Proposition 3. □

Lemma 2. *For $t \in [0,1]$ and $c_1 = \pi/2 - 1 < 0.5708$, we have*

$$0 \leq \arcsin(t) - t \leq c_1 t^3.$$

Proof. Let $f(t) = (\arcsin(t) - t)/t^3$. We see from the Taylor series that $f(t)$ is nonnegative and increasing in $[0,1]$. Thus, $\sup_{t\in[0,1]} f(t) = f(1) = \pi/2 - 1$. □

Lemma 3. *Suppose $b > 1$, $t \in [0,1]$, and c_1 as in Lemma 2. Then*

$$0 \leq b\arcsin(t/b) - t \leq c_1 t^3/b^2.$$

Proof. Replace t by t/b in Lemma 2 and multiply both sides of the resulting inequality by b. □

Proposition 7. *Suppose $b > 1$, $x > 0$, and $c_2 = (2 - 4/\pi)/3 < 0.2423$. Then*

$$|I(b,x) - J_0(x/b)| \leq c_2 x/b^3.$$

Proof. From the last integral of Proposition 3, we have

$$I(b,bx) = \frac{2}{\pi} \int_0^1 \cos\left(bx\arcsin\frac{t}{b}\right) \frac{dt}{\sqrt{1-t^2}}.$$

Also, from the integral representation (16) for J_0, we see that

$$J_0(x) = \frac{2}{\pi} \int_0^1 \cos(xt) \frac{dt}{\sqrt{1-t^2}}.$$

Thus, by subtraction,

$$I(b,bx) - J_0(x) = \frac{2}{\pi} \int_0^1 f(b,x,t) \frac{dt}{\sqrt{1-t^2}}, \tag{17}$$

where $f(b,x,t) = \cos(bx\arcsin(t/b)) - \cos(xt)$. Using $|\cos x - \cos y| \le |x-y|$, we have

$$|f(b,x,t)| \le |bx\arcsin(t/b) - xt|.$$

Thus, from Lemma 3,

$$|f(b,x,t)| \le c_1 t^3 x/b^2.$$

Taking norms in (17) gives

$$|I(b,bx) - J_0(x)| \le \frac{2c_1 x}{\pi b^2} \int_0^1 \frac{t^3 dt}{\sqrt{1-t^2}}. \tag{18}$$

The integral in (18) is easily seen to have the value $2/3$. Thus, replacing x by x/b in (18) completes the proof. □

Corollary 5. *If $b > 1$, $x > 0$, c_2 as in Proposition 7, and $c_3 = \sqrt{2/\pi} < 0.7979$, then*

$$|I(b,x)| \le c_2 x/b^3 + c_3(b/x)^{1/2}. \tag{19}$$

Proof. It is known [1, 9.2.28–9.2.31] that $|J_0(x)| \le \sqrt{2/(\pi x)}$ for real, positive x. Thus, the result follows from Proposition 7. □

Remark 6. The crossover point in Corollary 5 is for $b \approx x^{3/7}$: the first term in (19) dominates if $b \ll x^{3/7}$; the second term dominates if $b \gg x^{3/7}$.

Corollary 6. *If $x > 1$ and $b \ge x^{1/2}$, then*

$$|I(b,x)| \le c_3(b/x)^{1/2}(1 + c_5 b^{-1/2}),$$

where c_2, c_3 are as above and $c_5 = c_2/c_3 < 0.3037$.

Proof. From Corollary 5 we have

$$|I(b,x)| \le c_3(b/x)^{1/2}(1 + c_5 x^{3/2}/b^{7/2}).$$

The condition $b \ge x^{1/2}$ implies that $x^{3/2}/b^{7/2} \le b^{-1/2}$. □

For the remainder of this section we write $\beta := \arcsin(1/b)$.

Proposition 8. *For $b > 1$ and real positive x, we have*

$$I(b,x) = -\Re\left(\frac{2ie^{ix\beta}}{\pi}\int_0^\infty e^{-xu}\frac{\sqrt{b^2-1}\cosh u - i\sinh u}{\sqrt{1-(\cosh u + i\sqrt{b^2-1}\sinh u)^2}}\,du\right). \quad (20)$$

Proof. From the second integral in Proposition 3 we get $I(b,x) = \Re J(b,x)$, where

$$J(b,x) = \frac{2b}{\pi}\int_0^\beta e^{ixt}\frac{\cos t}{\sqrt{1-b^2\sin^2 t}}\,dt.$$

The function $1-b^2\sin^2 t$ has zeros at $t = \pm\beta + k\pi$ with $k \in \mathbb{Z}$ and only at these points. Also, $\beta = \arcsin(1/b) \in (0,\pi/2)$. Hence, if Ω denotes the complex plane $\mathbb{C}$ with two cuts along the half-lines $(-\infty, -\beta]$ and $[\beta, +\infty)$, then the function $(\cos t)/\sqrt{1-b^2\sin^2 t}$ is analytic on Ω. We consider the branch that is real and positive in the interval $(0,\beta)$. We apply Cauchy's theorem to the half strip $\Im t > 0$, $0 < \Re t < \beta$, obtaining

$$J(b,x) = \frac{i}{\pi}\int_0^\infty e^{-xu}\frac{2b\cosh u}{\sqrt{1+b^2\sinh^2 u}}\,du - \frac{ie^{ix\beta}}{\pi}\int_0^\infty e^{-xu}\frac{2b\cos(\beta + iu)}{\sqrt{1-b^2\sin^2(\beta + iu)}}\,du.$$

The first integral does not contribute to the real part. Taking the real part of the second integral and simplifying gives (20). □

In the following theorem we give an asymptotic expansion of $I(b,x)$.

Theorem 3. *For $b > 1$ fixed and real $x \to +\infty$, there is an asymptotic expansion of $I(b,x)$. If $\beta = \arcsin(1/b)$, the first three terms are given by*

$$I(b,x) = \frac{2}{\sqrt{2\pi}}\frac{(b^2-1)^{1/4}}{x^{1/2}}\cos(x\beta - \pi/4)$$

$$+ \frac{(b^2+2)}{4\sqrt{2\pi}}\frac{(b^2-1)^{-1/4}}{x^{3/2}}\sin(x\beta - \pi/4)$$

$$- \frac{(9b^4 - 28b^2 + 4)}{64\sqrt{2\pi}}\frac{(b^2-1)^{-3/4}}{x^{5/2}}\cos(x\beta - \pi/4) + \mathcal{O}\left(\frac{1}{x^{7/2}}\right).$$

Proof. We apply the Laplace method and Watson's lemma [25, Ch. 3, pg. 71] to the representation (20). □

Corollary 7. *For fixed $b > 1$, the function $I(b,x)$ has infinitely many real zeros.*

Proof. This is immediate from the first term of the asymptotic expansion above. The zeros are near the points $\pm\left(\frac{3\pi}{4} + k\pi\right)/\beta$ for $k \in \mathbb{Z}_{\geq 0}$. □

Corollary 8. *For fixed $\sigma > \frac{1}{2}$, the function $\psi_\sigma(x)$ has infinitely many real zeros.*

Proof. This is immediate from Proposition 2 and Corollary 7.   □

Corollary 9. *If $b > 1$ and $\beta = \arcsin(1/b)$, then for real $x \to +\infty$, we have*

$$I(b,x) = \beta^{1/2}(b^2 - 1)^{1/4}J_0(\beta x) + \mathcal{O}(x^{-3/2}).$$

Proof. The Bessel function $J_0(x)$ has an asymptotic expansion which gives

$$J_0(x) = \left(\frac{2}{\pi x}\right)^{1/2}\left(\cos(x - \pi/4) + \frac{1}{8x}\sin(x - \pi/4) + \mathcal{O}\left(\frac{1}{x^2}\right)\right).$$

Therefore, from Theorem 3, the difference $I(b,x) - \beta^{1/2}(b^2 - 1)^{1/4}J_0(\beta x)$ is of the order indicated.   □

Now we give a bound on the function $I(b,x)$ which is sharper than Corollary 5 in the region $x \gg b^{7/3}$.

Proposition 9. *For $b \geq \sqrt{2}$ and real $x \geq 5$, we have*

$$|I(b,x)| \leq 1.1512\sqrt{b/x}. \tag{21}$$

Proof. We consider the representation (20). Take $A := \sqrt{b^2 - 1}$ so the condition $b \geq \sqrt{2}$ implies that $A \geq 1$. It can be shown that, for $A \geq 1$ and real $u > 0$, the inequality

$$\left|\frac{A\cosh u - i\sinh u}{\sqrt{1 - (\cosh u + iA\sinh u)^2}}\right| \leq \frac{c_4\sqrt{A}}{\min(u^{1/2}, 1)}$$

holds. Here, the optimal constant is $c_4 = \sqrt{\coth(2)} < 1.0185$, attained at $A = u = 1$. (We omit details of the proof, which is elementary but tedious.) Hence, from (20),

$$|I(b,x)| \leq \frac{2c_4}{\pi}(b^2 - 1)^{1/4}\left\{\int_0^1 u^{-1/2}e^{-xu}\,du + \int_1^\infty e^{-xu}\,du\right\}$$

$$< \frac{2c_4\sqrt{b}}{\pi}\left(\sqrt{\pi/x} + e^{-x}/x\right).$$

For $x \geq 5$ we have $(2c_4/\pi)(\sqrt{\pi/x} + e^{-x}/x) \leq 1.1512/\sqrt{x}$.   □

Remark 7. The constant 1.1512 in (21) can be reduced if we do not ask for uniformity in b. From Theorem 3, we have

$$|I(b,x)| \leq c_3(1 - b^{-2})^{1/4}(b/x)^{1/2} + \mathcal{O}\left(x^{-3/2}\right) \quad \text{as } x \to +\infty,$$

so the constant can be reduced to $c_3 = (2/\pi)^{1/2} < 0.7979$ for all $x \geq x_0(b)$.

The following conjecture is consistent with our analytic results, for example, Corollary 5 and Theorem 3, and with extensive numerical evidence.

Conjecture 1. For all $b > 1$ and $x > 0$, we have $|I(b,x)| < \sqrt{\dfrac{2b}{\pi x}}$.

To conclude this section, we give a bound on $\psi_\sigma(x)$.

Theorem 4. *Let $\sigma > \frac{1}{2}$ be fixed. Then $|\psi_\sigma(x)| \leq 1$ for all $x \in \mathbb{R}$. Also, there exists a positive constant $c \geq 0.47$ and $x_0(\sigma)$ such that*

$$|\psi_\sigma(x)| \leq \exp\left(-\frac{cx^{1/\sigma}}{\log(x^{1/\sigma})}\right) \quad \text{for all real } x \geq x_0(\sigma).$$

Proof. The first inequality is immediate from the definition of $\psi_\sigma(x)$ as the characteristic function of a random variable.

To prove the last inequality, it is convenient to write $y := x^{1/\sigma}$. Let $\mathcal{P}(y)$ be the set of primes p in the interval $(y^{1/2}, y]$. We can assume that $\psi_\sigma(x) \neq 0$, because otherwise the inequality is trivial. From Proposition 6 and Corollary 6, we have

$$|\psi_\sigma(x)| \leq \prod_{p \in \mathcal{P}(y)} |I(p^\sigma, x)| \leq \prod_{p \in \mathcal{P}(y)} \left(c_3(p/y)^{\sigma/2}(1 + c_5 p^{-\sigma/2})\right),$$

which implies

$$-\log|\psi_\sigma(x)| \geq \sum_{p \in \mathcal{P}(y)} \left(-\log(c_3) + \tfrac{\sigma}{2}(\log y - \log p)\right) + \mathcal{O}(y^{1-\sigma/2}).$$

Using $\log(c_3) < -0.22$ and $\sigma > 1/2$ gives

$$-\log|\psi_\sigma(x)| \geq (\pi(y) - \pi(y^{1/2}))(\tfrac{\sigma}{2}\log y + 0.22) - \tfrac{\sigma}{2}\sum_{p \in \mathcal{P}(y)} \log p + \mathcal{O}(y^{3/4}), \quad (22)$$

where, as usual, $\pi(y)$ denotes the number of primes in the interval $[1, y]$.

From standard results on the distribution of primes [28], we have

$$\pi(y) = \frac{y}{\log y} + \frac{y}{\log^2 y} + \mathcal{O}\left(\frac{y}{\log^3 y}\right) \quad \text{and} \quad \sum_{p \in \mathcal{P}(y)} \log p = y + \mathcal{O}\left(\frac{y}{\log^2 y}\right).$$

Substituting in (22), we see that the leading terms of order y cancel, leaving

$$-\log|\psi_\sigma(x)| \geq (\tfrac{\sigma}{2} + 0.22)\frac{y}{\log y} + \mathcal{O}\left(\frac{y}{\log^2 y}\right).$$

Since $\tfrac{\sigma}{2} + 0.22 > 0.47$, the theorem follows, provided y is sufficiently large. $\square$

Remark 8. We find numerically that, for $\sigma \in (0.5, 1.1)$, we can take $c = 1$ and $x_0 = 5$ in Theorem 4.

8 An Algorithm for Computing $\psi_\sigma(x)$

There is a well-known technique, going back at least to Wrench [33], for accurately computing certain sums/products over primes. The idea is to express what we want to compute in terms of the *prime zeta function*

$$P(s) := \sum_p p^{-s} \quad (\Re(s) > 1).$$

The prime zeta function can be computed from $\log \zeta(s)$ using Möbius inversion:

$$P(s) = \sum_{r=1}^{\infty} \frac{\mu(r)}{r} \log \zeta(rs). \tag{23}$$

In fact, (23) gives the analytic continuation of $P(s)$ in the half-plane $\Re s > 0$ (see Titchmarsh [29, §9.5]), but we only need to compute $P(s)$ for real $s > 1$.

To illustrate the technique, temporarily ignore questions of convergence. From Theorem 2, we have

$$\log I(p^\sigma, x) = -\sum_{n=1}^{\infty} \frac{Q_n(x/2)}{n!^2} p^{-2n\sigma}.$$

Thus, taking logarithms in (2),

$$\log \psi_\sigma(x) = -\sum_p \sum_{n=1}^{\infty} \frac{Q_n(x/2)}{n!^2} p^{-2n\sigma} = -\sum_{n=1}^{\infty} \frac{Q_n(x/2)}{n!^2} P(2n\sigma). \tag{24}$$

Unfortunately, this approach fails, because $\psi_\sigma(x)$ has (infinitely many) real zeros – see Corollary 8. In fact, the series (24) converges for $|x| < |x_1(\sigma)|$, where $x_1(\sigma)$ is the zero of $\psi_\sigma(x)$ closest to the origin, and diverges for $|x| > |x_1(\sigma)|$.

Fortunately, a simple modification of the approach avoids this difficulty. Instead of considering a product over all primes, we consider the product over sufficiently large primes, say $p > p_0(x, \sigma)$. Corollary 2 guarantees that $I(p^\sigma, x)$ has no zeros in the disc $|x| < 2p^\sigma$. Thus, to evaluate $\psi_\sigma(x)$ for given σ and x, we should choose $2p_0^\sigma > |x|$, that is, $p_0 > |x/2|^{1/\sigma}$. In practice, to ensure rapid convergence, we might choose p_0 somewhat larger, say $p_0 \approx |4x|^{1/\sigma}$.

For the primes $p \le p_0$, we avoid logarithms and compute $I(p^\sigma, x)$ directly from the hypergeometric series of Proposition 4.

To summarise, the algorithm for computing $\psi_\sigma(x)$ with absolute error $\mathcal{O}(\varepsilon)$, for $x \in \mathbb{R}$, is as follows.

Algorithm for the Characteristic Function $\psi_\sigma(x)$

1. $p_0 \leftarrow \lceil |4x|^{1/\sigma} \rceil$.

2. $A \leftarrow \prod_{p \leq p_0} \left(1 + \sum_{n=1}^{N} \frac{1}{p^{2n\sigma} n!^2} \prod_{j=0}^{n-1} (j^2 - (x/2)^2) \right)$, where N is sufficiently large that the error in truncating the sum is $\mathcal{O}(\varepsilon)$. [Here A is the product over primes $\leq p_0$.]

3. $B \leftarrow \exp \left(-\sum_{n=1}^{N'} \frac{Q_n(x/2)}{n!^2} \left\{ P(2n\sigma) - \sum_{p \leq p_0} p^{-2n\sigma} \right\} \right)$, where N' is sufficiently large that the error in truncating the sum is $\mathcal{O}(\varepsilon)$, and $Q_n(x/2)$ is evaluated using the recurrence (8). [Here B is the product over primes $> p_0$.]

4. Return $A \times B$.

Remarks on the Algorithm for $\psi_\sigma(x)$

1. At step 3, $P(2n\sigma)$ can be evaluated using Eq. (23); time can be saved by precomputing the required values $\zeta(rs)$.

2. It is assumed that the computation is performed in floating-point arithmetic with sufficiently high precision and exponent range [6, Ch. 3]. For efficiency the precision should be varied dynamically as required, for example, to compensate for cancellation when summing the hypergeometric series at step 2 or when computing the term $\{ P(2n\sigma) - \sum_{p \leq p_0} p^{-2n\sigma} \}$ at step 3.

3. At step 3, an alternative is to evaluate $Q_n(x/2)$ using a table of coefficients $q_{n,k}$; these can be computed in advance using the recurrence of Proposition 5. This saves time (especially if many evaluations of $\psi_\sigma(x)$ at different points x are required, as is the case when evaluating $d(\sigma)$), at the expense of space and the requirement to estimate N' in advance.

4. The algorithm runs in polynomial time, in the sense that the number of bit-operations required to compute $\psi_\sigma(x)$ with absolute error $O(\varepsilon)$ is bounded by a polynomial (depending on σ and x) in $\log(1/\varepsilon)$.

9 Evaluation of $d(\sigma)$ and $d_-(\sigma)$

In this section we show how the densities $d(\sigma)$ and $d_-(\sigma)$ of Sect. 3 can be expressed in terms of the characteristic function ψ_σ.

Proposition 10. *For $\sigma > 1$, the support of the measure μ_σ of Sect. 3 is contained in the compact interval $[-L(\sigma), L(\sigma)]$, where*

$$L(\sigma) := \sum_p \arcsin(p^{-\sigma}).$$

Proof. Recall that μ_σ is the distribution of the random variable $\Im S$ considered in Sect. 2. From (4), $\Im S$ is equal to the sum of terms $-\arctan((\sin t)/(p^\sigma - \cos t))$ whose values are contained in the interval $[-\arcsin p^{-\sigma}, \arcsin p^{-\sigma}]$. Therefore, the range of $\Im S$ is contained in the interval $[-L(\sigma), L(\sigma)]$. $\qquad\square$

Remark 9. It may be shown that the support of μ_σ is exactly the interval $[L(\sigma), L(\sigma)]$.

Remark 10. As in van de Lune [23], we define σ_0 to be the (unique) real root in $(1, +\infty)$ of the equation $L(\sigma) = \pi/2$ and σ_1 to be the real root in $(1, +\infty)$ of $L(\sigma) = 3\pi/2$. These constants are relevant in Sect. 10.

Proposition 11. *For $\sigma > \frac{1}{2}$,*

$$d(\sigma) = 1 - \frac{2}{\pi} \int_0^\infty \psi_\sigma(x) \sin\left(\frac{\pi x}{2}\right) \frac{dx}{x}. \tag{25}$$

Proof. Recall from Sect. 3 that $1 - d(\sigma) = \mu_\sigma([-\pi/2, \pi/2])$. Since ψ_σ is the characteristic function associated to the distribution μ_σ, a standard result[3] in probability theory gives

$$1 - d(\sigma) = \frac{1}{2\pi} \lim_{X\to\infty} \int_{-X}^X \frac{\exp(ix\pi/2) - \exp(-ix\pi/2)}{ix} \psi_\sigma(x)\, dx.$$

Since $\psi_\sigma(x)$ is an even function, we obtain (25). $\qquad\square$

To evaluate $d(\sigma)$ numerically from (25), we have to perform a numerical integration. The following theorem shows that the integral may be replaced by a rapidly converging sum if $\sigma > 1$.

Theorem 5. *Let $\sigma > 1$ and $\ell > \max(\pi/2, L(\sigma))$. Then we have*

$$d(\sigma) = 1 - \frac{\pi}{2\ell} - \frac{2}{\pi} \sum_{n=1}^\infty \frac{1}{n} \psi_\sigma\left(\frac{\pi n}{\ell}\right) \sin\left(\frac{n\pi^2}{2\ell}\right). \tag{26}$$

Proof. Consider the function $\tilde{\rho}(x)$ equal to $\rho_\sigma(x)$ in the interval $[-\ell, \ell]$. Now extend $\tilde{\rho}(x)$ to the real line $\mathbb{R}$, making it periodic with period 2ℓ. Thus,

$$\tilde{\rho}(x) = \sum_{n \in \mathbb{Z}} f_n \exp\left(\frac{\pi i n x}{\ell}\right), \quad \text{where} \quad f_n = \frac{1}{2\ell} \int_{-\ell}^\ell \tilde{\rho}(x) \exp\left(-\frac{\pi i n x}{\ell}\right) dx.$$

Now $\tilde{\rho}(x) = \rho_\sigma(x)$ for $|x| \le \ell$ and $\rho_\sigma(x) = 0$ for $|x| > \ell$. Therefore,

$$f_n = \frac{1}{2\ell} \int_{-\ell}^\ell \rho_\sigma(x) \exp\left(-\frac{\pi i n x}{\ell}\right) dx = \frac{1}{2\ell} \int_{\mathbb{R}} \rho_\sigma(x) \exp\left(-\frac{\pi i n x}{\ell}\right) dx = \frac{1}{2\ell} \psi_\sigma\left(\frac{\pi n}{\ell}\right).$$

---

[3] Attributed to Paul Lévy.

Since $\psi_\sigma(x)$ is an even function,

$$\tilde{\rho}(x) = \frac{1}{2\ell} \sum_{n \in \mathbb{Z}} \psi_\sigma \left(\frac{\pi n}{\ell} \right) \exp \left(\frac{\pi i n x}{\ell} \right) = \frac{1}{2\ell} + \frac{1}{\ell} \sum_{n=1}^{\infty} \psi_\sigma \left(\frac{\pi n}{\ell} \right) \cos \frac{\pi n x}{\ell}. \qquad (27)$$

Now $d(\sigma) = 1 - \mu_\sigma([-\pi/2, \pi/2]) = 1 - \int_{-\pi/2}^{\pi/2} \rho_\sigma(t)\, dt$. Since $\pi/2 \le \ell$, we may replace $\rho_\sigma(t)$ by $\tilde{\rho}(t)$ in the integral. Hence, multiplying the equality (27) by the characteristic function of $[-\pi/2, \pi/2]$ and integrating, we get (26). $\qquad \square$

Remark 11. The sum in (26) can be seen as a numerical quadrature to approximate the integral in (25), taking a Riemann sum with stepsize $h = \pi/\ell$. However, we emphasise that (26) is *exact* under the conditions stated in Theorem 5. This is a consequence of the measure μ_σ having finite support when $\sigma > 1$. If $\sigma \in (\frac{1}{2}, 1]$, then μ_σ no longer has finite support and (26) only gives an approximation; however, this approximation converges rapidly to the exact result as $\ell \to \infty$, because μ_σ is well-approximated by measures with finite support.

Remark 12. If we take $m := 4\ell/\pi$ in the Theorem 5, we get the slightly simpler form

$$d(\sigma) = 1 - \frac{2}{m} - \frac{2}{\pi} \sum_{n=1}^{\infty} \frac{1}{n} \psi_\sigma \left(\frac{4n}{m} \right) \sin \left(\frac{2\pi n}{m} \right) \qquad (28)$$

for $m > \max(2, M(\sigma))$, where $M(\sigma) = 4L(\sigma)/\pi$. A good choice if $L(\sigma) < \pi$ is $m = 4$; then only the odd terms in the sum (28) contribute.

Computation of $d_-(\sigma)$

Recall that $d_-(\sigma)$ is the probability that $\Re\zeta(\sigma + it) < 0$. Let $a_k = a_k(\sigma)$ be the probability that $|\arg \zeta(\sigma + it)| > (2k+1)\pi/2$, that is,

$$a_k := 1 - \mu_\sigma([-(k + \tfrac{1}{2})\pi, (k + \tfrac{1}{2})\pi]).$$

Then

$$d_-(\sigma) = \sum_{k=0}^{\infty} (a_{2k} - a_{2k+1}) = \sum_{k=0}^{\infty} (-1)^k a_k. \qquad (29)$$

We have seen that, for $\sigma > 1$ and $m > \max(2, 4L(\sigma)/\pi)$, Eq. (28) gives $a_0 = d(\sigma)$. Similarly, under the same conditions, we have

$$a_k = 1 - \frac{4k+2}{m} - \frac{2}{\pi} \sum_{n=1}^{\infty} \frac{1}{n} \psi_\sigma \left(\frac{4n}{m} \right) \sin \left(\frac{(4k+2)\pi n}{m} \right). \qquad (30)$$

Using (29) and (30) in conjunction with an algorithm for the computation of ψ_σ, we can compute $d_-(\sigma)$ and also, of course, $d_+(\sigma) = 1 - d_-(\sigma)$. If $\sigma \in (\frac{1}{2}, 1]$, then we can take the limit of (30) as $m \to \infty$, or use an analogue of Proposition 11, to evaluate the constants a_k.

10 Numerical Results

In [3] we described a computation of the first 50 intervals ($t > 0$) on which $\Re\zeta(1 + it)$ takes negative values. The first such interval occurs for $t \approx 682{,}112.9$, and has length ≈ 0.05. From the lengths of the first 50 intervals we estimated that $d_-(1) \approx 3.85 \times 10^{-7}$. We also mentioned a Monte Carlo computation which gave $d_-(1) \approx 3.80 \times 10^{-7}$. The correct value is $3.7886\ldots \times 10^{-7}$. The difficulty of improving the accuracy of these computations or of extending them to other values of σ was one motivation for the analytic approach of the present paper.

The algorithm of Sect. 8 was implemented independently by two of us, using in one case Mathematica and in the other Magma. The Mathematica implementation precomputes a table of coefficients $q_{n,k}$; the Magma implementation uses the recurrence for the polynomials Q_n directly. The results obtained by both implementations are in agreement and also agree (up to the expected statistical error) with the results obtained by the Monte Carlo method in the region $0.6 \leq \sigma \leq 1.1$ where the latter method is feasible.

Table 2 gives some computed values of $d(\sigma)$ for $\sigma \in (0.5, 1.165]$. From van de Lune [23] we know that $d(\sigma) = d_-(\sigma) = 0$ for $\sigma \geq \sigma_0 \approx 1.19234$. Table 2 shows that $d(\sigma)$ is very small for σ close to σ_0. For example, $d(\sigma) < 10^{-100}$ for $\sigma \geq 1.15$. The small size of $d(\sigma)$ makes the computation difficult for $\sigma \geq 1.15$. We need to compute $\psi_\sigma(4n/m)$ to more than 100 decimal places to compensate for cancellation in the sum (28), in order to get any significant figures in $d(\sigma)$.

Selberg [26] (see also [16, 18, 30]) showed that, for $t \sim \mathrm{unif}(T, 2T)$,

$$\frac{\log\zeta(1/2 + it)}{\sqrt{\frac{1}{2}\log\log T}} \overset{d}{\to} X + iY \tag{31}$$

as $T \to \infty$, with $X, Y \sim N(0, 1)$. This implies that $d(1/2) = 1$ but gives no indication of the speed of convergence of $d(\sigma)$ as $\sigma \downarrow \frac{1}{2}$. Table 2 shows that convergence is very slow – for $\sigma - \frac{1}{2} \geq 10^{-11}$, we have $d(\sigma) < \frac{2}{3}$.

Table 2 $d(\sigma)$ for various $\sigma \in (0.5, 1.165]$

σ	$d(\sigma)$
$0.5 + 10^{-11}$	0.6533592249148917497
$0.5 + 10^{-5}$	0.4962734204446697434
0.6	$7.9202919267432753125 \times 10^{-2}$
0.7	$2.5228782796068962969 \times 10^{-2}$
0.8	$5.1401888600187247641 \times 10^{-3}$
0.9	$3.1401743610642112427 \times 10^{-4}$
1.0	$3.7886623606688718671 \times 10^{-7}$
1.1	$6.3088749952505014038 \times 10^{-22}$
1.15	$1.3815328080907034247 \times 10^{-103}$
1.16	$1.1172074815779368125 \times 10^{-194}$
1.165	$1.2798207752318534603 \times 10^{-283}$

Table 3 The difference
$d(\sigma) - d_-(\sigma)$

σ	$d(\sigma) - d_-(\sigma)$
$0.5 + 10^{-11}$	0.1547533823
0.6	$8.073328981 \times 10^{-11}$
0.7	$2.676004882 \times 10^{-32}$
0.8	$7.655052120 \times 10^{-210}$

It appears from numerical computations that $\Re\zeta(1/2 + it)$ is "usually positive" for those values of t for which computation is feasible. This is illustrated by several of the figures in [2]. Because the function $\sqrt{\log\log T}$ grows so slowly, the region that is feasible for computation may not show the typical behaviour of $\zeta(\sigma + it)$ for large t on or close to the critical line $\sigma = \frac{1}{2}$.

Table 3 gives the difference $d(\sigma) - d_-(\sigma)$. For $\sigma > 0.8$, there is no appreciable difference between $d(\sigma)$ and $d_-(\sigma)$. This is because the probability that $|\arg\zeta(\sigma + it)| > 3\pi/2$ is very small in this region. Indeed, $d(\sigma) = d_-(\sigma)$ for all $\sigma \geq \sigma_1 \approx 1.0068$, where σ_1 is the positive real root of $L(\sigma) = 3\pi/2$.

There is an appreciable difference between $d(\sigma)$ and $d_-(\sigma)$ very close to the critical line. For example, $d_-(0.5 + 10^{-11}) \approx 0.4986058426$, but $d(0.5 + 10^{-11}) \approx 0.6533592249$. Our numerical results suggest that $\lim_{\sigma\downarrow 1/2} d_-(\sigma) = 1/2$.

It is plausible that $d_-(\frac{1}{2}) = d_+(\frac{1}{2}) = \frac{1}{2}$, but Selberg's result (31) does not seem to be strong enough to imply this.

11 Conclusion

We have shown a precise sense in which $\Re\zeta(s)$ is "usually positive" in the half-plane $\sigma = \Re(s) > \frac{1}{2}$, given an explicit expression for the characteristic function ψ_σ, and given a feasible algorithm for the accurate computation of ψ_σ and consequently for the computation of the densities $d(\sigma)$ and $d_-(\sigma)$.

Our results could be generalised to cover Dirichlet L-functions because the character $\chi(p)$ in the Euler product

$$L(s, \chi) = \prod_p (1 - \chi(p)p^{-s})^{-1}$$

can be absorbed into the random variable z_p whenever $|\chi(p)| = 1$. Thus, it would only be necessary to omit, from sums/products over primes, all primes p for which $\chi(p)$ is zero, i.e. the finite number of primes that divide the modulus of the L-function. This would, of course, change the numerical results. Nevertheless, we expect $\Re L(s, \chi)$ to be "usually positive" for $\Re(s) > \frac{1}{2}$.

References

1. M. Abramowitz, I.A. Stegun, *Handbook of Mathematical Functions* (Dover, New York, 1973)
2. J. Arias de Reyna, X-ray of Riemann's zeta-function (26 Sept 2003), 42 pp. arXiv:math/0309433v1
3. J. Arias de Reyna, R.P. Brent, J. van de Lune, A note on the real part of the Riemann zeta-function, in *Herman te Riele Liber Amicorum* (CWI, Amsterdam, 2011). arXiv:1112.4910v1
4. H. Bohr, B. Jessen, Über die Werteverteilung der Riemannschen Zetafunktion. Acta Math. **54**, 1–35 (1930)
5. H. Bohr, B. Jessen, Über die Werteverteilung der Riemannschen Zetafunktion. Acta Math. **58**, 1–55 (1931)
6. R.P. Brent, P. Zimmermann, *Modern Computer Arithmetic* (Cambridge University Press, Cambridge, 2011)
7. L. Carlitz, A sequence of integers related to the Bessel function. Proc. Am. Math. Soc. **14**, 1–9 (1963)
8. A.B.O. Daalhuis, Hypergeometric function, in *NIST Handbook of Mathematical Functions*, ed. by F.W.J. Olver, D.M. Lozier, R.F. Boisvert et al. (Cambridge University Press, Cambridge, 2010)
9. H.M. Edwards, *Riemann's Zeta Function* (Academic, New York, 1974)
10. P.D.T.A. Elliott, On the distribution of $\arg L(s,\chi)$ on the half-plane $\sigma > \frac{1}{2}$, Acta Arith. **20**, 155–169 (1972)
11. J.-P. Gram, Note sur les zéros de la fonction $\zeta(s)$ de Riemann. Acta Math. **27**, 289–304 (1903)
12. P.R. Halmos, *Measure Theory* (Springer, New York, 1974)
13. J.I. Hutchinson, On the roots of the Riemann zeta function. Trans. Am. Math. Soc. **27**, 49–60 (1925)
14. E.L. Ince, *Ordinary Differential Equations* (Longman, Green and Co, London, 1926)
15. H. Iwaniec, E. Kowalski, Analytic number theory. Am. Math. Soc. Coll. Publ. **53** (2004)
16. D. Joyner, *Distribution Theorems of L-Functions*. Pitman Research Notes in Mathematics, vol 142 (Wiley, New York, 1986)
17. J. Kalpokas, J. Steuding, On the value-distribution of the Riemann zeta-function on the critical line. Moscow J. Combin. Number Theory **2**, 26–42 (2011). Also arXiv:0907.1910v1
18. P. Kühn, On Selberg's central limit theorem. Master's Thesis, Department of Mathematics, ETH Zürich, March 2011
19. Y. Lamzouri, The two dimensional distribution of values of $\zeta(1+it)$. Int. Math. Res. Notices, IMRN 2008, paper 106, 48 pp. Also arXiv:0801.3692v2
20. Y. Lamzouri, Extreme values of $\arg L(1,\chi)$. Acta Arith. **146**, 335–354 (2011). Also arXiv:1005.4425v1
21. Y. Lamzouri, On the distribution of extreme values of zeta and L-functions in the strip $\frac{1}{2} < \sigma < 1$. Int. Math. Res. Notices, IMRN 2011, no. 23, pp. 5449–5503. Also arXiv:1005.4640v2
22. A. Laurinčikas, Limit theorems for the Riemann zeta-function, in *Mathematics and Its Applications*, vol 352 (Kluwer Academic, Dordrecht, 1996)
23. J. van de Lune, Some observations concerning the zero-curves of the real and imaginary parts of Riemann's zeta function. Afdeling Zuivere Wiskunde, Report ZW 201/83. Mathematisch Centrum, Amsterdam, 1983, i+25 pp. http://oai.cwi.nl/oai/asset/6554/6554A.pdf
24. D.C. Milioto, A method for zeroing-in on $\Re\zeta(\sigma+it) < 0$ in the half-plane $\sigma > 1$ (20 Jan 2010), 13 pp. arXiv:1001.2962v3
25. F.W.J. Olver, *Introduction to Asymptotics and Special Functions* (Academic, New York, 1974)
26. A. Selberg, Contributions to the theory of the Riemann zeta-function. Arch. Math. Naturvid. **48**(5), 89–155 (1946)
27. J. Steuding, *Value-Distribution of L-Functions*. Lecture Notes in Mathematics, vol 1877 (Springer, Berlin and Heidelberg, 2007)

28. G. Tenenbaum, M. Mendès France, *The Prime Numbers and Their Distribution* (American Mathematical Society, Providence, 2000)

29. E.C. Titchmarsh, *The Theory of the Riemann Zeta-Function*, 2nd edn, ed. by D.R. Heath-Brown (The Clarendon Press, Oxford, 1986)

30. K. Tsang, The distribution of the values of the Riemann zeta-function. PhD Thesis, Department of Mathematics, Princeton University, October 1984

31. S.M. Voronin, Theorem on the "universality" of the Riemann zeta-function. Izv. Akad. Nauk SSSR, Ser. Matem. **39**, 475–486 (1975) (Russian); Math. USSR Izv. **9**, 443–445 (1975)

32. G.N. Watson, *A Treatise on the Theory of Bessel Functions* (Cambridge University Press, Cambridge, 1922/1995)

33. J.W. Wrench, Evaluation of Artin's constant and the twin prime constant. Math. Comp. **15**, 396–398 (1961)

Additive Combinatorics: With a View Towards Computer Science and Cryptography—An Exposition

Khodakhast Bibak

Abstract Recently, additive combinatorics has blossomed into a vibrant area in mathematical sciences. But it seems to be a difficult area to define—perhaps because of a blend of ideas and techniques from several seemingly unrelated contexts which are used there. One might say that additive combinatorics is a branch of mathematics concerning the study of combinatorial properties of algebraic objects, for instance, Abelian groups, rings, or fields. This emerging field has seen tremendous advances over the last few years and has recently become a focus of attention among both mathematicians and computer scientists. This fascinating area has been enriched by its formidable links to combinatorics, number theory, harmonic analysis, ergodic theory, and some other branches; all deeply cross-fertilize each other, holding great promise for all of them! In this exposition, we attempt to provide an overview of some breakthroughs in this field, together with a number of seminal applications to sundry parts of mathematics and some other disciplines, with emphasis on computer science and cryptography.

1 Introduction

Additive combinatorics is a compelling and fast-growing area of research in mathematical sciences, and the goal of this paper is to survey some of the recent developments and notable accomplishments of the field, focusing on both pure results and applications with a view towards computer science and cryptography. See [321] for a book on additive combinatorics, [237,238] for two books on additive number theory, and [330, 339] for two surveys on additive combinatorics. About

K. Bibak (✉)
Department of Combinatorics and Optimization, University of Waterloo,
Waterloo, ON, Canada N2L 3G1
e-mail: kbibak@uwaterloo.ca

J.M. Borwein et al. (eds.), *Number Theory and Related Fields: In Memory of Alf van der Poorten*, Springer Proceedings in Mathematics & Statistics 43, DOI 10.1007/978-1-4614-6642-0_4, © Springer Science+Business Media New York 2013

additive combinatorics over finite fields and its applications, the reader is referred to the very recent and excellent survey by Shparlinski [291].

One might say that additive combinatorics studies combinatorial properties of algebraic objects, for example, Abelian groups, rings, or fields, and in fact, focuses on the interplay between combinatorics, number theory, harmonic analysis, ergodic theory, and some other branches. Green [151] describes additive combinatorics as the following: "additive combinatorics is the study of *approximate mathematical structures* such as approximate groups, rings, fields, polynomials and homomor- phisms." *Approximate groups* can be viewed as finite subsets of a group with the property that they are *almost* closed under multiplication. Approximate groups and their applications (e.g., to expander graphs, group theory, probability, model theory, and so on) form a very active and promising area of research in additive combinatorics; the papers [64, 65, 67, 68, 151, 190, 315] contain many developments on this area. Gowers [145] describes additive combinatorics as the following: "ad- ditive combinatorics focuses on three classes of theorems: *decomposition theorems, approximate structural theorems,* and *transference principles."* These descriptions seem to be mainly inspired by new directions of this area. Techniques and approaches applied in additive combinatorics are often extremely sophisticated and may have roots in several unexpected fields of mathematical sciences. For instance, Hamidoune [172], through ideas from connectivity properties of graphs, established the so-called *isoperimetric method,* which is a strong tool in additive combinatorics; see also [173–175] and the nice survey [278]. As another example, Nathanson [239] employed *König's infinity lemma* on the existence of infinite paths in certain infinite graphs and introduced a new class of additive bases and also a generalization of the Erdős–Turán conjecture about additive bases of the positive integers. In [245] the authors employed tools from coding theory to estimating Davenport constants. Also, in [15, 230, 231, 268, 319] information-theoretic techniques are used to study sumset inequalities. Very recently, Alon et al. [8, 9], using graph-theoretic methods, studied sum-free sets of order m in finite Abelian groups and, also, sum-free subsets of the set $[1, n]$. Additive combinatorics problems in matrix rings is another active area of research [52, 60, 67, 82, 83, 114, 125, 133, 134, 183, 184, 206, 298].

A celebrated result by Szemerédi, known as Szemerédi's theorem (see [11, 12, 29, 122, 143, 144, 158, 161, 236, 246, 255, 256, 259, 260, 304, 310, 311, 329] for different proofs of this theorem), states that every subset A of the integers with positive upper density, that is, $\limsup_{N \to \infty} |A \cap [1, N]|/N > 0$, has arbitrary long arithmetic progressions. A stunning breakthrough of Green and Tao [154] (that answers a long- standing and folkloric conjecture by Erdős on arithmetic progressions, in a special case: the primes) says that primes contain arbitrary long arithmetic progressions. The fusion of methods and ideas from combinatorics, number theory, harmonic analysis, and ergodic theory used in its proof is very impressive.

Additive combinatorics has recently found a great deal of remarkable applica- tions to computer science and cryptography, for example, to expanders [20, 21, 39, 44, 51, 52, 55, 56, 59, 60, 66, 103, 104, 167, 206, 283, 342], extractors [19, 21, 26, 27, 39, 41, 105–107, 167, 182, 215, 346, 350], pseudorandomness [33, 223, 226, 227, 301] (also, [331, 334] are two surveys and [335] is a monograph on pseudorandomness), property testing [30, 176, 177, 181, 203, 204, 270, 281, 332] (see also [137]), complexity theory [27, 28, 31, 38, 244, 350], hardness amplification [301, 338, 340], probabilistic checkable proofs (PCPs) [271], information theory [15, 230, 231, 268, 319], discrete logarithm-based range protocols [77], noninteractive zero-knowledge (NIZK) proofs [220], compression functions [192], hidden shifted power problem [62], and Diffie–Hellman distributions [36, 37, 75, 121]. Additive combinatorics also has important applications in e-voting [77, 220]. Recently, Bourgain et al. [61] gave a new explicit construction of matrices satisfying the *restricted isometry property* (RIP) using ideas from additive combinatorics. RIP is related to the matrices whose behavior is nearly orthonormal (at least when acting on sufficiently sparse vectors); it has several applications, in particular, in compressed sensing [71, 72, 74]. Methods from additive combinatorics provide strong techniques for studying the so-called *threshold phenomena*, which is itself of significant importance in combinatorics, computer science, discrete probability, statistical physics, and economics [2, 34, 54, 119, 120, 198]. There are also very strong connections between ideas of additive combinatorics and the theory of *random matrices* (see, e.g., [322, 323, 341] and the references therein); the latter themselves have several applications in many areas of number theory, combinatorics, computer science, mathematical and theoretical physics, chemistry, and so on [1, 69, 73, 141, 241, 265, 323, 324]. This area also has many applications to group theory, analysis, exponential sums, expanders, complexity theory, discrete geometry, dynamical systems, and various other scientific disciplines.

Additive combinatorics has seen very fast advancements in the wake of extremely deep work on Szemerédi's theorem, the proof of the existence of long APs in the primes by Green and Tao, and generalizations and applications of the sum-product problem and continues to see significant progress (see [97] for a collection of open problems in this area). In the next section, we review Szemerédi's and Green–Tao theorems (and their generalizations), two cornerstone breakthroughs in additive combinatorics. In the third section, we will deal with the sum-product problem: yet another landmark achievement in additive combinatorics, and consider its generalizations and applications, especially to computer science and cryptography.

2   Szemerédi's and Green–Tao Theorems and Their Generalizations

Ramsey theory is concerned with the phenomenon that if a sufficiently large structure (complete graphs, arithmetic progressions, flat varieties in vector spaces, etc.) is partitioned arbitrarily into finitely many substructures, then at least one substructure has necessarily a particular property, and so total disorder is impossible. In fact, Ramsey theory seeks general conditions to guarantee the existence of substructures with regular properties. This theory has many applications, for example, in number theory, algebra, geometry, topology, functional analysis (in particular, in Banach space theory), set theory, logic, ergodic theory, information theory, and theoretical computer science (see, e.g., [263] and the references therein). *Ramsey's theorem* says that in any edge coloring of a sufficiently large complete graph, one can find monochromatic complete subgraphs. A nice result of the same spirit is *van der Waerden's theorem*: For a given k and r, there exists a number $N = N(k, r)$ such that if the integers in $[1, N]$ are colored using r colors, then there is a nontrivial monochromatic k-term arithmetic progression (k-AP). Intuitively, this theorem asserts that in any finite coloring of a structure, one will find a substructure of the same type at least in one of the color classes. Note that the finitary and infinitary versions of the van der Waerden's theorem are equivalent, through a compactness argument.

One landmark result in Ramsey theory is the Hales–Jewett theorem [171], which was initially introduced as a tool for analyzing certain kinds of games. Before stating this theorem, we need to define the concept of combinatorial line. A *combinatorial line* is a k-subset in the n-dimensional grid $[1, k]^n$ yielded from some template in $([1, k] \cup \{*\})^n$ by replacing the symbol $*$ with $1, \ldots, k$ in turn. The Hales–Jewett theorem states that for every r and k, there exists n such that every r-coloring of the n-dimensional grid $[1, k]^n$ contains a combinatorial line. Roughly speaking, it says that for every multidimensional grid whose faces are colored with a number of colors, there must necessarily be a line of faces of all the same color, if the dimension is sufficiently large (depending on the number of sides of the grid and the number of colors). Note that instead of seeking arithmetic progressions, the Hales–Jewett theorem seeks combinatorial lines. This theorem has many interesting consequences in Ramsey theory, two of which are van der Waerden's theorem and its multidimensional version, i.e., the Gallai–Witt theorem (see, e.g., [132, 149, 197] for further information).

Erdős and Turán [112] proposed a very strong form of van der Waerden's theorem—the density version of van der Waerden's theorem. They conjectured that arbitrarily long APs appear not only in finite partitions but also in every sufficiently dense subset of positive integers. More precisely, the Erdős–Turán conjecture states that if δ and k are given, then there is a number $N = N(k, \delta)$ such that any set $A \subseteq [1, N]$ with $|A| \geq \delta N$ contains a non-trivial k-AP. Roth [264] employed methods from Fourier analysis (or more specifically, the *Hardy–Littlewood circle method*) to prove the $k = 3$ case of the Erdős–Turán conjecture (see also [32, 35, 43, 98, 110, 165, 213, 234, 242, 272, 274, 287, 296]). Szemerédi [303] verified the Erdős–Turán conjecture for arithmetic progressions of length four. Finally, Szemerédi [304] by a tour de force of sophisticated combinatorial arguments proved the conjecture, now known as *Szemerédi's theorem*—one of the milestones of combinatorics. Roughly speaking, this theorem states that long arithmetic progressions are very widespread, and in fact it is not possible to completely get rid of them from a set of positive integers unless we can contract the set (sufficiently) to make it of density zero. Łaba and Pramanik [211] (also see [247]) proved that every compact set of reals with Lebesgue measure zero supporting a probabilistic measure satisfying appropriate dimensionality and Fourier decay conditions must contain nontrivial 3APs.

Conlon and Gowers [93] considered Szemerédi's theorem, and also several other combinatorial theorems such as Turán's theorem and Ramsey's theorem in sparse random sets. Also, Szemerédi-type problems in various structures other than integers have been a focus of significant amount of work. For instance, [150, 219] consider these kinds of problems in the finite field setting. Very recently, Bateman and Katz [23] (also see [22]) achieved new bounds for the *cap set problem*, which is basically Roth's problem, but in a vector space over finite fields (a set $A \subset \mathbb{F}_3^N$ is called a *cap set* if it contains no lines).

A salient ingredient in Szemerédi's proof (in addition to van der Waerden's theorem) is the *Szemerédi regularity lemma*. This lemma was conceived specifically for the purpose of this proof, but is now, by itself, one of the most powerful tools in extremal graph theory (see, e.g., [205, 258], which are two surveys on this lemma and its applications). Roughly speaking, it asserts that the vertex set of every (large) graph can be partitioned into relatively few parts such that the subgraphs between the parts are random-like. Indeed, this result states that each large dense graph may be decomposed into a low-complexity part and a pseudorandom part (note that Szemerédi's regularity lemma is the archetypal example of the *dichotomy between structure and randomness* [312]). The lemma has found numerous applications not only in graph theory but also in discrete geometry, additive combinatorics, and computer science. For example, as Trevisan [330] mentions, to solve a computational problem on a given graph, it might be easier to first construct a Szemerédi approximation—this resulted approximating graph has a simpler configuration and would be easier to treat. Note that the significance of Szemerédi's regularity lemma goes beyond graph theory: It can be reformulated as a result in information theory, approximation theory, as a compactness result on the completion of the space of finite graphs, etc. (see [222] and the references therein). Very recently, Tao and Green [158] established an *arithmetic regularity lemma* and a complementary *arithmetic counting lemma* that have several applications, in particular, an astonishing proof of Szemerédi's theorem.

The *triangle removal lemma* established by Ruzsa and Szemerédi [269] is one of the most notable applications of Szemerédi's regularity lemma. It asserts that each graph of order n with $o(n^3)$ triangles can be made triangle-free by removing $o(n^2)$ edges. In other words, if a graph has asymptotically few triangles, then it is asymptotically close to being triangle-free. As a clever application of this lemma, Ruzsa and Szemerédi [269] obtained a new proof of Roth's theorem (see also [296], in which the author using the triangle removal lemma proves Roth type theorems in finite groups). Note that a generalization of the triangle removal lemma, known as *simplex removal lemma*, can be used to deduce Szemerédi's theorem (see [144, 259, 260, 311]). The triangle removal lemma was extended by Erdős, Frankl, and Rödl [113] to the *graph removal lemma*, which roughly speaking, asserts that if a given graph does not contain too many subgraphs of a given type, then all the subgraphs of this type can be removed by deleting a few edges. More precisely, given a fixed graph H of order k, any graph of order n with $o(n^k)$ copies of H can be made H-free by removing $o(n^2)$ edges. Fox [115] gave a proof of the graph removal lemma which avoids applying Szemerédi's regularity lemma and gives a better bound (also see [92]). The graph removal lemma has many applications in graph theory, additive combinatorics, discrete geometry, and theoretical computer science. One surprising application of this lemma is to the area of *property testing*, which is now a very dynamic area in computer science [4–7, 13, 137, 138, 187, 254, 257, 261, 262, 270]. Property testing typically refers to the existence of sublinear time probabilistic algorithms (called testers), which distinguish between objects G (e.g., a graph) having a given property P (e.g., bipartiteness) and those being far away (in an appropriate metric) from P. Property testing algorithms have been recently designed and utilized for many kinds of objects and properties, in particular, discrete properties (e.g., graph properties, discrete functions, and sets of integers), geometric properties, algebraic properties, etc.

There is a (growing) number of proofs of Szemerédi's theorem, arguably seventeen proofs to this date. One such elegant proof that uses ideas from model theory was given by Towsner [329]. For another model theory based perspective, see [232], in which the authors give stronger regularity lemmas for some classes of graphs.

In fact, one might claim that many of these proofs have themselves opened up a new field of research. Furstenberg [122] by rephrasing it as a problem in *dynamical systems* and then applying several powerful techniques from ergodic theory achieved a nice proof of the Szemerédi's theorem. In fact, Furstenberg presented a correspondence between problems in the subsets of positive density in the integers and recurrence problems for sets of positive measure in a *probability measure-preserving system*. This observation is now known as the *Furstenberg correspondence principle*. *Ergodic theory* is concerned with the long-term behavior in dynamical systems from a statistical point of view (see, e.g., [108]). This area and its formidable way of thinking have made many strong connections with several branches of mathematics, including combinatorics, number theory, coding theory, group theory, and harmonic analysis; see, for example, [207–210] and the references therein for some connections between ergodic theory and additive combinatorics.

This ergodic-theoretic method is one of the most flexible known proofs and has been very successful at reaching considerable generalizations of Szemerédi's theorem. Furstenberg and Katznelson [123] obtained the *multidimensional Szemerédi theorem*. Their proof relies on the concept of *multiple recurrence*, a powerful tool in the interaction between ergodic theory and additive combinatorics. A purely combinatorial proof of this theorem was obtained roughly in parallel by Gowers [144] and Nagle et al. [236, 255, 256, 259, 260] and subsequently by Tao [311], via establishing a *hypergraph removal lemma* (see also [257, 314]). Also, Austin [11] proved the theorem via both ergodic-theoretic and combinatorial approaches. The multidimensional Szemerédi theorem was significantly generalized by Furstenberg and Katznelson [124] (via ergodic-theoretic approaches) and Austin [12] (via both ergodic-theoretic and combinatorial approaches), to the *density Hales–Jewett theorem*. The density Hales–Jewett theorem states that for every $\delta > 0$, there is some $N_0 \geq 1$ such that whenever $A \subseteq [1, k]^N$ with $N \geq N_0$ and $|A| \geq \delta k^N$, A contains a combinatorial line. Recently, in a massively collaborative online project, namely *Polymath 1* (a project that originated in Gowers's blog), the Polymath team found a purely combinatorial proof of the density Hales–Jewett theorem, which is also the first one providing explicit bounds for how large n needs to be [246] (also see [240]). Such bounds could not be obtained through the ergodic-theoretic methods, since these proofs rely on the Axiom of Choice. It is worth mentioning that this project was selected as one of the TIME Magazine's Best Ideas of 2009.

Furstenberg's proof gave rise to the field of *ergodic Ramsey theory*, in which arithmetical, combinatorial, and geometrical configurations, preserved in (sufficiently large) substructures of a structure, are treated via ideas and techniques from ergodic theory (or more specifically, multiple recurrence). Ergodic Ramsey theory has since produced a high number of combinatorial results, some of which have yet to be obtained by other means, and has also given a deeper understanding of the structure of measure-preserving systems. In fact, ergodic theory has been used to solve problems in Ramsey theory, and reciprocally, Ramsey theory has led to the discovery of new phenomena in ergodic theory. However, the ergodic-theoretic methods and the infinitary nature of their techniques have some limitations. For example, these methods do not provide any effective bound, since, as we already mentioned, they rely on the Axiom of Choice. Also, despite van der Waerden's theorem is not directly used in Furstenberg's proof, probably any effort to make the proof quantitative would result in rapidly growing functions. Furthermore, the ergodic-theoretic methods, to this day, have the limitation of only being able to deal with sets of positive density in the integers, although this density is allowed to be arbitrarily small. However, Green and Tao [154] discovered a *transference principle* which allowed one to reduce problems on structures in special sets of zero density (such as the primes) to problems on sets of positive density in the integers. It is worth mentioning that Conlon, Fox, and Zhao [94] established a *transference principle* extending several classical extremal graph-theoretic results, including the removal lemmas for graphs and groups (the latter leads to an extension of Roth's theorem), the Erdős–Stone–Simonovits theorem and Ramsey's theorem, to sparse pseudorandom graphs.

Gowers [143] generalized the arguments previously studied in [142, 264], in a substantial way. In fact, he employed combinatorics, generalized Fourier analysis, and inverse arithmetic combinatorics (including multilinear versions of *Freiman's theorem* on sumsets and the *Balog–Szemerédi theorem*) to reprove Szemerédi's theorem with explicit bounds. Note that Fourier analysis has a wide range of applications, in particular, to cryptography, hardness of approximation, signal processing, threshold phenomena for probabilistic models such as random graphs and percolations, and many other disciplines. Gowers's article introduced a kind of *higher-degree Fourier analysis*, which has been further developed by Green and Tao. Indeed, Gowers initiated the study of a new measure of functions, now referred to as *Gowers (uniformity) norms*, that resulted in a better understanding of the notion of *pseudorandomness*. The Gowers norm, which is an important special case of noise correlation (intuitively, the *noise correlation* between two functions f and g measures how much $f(x)$ and $g(y)$ correlate on random inputs x and y which are correlated), enjoys many properties and applications, and is now a very dynamic area of research in mathematical sciences; see [14, 153, 157, 159, 163, 164, 176, 181,

203,221,224,225,233,270,271] for more properties and applications of the Gowers norm. Also, the best known bounds for Szemerédi's theorem are obtained through the so-called *inverse theorems* for Gowers norms. Recently, Green and Tao [161] (see also [320]), using the density-increment strategy of Roth [264] and Gowers [142, 143], derived Szemerédi's theorem from the *inverse conjectures GI(s)* for the Gowers norms, which were recently established in [164].

To the best of my knowledge, there are two types of inverse theorems in additive combinatorics, namely, the *inverse sumset theorems of Freiman type* (see, e.g., [78, 116–118, 152, 155, 238, 275, 299, 306–308, 318]) and *inverse theorems for the Gowers norms* (see, e.g., [143, 146–148, 153, 156, 159, 162–164, 176, 181, 189, 203, 224, 225, 233, 270, 326, 332]). It is interesting that the inverse conjecture leads to a finite field version of Szemerédi's theorem [320]: Let $\mathbb{F}_p$ be a finite field. Suppose that $\delta > 0$ and $A \subset \mathbb{F}_p^n$ with $|A| \geq \delta |\mathbb{F}_p^n|$. If n is sufficiently large depending on p and δ, then A contains a (affine) line $\{x, x+r, \ldots, x+(p-1)r\}$ for some $x, r \in \mathbb{F}_p^n$ with $r \neq 0$ (actually, A contains an affine k-dimensional subspace, $k \geq 1$).

Suppose $r_k(N)$ is the cardinality of the largest subset of $[1, N]$ containing no nontrivial k-APs. Giving asymptotic estimates on $r_k(N)$ is an important inverse problem in additive combinatorics. Behrend [24] proved that

$$r_3(N) = \Omega\left(\frac{N}{2^{2\sqrt{2}\sqrt{\log_2 N}} \cdot \log^{1/4} N}\right).$$

Rankin [249] generalized Behrend's construction to longer APs. Roth proved that $r_3(N) = o(N)$. In fact, he proved the first nontrivial upper bound:

$$r_3(N) = O\left(\frac{N}{\log\log N}\right).$$

Bourgain [35, 43] improved Roth's bound. In fact, Bourgain [43] gave the upper bound:

$$r_3(N) = O\left(\frac{N(\log\log N)^2}{\log^{2/3} N}\right).$$

Sanders [274] proved the following upper bound which is the state of the art:

$$r_3(N) = O\left(\frac{N(\log\log N)^5}{\log N}\right).$$

Bloom [32], through the nice technique "translation of a proof in $\mathbb{F}_q[t]$ to one in $\mathbb{Z}/N\mathbb{Z}$," extends Sanders' proof to 4 and 5 variables. As Bloom mentions in his paper, many problems of additive combinatorics might be easier to attack via the approach "translating from $\mathbb{F}_p^N$ to $\mathbb{F}_q[t]$ and hence to $\mathbb{Z}/N\mathbb{Z}$."

Elkin [110] managed to improve Behrend's 62-year-old lower bound by a factor of $\Theta(\log^{1/2} N)$. Actually, Elkin showed that

$$r_3(N) = \Omega\left(\frac{N}{2^{2\sqrt{2}\sqrt{\log_2 N}}} \cdot \log^{1/4} N \right).$$

See also [165] for a short proof of Elkin's result and [242] for constructive lower bounds for $r_k(N)$. Schoen and Shkredov [277], using ideas from the paper of Sanders [273] and also the new probabilistic technique established by Croot and Sisask [98], obtained Behrend-type bounds for linear equations involving six or more variables. Thanks to this result, one may see that perhaps the Behrend-type constructions are not too far from being best possible.

Almost all the known proofs of Szemerédi's theorem are based on a *dichotomy between structure and randomness* [312, 316], which allows many mathematical objects to be split into a "structured part" (or "low-complexity part") and a "random part" (or "discorrelated part"). Tao [310] best describes almost all known proofs of Szemerédi's theorem collectively as the following: "Start with the set A (or some other object which is a proxy for A, e.g., a graph, a hypergraph, or a measure-preserving system). For the object under consideration, define some concept of randomness (e.g., ε-regularity, uniformity, small Fourier coefficients, or weak mixing), and some concept of structure (e.g., a nested sequence of arithmetically structured sets such as progressions or Bohr sets, or a partition of a vertex set into a controlled number of pieces, a collection of large Fourier coefficients, a sequence of almost periodic functions, a tower of compact extensions of the trivial factors). Obtain some sort of structure theorem that splits the object into a structured component, plus an error which is random relative to that structured component. To prove Szemerédi's theorem (or a variant thereof), one then needs to obtain some sort of *generalized von Neumann theorem* [154] to eliminate the random error, and then some sort of *structured recurrence theorem* for the structured component."

Erdős's famous conjecture on APs states that a set $A = \{a_1, a_2, \ldots, a_n, \ldots\}$ of positive integers, where $a_i < a_{i+1}$ for all i, with the divergent sum $\sum_{n \in \mathbb{Z}^+} \frac{1}{a_n}$, contains arbitrarily long APs. If true, the theorem includes both Szemerédi's and Green–Tao theorems as special cases. This conjecture seems to be too strong to hold and, in fact, might be very difficult to attack—it is not even known whether such a set must contain a 3AP! So, let us mention an equivalent statement for Erdős's conjecture that may be helpful. Let N be a positive integer. For a positive integer k, define $a_k(N) := r_k(N)/N$ (note that Szemerédi's theorem asserts that $\lim_{N \to \infty} a_k(N) = 0$, for all k). It can be proved (see [286]) that Erdős's conjecture is true if and only if the series $\sum_{i=1}^{\infty} a_k 4^i$ converges for any integer $k \geq 3$. So, to prove Erdős's conjecture, it suffices to obtain the estimate $a_k(N) \ll 1/(\log N)^{1+\varepsilon}$, for any $k \geq 3$ and for some $\varepsilon > 0$.

Szemerédi's theorem plays an important role in the proof of the Green–Tao theorem [154]: The primes contain arithmetic progressions of arbitrarily large length (note that the same result is valid for every subset of the primes with positive relative upper density). Green and Tao [160] also proved that there is a k-AP of primes all of whose terms are bounded by

which shows that how far out in the primes one must go to warrant finding a k-AP. A conjecture (see [207]) asserts that there is a k-AP in the primes all of whose terms are bounded by $k! + 1$.

There are three fundamental ingredients in the proof of the Green–Tao theorem (in fact, there are many similarities between Green and Tao's approach and the ergodic-theoretic method, see [188]). The first is Szemerédi's theorem itself. Since the primes do not have positive upper density, Szemerédi's theorem cannot be directly applied. The second major ingredient in the proof is a certain *transference principle* that allows one to use Szemerédi's theorem in a more general setting (a generalization of Szemerédi's theorem to the *pseudorandom sets*, which can have zero density). The last major ingredient is applying some notable features of the primes and their distribution through results of Goldston and Yildirim [139, 140], and proving the fact that this generalized Szemerédi theorem can be efficiently applied to the primes, and indeed, the set of primes will have the desired pseudorandom properties.

In fact, Green and Tao's proof employs the techniques applied in several known proofs of Szemerédi's theorem and exploits a dichotomy between structure and randomness. This proof is based on ideas and results from several branches of mathematics, for example, combinatorics, analytical number theory, pseudorandomness, harmonic analysis, and ergodic theory. Reingold et al. [252], and Gowers [145], independently obtained a short proof for a fundamental ingredient of this proof.

Tao and Ziegler [325] (see also [252]), via a transference principle for polynomial configurations, extended the Green–Tao theorem to cover polynomial progressions: Let $A \subset \mathcal{P}$ be a set of primes of positive relative upper density in the primes, i.e., $\limsup_{N \to \infty} |A \cap [1, N]| / |\mathcal{P} \cap [1, N]| > 0$. Then, given any integer-valued polynomials $P_1, \ldots, P_k$ in one unknown m with vanishing constant terms, the set A contains infinitely many progressions of the form $x + P_1(m), \ldots, x + P_k(m)$ with $m > 0$ (note that the special case when the polynomials are $m, 2m, \ldots, km$ implies the previous result that there are k-APs of primes). Tao [313] proved the analogue in the Gaussian integers. Green and Tao (in view of the parallelism between the integers and the polynomials over a finite field) thought that the analogue of their theorem should be held in the setting of function fields; a result that was proved by Lê [212]: Let $\mathbb{F}_q$ be a finite field over q elements. Then for any $k > 0$, one can find polynomials $f, g \in \mathbb{F}_q[t]$, $g \not\equiv 0$ such that the polynomials $f + Pg$ are all irreducible, where P runs over all polynomials $P \in \mathbb{F}_q[t]$ of degree less than k. Moreover, such structures can be found in every set of positive relative upper density among the irreducible polynomials. The proof of this interesting theorem follows the ideas of the proof of the Green–Tao theorem very closely.

3 Sum-Product Problem: Its Generalizations and Applications

The *sum-product problem* and its generalizations constitute another vibrant area in additive combinatorics and have led to many seminal applications to number theory, Ramsey theory, computer science, and cryptography.

Let us start with the definition of sumset, product set, and some preliminaries. We will follow closely the presentation of Tao [316]. Let A be a finite nonempty set of elements of a ring R. We define the *sumset* $A + A = \{a + b : a, b \in A\}$ and the *product set* $A \cdot A = \{a \cdot b : a, b \in A\}$. Suppose that no $a \in A$ is a zero divisor (otherwise, $A \cdot A$ may become very small, which lead to degenerate cases). Then one can easily show that $A + A$ and $A \cdot A$ will be at least as large as A. The set A may be almost closed under addition, which, for example, occurs when A is an arithmetic progression or an additive subgroup in the ring R (e.g., if $A \subset \mathbb{R}$ is an AP, then $|A + A| = 2|A| - 1$, and $|A \cdot A| \geq c|A|^{2-\varepsilon}$), or it may be almost closed under multiplication, which, for example, occurs when A is a geometric progression or a multiplicative subgroup in the ring R (e.g., if $N \subset \mathbb{R}$ is an AP and $A = \{2^n : n \in N\}$, then $|A \cdot A| = 2|A| - 1$, and $|A + A| \approx |A|^2$). Note that even if A is a dense subset of an arithmetic progression or additive subgroup (or a dense subset of an geometric progression or multiplicative subgroup), then $A + A$ (or $A \cdot A$, respectively) is still comparable in size to A. But it is difficult for A to be almost closed under addition and multiplication simultaneously, unless it is very close to a subring. The *sum-product phenomenon* says that if a finite set A is not close to a subring, then either the sumset $A + A$ or the product set $A \cdot A$ must be considerably larger than A. The reader can refer to [276] and the references therein to see some lower bounds on $|C - C|$ and $|C + C|$, where C is a convex set (a set of reals $C = \{c_1, \ldots, c_n\}$ is called *convex* if $c_{i+1} - c_i > c_i - c_{i-1}$, for all i). In the reals setting, does there exist an $A \subset \mathbb{R}$ for which $\max\{|A + A|, |A \cdot A|\}$ is "small"? Erdős and Szemerédi [111] gave a negative answer to this question. Actually, they proved the inequality $\max\{|A + A|, |A \cdot A|\} \geq c|A|^{1+\varepsilon}$ for a small but positive ε, where A is a subset of the reals. They also conjectured that $\max\{|A + A|, |A \cdot A|\} \geq c|A|^{2-\delta}$, for any positive δ. Much efforts have been made towards the value of ε. Elekes [109] observed that the sum-product problem has interesting connections to problems in incidence geometry. In particular, he applied the so-called *Szemerédi–Trotter theorem* and showed that $\varepsilon \geq 1/4$, if A is a finite set of real numbers. Elekes's result was extended to complex numbers in [292]. In the case of reals, the state of the art is due to Solymosi [295]: One can take ε arbitrarily close to $1/3$. For complex numbers, Solymosi [293], using the Szemerédi–Trotter theorem, proved that one can take ε arbitrarily close to $3/11$. Very recently, Rudnev [267], again using the Szemerédi–Trotter theorem, obtained a bound which is the state of the art in the case of complex numbers: One can take ε arbitrarily close to $19/69$.

Solymosi and Vu [298] proved a sum-product estimate for a special finite set of square matrices with complex entries, where that set is *well conditioned* (i.e., its matrices are far from being singular). Note that if we remove the latter condition (i.e., well conditioned!), then the theorem will not be true; see [298, Example 1.1]. Wolff [345], motivated by the "finite field Kakeya conjecture," formulated the finite field version of sum-product problem. The *Kakeya conjecture* says that the Hausdorff dimension of any subset of $\mathbb{R}^n$ that contains a unit line segment in every direction is equal to n; it is open in dimensions at least three. The *finite field Kakeya conjecture* asks for the smallest subset of $\mathbb{F}_q^n$ that contains a line in each direction. This conjecture was proved by Dvir [102] using a clever application of the so-called *polynomial method*; see also [101] for a nice survey on this problem and its applications especially in the area of randomness extractors. The polynomial method, which has proved to be very useful in additive combinatorics, is roughly described as the following: Given a field $\mathbb{F}$ and a finite subset $S \subset \mathbb{F}^n$, multivariate polynomials over $\mathbb{F}$ which vanish on all points of S usually get some combinatorial properties about S. (This has some similarities with what we usually do in algebraic geometry!) See, e.g., [197, Chapter 16] for some basic facts about the polynomial method, [102, 168–170] for applications in additive combinatorics, and [105, 167, 300] for applications in computer science.

Actually, the finite field version (of sum-product problem) becomes more difficult, because we will encounter with some difficulties in applying the Szemerédi–Trotter incidence theorem in this setting. In fact, the *crossing lemma*, which is an important ingredient in the proof of Szemerédi–Trotter theorem [302], relies on *Euler's formula* (and so on the topology of the plane) and consequently does not work in finite fields. Note that the proof that Szemerédi and Trotter presented for their theorem was somewhat complicated, using a combinatorial technique known as *cell decomposition* [305].

When working with finite fields, it is important to consider fields whose order is prime and not the power of a prime; because in the latter case, we can take A to be a subring which leads to the degenerate case $|A| = |A + A| = |A \cdot A|$. A stunning result in the case of finite field $\mathbb{F}_p$, with p prime, was proved by Bourgain, Katz, and Tao [57]. They proved the following:

If $A \subset \mathbb{F}_p$, and $p^\delta \leq |A| \leq p^{1-\delta}$ for some $\delta > 0$, then there exists $\varepsilon = \varepsilon(\delta) > 0$ such that $\max\{|A + A|, |A \cdot A|\} \geq c|A|^{1+\varepsilon}$.

This result is now known as the *sum-product theorem* for $\mathbb{F}_p$. In fact, this theorem holds if A is not too close to be the whole field. The condition $|A| \geq p^\delta$ in this theorem was removed by Bourgain, Glibichuk, and Konyagin in [58]. Also, note that the condition $|A| \leq p^{1-\delta}$ is necessary (e.g., if we consider a set A consisting of all elements of the field except one, then $\max\{|A + A|, |A \cdot A|\} = |A| + 1$). The idea for the proof of this theorem is by contradiction; assume that $|A + A|$ and $|A \cdot A|$ are close to $|A|$ and conclude that A is behaving very much like a subfield of $\mathbb{F}_p$. Sum-product estimates for rational functions (i.e., the results that one of $A + A$ or $f(A)$ is substantially larger than A, where f is a rational function) have also been treated (see, e.g., [19, 70]). Also, note that problems of the kind "interaction of summation and addition" are very important in various contexts of additive combinatorics and have very interesting applications (see [16, 130, 135, 136, 170, 191, 243, 253]).

Garaev [126] proved the first quantitative sum-product estimate for fields of prime order: Let $A \subset \mathbb{F}_p$ such that $1 < |A| < p^{7/13} \log^{-4/13} p$. Then

$$\max\{|A+A|, |A \cdot A|\} \gg \frac{|A|^{15/14}}{\log^{2/7} |A|}.$$

Garaev's result was extended and improved by several authors. Rudnev [266] proved the following: Let $A \subset \mathbb{F}_p^*$ with $|A| < \sqrt{p}$ and p large. Then

$$\max\{|A+A|, |A \cdot A|\} \gg \frac{|A|^{12/11}}{\log^{4/11} |A|}.$$

Li and Roche–Newton [217] proved a sum-product estimate for subsets of a finite field whose order is not prime: Let $A \subset \mathbb{F}_{p^n}$ with $|A \cap cG| \leq |G|^{1/2}$ for any subfield G of $\mathbb{F}_{p^n}$ and any element $c \in \mathbb{F}_{p^n}$. Then

$$\max\{|A+A|, |A \cdot A|\} \gg \frac{|A|^{12/11}}{\log_2^{5/11} |A|}.$$

See also [53, 127, 128, 201, 202, 214, 282, 284] for other generalizations and improvements of Garaev's result. As an application, Shparlinski [290], using Rudnev's result [266], estimates the cardinality, $\#\Gamma_p(T)$, of the set

$$\Gamma_p(T) = \{\gamma \in \mathbb{F}_p : \operatorname{ord}\gamma \leq T \text{ and } \operatorname{ord}(\gamma + \gamma^{-1}) \leq T\},$$

where $\operatorname{ord}\gamma$ (multiplicative order of γ) is the smallest positive integer t with $\gamma^t = 1$.

Tóth [328] generalized the Szemerédi–Trotter theorem to complex points and lines in $\mathbb{C}^2$ (also see [347] for a different proof of this result, and a sharp result in the case of $\mathbb{R}^4$). As another application of the sum-product theorem, Bourgain, Katz, and Tao [57] (also see [39]) derived an important Szemerédi–Trotter type theorem in prime finite fields (but did not quantify it):

If $\mathbb{F}_p$ is a prime field, and $\mathcal{P}$ and $\mathcal{L}$ are points and lines in the projective plane over $\mathbb{F}_p$ with cardinality $|\mathcal{P}|, |\mathcal{L}| \leq N < p^\alpha$ for some $0 < \alpha < 2$, then $\left|\{(p, l) \in \mathcal{P} \times \mathcal{L} : p \in l\}\right| \leq CN^{3/2-\varepsilon}$, for some $\varepsilon = \varepsilon(\alpha) > 0$. Note that it is not difficult to generalize this theorem from prime finite fields to every finite field that does not contain a large subfield. The first quantitative Szemerédi–Trotter type theorem in prime finite fields was obtained by Helfgott and Rudnev [185]. They showed that $\varepsilon \geq 1/10678$, when $|\mathcal{P}| = |\mathcal{L}| < p$. (Note that this condition prevents P from being the entire plane $\mathbb{F}_p^2$.) This result was extended to general finite fields (with a slightly weaker exponent) by Jones [193]. Also, Jones [194, 196] improved the result of Helfgott and Rudnev [185] by replacing $1/10678$ with $1/806 - o(1)$ and $1/662 - o(1)$, respectively. A near-sharp generalization of the Szemerédi–Trotter theorem to higher dimensional points and varieties was obtained in [297].

Using ideas from additive combinatorics (in fact, combining the techniques of cell decomposition and polynomial method in a novel way), Guth and Katz [170], achieved a near-optimal bound for the *Erdős distinct distance problem* in the plane. They proved that a set of N points in the plane has at least $c\frac{N}{\log N}$ distinct distances (see also [131] for some techniques and ideas related to this problem).

Hart, Iosevich, and Solymosi [178] obtained a new proof of the sum-product theorem based on incidence theorems for hyperbolas in finite fields which is achieved through some estimates on Kloosterman exponential sums. Some other results related to the incidence theorems can be found in [95, 179].

The sum-product theorem has a plethora of deep applications to various areas such as incidence geometry [31, 39, 57, 84, 86, 191, 195, 253, 309, 342, 343], analysis [79, 125, 180, 309], PDE [309], group theory [55, 67, 133, 134, 183, 184, 190, 206], exponential and character sums [18, 36–40, 42, 45–49, 58, 62, 63, 81, 129], number theory [17, 18, 42, 46, 59, 60, 81, 85, 89–91, 243], combinatorics [21, 52, 55, 309], expanders [39, 51, 52, 55, 59, 60, 66, 206, 283, 342], extractors [19, 21, 39, 41, 88, 106, 182, 346], dispersers [21], complexity theory [38], pseudorandomness [33, 39, 227], property testing [137, 270], hardness amplification [338, 340], PCPs [271], and cryptography [36, 37, 75, 121].

A sum-product problem associated with a graph was initiated by Erdős and Szemerédi [111]. Alon et al. [10] studied the sum-product theorems for sparse graphs and obtained some nice results when the graph is a matching.

Let us ask, does there exist any connection between the "sum-product problem" and "spectral graph theory"? Surprisingly, the answer is yes! In fact, the first paper that introduced and applied the spectral methods to estimate sum-product problems (and even more general problems) is the paper by Vu [342] (see also [180], in which Fourier analytic methods were used to generalize the results by Vu). In his elegant paper, Vu relates the sum-product bound to the expansion of certain graphs, and then via the relation of the spectrum (second eigenvalue) and expansion, one can deduce a rather strong bound. Vinh [336] (also see [337]), using ideas from spectral graph theory, derived a Szemerédi–Trotter-type theorem in finite fields and from there obtained a different proof of Garaev's result [128] on sum-product estimate for large subsets of finite fields. Also, Solymosi [294] applied techniques from spectral graph theory and obtained estimates similar to those of Garaev [126] that already followed via tools from exponential sums and Fourier analysis. One important ingredient in Solymosi's method [294] is the well-known *expander mixing lemma* (see, e.g., [186]), which, roughly speaking, states that on graphs with good expansion, the edges of the graphs are well distributed, and in fact, the number of edges between any two vertex subsets is about what one would expect for a random graph of that edge density.

The generalizations of the sum-product problem to polynomials, elliptic curves, and also the exponentiated versions of the problem in finite fields were obtained in [3, 19, 70, 96, 130, 285, 288, 289, 342]. Also, the problem in the commutative integral domain (with characteristic zero) setting was considered in [343]. Some other generalizations to algebraic division algebras and algebraic number fields were treated in [50, 80]. Tao [317] settled the sum-product problem in arbitrary rings.

As we already mentioned, the sum-product theorem is certainly not true for matrices over $\mathbb{F}_p$. However, Helfgott [183] proves that the theorem is true for $A \subset SL_2(\mathbb{F}_p)$. In particular, the set $A \cdot A \cdot A$ is much larger than A (more precisely, $|A \cdot A \cdot A| > |A|^{1+\varepsilon}$, where $\varepsilon > 0$, is an absolute constant), unless A is contained in a proper subgroup. Helfgott's theorem has found several applications, for instance, in some nonlinear sieving problems [60], in the spectral theory of Hecke operators [125] and in constructing expanders via Cayley graphs [52]. Underlying this theorem is the sum-product theorem. Very recently, Kowalski [206] obtained explicit versions of Helfgott's growth theorem for SL_2. Helfgott [184] proves his result when $A \subset SL_3(\mathbb{F}_p)$, as well. Gill and Helfgott [133] generalized Helfgott's theorem to $SL_n(\mathbb{F}_p)$, when A is small, that is, $|A| \leq p^{n+1-\delta}$, for some $\delta > 0$. The study of growth inside solvable subgroups of $GL_r(\mathbb{F}_p)$ is done in [134]. Breuillard, Green, and Tao [67], and also Pyber and Szabó [248], independently and simultaneously generalized Helfgott's theorem to $SL_n(\mathbb{F})$ (n arbitrary, $\mathbb{F}$ arbitrary finite field) and also to some other simple groups, as part of a more general result for groups of bounded Lie rank; see also [190].

Let us ask does there exist a "sum-division theory"? Solymosi [295] using the concept of *multiplicative energy* proved the following: If A is a finite set of positive real numbers, then

$$|A+A|^2|A \cdot A| \geq \frac{|A|^4}{4\lceil \log_2 |A| \rceil}.$$

Solymosi's result also gives

$$|A+A|^2|A/A| \geq \frac{|A|^4}{4\lceil \log_2 |A| \rceil}.$$

Li and Shen [218] removed the term $\lceil \log_2 |A| \rceil$ in the denominator. In fact, they proved the following: If A is a finite set of positive real numbers, then

$$|A+A|^2|A/A| \geq \frac{|A|^4}{4},$$

which concludes that

$$\max\{|A+A|, |A/A|\} \geq \frac{|A|^{4/3}}{2}.$$

One may ask about a "difference-product theory." The work of Solymosi [293] considers this type, but the state of the art not only for this type but also for all combinations of addition, multiplication, subtraction, and division in the case of complex numbers is due to Rudnev [267].

As we already mentioned, Rudnev [266] proved

$$\max\{|A+A|, |A \cdot A|\} \gg \frac{|A|^{12/11}}{\log^{4/11} |A|},$$

where $A \subset \mathbb{F}_p^*$ with $|A| < \sqrt{p}$ and p large. In Remark 2 of his paper, Rudnev [266], mentions an interesting fact: "one can replace either one or both the product set $A \cdot A$ with the ratio set A/A—in which case the logarithmic factor disappears—and the sumset $A + A$ with the difference set $A - A$"!

The sum-product problem can be applied efficiently to construct *randomness extractors* [19, 21, 39, 41, 88, 106, 182, 346]. Inspired by this fact, we are going to discuss some properties and applications of randomness extractors here. First, note that all cryptographic protocols, and in fact, many problems that arise in cryptography, algorithm design, distributed computing, and so on, rely completely on randomness and indeed are impossible to solve without access to it.

A *randomness extractor* is a deterministic polynomial-time computable algorithm that computes a function $\mathsf{Ext} : \{0,1\}^n \rightarrow \{0,1\}^m$, with the property that for any defective source of randomness X satisfying minimal assumptions, $\mathsf{Ext}(X)$ is close to being uniformly distributed. In other words, a randomness extractor is an algorithm that transforms a weak random source into an almost uniformly random source. Randomness extractors are interesting in their own right as combinatorial objects that "appear random" in many strong ways. They fall into the class of "pseudorandom" objects. *Pseudorandomness* is the theory of efficiently generating objects that "appear random" even though they are constructed with little or no true randomness; see [331, 334, 335] (and also the surveys [279, 280]). Error correcting codes, hardness amplifiers, epsilon-biased sets, pseudorandom generators, expander graphs, and Ramsey graphs are of other such objects. (Roughly speaking, an *expander* is a highly connected sparse finite graph, i.e., every subset of its vertices has a large set of neighbors. Expanders have a great deal of seminal applications in many disciplines such as computer science and cryptography; see [186, 229] for two excellent surveys on this area and its applications.) Actually, when studying large combinatorial objects in additive combinatorics, a helpful (and easier) procedure is to decompose them into a "structured part" and a "pseudorandom part".

Constructions of randomness extractors have been used to get constructions of communication networks and good expander graphs [76, 344], error correcting codes [166, 327], cryptographic protocols [228, 333], data structures [235], and samplers [349]. Randomness extractors are widely used in cryptographic applications (see, e.g., [25, 87, 99, 100, 199, 200, 216, 348]). This includes applications in construction of pseudorandom generators from one-way functions, design of cryptographic functionalities from noisy and weak sources, construction of key derivation functions, and extracting many private bits even when the adversary knows all except $\log^{\Omega(1)} n$ of the n bits [251] (see also [250]). They also have remarkable applications to quantum cryptography, where photons are used by the randomness extractor to generate secure random bits [279].

Ramsey graphs (i.e., graphs that have no large clique or independent set) have strong connections with extractors for two sources. Using this approach, Barak et al. [20] presented an explicit Ramsey graph that does not have cliques and independent sets of size $2^{\log^{o(1)} n}$, and ultimately beating the Frankl–Wilson construction!

Acknowledgements The author would like to thank Igor Shparlinski for many invaluable comments and for his inspiration throughout the preparation of this survey and his unending encouragement. I also thank Antal Balog, Emmanuel Breuillard, Ernie Croot, Harald Helfgott, Sergei Konyagin, Liangpan Li, Helger Lipmaa, Devanshu Pandey, Alain Plagne, László Pyber, Jeffrey Shallit, Emanuele Viola, and Van Vu for useful comments on this manuscript and/or sending some papers to me. Finally, I am grateful to the anonymous referees for their suggestions to improve this paper.

References

1. D. Achlioptas, Random matrices in data analysis. In *Proc. of ECML '04* (2004), pp. 1–8
2. D. Achlioptas, A. Naor, Y. Peres, Rigorous location of phase transitions in hard optimization problems. Nature **435**, 759–764 (2005)
3. O. Ahmadi, I.E. Shparlinski, Bilinear character sums and the sum-product problem on elliptic curves. Proc. Edinburgh Math. Soc. **53**, 1–12 (2010)
4. N. Alon, A. Shapira, Homomorphisms in graph property testing - A survey. In *Topics in Discrete Mathematics*, Algorithms and Combinatorics, vol. 26 (Springer, Berlin, 2006) pp. 281–313
5. N. Alon, A. Shapira, A characterization of the (natural) graph properties testable with one-sided error. SIAM J. Comput. **37**(6), 1703–1727 (2008)
6. N. Alon, A. Shapira, Every monotone graph property is testable. SIAM J. Comput. **38**(2), 505–522 (2008)
7. N. Alon, E. Fischer, I. Newman, A. Shapira, A combinatorial characterization of the testable graph properties: It's all about regularity. SIAM J. Comput. **39**(1), 143–167 (2009)
8. N. Alon, J. Balogh, R. Morris, W. Samotij, A refinement of the Cameron-Erdős conjecture (2012), arXiv/1202.5200
9. N. Alon, J. Balogh, R. Morris, W. Samotij, Counting sum-free sets in Abelian groups (2012), arXiv/1201.6654
10. N. Alon, O. Angel, I. Benjamini, E. Lubetzky, Sums and products along sparse graphs. Israel J. Math. **188**, 353–384 (2012)
11. T. Austin, Deducing the multidimensional Szemerédi theorem from an infinitary removal lemma. J. Anal. Math. **111**, 131–150 (2010)
12. T. Austin, Deducing the density Hales-Jewett theorem from an infinitary removal lemma. J. Theoret. Probab. **24**(3), 615–633 (2011)
13. T. Austin, T. Tao, Testability and repair of hereditary hypergraph properties. Random Structures **36**(4), 373–463 (2010)
14. P. Austrin, E. Mossel, Noise correlation bounds for uniform low degree functions. Ark. Mat. (to appear)
15. P. Balister, B. Bollobás, Projections, entropy and sumsets. Combinatorica **32**(2), 125–141 (2012)
16. A. Balog, Another sum-product estimate in finite fields. *Sovrem. Probl. Mat.* **16**, 31–37 (2012)
17. A. Balog, K.A. Broughan, I.E. Shparlinski, On the number of solutions of exponential congruences. Acta Arith. **148**(1), 93–103 (2011)
18. A. Balog, K.A. Broughan, I.E. Shparlinski, Sum-products estimates with several sets and applications. *Integers* **12**(5), 895–906 (2012)
19. B. Barak, R. Impagliazzo, A. Wigderson, Extracting randomness using few independent sources. SIAM J. Comput. **36**(4), 1095–1118 (2006)
20. B. Barak, A. Rao, R. Shaltiel, A. Wigderson, 2-source dispersers for sub-polynomial entropy and Ramsey graphs beating the Frankl-Wilson construction. In *Proc. 38th Annu. ACM Symp. Theory Comput (STOC '06)* (2006), pp. 671–680

21. B. Barak, G. Kindler, R. Shaltiel, B. Sudakov, A. Wigderson, Simulating independence: new constructions of condensers, Ramsey graphs, dispersers, and extractors. J. ACM **57**(4), Art. 20, 52 (2010)
22. M. Bateman, N.H. Katz, Structure in additively nonsmoothing sets (2011), arXiv/1104.2862
23. M. Bateman, N.H. Katz, New bounds on cap sets. J. Amer. Math. Soc. **25**, 585–613 (2012)
24. F.A. Behrend, On the sets of integers which contain no three in arithmetic progression. Proc. Nat. Acad. Sci. **23**, 331–332 (1946)
25. M. Bellare, S. Tessaro, A. Vardy, A cryptographic treatment of the wiretap channel (2012), arXiv/1201.2205
26. E. Ben-Sasson, S. Kopparty, Affine dispersers from subspace polynomials. In *Proc. 41st Annu. ACM Symp. Theory Comput (STOC '09)* (2009), pp. 65–74.
27. E. Ben-Sasson, N. Zewi, From affine to two-source extractors via approximate duality. In *Proc. 43rd Annu. ACM Symp. Theory Comput. (STOC '11)* (2011), pp. 177–186
28. E. Ben-Sasson, S. Lovett, N. Zewi, An additive combinatorics approach to the log-rank conjecture in communication complexity (2011), arXiv/1111.5884
29. V. Bergelson, A. Leibman, E. Lesigne, Intersective polynomials and the polynomial Szemerédi theorem. Adv. Math. **219**, 369–388 (2008)
30. A. Bhattacharyya, E. Grigorescu, A. Shapira, A unified framework for testing linear-invariant properties. In *Proc. 51st Annu. IEEE Symp. Found. Comput. Sci. (FOCS '10)* (2010), pp. 478–487
31. A. Bhattacharyya, Z. Dvir, A. Shpilka, S. Saraf, Tight lower bounds for 2-query LCCs over finite fields. In *Proc. 52nd Annu. IEEE Symp. Found. Comput. Sci. (FOCS '11)* (2011), pp. 638–647
32. T. Bloom, Translation invariant equations and the method of Sanders (2011), arXiv/1107.1110
33. A. Bogdanov and E. Viola, Pseudorandom bits for polynomials. In *Proc. 48th Annu. IEEE Symp. Found. Comput. Sci. (FOCS '07)* (2007), pp. 41–51
34. B. Bollobás, O. Riordan, The critical probability for random Voronoi percolation in the plane is 1/2. Probab. Theor. Related Fields **136**, 417–468 (2006)
35. J. Bourgain, On triples in arithmetic progression. Geom. Funct. Anal. **9**, 968–984 (1999)
36. J. Bourgain, New bounds on exponential sums related to the Diffie-Hellman distributions. C. R. Math. Acad. Sci. Paris **338**(11), 825–830 (2004)
37. J. Bourgain, Estimates on exponential sums related to the Diffie-Hellman distributions. Geom. Funct. Anal. **15**, 1–34 (2005)
38. J. Bourgain, Estimation of certain exponential sums arising in complexity theory. C. R. Math. Acad. Sci. Paris **340**(9), 627–631 (2005)
39. J. Bourgain, More on sum-product phenomenon in prime fields and its applications. Int. J. Number Theory **1**, 1–32 (2005)
40. J. Bourgain, Mordell's exponential sum estimate revisited. J. Amer. Math. Soc. **18**(2), 477–499 (2005)
41. J. Bourgain, On the construction of affine extractors. Geom. Funct. Anal. **17**(1) (2007), 33–57.
42. J. Bourgain, Some arithmetical applications of the sum-product theorems in finite fields. In *Geometric Aspects of Functional Analysis*, Lecture Notes in Math (Springer, 2007), pp. 99–116
43. J. Bourgain, Roth's theorem on progressions revisited. J. Anal. Math. **104**, 155–192 (2008)
44. J. Bourgain, Expanders and dimensional expansion. C. R. Acad. Sci. Paris, Ser. I **347**, 357–362 (2009)
45. J. Bourgain, Multilinear exponential sums in prime fields under optimal entropy condition on the sources. Geom. Funct. Anal. **18**(5), 1477–1502 (2009)
46. J. Bourgain, The sum-product phenomenon and some of its applications. In *Analytic Number Theory* (Cambridge University Press, Cambridge, 2009), pp. 62–74
47. J. Bourgain, Sum-product theorems and applications. In *Additive Number Theory* (Springer, New York, 2010), pp. 9–38
48. J. Bourgain, M.-C. Chang, Sum-product theorem and exponential sum estimates. C. R. Acad. Sci. Paris, Ser. I **339**, 463–466 (2004)

49. J. Bourgain, M.-C. Chang, A Gauss sum estimate in arbitrary finite fields. C.R. Math. Acad. Sci. Paris **342**(9), 643–646 (2006)
50. J. Bourgain, M.-C. Chang, Sum-product theorems in algebraic number fields. J. Anal. Math. **109**, 253–277 (2009)
51. J. Bourgain, A. Gamburd, New results on expanders. Comptes Rendus Acad. Sci. Paris Ser. I **342**, 717–721 (2006)
52. J. Bourgain, A. Gamburd, Uniform expansion bounds for Cayley graphs of $SL_2(\mathbb{F}_p)$. Ann. of Math. **167**(2), 625–642 (2008)
53. J. Bourgain, M. Garaev, On a variant of sum-product estimates and exponential sum bounds in prime fields. Math. Proc. Cambridge Philos. Soc. **146**, 1–21 (2009)
54. J. Bourgain, G. Kalai, Influences of variables and threshold intervals under group symmetries. Geom. Funct. Anal. **7**, 438–461 (1997)
55. J. Bourgain, P.P. Varjú, Expansion in $SL_d(\mathbb{Z}/q\mathbb{Z})$, q arbitrary. Invent. Math. **188**(1), 151–173 (2012)
56. J. Bourgain, A. Yehudayoff, Monotone expansion. In *Proc. 44th Annu. ACM Symp. Theory Comput (STOC '12)* (2012), pp. 1061–1078
57. J. Bourgain, N. Katz, T. Tao, A sum-product estimate in finite fields, and applications. Geom. Funct. Anal. **14**, 27–57 (2004)
58. J. Bourgain, A.A. Glibichuk, S.V. Konyagin, Estimates for the number of sums and products and for exponential sums in fields of prime order. J. London Math. Soc. **73**(2), 380–398 (2006)
59. J. Bourgain, A. Gamburd, P. Sarnak, Sieving and expanders. Comptes Rendus Acad. Sci. Paris Ser. I **343**, 155–159 (2006)
60. J. Bourgain, A. Gamburd, P. Sarnak, Affine linear sieve, expanders, and sum-product. Invent. Math. **179**(3), 559–644 (2010)
61. J. Bourgain, S. Dilworth, K. Ford, S. Konyagin, D. Kutzarova, Explicit constructions of RIP matrices and related problems. Duke Math. J. **159**(1), 145–185 (2011)
62. J. Bourgain, M.Z. Garaev, S.V. Konyagin, I.E. Shparlinski, On the hidden shifted power problem (2011), arXiv/1110.0812
63. J. Bourgain, M.Z. Garaev, S.V. Konyagin, I.E. Shparlinski, On congruences with products of variables from short intervals and applications (2012), arXiv/1203.0017
64. E. Breuillard, B. Green, Approximate groups I: The torsion-free nilpotent case. J. Inst. Math. Jussieu **10**(1), 37–57 (2011)
65. E. Breuillard, B. Green, Approximate groups II: The solvable linear case. Quart. J. Math. **62**(3), 513–521 (2011)
66. E. Breuillard, B. Green, T. Tao, Suzuki groups as expanders. Groups Geom. Dyn. **5**, 281–299 (2011)
67. E. Breuillard, B. Green, T. Tao, Approximate subgroups of linear groups. Geom. Funct. Anal. **21**(4), 774–819 (2011)
68. E. Breuillard, B. Green, T. Tao, The structure of approximate groups (2011), arXiv/1110.5008
69. É. Brezin, V. Kazakov, D. Serban, P. Wiegmann, A. Zabrodin (eds.), in *Applications of Random Matrices in Physics*. (NATO Science Series II, Mathematics, physics, and chemistry), vol. 221 (Springer, Dordrecht, 2006)
70. B. Bukh, J. Tsimerman, Sum-product estimates for rational functions. Proc. Lond. Math. Soc. **104**(1), 1–26 (2012)
71. E.J. Candès, The restricted isometry property and its implications for compressed sensing. C. R. Math. Acad. Sci. Paris **346**, 589–592 (2008)
72. E.J. Candès, T. Tao, Decoding by linear programming. IEEE Trans. Inform. Theory **51**, 4203–4215 (2005)
73. E. Candès, T. Tao, Near-optimal signal recovery from random projections: universal encoding strategies? IEEE Trans. Inform. Theory **52**, 5406–5425 (2006)
74. E.J. Candès, J. K. Romberg, T. Tao, Stable signal recovery from incomplete and inaccurate measurements. Comm. Pure Appl. Math. **59**, 1207–1223 (2006)
75. R. Canetti, J. Friedlander, S. Konyagin, M. Larsen, D. Lieman, I. E. Shparlinski, On the statistical properties of Diffie-Hellman distributions. Israel J. Math. part A **120**, 23–46 (2000)

76. M. Capalbo, O. Reingold, S. Vadhan, A. Wigderson, Randomness conductors and constant-degree lossless expanders. In *Proc. 34th Annu. ACM Symp. Theory Comput (STOC '02)* (2002), pp. 659–668

77. R. Chaabouni, H. Lipmaa, A. Shelat, Additive combinatorics and discrete logarithm based range protocols. In *Proc. ACISP '10*, Sydney, Australia, 5–7 July 2010. Lecture Notes in Computer Science, vol. 6168 (Springer, 2010), pp. 336–351

78. M.-C. Chang, A polynomial bound in Freiman's theorem. Duke Math. J. **113**(3), 399–419 (2002)

79. M.-C. Chang, A sum-product theorem in semi-simple commutative Banach algebras. J. Funct. Anal. **212**, 399–430 (2004)

80. M.-C. Chang, A sum-product estimate in algebraic division algebras. Israel J. Math. **150**, 369–380 (2005)

81. M.-C. Chang, On sum-product representations in $\mathbb{Z}_q$. J. Eur. Math. Soc. (JEMS) **8**(3), 435–463 (2006)

82. M.-C. Chang, Additive and multiplicative structure in matrix spaces. Combin. Probab. Comput. **16**(2), 219–238 (2007)

83. M.-C. Chang, On a matrix product question in cryptography (2012) (preprint)

84. M.-C. Chang, J. Solymosi, Sum-product theorems and incidence geometry. J. Eur. Math. Soc. (JEMS) **9**(3), 545–560 (2007)

85. M.-C. Chang, J. Cilleruelo, M.Z. Garaev, J. Hernández, I.E. Shparlinski, A. Zumalacárregui, Points on curves in small boxes and applications (2011), arXiv/1111.1543

86. J. Chapman, M. B. Erdoğan, D. Hart, A. Iosevich, D. Koh, Pinned distance sets, k-simplices, Wolff's exponent in finite fields and sum-product estimates. Math. Z. **271**(1–2), 63–93 (2012)

87. C. Chevalier, P.-A. Fouque, D. Pointcheval, S. Zimmer, Optimal randomness extraction from a Diffie-Hellman element. In *Advances in Cryptology Proc. of Eurocrypt '09*, ed. by A. Joux. Lecture Notes in Computer Science, vol. 5479 (Springer, 2009), pp. 572–589

88. B. Chor, O. Goldreich, Unbiased bits from sources of weak randomness and probabilistic communication complexity. SIAM J. Comput. **17**(2), 230–261 (1988)

89. J. Cilleruelo, M.Z. Garaev, Concentration of points on two and three dimensional modular hyperbolas and applications. Geom. Func. Anal. **21**, 892–904 (2011)

90. J. Cilleruelo, M.Z. Garaev, A. Ostafe, I.E. Shparlinski, On the concentration of points of polynomial maps and applications. *Math. Z.* **272**(3–4), 825–837 (2012)

91. J. Cilleruelo, I.E. Shparlinski, A. Zumalacárregui, Isomorphism classes of elliptic curves over a finite field in some thin families. *Math. Res. Lett.* **19**(2), 335–343 (2012)

92. D. Conlon, J. Fox, Bounds for graph regularity and removal lemmas. *Geom. Funct. Anal.* **22**(5), 1191–1256 (2012)

93. D. Conlon, W.T. Gowers, Combinatorial theorems in sparse random sets (2010), arXiv/1011.4310

94. D. Conlon, J. Fox, Y. Zhao, Extremal results in sparse pseudorandom graphs (2012), arXiv/1204.6645

95. D. Covert, D.Hart, A. Iosevich, D. Koh, M. Rudnev, Generalized incidence theorems, homogeneous forms and sum-product estimates in finite fields. European J. Combin. **31**(1), 306–319 (2010)

96. E. Croot, D. Hart, On sums and products in $\mathbb{C}[x]$. Ramanujan J. **22**, 33–54 (2010)

97. E. Croot, V. F. Lev, Open problems in additive combinatorics. In *Additive Combinatorics*, CRM Proc. Lecture Notes, Amer. Math. Soc., vol. 43 (2007), pp. 207–233

98. E. Croot, O. Sisask, A probabilistic technique for finding almost-periods of convolutions. Geom. Funct. Anal. **20**, 1367–1396 (2010)

99. D. Dachman-Soled, R. Gennaro, H. Krawczyk, T. Malkin, Computational extractors and pseudorandomness. In *Proc. 9th Int. Conf. Theory Cryptogr. (TCC '12)* (2012), pp. 383–403

100. Y. Dodis, D. Wichs, Non-malleable extractors and symmetric key cryptography from weak secrets. In *Proc. 41st Annu. ACM Symp. Theory Comput (STOC '09)* (2009), pp. 601–610

101. Z. Dvir, From randomness extraction to rotating needles. SIGACT News **40**(4), 46–61 (2009)

102. Z. Dvir, On the size of Kakeya sets in finite fields. J. Amer. Math. Soc. **22**(4), 1093–1097 (2009)
103. Z. Dvir, A. Shpilka, Towards dimension expanders over finite fields. Combinatorica **31**(3), 305–320 (2011)
104. Z. Dvir, A. Wigderson, Monotone expanders: Constructions and applications. Theory Comput. **6**(1), 291–308 (2010)
105. Z. Dvir, A. Wigderson, Kakeya sets, new mergers, and old extractors. SIAM J. Comput. **40**(3), 778–792 (2011)
106. Z. Dvir, A. Gabizon, A. Wigderson, Extractors and rank extractors for polynomial sources. Comput. Complex. **18**, 1–58 (2009)
107. Z. Dvir, S. Kopparty, S. Saraf, M. Sudan, Extensions to the method of multiplicities, with applications to kakeya sets and mergers. In *Proc. 50th Annu. IEEE Symp. Found. Comput. Sci. (FOCS '09)* (2009), pp. 181–190
108. M. Einsiedler, T. Ward, *Ergodic Theory: With a View Towards Number Theory* (Springer, Berlin, 2010)
109. G. Elekes, On the number of sums and products. Acta Arith. **81**, 365–367 (1997)
110. M. Elkin, An improved construction of progression-free sets. Israel J. Math. **184**, 93–128 (2011)
111. P. Erdős, E. Szemerédi, On sums and products of integers. In *Studies in Pure Mathematics* (Birkhäuser, Basel, 1983), pp. 213–218
112. P. Erdős, P. Turán, On some sequences of integers. J. Lond. Math. Soc. **11**, 261–264 (1936)
113. P. Erdős, P. Frankl, V. Rödl, The asymptotic number of graphs not containing a fixed subgraph and a problem for hypergraphs having no exponent. Graphs Combin. **2**, 113–121 (1986)
114. R. Ferguson, C. Hoffman, F. Luca, A. Ostafe, I.E. Shparlinski, Some additive combinatorics problems in matrix rings. Rev. Mat. Complut. **23**, 1–13 (2010)
115. J. Fox, A new proof of the graph removal lemma. Ann. Math. **174**(1), 561–579 (2011)
116. G.A. Freiman, *Foundations of a Structural Theory of Set Addition*, Translated from the Russian. Translations of Mathematical Monographs, vol. 37, Amer. Math. Soc., Providence, RI, 1973
117. G.A. Freiman, D. Grynkiewicz, O. Serra, Y.V. Stanchescu, Inverse additive problems for Minkowski sumsets I. *Collect. Math.* **63**(3), 261–286 (2012)
118. G.A. Freiman, D. Grynkiewicz, O. Serra, Y.V. Stanchescu, Inverse additive problems for Minkowski sumsets II. *J. Geom. Anal.* **23**(1), 395–414 (2013)
119. E. Friedgut, Hunting for sharp thresholds. Random Structures Algorithms **26**(1–2), 37–51 (2005)
120. E. Friedgut, V. Rödl, A. Ruciński, P. Tetali, A sharp threshold for random graphs with a monochromatic triangle in every edge coloring. Mem. Amer. Math. Soc. **179**(845), vi+66 (2006)
121. J.B. Friedlander, S. Konyagin, I.E. Shparlinski, Some doubly exponential sums over $\mathbb{Z}_m$. Acta Arith. **105**(4), 349–370 (2002)
122. H. Furstenberg, Ergodic behaviour of diagonal measures and a theorem of Szemerédi on arithmetic progressions. J. Anal. Math. **31**, 204–256 (1977)
123. H. Furstenberg, Y. Katznelson, An ergodic Szemerédi theorem for commuting transformations. J. Anal. Math. **34**, 275–291 (1978)
124. H. Furstenberg, Y. Katznelson, A density version of the Hales-Jewett theorem. J. Anal. Math. **57**, 64–119 (1991)
125. A. Gamburd, D. Jakobson, P. Sarnak, Spectra of elements in the group ring of $SU(2)$. J. Eur. Math. Soc. **1**, 51–85 (1999)
126. M.Z. Garaev, An explicit sum-product estimate in $\mathbb{F}_p$. Int. Math. Res. Not. **11**, 1-11 (2007) Article ID rnm035
127. M.Z. Garaev, A quantified version of Bourgain's sum-product estimate in $\mathbb{F}_p$ for subsets of incomparable sizes. Electron. J. Combin. **15**, (2008) Article #R58
128. M.Z. Garaev, The sum-product estimate for large subsets of prime fields. Proc. Amer. Math. Soc. **136**, 2735–2739 (2008)

129. M.Z. Garaev, Sums and products of sets and estimates of rational trigonometric sums in fields of prime order. Russ. Math. Surv. **65**(4), 599–658 (2010)
130. M.Z. Garaev, C.-Y. Shen, On the size of the set $A(A+1)$. Math. Z. **265**, 125–132 (2010)
131. J. Garibaldi, A. Iosevich, S. Senger, *The Erdős distance problem*, Amer. Math. Soc., Providence, RI, 2011
132. W. Gasarch, C. Kruskal, A Parrish, Purely combinatorial proofs of van der Waerden-type theorems, Draft book (2010)
133. N. Gill, H. Helfgott, Growth of small generating sets in $SL_n(\mathbb{Z}/p\mathbb{Z})$. Int. Math. Res. Not. **2010**, 26 (2010) Article ID rnq244
134. N. Gill, H. Helfgott, Growth in solvable subgroups of $GL_r(\mathbb{Z}/p\mathbb{Z})$ (2010), arXiv/1008.5264
135. A.A. Glibichuk, Sums of powers of subsets of an arbitrary finite field. Izv. Math. **75**(2), 253–285 (2011)
136. A.A. Glibichuk, M. Rudnev, On additive properties of product sets in an arbitrary finite field. J. Anal. Math. **108**, 159–170 (2009)
137. O. Goldreich (ed.), *Property Testing: Current Research and surveys*. Lecture Notes in Computer Science, vol. 6390 (Springer, 2010)
138. O. Goldreich, S. Goldwasser, D. Ron, Property testing and its connection to learning and approximation. J. ACM **45**(4), 653–750 (1998)
139. D.A. Goldston, C.Y. Yildirim, Higher correlations of divisor sums related to primes I: Triple correlations. Integers **3**, 66 (2003) paper A5
140. D.A. Goldston, C.Y. Yildirim, Higher correlations of divisor sums related to primes III: small gaps between primes. Proc. London Math. Soc. **95**(3), 653–686 (2007)
141. G. Golub, M. Mahoney, P. Drineas, L.H. Lim, Bridging the gap between numerical linear algebra, theoretical computer science, and data applications. SIAM News **9**(8), (2006)
142. W.T. Gowers, A new proof of Szemerédi's theorem for arithmetic progressions of length four. Geom. Funct. Anal. **8**, 529–551 (1998).
143. W.T. Gowers, A new proof of Szemerédi's theorem. Geom. Funct. Anal. **11**(3), 465–588 (2001)
144. W.T. Gowers, Hypergraph regularity and the multidimensional Szemerédi theorem. Ann. Math. (2) **166**(3), 897–946 (2007)
145. W.T. Gowers, Decompositions, approximate structure, transference, and the Hahn-Banach theorem. Bull. London Math. Soc. **42**, 573–606 (2010)
146. W.T. Gowers, J. Wolf, Linear forms and higher-degree uniformity for functions on $\mathbb{F}_p^n$. Geom. Funct. Anal. **21**, 36–69 (2011)
147. W.T. Gowers, J. Wolf, Linear forms and quadratic uniformity for functions on $\mathbb{Z}_N$. J. Anal. Math. **115**, 121–186 (2011)
148. W.T. Gowers, J. Wolf, Linear forms and quadratic uniformity for functions on $\mathbb{F}_p^n$. Mathematika **57**(2), 215–237 (2012)
149. R.L. Graham, B.L. Rothschild, J.H. Spencer, *Ramsey Theory*, 2nd edn. (Wiley, New York, 1990)
150. B. Green, Finite field models in arithmetic combinatorics. In *Surveys in Combinatorics*, London Math. Soc. Lecture Notes, vol. 327 (2005), pp. 1–27
151. B. Green, Approximate groups and their applications: work of Bourgain, Gamburd, Helfgott and Sarnak. In *Current Events Bulletin of the AMS* (2010), p. 25
152. B. Green, I. Ruzsa, Freiman's theorem in an arbitrary Abelian group. J. Lond. Math. Soc. **75**, 163–175 (2007)
153. B. Green, T. Tao, An inverse theorem for the Gowers U^3-norm, with applications. Proc. Edinburgh Math. Soc. **51**(1), 73–153 (2008)
154. B. Green, T. Tao, The primes contain arbitrarily long arithmetic progressions. Ann. Math. **167**, 481–547 (2008)
155. B. Green, T. Tao, Freiman's theorem in finite fields via extremal set theory. Combin. Probab. Comput. **18**, 335–355 (2009)

156. B. Green, T. Tao, New bounds for Szemerédi's theorem, II: a new bound for $r_4(N)$. In *Analytic Number Theory: Essays in Honour of Klaus Roth* (Cambridge University Press, Cambridge, 2009), pp. 180–204

157. B. Green, T. Tao, The distribution of polynomials over finite fields, with applications to the Gowers norms. Contrib. Discrete Math. **4**(2), 1–36 (2009)

158. B. Green, T. Tao, An arithmetic regularity lemma, associated counting lemma, and applications. In *An Irregular Mind: Szemerédi is 70*, Bolyai Soc. Math. Stud., vol. 21 (2010), pp. 261–334

159. B. Green, T. Tao, An equivalence between inverse sumset theorems and inverse conjectures for the U^3-norm. Math. Proc. Camb. Phil. Soc. **149**(1), 1–19 (2010)

160. B. Green, T. Tao, Linear equations in primes. Ann. Math. **171**(3), 1753–1850 (2010)

161. B. Green, T. Tao, Yet another proof of Szemerédi's theorem. In *An irregular mind: Szemerédi is 70*, Bolyai Soc. Math. Stud., vol. 21 (2010), pp. 335–342

162. B. Green, T. Tao, New bounds for Szemerédi's theorem, Ia: Progressions of length 4 in finite field geometries revisited (2012), arXiv/1205.1330

163. B. Green, T. Tao, T. Ziegler, An inverse theorem for the Gowers U^4-norm. Glasgow Math. J. **53**, 1–50 (2011)

164. B. Green, T. Tao, T. Ziegler, An inverse theorem for the Gowers $U^{s+1}[N]$-norm. *Ann. Math.* **176**(2), 1231–1372 (2012)

165. B. Green, J. Wolf, A note on Elkin's improvement of Behrend's construction. In *Additive Number Theory* (Springer, New York, 2010), pp. 141–144

166. V. Guruswami, Better extractors for better codes? In *Proc. 36th Annu. ACM Symp. Theory Comput. (STOC '04)* (2004), pp. 436–444

167. V. Guruswami, C. Umans, S. P. Vadhan, Unbalanced expanders and randomness extractors from Parvaresh-Vardy codes. J. ACM **56**(4), 1–34 (2009)

168. L. Guth, The endpoint case of the Bennett-Carbery-Tao multilinear Kakeya conjecture. Acta Math. **205**(2), 263–286 (2010)

169. L. Guth, N.H. Katz, Algebraic methods in discrete analogs of the Kakeya problem. Adv. Math. **225**(5), 2828–2839 (2010)

170. L. Guth, N.H. Katz, On the Erdős distinct distance problem in the plane (2010), arXiv/1011.4105

171. A.W. Hales, R.I. Jewett, Regularity and positional games, Trans. Amer. Math. Soc. **106**, 222–229 (1963)

172. Y.O. Hamidoune, An isoperimetric method in additive theory. J. Algebra **179**(2), 622–630 (1996)

173. Y.O. Hamidoune, Some additive applications of the isoperimetric approach. Ann. Inst. Fourier (Grenoble) **58**(6), 2007–2036 (2008)

174. Y.O. Hamidoune, The isoperimetric method. In *Combinatorial Number Theory and Additive Group Theory*, Adv. Courses Math, CRM Barcelona (Birkhäuser, Basel, 2009), pp. 241–252.

175. Y.O. Hamidoune, Topology of Cayley graphs applied to inverse additive problems. In *Proc. 3rd Int. Workshop Optimal Netw. Topol. (IWONT 2010)* (Barcelona, 2011), pp. 265–283

176. E. Haramaty, A. Shpilka, On the structure of cubic and quartic polynomials. In *Proc. 42nd Annu. ACM Symp. Theory Comput (STOC '10)* (2010), pp. 331–340

177. E. Haramaty, A. Shpilka, M. Sudan, Optimal testing of multivariate polynomials over small prime fields. In *Proc. 52nd Annu. IEEE Symp. Found. Comput. Sci. (FOCS '11)* (2011), pp. 629–637

178. D. Hart, A. Iosevich, J. Solymosi, Sum-product estimates in finite fields via Kloosterman sums. Int. Math. Res. Notices **2007**, 14 (2007) Article ID rnm007

179. D. Hart, A. Iosevich, D. Koh, M. Rudnev, Averages over hyperplanes, sum-product theory in vector spaces over finite fields and the Erdős-Falconer distance conjecture. Trans. Amer. Math. Soc. **363**, 3255–3275 (2011)

180. D. Hart, L. Li, C.-Y. Shen, Fourier analysis and expanding phenomena in finite fields. *Proc. Amer. Math. Soc.* **141**, 461–473 (2013)

181. H. Hatami, S. Lovett, Correlation testing for affine invariant properties on F_p^n in the high error regime. In *Proc. 43rd Annu. ACM Symp. Theory Comput (STOC '11)* (2011), pp. 187–194
182. N. Hegyvári, F. Hennecart, Explicit constructions of extractors and expanders. Acta Arith. **140**, 233–249 (2009)
183. H. Helfgott, Growth and generation in $SL_2(\mathbb{Z}/p\mathbb{Z})$. Ann. Math. **167**(2), 601–623 (2008)
184. H. Helfgott, Growth in $SL_3(\mathbb{Z}/p\mathbb{Z})$. J. Eur. Math. Soc. **13**(3), 761–851 (2011)
185. H. Helfgott, M. Rudnev, An explicit incidence theorem in $\mathbb{F}_p$. Mathematika **57**(1), 135–145 (2011)
186. S. Hoory, N. Linial, A. Wigderson, Expander graphs and their applications. Bull. Amer. Math. Soc. **43**, 439–561 (2006)
187. C. Hoppen, Y. Kohayakawa, C.G. Moreira, R.M. Sampaio, Property testing and parameter testing for permutations. In *Proc. 21th Annu. ACM-SIAM Symp. Discrete Algorithms* (2010), pp. 66–75
188. B. Host, B. Kra, Nonconventional ergodic averages and nilmanifolds. Ann. Math. (2) **161**(1), 397–488 (2005)
189. B. Host, B. Kra, A point of view on Gowers uniformity norms. New York J. Math. **18**, 213–248 (2012)
190. E. Hrushovski, Stable group theory and approximate subgroups. J. Amer. Math. Soc. **25**, 189–243 (2012)
191. A. Iosevich, O. Roche-Newton, M. Rudnev, On an application of Guth-Katz theorem. Math. Res. Lett. **18**(4), 1–7 (2011)
192. D. Jetchev, O. Özen, M. Stam, Collisions are not incidental: a compression function exploiting discrete geometry. In *Proc. 9th Int. Conf. Theory Cryptogr. (TCC '12)* (2012), pp. 303–320
193. T. Jones, Explicit incidence bounds over general finite fields. Acta Arith. **150**, 241–262 (2011)
194. T. Jones, An improved incidence bound for fields of prime order (2011), arXiv/1110.4752
195. T. Jones, New results for the growth of sets of real numbers (2012), arXiv/1202.4972
196. T. Jones, Further improvements to incidence and Beck-type bounds over prime finite fields (2012), arXiv/1206.4517
197. S. Jukna, *Extremal Combinatorics: With Applications in Computer Science*, 2nd edn. (Springer, Heidelberg, 2011)
198. G. Kalai, S. Safra, Threshold phenomena and infl uence: perspectives from mathematics, computer science, and economics. In *Computational Complexity and Statistical Physics* (Oxford University Press, New York, 2006), pp. 25–60
199. Y.T. Kalai, X. Li, A. Rao, 2-Source extractors under computational assumptions and cryptography with defective randomness. In *Proc. 50th Annu. IEEE Symp. Found. Comput. Sci. (FOCS '09)* (2009), pp. 617–626.
200. J. Kamp, A. Rao, S. Vadhan, D. Zuckerman, Deterministic extractors for small-space sources. J. Comput. System Sci. **77**, 191–220 (2011)
201. N. Katz, C.-Y. Shen, A slight improvement to garaevs sum product estimate. Proc. Amer. Math. Soc. **136**(7), 2499–2504 (2008)
202. N. Katz, C.-Y. Shen, Garaev's inequality in finite fields not of prime order. Online J. Anal. Comb. **3**(3), 6 (2008)
203. T. Kaufman, S. Lovett, Worst case to average case reductions for polynomials. In *Proc. 49th Annu. IEEE Symp. Found. Comput. Sci. (FOCS '08)* (2008), pp. 166–175
204. T. Kaufman, S. Lovett, New extension of the Weil bound for character sums with applications to coding. In *Proc. 52nd Annu. IEEE Symp. Found. Comput. Sci. (FOCS '11)* (2011), pp. 788–796
205. J. Komlós, M. Simonovits, Szemerédi's regularity lemma and its applications in graph theory. In *Combinatorics, Paul Erdős is eighty*, Bolyai Soc. Math. Stud. 2, vol. 2 (1996), pp. 295–352
206. E. Kowalski, Explicit growth and expansion for SL_2 (2012), arXiv/1201.1139
207. B. Kra, The Green-Tao Theorem on arithmetic progressions in the primes: an ergodic point of view. Bull. Amer. Math. Soc. **43**(1), 3–23 (2005)
208. B. Kra, From combinatorics to ergodic theory and back again. Proc. Int. Congr. Math. **III**, 57–76 (2006)

209. B. Kra, Ergodic methods in additive combinatorics. In *Additive combinatorics*, CRM Proc. Lecture Notes, Amer. Math. Soc., vol. 43 (2007), pp. 103–144

210. B. Kra, Poincaré recurrence and number theory: thirty years later. Bull. Amer. Math. Soc. **48**, 497–501 (2011)

211. I. Łaba, M. Pramanik, Arithmetic progressions in sets of fractional dimension. Geom. Funct. Anal. **19**(2), 429–456 (2009)

212. T.H. Lê, Green-Tao theorem in function fields Acta Arith. **147**(2), 129–152 (2011)

213. V.F. Lev, Character-free approach to progression-free sets. Finite Fields Appl. **18**(2), 378–383 (2012)

214. L. Li, Slightly improved sum-product estimates in fields of prime order. Acta Arith. **147**(2), 153–160 (2011)

215. X. Li, A new approach to affine extractors and dispersers. In *Proc. 2011 IEEE 26th Annu. Conf. Comput. Complexity (CCC '11)* (2011), pp. 137–147

216. X. Li, Non-malleable extractors, two-source extractors and privacy amplification, In *Proceeding 53rd Annual IEEE Symposium Foundation Computer Science (FOCS '12)*, (IEEE, New york, 2012), pp. 688–697

217. L. Li, O. Roche-Newton, An improved sum-product estimate for general finite fields. SIAM J. Discrete Math. **25**(3), 1285–1296 (2011)

218. L. Li, J. Shen, A sum-division estimate of reals. Proc. Amer. Math. Soc. **138**(1), 101–104 (2010)

219. Y. Lin, J. Wolf, On subsets of $\mathbb{F}_q^n$ containing no k-term progressions. European J. Combin. **31**(5), 1398–1403 (2010)

220. H. Lipmaa, Progression-free sets and sublinear pairing-based non-interactive zero-knowledge arguments. In *Proc. 9th Int. Conf. Theory Cryptogr. (TCC '12)* (2012), pp. 169–189

221. H. Liu, Gowers uniformity norm and pseudorandom measures of the pseudorandom binary sequences. Int. J. Number Theory **7**(5), 1279–1302 (2011)

222. L. Lovász, B. Szegedy, Regularity partitions and the topology on graphons. In *An irregular mind: Szemeredi is 70*, Bolyai Soc. Math. Stud., vol. 21 (2010), pp. 415–446

223. S. Lovett, Unconditional pseudorandom generators for low degree polynomials. Theory Comput. **5**(1), 69–82 (2009)

224. S. Lovett, Equivalence of polynomial conjectures in additive combinatorics. *Combinatorica* **32**(5), 607–618 (2012)

225. S. Lovett, R. Meshulam, A. Samorodnitsky, Inverse conjecture for the Gowers norm is false. In *Proc. 40th Annu. ACM Symp. Theory Comput. (STOC '08)* (2008), pp. 547–556

226. S. Lovett, P. Mukhopadhyay, A. Shpilka, Pseudorandom generators for $CC_0[p]$ and the Fourier spectrum of low-degree polynomials over finite fields. In *Proc. 51st Annu. IEEE Symp. Found. Comput. Sci. (FOCS '10)* (2010), pp. 695–704

227. S. Lovett, O. Reingold, L. Trevisan, S. Vadhan, Pseudorandom bit generators that fool modular sums. In *Proc. APPROX '09 / RANDOM '09*, Lecture Notes in Computer Science, vol. 5687 (2009), pp. 615–630

228. C.J. Lu, Encryption against storage-bounded adversaries from on-line strong extractors. J. Cryptology **17**(1), 27–42 (2004)

229. A. Lubotzky, Expander graphs in pure and applied mathematics. Bull. Amer. Math. Soc. **49**, 113–162 (2012)

230. M. Madiman, A. Marcus, P. Tetali, Information-theoretic inequalities in additive combinatorics. In *Proc. of the 2010 IEEE Information Theory Workshop* (2010), pp. 1–4

231. M. Madiman, A. Marcus, P. Tetali, Entropy and set cardinality inequalities for partition-determined functions. Random Structures Algorithms **40**(4), 399–424 (2012)

232. M. Malliaris, S. Shelah, Regularity lemmas for stable graphs. Trans. Amer. Math. Soc. (to appear)

233. L. Matthiesen, Linear correlations amongst numbers represented by positive definite binary quadratic forms. *Acta Arith.* **154**, 235–306 (2012)

234. R. Meshulam, On subsets of finite Abelian groups with no 3-term arithmetic progressions. J. Combin. Theory, Ser. A **71**(1), 168–172 (1995)

235. P. Miltersen, N. Nisan, S. Safra, A. Wigderson, On data structures and asymmetric communication complexity. J. Comput. Syst. Sci. **57**(1), 37–49 (1998)
236. B. Nagle, V. Rödl, M. Schacht, The counting lemma for regular k-uniform hypergraphs. Random Structures Algorithms **28**, 113–179 (2006)
237. M.B. Nathanson, *Additive Number Theory: The Classical Bases*, vol. 164 (Springer, New York, 1996)
238. M.B. Nathanson, *Additive Number Theory: Inverse Problems and the Geometry of Sumsets*, vol. 165 (Springer, New York, 1996)
239. M.B. Nathanson, Generalized additive bases, König's lemma, and the Erdős-Turán conjecture. J. Number Theory **106**(1), 70–78 (2004)
240. M.A. Nielsen, Introduction to the Polymath project and "density Hales-Jewett and Moser numbers", In *An Irregular Mind: Szemerédi is 70*, Bolyai Soc. Math. Stud., vol. 21 (2010), pp. 651–657
241. J.I. Novak, Topics in combinatorics and random matrix theory, PhD thesis, Queen's University, 2009
242. K. O'Bryant, Sets of integers that do not contain long arithmetic progressions. Electron. J. Combin. **18**, (2011) Article #P59
243. A. Ostafe, I.E. Shparlinski, On the Waring problem with Dickson polynomials in finite fields. Proc. Amer. Math. Soc. **139**, 3815–3820 (2011)
244. A. Plagne, On threshold properties of k-SAT: An additive viewpoint. European J. Combin. **27**, 1186–1198 (2006)
245. A. Plagne, W.A. Schmid, An application of coding theory to estimating Davenport constants. Des. Codes Cryptogr. **61**, 105–118 (2011)
246. D.H.J. Polymath, A new proof of the density Hales-Jewett theorem. Ann. Math. **175**(3), 1283–1327 (2012)
247. P. Potgieter, Arithmetic progressions in Salem-type subsets of the integers. J. Fourier Anal. Appl. **17**, 1138–1151 (2011)
248. L. Pyber and E. Szabó, Growth in finite simple groups of Lie type of bounded rank (2010), arXiv/1005.1858
249. R.A. Rankin, Sets of integers containing not more than a given number of terms in arithmetical progression. Proc. Roy. Soc. Edinburgh Sect. A **65**, 332–344 (1962)
250. A. Rao, Extractors for a constant number of polynomially small min-entropy independent sources. SIAM J. Comput. **39**(1), 168–194 (2009)
251. A. Rao, Extractors for low-weight affine sources. In *Proc. 24th Annu. IEEE Conf. Comput. Complexity (CCC '09)* (2009), pp. 95–101.
252. O. Reingold, L. Trevisan, M. Tulsiani, S. Vadhan, Dense subsets of pseudorandom sets. Electron. Colloq. Comput. Complexity (ECCC), Technical Report TR08-045 (2008)
253. O. Roche-Newton, M. Rudnev, Areas of rectangles and product sets of sum sets (2012), arXiv/1203.6237
254. V. Rödl, M. Schacht, Property testing in hypergraphs and the removal lemma. In *Proc. 39th Annu. ACM Symp. Theory Comput. (STOC '07)* (2007), pp. 488–495
255. V. Rödl, M. Schacht, Regular partitions of hypergraphs: Regularity lemmas. Combin. Probab. Comput. **16**(6), 833–885 (2007)
256. V. Rödl, M. Schacht, Regular partitions of hypergraphs: Counting lemmas. Combin. Probab. Comput. **16**(6), 887–901 (2007)
257. V. Rödl, M. Schacht, Generalizations of the removal lemma. Combinatorica **29**(4), 467–501 (2009)
258. V. Rödl, M. Schacht, Regularity lemmas for graphs. In *Fete of Combinatorics and Computer Science*, Bolyai Soc. Math. Stud., vol. 20 (2010), pp. 287–325
259. V. Rödl, J. Skokan, Regularity lemma for k-uniform hypergraphs. Random Structures Algorithms **25**, 1–42 (2004)
260. V. Rödl, J. Skokan, Applications of the regularity lemma for uniform hypergraphs. Random Struct. Algorithms **28**, 180–194 (2006)

261. D. Ron, Property testing: A learning theory perspective. Found. Trends Mach. Learn. **1**(3), 307–402 (2008)
262. D. Ron, Algorithmic and analysis techniques in property testing. Found. Trends Theor. Comput. Sci. **5**(2), 73–205 (2009)
263. V. Rosta, Ramsey theory applications. Electron. J. Combin. **11**, 43 (2004), Article #DS13
264. K. F. Roth, On certain sets of integers. J. Lond. Math. Soc. **28**, 104–109 (1953)
265. M. Rudelson, R. Vershynin, Non-asymptotic theory of random matrices: extreme singular values. In *Proc. Internat. Congr. Math.*, vol. III (Hindustan Book Agency, New Delhi, 2011), pp. 1576–1602
266. M. Rudnev, An improved sum-product inequality in fields of prime order. *Int. Math. Res. Not.* (**16**), 3693–3705 (2012)
267. M. Rudnev, On new sum-product type estimates (2011), arXiv/1111.4977
268. I.Z. Ruzsa, Sumsets and Entropy. Random Structures Algorithms **34**(1), 1–10 (2009)
269. I.Z. Ruzsa, E. Szemerédi, Triple systems with no six points carrying three triangles, in Combinatorics. Colloq. Math. Soc. J. Bolyai **18**, 939–945. (1978)
270. A. Samorodnitsky, Low-degree tests at large distances. In *Proc. 39th Annu. ACM Symp. Theory Comput. (STOC '07)* (2007), pp. 506–515
271. A. Samorodnitsky, L. Trevisan, Gowers uniformity, influence of variables, and PCPs. SIAM J. Comput. **39**(1), 323–360 (2009)
272. T. Sanders, Appendix to 'Roth's theorem on progressions revisited,' by J. Bourgain. J. Anal. Math. **104**, 193–206 (2008)
273. T. Sanders, On the Bogolyubov-Ruzsa lemma (2010), arXiv/1011.0107v1
274. T. Sanders, On Roth's theorem on progressions. Ann. Math. **174**(1), 619–636 (2011)
275. T. Schoen, Near optimal bounds in Freiman's theorem. Duke Math. J. **158**(1), 1–12 (2011)
276. T. Schoen, I.D. Shkredov, On sumsets of convex sets. Combin. Probab. Comput. **20**, 793–798 (2011)
277. T. Schoen, I.D. Shkredov, Roth's theorem in many variables (2011), arXiv/1106.1601
278. O. Serra, An isoperimetric method for the small sumset problem. In *Surveys in combinatorics*, London Math. Soc. Lecture Note Ser., vol. 327 (2005), pp. 119–152
279. R. Shaltiel, Recent developments in explicit construction of extractors. Bull. Eur. Assoc. Theoret. Comput. Sci. **77**, 67–95 (2002)
280. R. Shaltiel, An introduction to randomness extractors. In *ICALP 2011, Part II*, ed. by L. Aceto, M. Henzinger, J. Sgall. Lecture Notes in Computer Science, vol. 6756 (2011), pp. 21–41
281. A. Shapira, Green's conjecture and testing linear-invariant properties. In *Proc. 41st Annu. ACM Symp. Theory Comput (STOC '09)* (2009), pp. 159–166.
282. C.-Y. Shen, An extension of Bourgain and Garaev's sum-product estimates. Acta Arith. **135**(4), 351–356 (2008)
283. C.-Y. Shen, On the sum product estimates and two variables expanders. Publ. Mat. **54**(1), 149–157 (2010)
284. C.-Y. Shen, Sum-product phenomenon in finite fields not of prime order. Rocky Mountain J. Math. **41**(3), 941–948 (2011)
285. C.-Y. Shen, Algebraic methods in sum-product phenomena. Israel J. Math. **188**(1), 123–130 (2012)
286. I.D. Shkredov, Szemerédi's theorem and problems on arithmetic progressions. Russian Math. Surveys **61**(6), 1101–1166 (2006)
287. I.D. Shkredov, On a two-dimensional analogue of Szemerédi's theorem in Abelian groups. Izv. Math. **73**(5), 1033–1075 (2009)
288. I.E. Shparlinski, On the elliptic curve analogue of the sum-product problem. Finite Fields and Their Appl. **14**, 721–726 (2008)
289. I.E. Shparlinski, On the exponential sum-product problem. Indag. Math. **19**, 325–331 (2009)
290. I.E. Shparlinski, Sum-product estimates and multiplicative orders of γ and $\gamma + \gamma^{-1}$ in finite fields. Bull. Aust. Math. Soc. **85**, 505–508 (2012)

291. I.E. Shparlinski, Additive combinatorics over finite fields: New results and applications. In *Proc. RICAM-Workshop on Finite Fields and Their Applications: Character Sums and Polynomials*, De Gruyter (to appear)

292. J. Solymosi, On sum-sets and product-sets of complex numbers. J. Théor. Nombres Bordeaux **17**, 921–924 (2005)

293. J. Solymosi, On the number of sums and products. Bull. Lond. Math. Soc. **37**, 491–494 (2005)

294. J. Solymosi, Incidences and the spectra of graphs. In *Building bridges: between mathematics and computer science*, Bolyai Soc. Math. Stud., vol. 19 (2008), pp. 499–513.

295. J. Solymosi, Bounding multiplicative energy by the sumset. Adv. Math. **222**(2), 402–408 (2009)

296. J. Solymosi, Roth type theorems in finite groups, arXiv/1201.3670, (2012).

297. J. Solymosi, T. Tao, An incidence theorem in higher dimensions. *Discrete Comput. Geom.* **48**(2), 255–280 (2012)

298. J. Solymosi, V. Vu, Sum-product estimates for well-conditioned matrices. Bull. London Math. Soc. **41**, 817–822 (2009)

299. B. Sudakov, E. Szemerédi, V. Vu, On a question of Erdős and Moser. Duke Math. J. **129**, 129–155 (2005)

300. M. Sudan, Decoding of Reed-Solomon codes beyond the errorcorrection bound. J. Complexity **13**(1), 180–193 (1997)

301. M. Sudan, L. Trevisan, S.P. Vadhan, Pseudorandom generators without the XOR lemma. J. Comput. Syst. Sci. **62**(2), 236–266 (2001)

302. L. Székely, Crossing numbers and hard Erdős problems in discrete geometry. Combin. Probab. Comput. **6**, 353–358 (1997)

303. E. Szemerédi, On sets of integers containing no four elements in arithmetic progression. Acta Math. Acad. Sci. Hung. **20**, 89–104 (1969)

304. E. Szemerédi, On sets of integers containing no k elements in arithmetic progression. Acta Arith. **27**, 199–245 (1975)

305. E. Szemerédi, W.T. Trotter Jr., Extremal problems in discrete geometry. Combinatorica **3**, 381–392 (1983)

306. E. Szemerédi, V. Vu, Long arithmetic progressions in sum-sets and the number of x-sum-free sets. Proc. Lond. Math. Soc. **90**(2), 273–296 (2005)

307. E. Szemerédi, V. Vu, Finite and infinite arithmetic progressions in sumsets. Ann. Math. (2) **163**(1), 1–35 (2006)

308. E. Szemerédi, V. Vu, Long arithmetic progressions in sumsets: Thresholds and bounds. J. Amer. Math. Soc. **19**, 119–169 (2006)

309. T. Tao, From rotating needles to stability of waves: emerging connections between combinatorics, analysis, and PDE. Notices Amer. Math. Soc. **48**(3), 294–303 (2001)

310. T. Tao, A quantitative ergodic theory proof of Szemerédi's theorem. Electron. J. Combin. **13**, (2006) Article #R99

311. T. Tao, A variant of the hypergraph removal lemma. J. Combin. Theory Ser. A **113**, 1257–1280 (2006)

312. T. Tao, The dichotomy between structure and randomness, arithmetic progressions, and the primes. In *Proc. Int. Congr Math.* (Madrid, Spain, 2006), pp. 581–608

313. T. Tao, The Gaussian primes contain arbitrarily shaped constellations. J. d'Analyse Math. **99**, 109–176 (2006)

314. T. Tao, A correspondence principle between (hyper)graph theory and probability theory, and the (hyper)graph removal lemma. J. Anal. Math. **103**, 1–45 (2007)

315. T. Tao, Product set estimates in noncommutative groups. Combinatorica **28**, 547–594 (2008)

316. T. Tao, *Structure and Randomness: Pages from Year One of a Mathematical Blog*, Amer. Math. Soc. (2008)

317. T. Tao, The sum-product phenomenon in arbitrary rings. Contrib. Discrete Math. **4**(2), 59–82 (2009)

318. T. Tao, Freiman's theorem for solvable groups. Contrib. Discrete Math. **5**(2), 137–184 (2010)
319. T. Tao, Sumset and inverse sumset theory for Shannon entropy. Combin. Probab. Comput. **19**, 603–639 (2010)
320. T. Tao, *Higher Order Fourier Analysis*, vol. 142 (American Mathematical Society, Providence, 2012)
321. T. Tao, V. Vu, *Additive Combinatorics*, vol. 105 (Cambridge University Press, Cambridge, 2006)
322. T. Tao, V. Vu, Random matrices: The circular law. Commun. Contemp. Math. **10**(2), 261–307 (2008)
323. T. Tao, V. Vu, From the Littlewood-Offord problem to the circular law: universality of the spectral distribution of random matrices. Bull. Amer. Math. Soc. **46**, 377–396 (2009)
324. T. Tao, V. Vu, Random matrices: Universality of local eigenvalue statistics up to the edge, Commun. Math. Phys. **298**(2), 549–572 (2010)
325. T. Tao, T. Ziegler, The primes contain arbitrarily long polynomial progressions. Acta Math. **201**, 213–305 (2008)
326. T. Tao, T. Ziegler, The inverse conjecture for the Gowers norm over finite fields via the correspondence principle. Anal. PDE **3**(1), 1–20 (2010)
327. A. Ta-Shma, D. Zuckerman, Extractor codes. IEEE Trans. Inform. Theory **50**(12), 3015–3025 (2004)
328. C. D. Tóth, The Szemerédi-Trotter theorem in the complex plane (2003), arXiv/0305283
329. H. Towsner, A model theoretic proof of Szemerédi's theorem (2010), arXiv/1002.4456
330. L. Trevisan, Additive combinatorics and theoretical computer science. ACM SIGACT News **40**(2), 50–66 (2009)
331. L. Trevisan, Pseudorandomness in computer science and in additive combinatorics. In *An Irregular Mind: Szemerédi is 70*, Bolyai Soc. Math. Stud., vol. 21 (2010), pp. 619–650.
332. M. Tulsiani, J. Wolf, Quadratic Goldreich-Levin theorems. In *Proc. 52th Annu. IEEE Symp. Found. Comput. Sci. (FOCS '11)* (2011), pp. 619–628.
333. S.P. Vadhan, Constructing locally computable extractors and cryptosystems in the bounded-storage model. J. Cryptology **17**(1), 43–77 (2004)
334. S.P. Vadhan, The unified theory of pseudorandomness. In *Proc. Internat. Congr. Math.*, vol. IV (Hindustan Book Agency, New Delhi, 2011), pp. 2723–2745.
335. S.P. Vadhan, Pseudorandomness. *Found. Trends Theor. Comput. Sci.* **7**(1–3), 1–336 (2012)
336. L.A. Vinh, The Szemerédi-Trotter type theorem and the sum-product estimate in finite fields. European J. Combin. **32**(8), 1177–1181 (2011)
337. L.A. Vinh, Sum and shifted-product subsets of product-sets over finite rings. Electron. J. Combin. **19**(2), (2012) Article #P33
338. E. Viola, New correlation bounds for $GF(2)$ polynomials using Gowers uniformity. Electron. Colloq. Comput. Complexity (ECCC), Technical Report TR06–097 (2006)
339. E. Viola, Selected results in additive combinatorics: An exposition. Theory Comput. Library, Graduate Surveys, **3**, 1–15 (2011)
340. E. Viola, A. Wigderson, Norms, XOR lemmas, and lower bounds for polynomials and protocols. Theory Comput. **4**, 137–168 (2008)
341. V. Vu, Random discrete matrices. In *Horizons of combinatorics*, Bolyai Soc. Math. Stud., vol. 17 (2008), pp. 257–280
342. V. Vu, Sum-product estimates via directed expanders. Math. Res. Lett. **15**(2), 375–388 (2008)
343. V. Vu, M. M. Wood, P.M. Wood, Mapping incidences. J. Lond. Math. Soc. **84**(2), 433–445 (2011)
344. A. Wigderson and D. Zuckerman, Expanders that beat the eigenvalue bound: Explicit construction and applications. Combinatorica **19**(1), 125–138 (1999)
345. T. Wolff, Recent work connected with the Kakeya problem, Prospects in Mathematics (Princeton, NJ, 1996), pp. 129–162. Amer. Math. Soc., Providence, RI, 1999
346. A. Yehudayoff, Affine extractors over prime fields. Combinatorica **31**(2), 245–256 (2011)
347. J. Zahl, A Szemerédi-Trotter type theorem in $\mathbb{R}^4$ (2012), arXiv/1203.4600

348. M. Zimand, Simple extractors via constructions of cryptographic pseudorandom generators. Theor. Comput. Sci. **411**, 1236–1250 (2010)
349. D. Zuckerman, Randomness-optimal oblivious sampling. Random Structures Algorithms **11**, 345–367 (1997)
350. D. Zuckerman, Linear degree extractors and the inapproximability of max clique and chromatic number. Theory Comput. **3**(1), 103–128 (2007)

Transcendence of Stammering Continued Fractions

Yann Bugeaud

To the memory of Alf van der Poorten

Abstract Let $\theta = [0; a_1, a_2, \ldots]$ be an algebraic number of degree at least three. Recently, we have established that the sequence of partial quotients $(a_\ell)_{\ell \geq 1}$ of θ is not too simple and cannot be generated by a finite automaton. In this expository paper, we point out the main ingredients of the proof and we briefly survey earlier works.

1 Introduction

It is widely believed that the continued fraction expansion of an irrational algebraic number

$$\theta = \lfloor \theta \rfloor + [0; a_1, a_2, \ldots, a_\ell, \ldots] = \lfloor \theta \rfloor + \cfrac{1}{a_1 + \cfrac{1}{a_2 + \cfrac{1}{\cdots}}}$$

either is eventually periodic (and we know that this is the case if, and only if, θ is a quadratic irrational) or contains arbitrarily large partial quotients. Here, and in all that follows, $\lfloor x \rfloor$ and $\lceil x \rceil$ denote, respectively, the integer part and the upper integer part of the real number x.

A preliminary step consists in providing explicit examples of transcendental continued fractions. The first result of this type goes back to Liouville [16], who

Y. Bugeaud (✉)
Université de Strasbourg, Mathématiques, 7 rue René Descartes, 67084 Strasbourg, France
e-mail: bugeaud@math.unistra.fr

J.M. Borwein et al. (eds.), *Number Theory and Related Fields: In Memory of Alf van der Poorten*, Springer Proceedings in Mathematics & Statistics 43, DOI 10.1007/978-1-4614-6642-0_5, © Springer Science+Business Media New York 2013

constructed transcendental real numbers with a very fast-growing sequence of partial quotients. His key tool is the so-called Liouville inequality which asserts that if θ is a real algebraic number of degree $d \geq 2$, then there exists a positive constant $c_1(\theta)$ such that

$$|\theta - p/q| \geq c_1(\theta)q^{-d}, \quad \text{for every rational number } p/q \text{ with } q \geq 1.$$

Subsequently, various authors used deeper transcendence criteria from Diophantine approximation to construct other classes of transcendental continued fractions. Of particular interest is the work of Maillet [19] (see also Section 34 of [22]), who was the first to give explicit examples of transcendental continued fractions with bounded partial quotients. A particular case of Maillet's result asserts that if $(a_\ell)_{\ell \geq 1}$ is a non-eventually periodic sequence of positive integers at most equal to M, and if there is an increasing sequence $(\ell_n)_{n \geq 1}$ such that

$$a_{\ell_n} = a_{\ell_n+1} = \ldots = a_{n\ell_n} = 1,$$

for $n \geq 1$, then the real number

$$\alpha = [0; a_1, a_2, \ldots]$$

is transcendental. His proof is based on a general form of the Liouville inequality which limits the approximation of real algebraic numbers θ of degree $d \geq 3$ by quadratic irrationals. More precisely, Maillet showed that there exists a positive constant $c_2(\theta)$ such that

$$|\theta - \gamma| \geq c_2(\theta)H(\gamma)^{-d}, \quad \text{for every real quadratic number } \gamma. \tag{1.1}$$

Here, and everywhere in the present text, $H(P)$ denotes the height of the integer polynomial $P(X)$, that is, the maximum of the absolute values of its coefficients; furthermore, $H(\gamma)$ denotes the height of the algebraic number γ, that is, the height of its minimal defining polynomial over $\mathbf{Z}$. A rapid (and rough) calculation shows that the height of the quadratic irrational real number

$$\alpha_n := [0; a_1, \ldots, a_{\ell_n-1}, \overline{1}],$$

where the notation $\overline{1}$ means that the partial quotient 1 is repeated infinitely many times, satisfies

$$H(\alpha_n) \leq \prod_{i=1}^{\ell_n}(a_i+2)^2 \leq (M+2)^{2\ell_n}. \tag{1.2}$$

This provides us with infinitely many very good approximations to α. Indeed, by construction, for $n \geq 1$, the first $n\ell_n$ partial quotients of α and α_n are the same; thus, we derive from (1.2) and (4.4) below that

$$|\alpha - \alpha_n| < 2^{2-n\ell_n} \leq 4H(\alpha_n)^{-n(\log 2)/(2\log(M+2))}. \tag{1.3}$$

It then follows from (1.1) that α cannot be algebraic of degree ≥ 3. As $(a_\ell)_{\ell \geq 1}$ is infinite and not eventually periodic, α is transcendental.

Baker [9] used in 1962 Roth's theorem for number fields obtained by LeVeque to strongly improve upon the results of Maillet and make them more explicit. He observed that, when infinitely many of the quadratic approximations found by Maillet lie in the same quadratic number field, one can replace the use of (1.1) by that of LeVeque's theorem, which asserts that, for any given real number field $\mathbf{K}$, any positive real number ε and any real algebraic number θ lying outside $\mathbf{K}$, there exists a positive constant $c_3(\theta, \varepsilon)$ such that

$$|\theta - \gamma| \geq c_3(\theta, \varepsilon) H(\gamma)^{-2-\varepsilon}, \quad \text{for every real algebraic number } \gamma \text{ in } \mathbf{K}. \quad (1.4)$$

This is clearly relevant for the example mentioned above, since all the α_n belong to the quadratic field $\mathbf{Q}(\sqrt{5})$. In particular, it follows from (1.3) and (1.4) that if $(a_\ell)_{\ell \geq 1}$ is a non-eventually periodic sequence of positive integers at most equal to M, and if there is an increasing sequence $(\ell_n)_{n \geq 1}$ such that

$$a_{\ell_n} = a_{\ell_n+1} = \ldots = a_{\lfloor \kappa \ell_n \rfloor} = 1,$$

for $n \geq 1$ and some real number $\kappa > 4(\log(M+2))/(\log 2)$, then the real number $\alpha = [0; a_1, a_2, \ldots]$ is transcendental.

Subsequently, further transcendence results have been obtained by applying a corollary to the Schmidt subspace theorem which states that, for any positive real number ε and any real algebraic number θ of degree at least 3, there exists a positive constant $c_4(\theta, \varepsilon)$ such that

$$|\theta - \gamma| \geq c_4(\theta, \varepsilon) H(\gamma)^{-3-\varepsilon}, \quad \text{for every real quadratic number } \gamma; \quad (1.5)$$

see Corollary 1. The difference between (1.4) and (1.5) is that one takes into account *every* real quadratic number in (1.5), while the approximants in (1.4) all belong to the same number field. By means of (1.5), Davison [14], Queffélec [23] and other authors [7, 15] established the transcendence of several families of continued fractions with bounded partial quotients. In particular, the real number whose sequence of partial quotients is the Thue–Morse sequence or any Sturmian or quasi-Sturmian sequence is transcendental [7, 23].

The next step, initiated in [1], has been the use of the Schmidt subspace theorem, instead of its corollary (1.5), to get several combinatorial transcendence criteria for continued fraction expansions [1, 3–5]. Recently in [10], we have shown how a slight modification of their proofs allows us to considerably improve two of these criteria. In the present survey, we focus on the new combinatorial transcendence criterion for stammering continued fractions established in [10] and explain the two main ingredients of its proof. We also point out some of its applications, including the proof of the Cobham–Loxton–van der Poorten conjecture for automatic continued fraction expansions.

2 Recent Results

Throughout this note, we identify a sequence $\mathbf{a} = (a_\ell)_{\ell \geq 1}$ of positive integers with the infinite word $a_1 a_2 \ldots a_\ell \ldots$, as well denoted by $\mathbf{a}$. This should not cause any confusion.

For $n \geq 1$, we denote by $p(n, \mathbf{a})$ the number of distinct blocks of n consecutive letters occurring in the word $\mathbf{a}$, that is,

$$p(n, \mathbf{a}) := \mathrm{Card}\{a_{\ell+1} \ldots a_{\ell+n} : \ell \geq 0\}.$$

The function $n \mapsto p(n, \mathbf{a})$ is called the complexity function of $\mathbf{a}$. A well-known result of Morse and Hedlund [20, 21] asserts that $p(n, \mathbf{a}) \geq n + 1$ for $n \geq 1$, unless $\mathbf{a}$ is ultimately periodic (in which case there exists a constant C such that $p(n, \mathbf{a}) \leq C$ for $n \geq 1$).

Let α be an irrational real number and write

$$\alpha = \lfloor \alpha \rfloor + [0; a_1, a_2, \ldots].$$

Let $\mathbf{a}$ denote the infinite word $a_1 a_2 \ldots$. A natural way to measure the intrinsic *complexity* of α is to count the number $p(n, \alpha) := p(n, \mathbf{a})$ of distinct blocks of given length n in the word $\mathbf{a}$.

Let α be a real algebraic number of degree at least three. A first step towards a proof that α has unbounded partial quotients would be to get a good lower bound for $p(n, \alpha)$. Theorem 1.1 of [10], reproduced below, asserts that the complexity function of an algebraic number of degree at least three cannot increase too slowly.

Theorem 1. *Let* $\mathbf{a} = (a_\ell)_{\ell \geq 1}$ *be a sequence of positive integers which is not ultimately periodic. If the real number*

$$\alpha := [0; a_1, a_2, \ldots, a_\ell, \ldots]$$

is algebraic, then

$$\lim_{n \to +\infty} \frac{p(n, \alpha)}{n} = +\infty. \tag{2.1}$$

Theorem 1 improves Theorem 7 from [7] and Theorem 4 from [1], where

$$\lim_{n \to +\infty} p(n, \alpha) - n = +\infty$$

was proved instead of (2.1). This gives a positive answer to Problem 3 of [1].

An infinite sequence $\mathbf{a} = (a_\ell)_{\ell \geq 1}$ is an automatic sequence if it can be generated by a finite automaton, that is, if there exists an integer $k \geq 2$ such that a_ℓ is a finite-state function of the representation of ℓ in base k, for every $\ell \geq 1$. We refer the reader to [8] for a more precise definition and examples of automatic sequences. Let $b \geq 2$ be an integer. In 1968, Cobham [12] asked whether a real number whose

b-ary expansion can be generated by a finite automaton is always either rational or transcendental. After several attempts by Cobham himself and by Loxton and van der Poorten [17], Loxton and van der Poorten [18] asserted in 1988 that the b-ary expansion of an irrational algebraic number cannot be generated by a finite automaton. The proof proposed in [18], which rests on a method introduced by Mahler, contains unfortunately a gap. A positive answer to Cobham's question was finally given in [2], by means of the combinatorial transcendence criterion established in [6]. Since the complexity function of every automatic sequence **a** satisfies $p(n, \mathbf{a}) = O(n)$ (this was proved by Cobham [13] in 1972), Theorem 1 implies straightforwardly the next result.

Theorem 2. *The continued fraction expansion of an algebraic number of degree at least three cannot be generated by a finite automaton.*

Before stating our combinatorial transcendence criterion for continued fractions, we introduce some notation. The length of a word W, that is, the number of letters composing W, is denoted by $|W|$. For any positive integer k, we write W^k for the word $W \ldots W$ (k times repeated concatenation of the word W). More generally, for any positive real number x, we denote by W^x the word $W^{\lfloor x \rfloor} W'$, where W' is the prefix of W of length $\lceil (x - \lfloor x \rfloor) |W| \rceil$.

Let $\mathbf{a} = (a_\ell)_{\ell \geq 1}$ be a sequence of positive integers. We say that **a** satisfies Condition ($*$) if **a** is not ultimately periodic and if there exist $w > 1$ and two sequences of finite words $(U_n)_{n \geq 1}$, $(V_n)_{n \geq 1}$ such that:

(a) For every $n \geq 1$, the word $U_n V_n^w$ is a prefix of the word **a**.
(b) The sequence $(|U_n|/|V_n|)_{n \geq 1}$ is bounded.
(c) The sequence $(|V_n|)_{n \geq 1}$ is increasing.

Equivalently, the word **a** satisfies Condition ($*$) if there exists a positive real number ε such that, for arbitrarily large integers N, the prefix $a_1 a_2 \ldots a_N$ of **a** contains two disjoint occurrences of a word of length $\lfloor \varepsilon N \rfloor$.

The key tool for the proofs of Theorems 1 and 2 is the following combinatorial transcendence criterion.

Theorem 3. *Let $\mathbf{a} = (a_\ell)_{\ell \geq 1}$ be a sequence of positive integers. Let $(p_\ell/q_\ell)_{\ell \geq 1}$ denote the sequence of convergents to the real number*

$$\alpha := [0; a_1, a_2, \ldots, a_\ell, \ldots].$$

*Assume that the sequence $(q_\ell^{1/\ell})_{\ell \geq 1}$ is bounded. If **a** satisfies Condition ($*$), then α is transcendental.*

Theorem 3 was established in [10]. Its proof uses the Schmidt subspace theorem; see Theorem 4. Consequently, the proofs of Theorems 1 and 2 rest ultimately on the Schmidt subspace theorem. This is also the case for the similar results on expansions of irrational algebraic numbers to an integer base; see [2, 6].

A simple combinatorial study (see, e.g. [10]) shows that if (2.1) does not hold for a real number $\alpha := [0; a_1, a_2, \ldots, a_\ell, \ldots]$, then the sequence $(a_\ell)_{\ell \geq 1}$ either is ultimately periodic or satisfies Condition $(*)$ above. In the latter case, Theorem 3 implies that α is transcendental. This shows that Theorem 1 is a consequence of Theorem 3.

3   The Schmidt Subspace Theorem

The proof of Theorem 3 rests on the Schmidt subspace theorem.

Theorem 4. *Let $m \geq 2$ be an integer. Let $L_1, \ldots, L_m$ be linearly independent linear forms in $\mathbf{x} = (x_1, \ldots, x_m)$ with algebraic coefficients. Let ε be a positive real number. Then, the set of solutions $\mathbf{x} = (x_1, \ldots, x_m)$ in $\mathbf{Z}^m$ to the inequality*

$$|L_1(\mathbf{x}) \cdots L_m(\mathbf{x})| \leq (\max\{|x_1|, \ldots, |x_m|\})^{-\varepsilon}$$

lies in finitely many proper subspaces of $\mathbf{Q}^m$.

Proof. See, e.g. [25, 26]. □

Roth's theorem (i.e., (1.4) with $\mathbf{K} = \mathbf{Q}$) is equivalent to the case $m = 2$ of Theorem 4. We point out an immediate consequence of the case $m = 3$ of Theorem 4, which extends Roth's theorem to approximation by quadratic numbers.

Corollary 1. *Let θ be a real algebraic number of degree at least 3. Let ε be a positive real number. Then, there are only finitely many integer polynomials $P(X)$ of degree at most 2 such that*

$$|P(\theta)| < H(P)^{-2-\varepsilon}.$$

Consequently, there exists a positive constant $c(\theta, \varepsilon)$ such that

$$|\theta - \gamma| > c(\theta, \varepsilon) H(\gamma)^{-3-\varepsilon},$$

for any algebraic number γ of degree at most 2.

Proof. By Theorem 4 applied with the three linear forms $X_2 \theta^2 + X_1 \theta + X_0$, X_1, X_2, the set of integer triples (x_0, x_1, x_2) satisfying

$$|x_2 \theta^2 + x_1 \theta + x_0| \cdot |x_1| \cdot |x_2| \leq (\max\{|x_0|, |x_1|, |x_2|\})^{-\varepsilon} \qquad (3.1)$$

lies in finitely many proper subspaces of $\mathbf{Q}^3$. If $x_1 x_2 = 0$, then, by Roth's theorem (i.e., (1.4) for $\mathbf{K} = \mathbf{Q}$), there are only finitely many integers y_0, y_2, z_0, z_1 such that $y_2 z_1 \neq 0$ and

$$|y_2| \cdot |y_2 \theta^2 + y_0| < (\max\{|y_0|, |y_2|\})^{-\varepsilon}, \quad |z_1| \cdot |z_1 \theta + z_0| < (\max\{|z_0|, |z_1|\})^{-\varepsilon}.$$

Consequently, we can assume that x_1 and x_2 are both non-zero.

Let $a_0 X_0 + a_1 X_1 + a_2 X_2 = 0$ denote a proper subspace of $\mathbf{Q}^3$, with a_0, a_1, a_2 in $\mathbf{Z}$ and $a_0 \neq 0$. If (3.1) and $a_0 x_0 + a_1 x_1 + a_2 x_2 = 0$ hold for an integer triple (x_0, x_1, x_2) with $x_1 x_2 \neq 0$, then

$$|x_2 \theta^2 + x_1 \theta + x_0| = |x_2(\theta^2 - a_2/a_0) + x_1(\theta - a_1/a_0)|.$$

By Roth's theorem, there are only finitely many integer pairs (x_1, x_2) such that $x_1 x_2 \neq 0$ and

$$|x_2(\theta^2 - a_2/a_0) + x_1(\theta - a_1/a_0)| \cdot |x_1| \cdot |x_2| \leq (\max\{|x_1|, |x_2|\})^{-\varepsilon}.$$

Consequently, the triple (x_0, x_1, x_2) is lying in a finite set, which depends on a_0, a_1, a_2. This proves the first statement of the corollary.

The second statement follows immediately since there is an absolute constant c such that, for any integer polynomial $P(X)$, we have

$$|P(\theta)| \leq c H(P) \cdot |\theta - \gamma|,$$

where γ is the root of $P(X)$ which is the closest to θ. □

4 Auxiliary Results on Continued Fractions

Classical references on the theory of continued fractions include [22, 26].

Let $\alpha := [0; a_1, a_2, \ldots]$ be a real irrational number. Set $p_{-1} = q_0 = 1$ and $q_{-1} = p_0 = 0$. For $\ell \geq 1$, set $p_\ell/q_\ell = [0; a_1, a_2, \ldots, a_\ell]$ and note that

$$q_\ell = a_\ell q_{\ell-1} + q_{\ell-2}.$$

The theory of continued fraction implies that

$$|q_\ell \alpha - p_\ell| < q_{\ell+1}^{-1}, \quad \text{for } \ell \geq 1, \tag{4.1}$$

and

$$q_{\ell+h} \geq q_\ell (\sqrt{2})^{h-1}, \quad \text{for } h, \ell \geq 1. \tag{4.2}$$

It follows from (4.1) that if two real irrational numbers α and α' have the same first ℓ partial quotients for some integer $\ell \geq 1$, then

$$|\alpha - \alpha'| \leq 2 q_\ell^{-2}, \tag{4.3}$$

where q_ℓ denotes the denominator of the ℓ-th convergent to α, and

$$|\alpha - \alpha'| \leq 2^{2-\ell}, \tag{4.4}$$

by (4.2).

In this and the next sections, we use the notation

$$[0; a_1, \ldots, a_r, \overline{a_{r+1}, \ldots, a_{r+s}}] := [0; U, \overline{V}],$$

where $U = a_1 \ldots a_r$ and $V = a_{r+1} \ldots a_{r+s}$, to indicate that the block of partial quotients $a_{r+1}, \ldots, a_{r+s}$ is repeated infinitely many times. We also denote by ζ' the Galois conjugate of a quadratic real number ζ. We reproduce below Lemma 6.1 from [11].

Lemma 1. *Let α be a quadratic real number with ultimately periodic continued fraction expansion*

$$\alpha = [0; a_1, \ldots, a_r, \overline{a_{r+1}, \ldots, a_{r+s}}],$$

with $r \geq 3$ and $s \geq 1$ and denote by α' its Galois conjugate. Let $(p_\ell/q_\ell)_{\ell \geq 1}$ denote the sequence of convergents to α. There exists an absolute constant κ such that if $a_r \neq a_{r+s}$, then we have

$$|\alpha - \alpha'| \leq \kappa a_r^2 \max\{a_{r-2}, a_{r-1}\} q_r^{-2}.$$

Lemma 1 is an easy consequence of the theorem of Galois (see [22, p. 83]) which states that the Galois conjugate of

$$[a_{r+1}; \overline{a_{r+2}, \ldots, a_{r+s}, a_{r+1}}]$$

is the quadratic number

$$-[0; \overline{a_{r+s}, \ldots, a_{r+2}, a_{r+1}}].$$

Although we do not use it in the computation (5.9), it can be considered as a key observation for the proof of Theorem 3.

5   Transcendence Criterion for Stammering Continued Fractions

In this section, we explain the main ingredients of the proof of Theorem 3. Let $\mathbf{a} = (a_\ell)_{\ell \geq 1}$ be a sequence of positive integers. Let w and w' be non-negative real numbers with $w > 1$. We say that $\mathbf{a}$ satisfies Condition $(*)_{w,w'}$ if $\mathbf{a}$ is not ultimately periodic and if there exist two sequences of finite words $(U_n)_{n \geq 1}$, $(V_n)_{n \geq 1}$ such that:

(a) For every $n \geq 1$, the word $U_n V_n^w$ is a prefix of the word $\mathbf{a}$.
(b) The sequence $(|U_n|/|V_n|)_{n \geq 1}$ is bounded from above by w'.
(c) The sequence $(|V_n|)_{n \geq 1}$ is increasing.

Theorem 5. *Let $\mathbf{a} = (a_\ell)_{\ell \geq 1}$ be a sequence of positive integers. Let $(p_\ell/q_\ell)_{\ell \geq 1}$ denote the sequence of convergents to the real number*

$$\alpha := [0; a_1, a_2, \ldots, a_\ell, \ldots].$$

Assume that the sequence $(q_\ell^{1/\ell})_{\ell \geq 1}$ *is bounded. If there exist non-negative real numbers w and w' with w > 1 such that* **a** *satisfies Condition* $(*)_{w,w'}$, *then* α *is transcendental.*

Theorem 5, established in [10], improves Theorem 2 from [1] and Theorem 3.1 from [5], where the assumption

$$w > ((2\log M / \log m) - 1)w' + 1 \qquad (5.1)$$

was required, with $M = \limsup_{\ell \to +\infty} q_\ell^{1/\ell}$ and $m = \liminf_{\ell \to +\infty} q_\ell^{1/\ell}$. Furthermore, it contains Theorem 3.2 from [3].

The reader is directed to [10] for a complete proof of Theorem 5. We content ourselves to explain how Theorem 4 and Corollary 1 can be applied to prove the transcendence of families of stammering continued fractions. We compare the various results obtained under the assumption that the sequence $(q_\ell^{1/\ell})_{\ell \geq 1}$ converges, which makes the comparisons easier.

Assume that the real numbers w and w' are fixed as well as the sequences $(U_n)_{n \geq 1}$ and $(V_n)_{n \geq 1}$ occurring in the definition of Condition $(*)_{w,w'}$. Set $r_n = |U_n|$ and $s_n = |V_n|$, for $n \geq 1$. We assume that the real number $\alpha := [0; a_1, a_2, \ldots]$ is algebraic of degree at least three. Throughout this section, the numerical constants implied in $\ll$ are absolute.

We observe that α admits infinitely many good quadratic approximants obtained by truncating its continued fraction expansion and completing by periodicity. With the above notation, for $n \geq 1$, the real number α is close to the quadratic number

$$\alpha_n = [0; U_n, \overline{V_n}].$$

Namely, since the first $r_n + \lfloor w s_n \rfloor$ partial quotients of α and of α_n are the same, we deduce from (4.3) that

$$|\alpha - \alpha_n| \leq 2q_{r_n + \lfloor w s_n \rfloor}^{-2}. \qquad (5.2)$$

Furthermore, α_n is root of the quadratic polynomial (see, e.g., [22])

$$\begin{aligned}
P_n(X) := &(q_{r_n-1}q_{r_n+s_n} - q_{r_n}q_{r_n+s_n-1})X^2 \\
&-(q_{r_n-1}p_{r_n+s_n} - q_{r_n}p_{r_n+s_n-1} + p_{r_n-1}q_{r_n+s_n} - p_{r_n}q_{r_n+s_n-1})X \\
&+(p_{r_n-1}p_{r_n+s_n} - p_{r_n}p_{r_n+s_n-1}),
\end{aligned}$$

and we deduce that

$$H(\alpha_n) \leq H(P_n) \leq 2q_{r_n}q_{r_n+s_n}.$$

Consequently,

$$|\alpha - \alpha_n| \ll H(\alpha_n)^{-2(\log q_{r_n + \lfloor w s_n \rfloor})/(\log q_{r_n}q_{r_n+s_n})}.$$

Assuming that $(q_\ell^{1/\ell})_{\ell \geq 1}$ converges, our assumption that α is algebraic contradicts the last assertion of Corollary 1 when

$$\limsup_{n \to +\infty} \frac{r_n + \lfloor w s_n \rfloor}{2r_n + s_n} > \frac{3}{2},$$

that is, when

$$w > 2w' + 3/2. \tag{5.3}$$

This is the approach followed in [7, 23]. It applies, for instance, when $(a_\ell)_{\ell \geq 1}$ is the Thue–Morse sequence

$$\mathbf{t} := 1221211221121221\ldots$$

on $\{1, 2\}$ defined as the fixed point beginning by 1 of the morphism τ defined by $\tau(1) = 12$ and $\tau(2) = 21$. Indeed, for each $n \geq 1$, the prefix of length $5 \cdot 2^n$ of $\mathbf{t}$ is equal to its prefix of length $3 \cdot 2^n$ raised to the power $5/3$. Thus, we can take $w = 5/3$ and $w' = 0$. The fact that the sequence $(q_\ell^{1/\ell})_{\ell \geq 1}$ converges in this case has been established in [24].

The main new ingredient in [1] is the use of Theorem 4 with $m = 4$, instead of Corollary 1, which is deduced from Theorem 4 with $m = 3$. Let us now explain to which system of four linear forms we apply Theorem 4. By (4.1), we have

$$
\begin{aligned}
&|(q_{r_n-1}q_{r_n+s_n} - q_{r_n}q_{r_n+s_n-1})\alpha - (q_{r_n-1}p_{r_n+s_n} - q_{r_n}p_{r_n+s_n-1})| \\
&\leq q_{r_n-1}|q_{r_n+s_n}\alpha - p_{r_n+s_n}| + q_{r_n}|q_{r_n+s_n-1}\alpha - p_{r_n+s_n-1}| \\
&\leq 2\,q_{r_n}q_{r_n+s_n}^{-1}
\end{aligned}
\tag{5.4}
$$

and, likewise,

$$
\begin{aligned}
&|(q_{r_n-1}q_{r_n+s_n} - q_{r_n}q_{r_n+s_n-1})\alpha - (p_{r_n-1}q_{r_n+s_n} - p_{r_n}q_{r_n+s_n-1})| \\
&\leq q_{r_n+s_n}|q_{r_n-1}\alpha - p_{r_n-1}| + q_{r_n+s_n-1}|q_{r_n}\alpha - p_{r_n}| \\
&\leq 2\,q_{r_n}^{-1}q_{r_n+s_n}.
\end{aligned}
\tag{5.5}
$$

Furthermore, we have

$$|P_n(\alpha)| \ll H(P_n) \cdot |\alpha - \alpha_n| \ll q_{r_n}q_{r_n+s_n}q_{r_n+\lfloor w s_n \rfloor}^{-2}. \tag{5.6}$$

We consider the four linearly independent linear forms:

$$
\begin{aligned}
L_1(X_1, X_2, X_3, X_4) &= \alpha^2 X_1 - \alpha(X_2 + X_3) + X_4, \\
L_2(X_1, X_2, X_3, X_4) &= \alpha X_1 - X_2, \\
L_3(X_1, X_2, X_3, X_4) &= \alpha X_1 - X_3, \\
L_4(X_1, X_2, X_3, X_4) &= X_1.
\end{aligned}
$$

Instead of treating the coefficient of X in $P_n(X)$ as a single variable, we cut it into two variables. Evaluating these linear forms on the quadruple

$$\underline{z}_n := (q_{r_n-1}q_{r_n+s_n} - q_{r_n}q_{r_n+s_n-1}, q_{r_n-1}p_{r_n+s_n} - q_{r_n}p_{r_n+s_n-1},$$
$$p_{r_n-1}q_{r_n+s_n} - p_{r_n}q_{r_n+s_n-1}, p_{r_n-1}p_{r_n+s_n} - p_{r_n}p_{r_n+s_n-1}),$$

it follows from (5.4)–(5.6) that

$$\prod_{1 \le j \le 4} |L_j(\underline{z}_n)| \ll q_{r_n}^2 q_{r_n+s_n}^2 q_{r_n+\lfloor ws_n \rfloor}^{-2}. \tag{5.7}$$

Again on the assumption that $(q_\ell^{1/\ell})_{\ell \ge 1}$ converges, we are able to apply Theorem 4 (and, with some additional work, deduce that α is transcendental) only when

$$\limsup_{n \to +\infty} \frac{r_n + ws_n}{2r_n + s_n} > 1,$$

that is, when

$$w > w' + 1. \tag{5.8}$$

This is precisely the inequality (5.1) with $m = M$.

The novelty in [10] is the observation that the estimate (5.6) can be considerably improved when r_n is large. Namely, using (5.2), (5.4) and (5.5), we get

$$
\begin{aligned}
|P_n(\alpha)| &= |P_n(\alpha) - P_n(\alpha_n)| \\
&= |(q_{r_n-1}q_{r_n+s_n} - q_{r_n}q_{r_n+s_n-1})(\alpha - \alpha_n)(\alpha + \alpha_n) \\
&\quad - (q_{r_n-1}p_{r_n+s_n} - q_{r_n}p_{r_n+s_n-1} + p_{r_n-1}q_{r_n+s_n} - p_{r_n}q_{r_n+s_n-1})(\alpha - \alpha_n)| \\
&= |\alpha - \alpha_n| \cdot |(q_{r_n-1}q_{r_n+s_n} - q_{r_n}q_{r_n+s_n-1})\alpha - (q_{r_n-1}p_{r_n+s_n} - q_{r_n}p_{r_n+s_n-1}) \\
&\quad + (q_{r_n-1}q_{r_n+s_n} - q_{r_n}q_{r_n+s_n-1})\alpha - (p_{r_n-1}q_{r_n+s_n} - p_{r_n}q_{r_n+s_n-1}) \\
&\quad + (q_{r_n-1}q_{r_n+s_n} - q_{r_n}q_{r_n+s_n-1})(\alpha_n - \alpha)| \\
&\ll |\alpha - \alpha_n| \cdot (q_{r_n}q_{r_n+s_n}^{-1} + q_{r_n}^{-1}q_{r_n+s_n} + q_{r_n}q_{r_n+s_n}|\alpha - \alpha_n|) \\
&\ll |\alpha - \alpha_n|q_{r_n}^{-1}q_{r_n+s_n} \\
&\ll q_{r_n}^{-1}q_{r_n+s_n}q_{r_n+\lfloor ws_n \rfloor}^{-2}.
\end{aligned} \tag{5.9}
$$

Compared to the estimate (5.6), which was used in [1], we gain a factor $q_{r_n}^{-2}$. As we will see below, this allows us eventually to replace the assumption (5.8) by (5.11) below. The improvement can be explained by Lemma 1. Indeed, since

$$|P_n(\alpha)| \le H(P_n) \cdot |\alpha - \alpha_n| \cdot |\alpha - \alpha_n'|,$$

where α_n' denotes the Galois conjugate of α_n, we get an improvement on (5.6) when α_n' is close to α, that is, when α_n' is close to α_n. And Lemma 1 precisely asserts that this situation holds when r_n is large.

Using (5.9) we can slightly improve (5.3) by applying the first statement of Corollary 1 instead of the second one. Indeed, we can conclude that α is transcendental when there exist $\varepsilon > 0$ and arbitrarily large integers n such that

$$q_{r_n}^{-1}q_{r_n+s_n}q_{r_n+\lfloor ws_n \rfloor}^{-2}q_{r_n}^2q_{r_n+s_n}^2 < (q_{r_n}q_{r_n+s_n})^{-\varepsilon}.$$

If $(q_\ell^{1/\ell})_{\ell \geq 1}$ converges, this shows that the assumption

$$w > w' + 3/2 \qquad (5.10)$$

is enough to deduce that α is transcendental.

By combining the use of Theorem 4 with $m = 4$ and (5.9), we are able to improve (5.8) in the same way as (5.10) improves (5.3). Namely, we have

$$\begin{aligned} \Pi_{1 \leq j \leq 4} |L_j(\underline{z}_n)| &\ll q_{r_n+s_n}^2 \, q_{r_n+\lfloor ws_n \rfloor}^{-2} \\ &\ll 2^{-(w-1)s_n} \\ &\ll (q_{r_n} q_{r_n+s_n})^{-\delta(w-1)s_n/(2r_n+s_n)}, \end{aligned}$$

if n is sufficiently large, where we have set

$$\delta = \frac{\log 2}{1 + \limsup_{\ell \to +\infty} q_\ell^{1/\ell}}.$$

Thus, with $\varepsilon = \delta(w-1)/(2w'+2)$, which is positive when

$$w > 1, \qquad (5.11)$$

we see that

$$\prod_{1 \leq j \leq 4} |L_j(\underline{z}_n)| \ll (q_{r_n} q_{r_n+s_n})^{-\varepsilon}$$

holds for any sufficiently large integer n.

We can then apply Theorem 4 to prove that α is transcendental. The details are given in [10].

6   Two Open Questions

We conclude this survey by two open questions.

Let $(a_\ell)_{\ell \geq 1}$ be the sequence defined by $a_\ell = 2$ if ℓ is a perfect square and $a_\ell = 1$ otherwise. Is the real number

$$[0; a_1, a_2, \ldots] = [0; 2, 1, 1, 2, 1, 1, \ldots]$$

transcendental? Theorem 3 cannot be applied in this case since the sequence of squares grows too slowly.

Theorem 2 asserts that automatic continued fractions are transcendental or quadratic. Conjecturally, the same holds for morphic continued fractions (see [8] for a precise definition). Since there exist morphic words $\mathbf{a} = (a_\ell)_{\ell \geq 1}$ whose complexity function $n \mapsto p(n, \mathbf{a})$ grows as fast as a constant times n^2, Theorem 3 is not strong enough to give a positive answer to this conjecture.

References

1. B. Adamczewski, Y. Bugeaud, On the complexity of algebraic numbers, II. Continued fractions. Acta Math. **195**, 1–20 (2005)
2. B. Adamczewski, Y. Bugeaud, On the complexity of algebraic numbers, I. Expansions in integer bases. Ann. Math. **165**, 547–565 (2007)
3. B. Adamczewski, Y. Bugeaud, On the Maillet–Baker continued fractions. J. Reine Angew. Math. **606**, 105–121 (2007)
4. B. Adamczewski, Y. Bugeaud, Palindromic continued fractions. Ann. Inst. Fourier (Grenoble) **57**, 1557–1574 (2007)
5. B. Adamczewski, Y. Bugeaud, L. Davison, Continued fractions and transcendental numbers. Ann. Inst. Fourier (Grenoble) **56**, 2093–2113 (2006)
6. B. Adamczewski, Y. Bugeaud, F. Luca. Sur la complexité des nombres algébriques. C. R. Acad. Sci. Paris **339**, 11–14 (2004)
7. J.-P. Allouche, J.L. Davison, M. Queffélec, L.Q. Zamboni, Transcendence of Sturmian or morphic continued fractions. J. Number Theory **91**, 39–66 (2001)
8. J.-P. Allouche, J. Shallit, *Automatic Sequences: Theory, Applications, Generalizations* (Cambridge University Press, Cambridge, 2003)
9. A. Baker, Continued fractions of transcendental numbers. Mathematika **9**, 1–8 (1962)
10. Y. Bugeaud, Automatic continued fractions are transcendental or quadratic. Ann. Sci. Ecole Norm. Sup. (to appear)
11. Y. Bugeaud, Continued fractions with low complexity: transcendence measures and quadratic approximation. Compos. Math. **148**, 718–750 (2012)
12. A. Cobham, On the Hartmanis–Stearns problem for a class of tag machines, in *Conference Record of 1968 Ninth Annual Symposium on Switching and Automata Theory*, Schenectady, New York, 1968, pp. 51–60
13. A. Cobham, Uniform tag sequences. Math. Syst. Theory **6**, 164–192 (1972)
14. J.L. Davison, A class of transcendental numbers with bounded partial quotients, in *Number Theory and Applications*, ed. by A.B. Banff. NATO Adv. Sci. Inst. Ser. C Math. Phys. Sci., vol 265 (Kluwer Academic, Dordrecht, 1988/1989), pp. 365–371
15. P. Liardet, P. Stambul, Séries de Engel et fractions continuées. J. Théor. Nombres Bordeaux **12**, 37–68 (2000)
16. J. Liouville, Sur des classes très étendues de quantités dont la valeur n'est ni algébrique, ni même réductible à des irrationelles algébriques. C. R. Acad. Sci. Paris **18**, 883–885, 910–911 (1844)
17. J.H. Loxton, A.J. van der Poorten, Arithmetic properties of the solutions of a class of functional equations. J. Reine Angew. Math. **330**, 159–172 (1982)
18. J.H. Loxton, A.J. van der Poorten, Arithmetic properties of automata: regular sequences. J. Reine Angew. Math. **392**, 57–69 (1988)
19. E. Maillet, *Introduction à la théorie des nombres transcendants et des propriétés arithmétiques des fonctions* (Gauthier-Villars, Paris, 1906)
20. M. Morse, G.A. Hedlund, Symbolic dynamics. Am. J. Math. **60**, 815–866 (1938)
21. M. Morse, G.A. Hedlund, Symbolic dynamics II. Am. J. Math. **62**, 1–42 (1940)
22. O. Perron, *Die Lehre von den Kettenbrüchen* (Teubner, Leibzig, 1929)
23. M. Queffélec, Transcendance des fractions continues de Thue–Morse. J. Number Theory **73**, 201–211 (1998)
24. M. Queffélec, Irrational numbers with automaton-generated continued fraction expansion, in *Dynamical Systems* (Luminy-Marseille, 1998), pp. 190–198 [World Sci. Publ., River Edge, 2000]
25. W.M. Schmidt, Norm form equations. Ann. Math. **96**, 526–551 (1972)
26. W.M. Schmidt, *Diophantine Approximation*. Lecture Notes in Mathematics, vol 785 (Springer, Berlin, 1980)

Algebraic Independence of Infinite Products and Their Derivatives

Peter Bundschuh

To the memory of Alf van der Poorten

Abstract For fixed rational integers $q > 1$, and for non-constant polynomials P with $P(0) = 1$ and with algebraic coefficients, we consider the infinite product $A_q(z) = \prod_{k \geq 0} P(z^{q^k})$. Using Mahler's transcendence method, we prove results on the algebraic independence over $\mathbb{Q}$ of the numbers $A_q(\alpha), A_q'(\alpha), A_q''(\alpha), \ldots$ at algebraic points α with $0 < |\alpha| < 1$. The basic analytic ingredient of the proof is the hypertranscendence of the function $A_q(z)$, and we provide sufficient criteria for it.

Key words Hypertranscendence • Mahler-type functional equations • transcendence and algebraic independence • Mahler's method

Mathematics Subject Classification (2010): 11J81, 11J91, 12H05, 34M15

1 Introduction and Main Results

Very recently, extending work of Coons [3], we studied in [2] arithmetic and related analytic questions concerning the generating function

$$A(z) := \sum_{n=0}^{\infty} a_{n+1} z^n \tag{1}$$

P. Bundschuh (✉)
Mathematisches Institut, Universität zu Köln, Weyertal 86-90, 50931 Köln, Germany
e-mail: pb@math.uni-koeln.de

J.M. Borwein et al. (eds.), *Number Theory and Related Fields: In Memory of Alf van der Poorten*, Springer Proceedings in Mathematics & Statistics 43, DOI 10.1007/978-1-4614-6642-0_6, © Springer Science+Business Media New York 2013

of Stern's sequence $(a_n)_{n=0,1,...}$ defined by $a_0 := 0, a_1 := 1$ and, for $n \in \mathbb{N}$, the set of all positive rational integers, by

$$a_{2n} := a_n, \quad a_{2n+1} := a_n + a_{n+1}. \tag{2}$$

More precisely, we proved the algebraic independence over $\mathbb{Q}$ of the numbers[1] $A(\alpha), A'(\alpha), A''(\alpha), \ldots$ for every complex algebraic number α with $0 < |\alpha| < 1$. Analytically, the crucial ingredient to this proof was the fact that the function $A(z)$, having the unit circle as its natural boundary, is hypertranscendental. Remember that an analytic function is called *hypertranscendental* if it satisfies no algebraic differential equation, that is, no finite collection of derivatives of the function is algebraically dependent over $\mathbb{C}(z)$. It should be noticed that our algebraic independence proof for the *numbers* $A^{(\mu)}(\alpha), \mu \in \mathbb{N}_0 := \mathbb{N} \cup \{0\}$, as well as our hypertranscendence proof for the *function* $A(z)$, heavily depended on the fact that this function satisfies the Mahler-type functional equation

$$A(z) = P(z)A(z^2) \quad \text{with} \quad P(z) := 1 + z + z^2. \tag{3}$$

Clearly, iteration of (3) yields the following product expansion for $A(z)$:

$$A(z) = \prod_{k=0}^{\infty} P(z^{2^k}). \tag{4}$$

It is the main aim of the present article to fairly generalize our results in [2] concerning the above particular function $A(z)$. Our first basic statement is the following hypertranscendence criterion for infinite products of type (4), where, *from now on, the hypothesis $q \in \mathbb{N} \setminus \{1\}$ will hold*.

Theorem 1. *Let $P \in \mathbb{C}[z]$ be nonconstant with $P(0) = 1$. If the functional equation*

$$w(z) - qw(z^q) = \frac{zP'(z)}{P(z)} \tag{5}$$

has no solution $w \in \mathbb{C}(z)$, then the infinite product

$$A_q(z) := \prod_{k=0}^{\infty} P(z^{q^k}) \tag{6}$$

is hypertranscendental.

Note here that the infinite product (6) is holomorphic at least in the unit disk $\mathbb{D} := \{z \in \mathbb{C} \,|\, |z| < 1\}$. Evidently, Theorem 1 reduces the hypertranscendence investigation to the simpler looking problem of rational insolvability of Eq. (5) which subsequently will be called *the adjoint functional equation of* (6).

---

[1] Of course, this means that every finite subset of these numbers is algebraically independent.

Our next two results will provide, under various conditions, criteria for rational insolvability of the adjoint functional equation of infinite products of type (6). The first one contains a necessary and sufficient condition and reads as follows.

Criterion 1. *For $P \in \mathbb{C}[z]$ satisfying $P(0) = 1$ and $1 \le \deg P < q$, the following conditions are equivalent:*

(i) *The functional equation (5) has a solution $w \in \mathbb{C}(z)$.*
(ii) *$P(z)$ is of the shape $\sum_{j=0}^{q-1}(z/\eta)^j$ with some $(q-1)$st root of unity η.*

Remark 1. Note that, in all of our applications of Criterion 1, we need only its implication (i) $\Rightarrow$ (ii). Note also that we could formulate (i) equivalently as "... has exactly one solution ...". Namely, $w(z) = qw(z^q)$ is just trivially solvable in $\mathbb{C}(z)$ as one easily sees by inserting the Laurent series expansion of $w(z)$ about $z = 0$ into this equation.

Whereas, in Criterion 1, the degree of P has to be less than the given q, the second one needs no such size condition. Instead, it concerns only quite particular polynomials connected with cyclotomic ones and has to suppose some coprimality condition.

Criterion 2. *Let $q, s \in \mathbb{N} \setminus \{1\}$ and $P(z)$ be $1 + z + \ldots + z^{s-1}$ or the sth cyclotomic polynomial $\Phi_s(z)$. If $(q, s) = 1$, then the adjoint functional equation (5) corresponding to (6) is rationally unsolvable.*

Remark 2. The case $q = 2, s = 3$ is at the heart of our proof of Theorem 1 in [2], and the proof of Criterion 2 will follow the reasoning given there [loc. cit., pp. 365–366].

The subsequent Theorem 2 is a criterion for the algebraic independence of infinite products of type (6) and their derivatives at algebraic points.

Theorem 2. *Let[2] $P \in \overline{\mathbb{Q}}[z]$ be nonconstant with $P(0) = 1$, and suppose that the function $A_q(z)$ defined in $\mathbb{D}$ according to (6) is hypertranscendental. Then, for every $\alpha \in \overline{\mathbb{Q}}^{\times} \cap \mathbb{D}$ with $P(\alpha^{q^j}) \neq 0$ for each $j \in \mathbb{N}_0$, the numbers $A_q(\alpha), A_q'(\alpha), A_q''(\alpha), \ldots$ are algebraically independent.*

It should be noted here that transcendence questions on infinite products have been studied already earlier. So Tachiya [14], e.g., considered products of type

$$\prod_{k=0}^{\infty} \frac{P_k(z^{q^k})}{Q_k(z^{q^k})}$$

with polynomials P_k, Q_k of uniformly bounded degrees and satisfying $P_k(0) = Q_k(0) = 1$ for any $k \in \mathbb{N}_0$. Furthermore their coefficients have to belong to some fixed algebraic number field, and the "denominators" as well as the "houses" of these coefficients are not allowed to increase too fast with k. Then Tachiya proved the transcendence of the above product at all points $\alpha \in \overline{\mathbb{Q}}^{\times} \cap \mathbb{D}$ satisfying $P_k(\alpha^{q^k})Q_k(\alpha^{q^k}) \neq 0$ for any $k \in \mathbb{N}_0$, if the product, as a function of z, is not rational.

---

[2]As usual, $\overline{\mathbb{Q}}$ denotes the field of all complex algebraic numbers.

Compared to Tachiya's products, we do not allow denominator polynomials on the right-hand side of (6), and moreover, we work only with a fixed numerator polynomial. On the other hand, we prove algebraic independence of the products and all its derivatives at α, not merely transcendence.

Whereas our proofs of Theorem 1, Criteria 1 and 2, and Theorem 2 will be given in Sects. 2, 3, and 4, respectively, to conclude our introduction, we will present only one immediate application of these results. A few more will follow in Sect. 5.

Corollary 1. *For $q,s \in \mathbb{N} \setminus \{1\}$, define, with $P_s(z) := 1 + z + \ldots + z^{s-1}$,*

$$C_{q,s}(z) := \prod_{k=0}^{\infty} P_s(z^{q^k}), \quad \text{and further} \quad \tilde{C}_{q,s}(z) := \prod_{k=0}^{\infty} \Phi_s(z^{q^k}) \qquad (7)$$

in the unit disk. If $(q,s) = 1$ holds in the case $s \geq q$, then $C_{q,s}(\alpha), C'_{q,s}(\alpha), C''_{q,s}(\alpha), \ldots$ are algebraically independent for every $\alpha \in \overline{\mathbb{Q}}^\times \cap \mathbb{D}$. The same statement holds for the numbers $\tilde{C}_{q,s}(\alpha), \tilde{C}'_{q,s}(\alpha), \tilde{C}''_{q,s}(\alpha), \ldots$.

Proof. If $s < q$ or $s \geq q$, then Criterion 1 or 2, respectively, tells us that the adjoint functional equation of both of the products in (7) is rationally unsolvable, whence, by Theorem 1, $C_{q,s}(z)$ and $\tilde{C}_{q,s}(z)$ are hypertranscendental. Then Theorem 2 gives us our arithmetical claim. $\square$

Remark 3. Note that the case $q = 2, s = 3$ is just Theorem 2 in [2]; hence, $C_{3,2} = \tilde{C}_{3,2}$ is just the particular A from (1), (3), (4). Note also that $C_{s,s}(z) = 1/(1-z)$ holds for any $s \geq 2$, by $P_s(z) = (1 - z^s)/(1 - z)$.

2 Proof of Theorem 1

This proof will be highly based on the subsequently quoted Theorem 3 in Nishioka's paper [9] which itself is deduced from a necessary condition for the existence of differentially algebraic solutions of certain types of functional equations.

Theorem N1. *Let C be a field of characteristic 0, and suppose that $f \in C[[z]]$ has the following two properties:*

(i) *For suitable $m \in \mathbb{N}_0$, the series $f, \delta f, \ldots, \delta^m f$ are algebraically dependent over $C(z)$, where δ denotes the differential operator $z \frac{d}{dz}$.*

(ii) *For suitable q, the series f satisfies the functional equation*

$$f(z^q) = u(z)f(z) + v(z), \qquad (8)$$

where $u, v \in C(z), u \neq 0$. If $u(z) = s_M z^M + \ldots$ with $M \in \mathbb{Z}, s_M \in C^\times$ define $Q := [M/(q-1)]$.

Then there exists some $w \in C(z)$ satisfying

$$w(z^q) = u(z)w(z) + v(z)$$

or

$$w(z^q) = u(z)w(z) + v(z) - \gamma \frac{u_1(z)z^{Qq}}{u_2(z)},$$

where $u_1(z) = u(z)/(s_M z^M), u_2 \in C(z) \setminus \{0\}$ fulfill the condition $u_2(z^q) = u_2(z)/u_1(z)$, and $\gamma \in C$ is the constant term in the z-expansion of $v(z)u_2(z)/(u_1(z)z^{Qq})$ in case $s_M = 1$ and $M = Q(q-1)$, but $\gamma = 0$ otherwise.

To prove now our Theorem 1 along the lines we followed in [2], we suppose that $A_q(z)$ from (6) is not hypertranscendental, i.e., that there is an $m \in \mathbb{N}_0$ such that $A_q, A'_q, \ldots, A_q^{(m)}$ are algebraically dependent over $\mathbb{C}(z)$. As it is easily seen by induction, the equation

$$\delta^\mu A_q = \sum_{\lambda=1}^{\mu} c_{\lambda,\mu} z^\lambda A_q^{(\lambda)}$$

holds for any $\mu \in \mathbb{N}$ with explicit $c_{1,\mu}, \ldots, c_{\mu,\mu} \in \mathbb{N}, c_{\mu,\mu} = 1$ which are known to be Stirling numbers of the second kind. Therefore, our above assumption on A_q and its derivatives is equivalent to the algebraic dependence of $A_q, \delta A_q, \ldots, \delta^m A_q$ over $\mathbb{C}(z)$, whence the transcendence degree

$$\mathrm{trdeg}_{\mathbb{C}} \mathbb{C}(z, A_q, \delta A_q, \delta^2 A_q, \ldots)$$

is finite.

Since $A_q(0) = 1$ we know that A_q is zero-free in some disk about the origin, whence $B := \frac{\delta A_q}{A_q}$ is holomorphic in this same disk. Furthermore, B belongs to the function field $\mathbb{C}(z, A_q, \delta A_q, \delta^2 A_q, \ldots)$ of finite transcendence degree over $\mathbb{C}$, whence the functions $B, \delta B, \delta^2 B, \ldots$ are algebraically dependent over $\mathbb{C}(z)$.

We now are going to apply Theorem N1 to $f = B$ and have only to check that our B satisfies a functional equation of type (8). To this purpose, we use $A_q(z) = P(z)A_q(z^q)$ coming from (6) and consider

$$(\delta A_q)(z) = z(P'(z)A_q(z^q) + qz^{q-1}P(z)A'_q(z^q)) = (\delta P)(z)A_q(z^q) + qP(z)z^q \frac{d}{dz^q}A_q(z^q).$$

From this and our above definition of B, we deduce

$$B(z) = \frac{\delta P(z)}{P(z)} + qB(z^q) \iff B(z^q) = \frac{1}{q}B(z) - \frac{\delta P(z)}{qP(z)}.$$

Therefore, (8) holds with $u(z) = 1/q, v(z) = -\delta P(z)/(qP(z))$; note here $M = 0, s_M = 1/q$, whence $\gamma = 0$. But then, according to Theorem N1, there is some $w \in \mathbb{C}(z)$ satisfying equation (5). Thus, Theorem 1 is established.

3 On the Rational Insolvability of Eq. (5)

In the subsequent proofs of Criteria 1 and 2, the following lemma will play a decisive role.

Lemma 1. *For any $P \in \mathbb{C}[z] \setminus \{0\}$, the formula*

$$\frac{zP'(z)}{P(z)} = s + \sum_{\sigma=1}^{s} \frac{z_\sigma}{z - z_\sigma}$$

holds with $s := \deg P$, and $z_1, \ldots, z_s$ the (not necessarily distinct) zeros of P.

Proof. The case of constant P is clear (but not needed later). Thus, let $z_1, \ldots, z_r$ denote the distinct zeros of P, of multiplicities $\omega_1, \ldots, \omega_r$, respectively. Logarithmic differentiation of the product representation of P then leads to

$$\frac{P'(z)}{P(z)} = \sum_{\rho=1}^{r} \frac{\omega_\rho}{z - z_\rho} \quad \left(= \sum_{\sigma=1}^{s} \frac{1}{z - z_\sigma} \right),$$

whence the result. $\qquad\qquad\qquad\qquad\qquad\qquad\qquad\qquad\qquad\qquad\qquad\qquad\qquad$ □

Proof of Criterion 1. For the implication (i) $\Rightarrow$ (ii), we work out a suggestion of Nishioka [12, p. 133]. For given q and P, we assume that (5) has a solution $w \in \mathbb{C}(z)$ which we may write as $w(z) = u(z)/v(z)$ with coprime $u, v \in \mathbb{C}[z] \setminus \{0\}$ (note $P' \neq 0$). Then (5) is equivalent to

$$P(z)\big(u(z)v(z^q) - qu(z^q)v(z)\big) = zP'(z)v(z)v(z^q) \qquad (9)$$

from which we conclude $v(z^q)|P(z)u(z^q)v(z)$, hence

$$v(z^q)|P(z)v(z) \qquad (10)$$

(divisibility in $\mathbb{C}[z]$) in virtue of the coprimality of $u(z^q), v(z^q)$. Relation (10) implies $(q-1)\deg v \leq \deg P < q$, by a hypothesis of Criterion 1, whence $\deg v \leq 1$. If $\deg v = 0$, i.e., $v = v_0 \in \mathbb{C}^\times$, say, then equation (9) reduces to $P(z)(u(z) - qu(z^q)) = v_0 zP'(z)$, whence u is also constant, and by taking $z = 0$, we find $(1-q)u = 0$, a contradiction. Therefore, we must have $\deg v = 1$, and without loss of generality, we may write $v(z) = z - \eta$ with some $\eta \in \mathbb{C}^\times$; note $\eta \neq 0$, by (10). Again from (10), we see $q - 1 \leq \deg P$, and the opposite inequality is a hypothesis of Criterion 1. Thus, the quotient

$$\frac{P(z)(z - \eta)}{z^q - \eta}$$

is a nonzero constant, in fact, equal to 1, by considering $z = 0$. In $P(z) = \sum_{j=0}^{q-1} p_j z^j$, we must have $p_{q-1} = 1$, $p_0 = 1$, and $p_{j-1} = \eta p_j$ for $j = 1, \ldots, q-1$. On induction, we finally conclude herefrom $p_j = \eta^{-j}$ for $j = 0, \ldots, q-1$, where η satisfies $\eta^{q-1} = 1$.

To prove the implication (ii) $\Rightarrow$ (i), we suppose that $P(z)$ is of the shape described in (ii). Then

$$P(z) = \frac{(z/\eta)^q - 1}{(z/\eta) - 1} = \prod_{j=1}^{q-1}\left(\frac{z}{\eta} - \xi^j\right) = \prod_{j=1}^{q-1}(z - \eta\xi^j)$$

with $\eta^{q-1} = 1$ and some fixed primitive qth root of unity ξ. From Lemma 1, we conclude, for the moment only if $z \in \mathbb{D}$,

$$\frac{zP'(z)}{P(z)} = q - 1 + \sum_{j=1}^{q-1}\frac{\eta\xi^j}{z - \eta\xi^j} = \sum_{j=1}^{q-1}\left(1 - \frac{1}{1 - z/(\eta\xi^j)}\right) = -\sum_{j=1}^{q-1}\sum_{v=1}^{\infty}\left(\frac{z}{\eta\xi^j}\right)^v$$

$$= -\sum_{v=1}^{\infty}\left(\frac{z}{\eta}\right)^v\left(\sum_{j=0}^{q-1}\xi^{-jv} - 1\right) = \sum_{v=1}^{\infty}\left(\frac{z}{\eta}\right)^v - q\sum_{\mu=1}^{\infty}\left(\frac{z}{\eta}\right)^{q\mu}.$$

Note that the last equality comes from the fact that $\sum_{j=0}^{q-1}\xi^{-jv}$ equals q if $q|v$ but 0 otherwise. Over all, we find, in virtue of $\eta^{q-1} = 1$,

$$\frac{zP'(z)}{P(z)} = \frac{z}{\eta - z} - q\frac{z^q}{\eta - z^q}.$$

Comparing this with equation (5), we immediately recognize $w(z) = z/(\eta - z)$ as its solution.   □

Proof of Criterion 2. To treat both cases of P simultaneously, we write

$$P(z) = \prod_{1 \le \sigma < s}^{*}(z - \zeta^\sigma) \tag{11}$$

with some fixed primitive sth root of unity ζ. Here $*$ indicates that the product (and soon also the sum) has to be taken, in the case $P = \Phi_s$, only over those $\sigma \in \{1,\ldots,s-1\}$ coprime to s. Assuming that (5) is rationally solvable is, by (11) and Lemma 1, equivalent to the existence of some $w \in \mathbb{C}(z)$ satisfying

$$w(z) - qw(z^q) = s^* + \sum_{1 \le \sigma < s}^{*}\frac{\zeta^\sigma}{z - \zeta^\sigma}, \tag{12}$$

where s^* denotes $s - 1$ or $\varphi(s)$ (Euler's totient function) in the first or second case, respectively.

By our hypothesis $(q,s) = 1$, the (multiplicative) order of q modulo s is defined; let us denote it by $r \in \mathbb{N}$. If $r \ge 2$, then we replace in (12) z by z^{q^ρ} with $\rho \in \{1,\ldots,r-1\}$ and obtain

$$w(z^{q^\rho}) - qw(z^{q^{\rho+1}}) = s^* + \sum_{1 \le \sigma < s}^{*}\frac{\zeta^\sigma}{z^{q^\rho} - \zeta^\sigma}. \tag{13}$$

We next want to investigate the behavior of the rational function on the right-hand side of (13) near $z = \zeta$. To this purpose, we note the following. By $(q, s) = 1$, the linear diophantine equation $as + bq^\rho = 1$ is solvable in $\mathbb{Z}$ for any $\rho \in \mathbb{N}$ and even with $(b, s) = 1$. Hence, there exists a unique b^{-1} in the prime residue class group modulo s, and we have $b^{-1} \equiv q^\rho \pmod{s}$. Note that this $b^{-1} = b^{-1}(\rho)$ may be taken from $\{1, \ldots, s - 1\}$. Thus, we see from (13)

$$w(z^{q^\rho}) - qw(z^{q^{\rho+1}}) = \frac{\zeta^{b^{-1}(\rho)}}{z^{q^\rho} - \zeta^{b^{-1}(\rho)}} + s^* + \sum_{\substack{1 \leq \sigma < s \\ \sigma \neq b^{-1}(\rho)}}{}^* \ldots = \frac{\zeta^{q^\rho}}{z^{q^\rho} - \zeta^{q^\rho}} + s^* + \sum_{\substack{1 \leq \sigma < s \\ \sigma \neq q^\rho}}{}^* \ldots$$

(14)

since $b^{-1}(\rho) \equiv q^\rho \pmod{s}$. By $\zeta^{q^\rho} - \zeta^\sigma = 0 \iff q^\rho \equiv \sigma \pmod{s}$, the sum on the right-hand side of (14) is holomorphic at $z = \zeta$. On the other hand, we have

$$\frac{\zeta^{q^\rho}}{z^{q^\rho} - \zeta^{q^\rho}} = \frac{\zeta^{q^\rho}}{(z - \zeta)(z^{q^\rho - 1} + \ldots + \zeta^{q^\rho - 1})} = \frac{\zeta^{q^\rho}}{q^\rho \zeta^{q^\rho - 1}(z - \zeta)} + \ldots = \frac{\zeta}{q^\rho(z - \zeta)} + \ldots$$

(15)

near $z = \zeta$, where $\ldots$ stands (here and up to (17)) for certain holomorphic functions not to be specified in more detail. Combining (14) and (15) we are lead to

$$q^\rho w(z^{q^\rho}) - q^{\rho+1} w(z^{q^{\rho+1}}) = \frac{\zeta}{z - \zeta} + \ldots$$

(16)

On adding (16) for $\rho = 0, \ldots, r - 1$, we find

$$w(z) - q^r w(z^{q^r}) = r \frac{\zeta}{z - \zeta} + \ldots;$$

(17)

note here that the case $\rho = 0$ of (16) comes from (12).

From (17) and $\zeta^{q^r} = \zeta$, we infer that our rational function $w(z)$ has a pole at $z = \zeta$ of order $\lambda \in \mathbb{N}$, say. Hence, $w(z) = c_\lambda(z - \zeta)^{-\lambda} + \ldots$ with $c_\lambda \neq 0$, where now $\ldots$ indicates the summands of higher-order in the Laurent expansion at $z = \zeta$. We next consider the beginning of the Laurent expansion of the left-hand side of (17) taking again $\zeta^{q^r} = \zeta$ into account:

$$\frac{c_\lambda}{(z - \zeta)^\lambda} + \ldots - q^r \left(\frac{c_\lambda}{(z^{q^r} - \zeta^{q^r})^\lambda} + \ldots \right) = \frac{c_\lambda}{(z - \zeta)^\lambda} + \ldots$$

$$- q^r \left(\frac{c_\lambda}{(z - \zeta)^\lambda (z^{q^r - 1} + \ldots)^\lambda} + \ldots \right) = \frac{c_\lambda(1 - q^{r(1-\lambda)})}{(z - \zeta)^\lambda} + \ldots$$

(18)

If $\lambda = 1$, then the numerator of the term indicated on the right-hand side of (18) vanishes, whence the left-hand side of (17) is holomorphic at $z = \zeta$. But if $\lambda \geq 2$, then the right-hand side of (18) has a pole of order λ at $z = \zeta$. Both of these conclusions contradict equation (17).   □

4   Proof of Theorem 2

Denoting by K the algebraic number field obtained by adjoining to $\mathbb{Q}$ the coefficients of P, we know $A_q \in K[[z]]$. Since A_q satisfies the functional equation $A_q(z) = P(z)A_q(z^q)$, we obtain that A_q fulfills

$$(D^\lambda A_q)(z) = \sum_{\mu=0}^{\lambda} \binom{\lambda}{\mu} (D^{\lambda-\mu}P)(z)D^\mu(A_q(z^q)), \tag{19}$$

where now D denotes the ordinary differential operator $\frac{d}{dz}$. For the function $A_q(z^q)$ and its derivatives with respect to z, we need the following auxiliary result.

Lemma 2. *For every $\mu \in \mathbb{N}$, one has*

$$D^\mu(A_q(z^q)) = Q_{\mu,1}(z)(DA_q)(z^q) + \ldots + Q_{\mu,\mu}(z)(D^\mu A_q)(z^q) \tag{20}$$

with explicit $Q_{\mu,1}, \ldots, Q_{\mu,\mu} \in \mathbb{Z}[z]$ and $Q_{\mu,\mu}(z) = (qz^{q-1})^\mu$.

Proof. We first note $\frac{d}{dz}A_q(z^q) = qz^{q-1}(DA_q)(z^q)$ giving the assertion for $\mu = 1$. Assume it is already proved for some $\mu \in \mathbb{N}$; then, applying the operator D to equation (20), we find

$$D^{\mu+1}(A_q(z^q)) = \sum_{v=1}^{\mu} Q'_{\mu,v}(z)(D^v A_q)(z^q) + \sum_{v=1}^{\mu} Q_{\mu,v}(z)qz^{q-1}(D^{v+1}A_q)(z^q).$$

This establishes our assertion for $\mu+1$ taking $Q_{\mu+1,\mu+1}(z) := qz^{q-1}Q_{\mu,\mu}(z) = (qz^{q-1})^{\mu+1}$ into account. $\square$

By inserting (20) into (19), we obtain for every $\lambda \in \mathbb{N}$

$$(D^\lambda A_q)(z) = (D^\lambda P)(z)A_q(z^q) + \sum_{\mu=1}^{\lambda} \binom{\lambda}{\mu}(D^{\lambda-\mu}P)(z)\sum_{v=1}^{\mu} Q_{\mu,v}(z)(D^v A_q)(z^q) \tag{21}$$

$$= \sum_{v=0}^{\lambda} R_{\lambda,v}(z)(D^v A_q)(z^q),$$

where we wrote

$$R_{\lambda,0}(z) := (D^\lambda P)(z), \ R_{\lambda,v}(z) := \sum_{\mu=v}^{\lambda} \binom{\lambda}{\mu}(D^{\lambda-\mu}P)(z)Q_{\mu,v}(z) \quad (v=1,\ldots,\lambda).$$

Notice here that all $R_{\lambda,v} \in K[z]$, and furthermore

$$R_{\lambda,\lambda}(z) = P(z)Q_{\lambda,\lambda}(z) = P(z)(qz^{q-1})^\lambda \quad (\lambda \in \mathbb{N}_0). \tag{22}$$

Coming back to the usual notation of (higher) derivatives, (21) says for every $m \in \mathbb{N}_0$, in matrix notation,

$$^\tau(A_q(z), A_q'(z), \ldots, A_q^{(m)}(z)) = \tilde{A}(z) \cdot {}^\tau(A_q(z^q), A_q'(z^q), \ldots, A_q^{(m)}(z^q)) \qquad (23)$$

with lower triangular $\tilde{A} \in \mathrm{Mat}_{m+1,m+1}(K[z]), \tau$ denoting the matrix transpose. Clearly, by (22) we get

$$\det \tilde{A}(z) = (qz^{q-1})^{m(m+1)/2} P(z)^{m+1},$$

and this does not vanish if $zP(z) \neq 0$.

Next we want to apply the following algebraic independence criterion of Nishioka ([12], Theorem 4.2.1; see also [11], Corollary 2).

Theorem N2. *Let K be an algebraic number field. Suppose that $f_0, \ldots, f_m \in K[[z]]$ converge in some disk $U \subset \mathbb{D}$ about the origin, where they satisfy the matrix functional equation*

$$^\tau(f_0(z^q), \ldots, f_m(z^q)) = \mathcal{A}(z) \cdot {}^\tau(f_0(z), \ldots, f_m(z)) \qquad (24)$$

with $\mathcal{A} \in \mathrm{Mat}_{m+1,m+1}(K(z))$. If $\alpha \in \overline{\mathbb{Q}}^\times \cap U$ is such that none of the α^{q^j} ($j \in \mathbb{N}_0$) is a pole of $\mathcal{A}$, then the following inequality holds for the transcendence degrees:

$$\mathrm{trdeg}_\mathbb{Q} \mathbb{Q}(f_0(\alpha), \ldots, f_m(\alpha)) \geq \mathrm{trdeg}_{K(z)} K(z)((f_0(z), \ldots, f_m(z)).$$

To conclude the proof of Theorem 2, we use Theorem N2 with $f_\mu := A_q^{(\mu)}$ ($\mu = 0, \ldots, m$) for arbitrary $m \in \mathbb{N}_0$ and with $U = \mathbb{D}$. By (23), we may choose $\mathcal{A}(z) := \tilde{A}(z)^{-1}$ in (24), implying

$$(qz^{q-1})^{m(m+1)/2} P(z)^{m+1} \mathcal{A}(z) \in \mathrm{Mat}_{m+1,m+1}(K[z]).$$

According to Theorem N2 we know, for every $\alpha \in \overline{\mathbb{Q}}^\times \cap \mathbb{D}$ with $P(\alpha^{q^j}) \neq 0$ ($j = 0, 1, \ldots$), that

$$\mathrm{trdeg}_\mathbb{Q} \mathbb{Q}(A_q(\alpha), \ldots, A_q^{(m)}(\alpha)) \geq \mathrm{trdeg}_{K(z)} K(z)(A_q(z), \ldots, A_q^{(m)}(z)). \qquad (25)$$

Since all Taylor coefficients of $A_q(z)$ (and hence of $A_q'(z), A_q''(z), \ldots$) about the origin lie in the number field K, the transcendence degree on the right-hand side of (25) equals

$$\mathrm{trdeg}_{\mathbb{C}(z)} \mathbb{C}(z)(A_q(z), \ldots, A_q^{(m)}(z))$$

By the hypertranscendence hypothesis in Theorem 2, this last transcendence degree equals $m + 1$, whence also the transcendence degree on the left-hand side of (25). Since our $m \in \mathbb{N}_0$ was arbitrary, Theorem 2 is established.

Remark 4. Analogously to our procedure in Remark 2 of [2], we could use a result of Amou [1, Theorem 1] instead of Theorem N2 to prove

$$\mathrm{trdeg}_{\mathbb{Q}}\mathbb{Q}(\alpha, A_q(\alpha), \ldots, A_q^{(m)}(\alpha)) \geq [m/2] + 1$$

for any transcendental $\alpha \in \mathbb{D}$ and any $m \in \mathbb{N}_0$.

5 Three More Applications

We begin with probably the most prominent infinite product of type (6).

Corollary 2. *Let E_q be defined in $\mathbb{D}$ by*

$$E_q(z) := \prod_{k=0}^{\infty}(1 - z^{q^k}).$$

Then, for any $\alpha \in \overline{\mathbb{Q}}^{\times} \cap \mathbb{D}$, the numbers $E_q(\alpha), E_q'(\alpha), E_q''(\alpha), \ldots$ are algebraically independent.

Proof. We apply Theorem 2 with $P(z) = 1 - z$. By (i) $\Rightarrow$ (ii) of Criterion 1, the adjoint functional equation (5) is rationally unsolvable; hence, $E_q(z)$ is hypertranscendental, by Theorem 1. Our claim then follows from Theorem 2. □

Remark 5. Notice that the transcendence of $E_2(\alpha)$ for $\alpha \in \overline{\mathbb{Q}}^{\times} \cap \mathbb{D}$ has been proved first by Mahler [6] and rediscovered several times later. For example, in [4], the reader may find a transcendence proof for just the so-called Thue–Morse number $E_2(\frac{1}{2})$.

Next, we describe a consequence of Corollary 2 concerning so-called Lambert series, i.e., infinite series of type $\sum_{n=1}^{\infty} a_n z^n/(1 - z^n)$ with $(a_n) \in \mathbb{C}^{\mathbb{N}}$. On denoting in $\mathbb{D}$

$$L_q(z) := \sum_{n \in q^{\mathbb{N}_0}} n \frac{z^n}{1 - z^n}, \tag{26}$$

we conclude $L_q(z) = -zE_q'(z)/E_q(z)$ in $\mathbb{D}$. Hence, *for any $\alpha \in \overline{\mathbb{Q}}^{\times} \cap \mathbb{D}$, the numbers $L_q(\alpha), L_q'(\alpha), L_q''(\alpha), \ldots$ are algebraically independent.*

Quite similar to (26) is the Lambert series

$$M_q(z) := \sum_{n \in q^{\mathbb{N}_0}} \frac{z^n}{1 - z^n} \quad (z \in \mathbb{D})$$

arithmetically investigated already by Mahler [7]. Based on the functional equation $M_q(z) = M_q(z^q) + z/(1 - z)$, he established the algebraic independence of the numbers $M_q(\alpha), M_q'(\alpha), M_q''(\alpha), \ldots$ for every $\alpha \in \overline{\mathbb{Q}}^{\times} \cap \mathbb{D}$.

Corollary 3. *Let* $P(z) := 1 + z - z^2$ *be the common companion polynomial of the Fibonacci and Lucas recurrences. On denoting*

$$F_q(z) := \prod_{k=0}^{\infty} P(z^{q^k}) \quad (z \in \mathbb{D}),$$

the numbers $F_q(\alpha), F_q'(\alpha), F_q''(\alpha), \ldots$ *are algebraically independent for any* $\alpha \in \overline{\mathbb{Q}}^{\times} \cap \mathbb{D}$ *with* $\alpha^{q^j} \neq (1 - \sqrt{5})/2$ $(j = 0, 1, \ldots)$.

Proof. For the hypertranscendence of $F_q(z)$, we rely on (i) $\Rightarrow$ (ii) of Criterion 1 if $q \geq 3$. But in the case $q = 2$ ($= \deg P$), we use Theorem 1 plus a divisibility argument similar to the one used in the proof of Criterion 1 to show that the adjoint functional equation (5), i.e.,

$$w(z) - 2w(z^2) = \frac{z(1 - 2z)}{1 + z - z^2}, \tag{27}$$

is rationally unsolvable.

Assume, on the contrary, that (27) has a solution $w \in \mathbb{C}(z)$. On writing it as $w = u/v$ with coprime $u, v \in \mathbb{C}[z] \setminus \{0\}$, we see that the right-hand side of (27) tends to some finite nonzero value (in fact, to 2) as $z \to \infty$, whence $\deg u = \deg v$. From (27), we further obtain

$$z(1 - 2z)v(z)v(z^2) = (1 + z - z^2)(u(z)v(z^2) - 2u(z^2)v(z))$$

and herefrom $v(z^2)|(1 + z - z^2)u(z^2)v(z)$. Since $u(z), v(z)$ are assumed to be coprime, the same holds for $u(z^2), v(z^2)$, whence the last divisibility relation leads to

$$v(z^2)|(1 + z - z^2)v(z) \tag{28}$$

implying $\deg v \leq 2$.

If $\deg v = 0$, then w is constant contradicting (27). If $\deg v = 1$ we may write $v(z) = z + a$ with $a \neq 0, -1$ (compare (28)). Again from (28), we deduce

$$\frac{(1 + z - z^2)(z + a)}{z^2 + a} = b - z,$$

where b must be 1 (take $z = 0$); then, taking $z = 1$, we get the desired contradiction. If finally $\deg v = 2$, write $v(z) = z^2 + az + b$ with $b \neq 0$ (by (28)); again (28) gives $(1 + z - z^2)(z^2 + az + b) = -(z^4 + az^2 + b)$ leading to $b = 0$.

Since $(1 - \sqrt{5})/2$ is the only zero of P in $\mathbb{D}$, Corollary 3 follows from Theorem 2. $\square$

To formulate our last application, we need a bit more notation. Assume $\ell \in \mathbb{N}, \ell < q$, and suppose that $d_1, \ldots, d_\ell \in \mathbb{N}$ are distinct and less than q. Putting $\underline{d} := (d_1, \ldots, d_\ell)$ we denote, for $n \in \mathbb{N}_0$,

$$g_{q,\underline{d}}(n) := \begin{cases} 1 & \text{if in the } q\text{-adic expansion of } n \text{ no digit } d_1,\ldots,d_\ell \text{ occurs,} \\ 0 & \text{otherwise.} \end{cases}$$

Clearly, $g_{q,\underline{d}}(0) = 1$ and furthermore, one can easily see that the sequence $(g_{q,\underline{d}}(n))_{n=0,1,\ldots}$ cannot be ultimately periodic unless we have $\ell = q - 1$, where $g_{q,\underline{d}}(n) = 0$ for any $n \in \mathbb{N}$. Hence, we suppose $q \geq 3, \ell \leq q - 2$ from now on, and $\underline{d}$ as before. Then we consider the generating power series

$$G_{q,\underline{d}}(z) := \sum_{n=0}^{\infty} g_{q,\underline{d}}(n) z^n \tag{29}$$

of $(g_{q,\underline{d}}(n))$ having convergence radius 1. By a classical result of Szegő [13], no $G_{q,\underline{d}}$ can be analytically continued beyond the unit circle, whence all these $G_{q,\underline{d}}$ are transcendental functions.

For the arithmetical applications of more importance is the fact that these $G_{q,\underline{d}}$ are hypertranscendental. To see this, we write any $n \in \mathbb{N}_0$ uniquely as $n = d_0 + n'q$ with $d_0 \in \{0,\ldots,q-1\}, n' \in \mathbb{N}_0$. Thus, for any fixed $\underline{d}$ as above, we easily see

$$g_{q,\underline{d}}(d_0 + n'q) = g_{q,\underline{d}}(d_0) \cdot g_{q,\underline{d}}(n')$$

which says that the number theoretic functions $g_{q,\underline{d}}$ are strongly q-multiplicative in the sense of Toshimitsu [15], a property which leads to

$$G_{q,\underline{d}}(z) = \sum_{d_0=0}^{q-1} \sum_{n'=0}^{\infty} g_{q,\underline{d}}(d_0) g_{q,\underline{d}}(n') z^{d_0+n'q} = \left(\sum_{d_0=0}^{q-1} g_{q,\underline{d}}(d_0) z^{d_0} \right) \left(\sum_{n'=0}^{\infty} g_{q,\underline{d}}(n') (z^q)^{n'} \right).$$

This is a Mahler-type functional equation of the shape

$$G_{q,\underline{d}}(z) = P_{q,\underline{d}}(z) G_{q,\underline{d}}(z^q) \quad \text{with} \quad P_{q,\underline{d}}(z) := \sum_{d_0=0}^{q-1} g_{q,\underline{d}}(d_0) z^{d_0}, \tag{30}$$

and our assertion is the following.

Corollary 4. *Suppose $q, \ell \in \mathbb{N}, q \geq 3, \ell \leq q - 2, \underline{d} := (d_1,\ldots,d_\ell)$ as above, and let $G_{q,\underline{d}}(z)$ denote the generating function (29) of the sequence $(g_{q,\underline{d}}(n))$. Then, for any $\alpha \in \overline{\mathbb{Q}}^{\times} \cap \mathbb{D}$ satisfying $P_{q,\underline{d}}(\alpha^{q^j}) \neq 0$ for $j = 0,1,\ldots$, the numbers $G_{q,\underline{d}}(\alpha), G'_{q,\underline{d}}(\alpha), G''_{q,\underline{d}}(\alpha),\ldots$ are algebraically independent.*

Proof. Cleary, we have the product expansion

$$\prod_{k=0}^{\infty} P_{q,\underline{d}}(z^{q^k}) \tag{31}$$

of $G_{q,\underline{d}}(z)$, by (30). Moreover, since $g_{q,\underline{d}}(0) = 1$ and, for any $d_0 \in \{1,\ldots,q-1\}$ we have $g_{q,\underline{d}}(d_0) = 0$ if and only if d_0 coincides with one of the given $d_1,\ldots,d_\ell$, the implication (i) $\Rightarrow$ (ii) of Criterion 1 tells us that the functional equation adjoint to (31) is rationally unsolvable, whence (31) is hypertranscendental. $\qquad\square$

Remark 6. In the case $\ell = 1$, Mahler [8] proved the transcendence of all $G_{q,d}(\alpha), d \in \{1, \ldots, q-1\}$ if $q \geq 3$, where $\alpha \in \overline{\mathbb{Q}}^{\times} \cap \mathbb{D}$ is not a zero of $G_{q,d}$. By the way, the question of these zeros, completely settled by Mahler in the case $\ell = 1$ will not be discussed here for $\ell > 1$. Notice that, at the end of his paper, he mentioned the possibility to extend his transcendence result to the more general case of ℓ.

It should also be pointed out that generalizations of this Mahler result in a different direction have been meanwhile established. Namely, Loxton and van der Poorten [5], and later Nishioka and Nishioka [10] proved algebraic independence results for functions of type $G_{q,\underline{d}}(z)$ with fixed q and different (admissible) $\underline{d}$'s as above, and of their values at algebraic points but with no derivatives included.

References

1. M. Amou, Algebraic independence of the values of certain functions at a transcendental number. Acta Arith. **59**, 71–82 (1991)
2. P. Bundschuh, Transcendence and algebraic independence of series related to Stern's sequence. Int. J. Number Theory **8**, 361–376 (2012)
3. M. Coons, The transcendence of series related to Stern's diatomic sequence. Int. J. Number Theory **6**, 211–217 (2010)
4. M. Dekking, Transcendance du nombre de Thue-Morse. C. R. Acad. Sci. Paris Sér. A **285**, 157–160 (1977)
5. J.H. Loxton, A.J. van der Poorten, Arithmetic properties of the solutions of a class of functional equations. J. Reine Angew. Math. **330**, 159–172 (1982)
6. K. Mahler, Arithmetische Eigenschaften der Lösungen einer Klasse von Funktionalgleichungen. Math. Ann. **101**, 342–366 (1929)
7. K. Mahler, Arithmetische Eigenschaften einer Klasse transzendental-transzendenter Funktionen. Math. Z. **32**, 545–585 (1930)
8. K. Mahler, On the generating function of the integers with a missing digit. J. Indian Math. Soc. (N.S.) A **15**, 33–40 (1951)
9. Ke. Nishioka, A note on differentially algebraic solutions of first order linear difference equations. Aequationes Math. **27**, 32–48 (1984)
10. Ke. Nishioka, Ku. Nishioka, Algebraic independence of functions satisfying a certain system of functional equations. Funkcial. Ekvac. **37**, 195–209 (1994)
11. Ku. Nishioka, New approach in Mahler's method. J. Reine Angew. Math. **407**, 202–219 (1990)
12. Ku. Nishioka, *Mahler Functions and Transcendence*. LNM, vol 1631 (Springer, Berlin, 1996)
13. G. Szegő, Über Potenzreihen mit endlich vielen verschiedenen Koeffizienten. Sitz.ber. preuß. Akad. Wiss. Math.-Phys. Kl. **1922**, 88–91 (1922)
14. Y. Tachiya, Transcendence of certain infinite products. J. Number Theory **125**, 182–200 (2007)
15. T. Toshimitsu, Strongly q-additive functions and algebraic independence. Tokyo J. Math. **21**, 107–113 (1998)

Small Representations by Indefinite Ternary Quadratic Forms

J.B. Friedlander and H. Iwaniec

Dedicated in memory of Alf van der Poorten

Abstract We study nonhomogeneous Pythagorean triples, that is, solutions of the diophantine equation $x^2 + y^2 - z^2 = D$ where D is fixed and non-zero. We are particularly concerned, for D large and positive, with counting the solutions having x, y, z, all small relative to D.

1 Introduction

There has been significant development in the theory of the representation of integers by ternary quadratic forms using methods of automorphic forms and ideas from ergodic theory. A broad overview of the subject matter is presented in the ICM Proceedings [4]. In the case of definite forms there are excellent accounts of the results in [2,5].

The case of indefinite forms has received far less coverage, and it does not follow in quite the same way, although it is likely that some modifications in these technologies will succeed to complete the picture. Our goal in this note is to show that the general results [1] about Weyl sums can be used to treat some nice special cases of indefinite forms. A few are already established in [1].

For illustration, we consider the representations of the integer D by perhaps the simplest such form, namely,

J.B. Friedlander (✉)
Department of Mathematics, University of Toronto, Toronto, ON, Canada M5S 2E4
e-mail: frdlndr@math.toronto.edu

H. Iwaniec
Department of Mathematics, Rutgers University, New Brunswick, NJ 08903, USA

J.M. Borwein et al. (eds.), *Number Theory and Related Fields: In Memory of Alf van der Poorten*, Springer Proceedings in Mathematics & Statistics 43,
DOI 10.1007/978-1-4614-6642-0_7, © Springer Science+Business Media New York 2013

$$x^2 + y^2 - z^2.$$

For $D = 0$ the solutions are just the Pythagorean triples, whose explicit description is given in almost every introductory book. We shall take D to be positive. We think of D as being large and want to count the solutions of

$$D = x^2 + y^2 - z^2 \tag{1.1}$$

where to make it more of a challenge, we ask that all of the variables x, y, z be small. It is obvious that one of x, y needs to exceed $\sqrt{D/2}$ so that the best we can hope for is to find solutions with x, y, z all about this small. Actually, we shall be able to find solutions to (1.1) with $x^2 + y^2$ close to D so that z is considerably smaller than $\sqrt{D}$. It turns out to be convenient for counting purposes if we attach weights to the three variables, a natural arithmetic weight to the pair (x, y) and an analytic one to the variable z. With this in mind we are led to study the following sum:

$$\sum_m r(m^2 + D) F(m). \tag{1.2}$$

Here, we assume that D is a large positive square-free number with $D \equiv 5 \pmod 8$, so D is a fundamental discriminant. As usual $r(n)$ is the number of representations of n as the sum of two squares. The function $F(x)$ is a smooth function supported on

$$M < x < 2M. \tag{1.3}$$

One could consider an arbitrary binary quadratic form in place of the sum of two squares, but we shall take advantage of the simpler structure of $r(n)$. The problem would not be much of a challenge if D is fixed and M is large. We succeed in getting the asymptotic formula for (1.2) when M is considerably small in comparison to D.

Theorem 1.1. *Let $F(x)$ be supported on the segment* (1.3) *with*

$$|F^{(\alpha)}(x)| \leqslant M^{-\alpha}, \qquad \alpha = 0, \ldots, 6 \tag{1.4}$$

Then

$$\sum_m r(m^2 + D) F(m) = \frac{48}{\pi} L(1, \chi) \int F(x) dx + O\left(D^{\frac{1}{6} - \frac{1}{3996}} M^{\frac{2}{3}}\right) \tag{1.5}$$

where $\chi = \chi_D$ is the real character of conductor D and the implied constant is absolute.

Note that we save a fixed power in the error term, and the result is meaningful if

$$M > D^{\frac{1}{2} - \frac{1}{1332}}. \tag{1.6}$$

The range $M \asymp D^{\frac{1}{2}}$ represents an important barrier. We have assumed that $D \equiv 5 \pmod 8$ to simplify the presentation. The case of positive $D \equiv 1 \pmod 8$ requires only small changes of our arguments, but other fundamental discriminants positive and negative would force us to substantially augment the material in [1] which we are reluctant to do.

We note that in work essentially simultaneous to ours, Liu and Masri have derived [3] a beautiful asymptotic formula for the corresponding sum in the same critical range, wherein $D < 0$ and $r(n)$ is replaced by the divisor function $\tau(n)$. Indeed, the subject of sums of arithmetic functions along quadratic progressions has a long history as well as a number of deep recent developments; see the paper [6] of Templier and Tsimerman and also the references therein.

2 Preparations

For $n = m^2 + D$ with $D \equiv 5 \pmod 8$ we write

$$r(n) = 4|\{u, v > 0;\ u^2 + v^2 = n\}|$$

which implies that m is even and one of u, v is even, the other being odd. Our sum (1.2) is equal to

$$S = 4 \sum \sum \sum_{u^2 + v^2 = m^2 + D} F(m).$$

Note that

$$0 < u, v < 2M + \sqrt{D} = X \text{ say, and } u^2 + v^2 \asymp X^2.$$

Hence, only one of u, v could be small, say smaller than Y, with Y to be chosen later subject to $1 \leqslant Y \leqslant X$. Such terms contribute to S at most (note that $m \neq u, m \neq v$):

$$8 \sum_{0 < u \leqslant Y} \tau(|D - u^2|) \ll Y X^\varepsilon.$$

Let $G(t)$ be a smooth function whose graph is

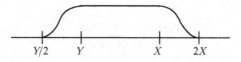

By the above considerations, taking u to be even, we obtain

$$S = S(Y) + O(Y X^\varepsilon) \tag{2.1}$$

with

$$S(Y) = 8 \sum_{\substack{u^2+v^2=m^2+D \\ u,m \text{ even}}} G(u)G(v)F(m). \tag{2.2}$$

We write the equation $u^2 + v^2 = m^2 + D$ in the form $D = u^2 + ac$ with $a = v - m$, $c = v + m$. Note that this change of variables v, m into a, c with c odd is a one-to-one correspondence in integers because of the parity assumptions and the choice of $D \equiv 5 \pmod 8$. Hence,

$$S(Y) = 8 \sum_{c \text{ odd}} \sum_{\substack{u^2 \equiv D \pmod c \\ u \text{ even}}} G(u) G\left(\frac{c}{2} + \frac{D-u^2}{2c}\right) F\left(\frac{c}{2} - \frac{D-u^2}{2c}\right).$$

Note that by the support of F and G, it follows that

$$M < c < 4X. \tag{2.3}$$

This information is redundant, but we keep it in mind.

Next we split the summation over u into residue classes modulo $2c$ where

$$u \equiv \bar{2}\, 2b \pmod{2c} \text{ with } b^2 \equiv D \pmod c,$$

and naturally, $\bar{2}$ is the multiplicative inverse of 2 modulo c. Having done this, we apply Poisson's formula, getting

$$\sum_u = \frac{1}{2c} \sum_{h \in \mathbb{Z}} e\left(\frac{\bar{2}hb}{c}\right) I_h(c),$$

where $I_h(c)$ is given by the Fourier integral:

$$I_h(c) = \int G(t) G\left(\frac{c}{2} + \frac{D-t^2}{2c}\right) F\left(\frac{c}{2} - \frac{D-t^2}{2c}\right) e\left(\frac{ht}{2c}\right) dt.$$

The resulting complete sum over b modulo c is the Weyl sum:

$$W_{\bar{2}h}(D;c) = \sum_{b^2 \equiv D \pmod c} e\left(\frac{\bar{2}hb}{c}\right);$$

see (1.1) of [1]. By (6.10) of [1], this is also equal to $\frac{1}{2}W_{2h}(D;4c)$. Hence,

$$S(Y) = \sum_h \sum_{c \text{ odd}} \frac{2}{c} W_{2h}(D; 4c) I_h(c) = \sum_h S_h(Y), \text{ say.} \tag{2.4}$$

3   The Main Term

The zero frequency $h = 0$ in (2.4) yields

$$S_0(Y) = 2 \sum_{c \text{ odd}} \frac{\rho(4c)}{c} I_0(c) \tag{3.1}$$

where $\rho(4c)$ is the number of solutions $b \pmod{4c}$ of $b^2 \equiv D \pmod{4c}$ (see Sect. 12.1 of [1]) and

$$I_0(c) = \int G(t) G\left(\frac{c}{2} + \frac{D - t^2}{2c}\right) F\left(\frac{c}{2} - \frac{D - t^2}{2c}\right) dt. \tag{3.2}$$

We are going to execute the summation over c before computing the integral. To this end we consider the Dirichlet generating series:

$$L(s) = \sum_{c \text{ odd}} \rho(4c) c^{-s}. \tag{3.3}$$

We have

$$L(s) = 4^s \sum_{c \equiv 0(4)} \rho(c) c^{-s}$$

because $\rho(c) = 0$ if $c \equiv 0 \pmod{8}$ for $D \equiv 5 \pmod{8}$. By (12.3) of [1], we get

$$L(s) = 2\zeta(s) L(s, \chi) / \zeta(2s). \tag{3.4}$$

Therefore, $L(s)$ is holomorphic in $\operatorname{Re} s \geqslant \frac{1}{2}$ except for a simple pole at $s = 1$ with

$$\operatorname{res}_{s=1} L(s) = 12\pi^{-2} L(1, \chi). \tag{3.5}$$

On the line $\operatorname{Re} s = \frac{1}{2}$, we get by the Burgess bound

$$L(s) \ll |s| D^{\frac{3}{16} + \varepsilon}. \tag{3.6}$$

Here we point out that the weaker exponent $\frac{1}{4}$ in place of $\frac{3}{16}$ would not be sufficient for our applications, so even for the evaluation of the main term, we need a subconvexity bound in the conductor aspect.

Using (3.1), (3.2), (3.5), and (3.6), we derive by a standard contour integration:

$$S_0(Y) = \frac{24}{\pi^2} L(1, \chi) I(G, F) + O(XM^{-\frac{1}{2}} D^{\frac{3}{16} + \varepsilon}), \tag{3.7}$$

where

$$I(G,F) = \iint G(t)\, G\left(\frac{x}{2} + \frac{D-t^2}{2x}\right) F\left(\frac{x}{2} - \frac{D-t^2}{2x}\right) \frac{dx}{x} dt.$$

Removing the cutoff functions G, we extend this integral to

$$I(F) = \iint_{\mathcal{B}(X)} F\left(\frac{x}{2} - \frac{D-t^2}{2x}\right) \frac{dx}{x} dt \tag{3.8}$$

up to an error term $O(Y \log X)$ by the same augments which gave (2.1). Here, the integration is over the box:

$$\mathcal{B}(X) = \{x, t;\ 0 < x < 4X,\ 0 < t < 2X\}. \tag{3.9}$$

Of course, there is one more restriction:

$$M < \frac{x}{2} - \frac{D-t^2}{2x} < 2M \tag{3.10}$$

given by the support of $F(y)$. Hence, we arrive at

$$S_0(Y) = \frac{24}{\pi^2} L(1,\chi) I(F) + O\big(Y \log D + X M^{-\frac{1}{2}} D^{\frac{3}{16}+\varepsilon}\big). \tag{3.11}$$

It remains to evaluate the complete integral $I(F)$. Changing the variable x into

$$y = \frac{x}{2} - \frac{D-t^2}{2x},$$

we get y with $M < y < 2M$ and compute as follows:

$$x^2 + t^2 - D = 2xy, \quad (x-y)^2 = D + y^2 - t^2 \geqslant 0,$$

and

$$x = y + \sqrt{D+y^2-t^2}, \quad x^{-1}dx = (D+y^2-t^2)^{-\frac{1}{2}} dy.$$

Hence,

$$I(F) = \int F(y)\left(\int (D+y^2-t^2)^{-\frac{1}{2}} dt\right) dy$$

$$= \left(\int F(y)\, dy\right) \int_0^1 (1-t^2)^{-\frac{1}{2}} dt = 2\pi \int F(y)\, dy = 2\pi \hat{F}(0).$$

Inserting this into (3.11) we conclude that

$$S_0(Y) = \frac{48}{\pi} L(1,\chi) \hat{F}(0) + O\big(Y \log D + X M^{-\frac{1}{2}} D^{\frac{3}{16}+\varepsilon}\big). \tag{3.12}$$

4   The Weyl Sums

Now we are going to estimate the sums $S_h(Y)$ for every $h \neq 0$ in (2.4). These are sums of Weyl sums over the moduli $4c$, and they constitute the core of our work. Recall that c ranges over odd integers in the segment (2.3) which restriction follows from the support of the functions G, F in the integral $I_h(c)$. Although this segment is quite well localized, we need to subdivide it into dyadic segments $C < c < 2C$ before we can appeal to the results of [1], specifically to Theorem 1.1. Of course, we subdivide by a smooth partition of unity, so the number of such partial sums (partition components) is bounded by $O(\log X)$. We need Theorem 1.1 of [1] for the test function:

$$f(x) = 2\frac{\phi(x)}{x}I_h(x) \tag{4.1}$$

where $\phi(y)$ is the partition component supported in $C < x < 2C$ with $\frac{1}{2}M < C < 4X$. On applying integration by parts zero or six times to

$$I_h(x) = \int G(t)G\left(\frac{x}{2} + \frac{D - t^2}{2x}\right)F\left(\frac{x}{2} - \frac{D - t^2}{2x}\right)e\left(\frac{ht}{2x}\right)dt \tag{4.2}$$

and then estimating trivially, we get the bound (for every $h > 0$, the case $h < 0$ begin similar):

$$f(x) \ll \frac{X}{C}\left(1 + \frac{hY}{C}\right)^{-6}. \tag{4.3}$$

If we differentiate (4.1) in x three times and integrate over t as before, then we get

$$x^3 f'''(x) \ll \frac{X}{C}\left(1 + \frac{hY}{C}\right)^{-3}. \tag{4.4}$$

Hence, our test function $f(x)$ satisfies the condition (1.3) of [1] up to a normalizing factor given in the upper bounds (4.3) and (4.4). Therefore, Theorem 1.1 of [1] yields

$$2\sum_{c \text{ odd}} \frac{\phi(c)}{c}W_{2h}(D; 4c)I_h(c) \ll \frac{X}{C}\left(1 + \frac{hY}{C}\right)^{-3}h^{\frac{1}{4}}\left(C + h\sqrt{D}\right)^{\frac{3}{4}}D^{\frac{1}{8} - \frac{1}{1331}}.$$

In fact Theorem 1.1 of [1] involves such a sum over all c divisible by a fixed integer q, so we derive the above estimate by using $q = 1$ and $q = 2$ and taking the difference. Adding the above estimate over the partition of unity, we get

$$S_h(Y) \ll \frac{X}{M}\left(1 + \frac{hY}{X}\right)^{-3}h^{\frac{1}{4}}\left(X + h\sqrt{D}\right)^{\frac{3}{4}}D^{\frac{1}{8} - \frac{1}{1331}}\log X.$$

for every $h > 0$. The same bound holds for negative frequencies with h replaced by $|h|$. Summing over all $h \neq 0$, we get

$$S(Y) - S_0(Y) \ll \frac{X}{M} \left(\frac{X}{Y}\right)^{\frac{5}{4}} \left(X + \frac{X}{Y}\sqrt{D}\right)^{\frac{3}{4}} D^{\frac{1}{8} - \frac{1}{1331}} \log X.$$

Now we apply (3.12) for the main term $S_0(Y)$ getting

$$S = \frac{48}{\pi} L(1, \chi) \hat{F}(0) +$$
$$O\left(\left(Y + XM^{-\frac{1}{2}} D^{\frac{3}{16}} + X^3 M^{-1} \left(Y^{-\frac{5}{4}} + Y^{-2} D^{\frac{3}{8}}\right) D^{\frac{1}{8} - \frac{1}{1331}}\right) X^\varepsilon\right), \tag{4.5}$$

provided $Y \leqslant X$. Recall that $X = 2M + \sqrt{D}$. We choose

$$Y = XM^{-\frac{1}{3}} D^{\frac{1}{6} - \frac{1}{3993}}$$

finding the error term in (4.5) to be bounded by the one in (1.5) provided that

$$M \geqslant D^{\frac{1}{2} - \frac{1}{1331}}. \tag{4.6}$$

However, (1.5) is trivial if (4.6) does not hold. This completes the proof of the theorem.

Acknowledgements JF was supported in part by NSERC grant A5123; HI was supported in part by NSF grant DMS-1101574.

References

1. W. Duke, J.B. Friedlander, H. Iwaniec, Weyl sums for quadratic roots, IMRN (2011)
2. W. Duke, R. Schulze-Pillot, Representation of integers by positive ternary quadratic forms and equidistribution of lattice points on ellipsoids. Invent. Math. **99**, 49–57 (1990)
3. S.-C. Liu, R. Masri, The average of the divisor function over values of a quadratic polynomial (2012, preprint)
4. P. Michel, A. Venkatesh, Equidistribution, L-functions and ergodic theory: on some problems of Yu. V. Linnik, in *Proc. ICM*, vol II, Zurich, 2006, pp. 421–457
5. P. Michel, A. Venkatesh, The subconvexity problems for GL_2. Pub. Math. IHES **111**, 171–271 (2010)
6. N. Templier, J. Tsimerman, Non-split sums of coefficients of $GL(2)$-automorphic forms. Israel J. Math. (2013, in press)

Congruences for Andrews' spt-Function Modulo 32760 and Extension of Atkin's Hecke-Type Partition Congruences

F.G. Garvan

Dedicated to the memory of A.J. (Alf) van der Poorten, my former teacher

Abstract New congruences are found for Andrews' smallest parts partition function spt(n). The generating function for spt(n) is related to the holomorphic part $\alpha(24z)$ of a certain weak Maass form $\mathcal{M}(z)$ of weight $\frac{3}{2}$. We show that a normalized form of the generating function for spt(n) is an eigenform modulo 72 for the Hecke operators $T(\ell^2)$ for primes $\ell > 3$ and an eigenform modulo p for $p = 5, 7$, or 13 provided that $(\ell, 6p) = 1$. The result for the modulus 3 was observed earlier by the author and considered by Ono and Folsom. Similar congruences for higher powers of p (namely 5^6, 7^4, and 13^2) occur for the coefficients of the function $\alpha(z)$. Analogous results for the partition function were found by Atkin in 1966. Our results depend on the recent result of Ono that $\mathcal{M}_\ell(z/24)$ is a weakly holomorphic modular form of weight $\frac{3}{2}$ for the full modular group where

$$\mathcal{M}_\ell(z) = \mathcal{M}(z)|T(\ell^2) - \left(\frac{3}{\ell}\right)(1 + \ell)\mathcal{M}(z).$$

Key words Andrews's spt-function • Weak Maass forms • Congruences • Partitions • Modular forms

Mathematics Subject Classification (2010): Primary 11P83, 11F33, 11F37; Secondary 11P82, 05A15, 05A17

The author was supported in part by NSA Grant H98230-09-1-0051. The first draft of this paper was written October 25, 2010.

F.G. Garvan (✉)
Department of Mathematics, University of Florida, Gainesville, FL 32611-8105, USA
e-mail: fgarvan@ufl.edu

J.M. Borwein et al. (eds.), *Number Theory and Related Fields: In Memory of Alf van der Poorten*, Springer Proceedings in Mathematics & Statistics 43, DOI 10.1007/978-1-4614-6642-0_8, © Springer Science+Business Media New York 2013

1 Introduction

Andrews [2] defined the function $\text{spt}(n)$ as the number of smallest parts in the partitions of n. For example, the partitions of 4 are

$$\overline{4}, \quad 3+\overline{1}, \quad \overline{2}+\overline{2}, \quad 2+\overline{1}+\overline{1}, \quad \overline{1}+\overline{1}+\overline{1}+\overline{1},$$

so that $\text{spt}(4) = 10$. Andrews related this function to the second-rank moment and proved some surprising congruences mod 5, 7, and 13. Rank and crank moments were introduced by Atkin and the author [3]. Bringmann [7] studied analytic, asymptotic, and congruence properties of the generating function for the second-rank moment as a quasi-weak Maass form. Further congruence properties of Andrews' spt-function were found by the author [12, 13], Folsom and Ono [11], and Ono [14]. In particular, Ono [14] proved that if $\left(\frac{1-24n}{\ell}\right) = 1$, then

$$\text{spt}(\ell^2 n - \tfrac{1}{24}(\ell^2 - 1)) \equiv 0 \pmod{\ell}, \tag{1.1}$$

for any prime $\ell \geq 5$. This amazing result was originally conjectured by the author[1]. Earlier special cases were observed by Tina Garrett (K.C. Garrett, October 18, 2007, private communication) and her students. Recently the author [13] has proved the following congruences for powers of 5, 7, and 13. For $a, b, c \geq 3$,

$$\text{spt}(5^a n + \delta_a) + 5\,\text{spt}(5^{a-2} n + \delta_{a-2}) \equiv 0 \pmod{5^{2a-3}}, \tag{1.2}$$

$$\text{spt}(7^b n + \lambda_b) + 7\,\text{spt}(7^{b-2} n + \lambda_{b-2}) \equiv 0 \pmod{7^{\lfloor \frac{1}{2}(3b-2)\rfloor}}, \tag{1.3}$$

$$\text{spt}(13^c n + \gamma_c) - 13\,\text{spt}(13^{c-2} n + \gamma_{c-2}) \equiv 0 \pmod{13^{c-1}}, \tag{1.4}$$

where δ_a, λ_b, and γ_c are the least nonnegative residues of the reciprocals of 24 mod 5^a, 7^b, and 13^c, respectively. Since this paper was first written, Ahlgren, Bringmann, and Lovejoy [1] have generalized Ono's congruence (1.1) to higher powers of ℓ. They have also obtained analogous results for other spt-like functions which were studied by Bringmann, Lovejoy, and Osburn [8, 9]. The congruences (1.2)–(1.4) do not follow from Ahlgren, Bringmann, and Lovejoy's results.

As in [13, 14], we define

$$\mathbf{a}(n) := 12\text{spt}(n) + (24n - 1)p(n), \tag{1.5}$$

for $n \geq 0$, and define

$$\alpha(z) := \sum_{n \geq 0} \mathbf{a}(n) q^{n - \frac{1}{24}}, \tag{1.6}$$

---

[1]The congruence (1.1) was first conjectured by the author in a colloquium given at the University of Newcastle, Australia on July 17, 2008.

where as usual $q = \exp(2\pi i z)$ and $\Im(z) > 0$. We note that spt$(0) = 0$ and $p(0) = 1$. Bringmann [7] showed that $\alpha(24z)$ is the holomorphic part of the weight $\frac{3}{2}$ weak Maass form $\mathcal{M}(z)$ on $\Gamma_0(576)$ with Nebentypus χ_{12}, where

$$\mathcal{M}(z) := \alpha(24z) - \frac{3i}{\pi\sqrt{2}} \int_{-\bar{z}}^{i\infty} \frac{\eta(24\tau)\,d\tau}{(-i(\tau+z))^{\frac{3}{2}}}, \tag{1.7}$$

$\eta(z) := q^{\frac{1}{24}} \prod_{n=1}^{\infty}(1 - q^n)$ is the Dedekind eta-function, the function $\alpha(z)$ is defined in (1.6), and

$$\chi_{12}(n) = \begin{cases} 1 & \text{if } n \equiv \pm 1 \pmod{12}, \\ -1 & \text{if } n \equiv \pm 5 \pmod{12}, \\ 0 & \text{otherwise.} \end{cases} \tag{1.8}$$

In Eq. (1.7) we have corrected sign errors in Bringmann's result [7, Eq. (1.6)]. Ono [14] showed that for $\ell \geq 5$ prime, the operator

$$T(\ell^2) - \chi_{12}(\ell)(1 + \ell) \tag{1.9}$$

annihilates the nonholomorphic part of $\mathcal{M}(z)$, and the function $\mathcal{M}_\ell(z/24)$ is a weakly holomorphic modular form of weight $\frac{3}{2}$ for the full modular group, where

$$\mathcal{M}_\ell(z) = \mathcal{M}(z)|T(\ell^2) - \chi_{12}(\ell)(1+\ell).\mathcal{M}(z) = \alpha(24z)|T(\ell^2) - \chi_{12}(\ell)(1+\ell)\alpha(24z). \tag{1.10}$$

In fact, he obtained

Theorem 1.1 (Ono [14]). *If $\ell \geq 5$ is prime, then the function*

$$\mathcal{M}_\ell(z/24)\,\eta(z)^{\ell^2} \tag{1.11}$$

is an entire modular form of weight $\frac{1}{2}(\ell^2 + 3)$ for the full modular group $\Gamma(1)$.

Applying this theorem Ono obtained

$$\mathcal{M}_\ell(z) \equiv 0 \pmod{\ell}. \tag{1.12}$$

The congruence (1.1) then follows easily.

Folsom and Ono [11] sketched the proof of the following

Theorem 1.2 (Folsom and Ono). *If $\ell \geq 5$ is prime, then*

$$\text{spt}(\ell^2 n - s_\ell) + \chi_{12}(\ell)\left(\frac{1 - 24n}{\ell}\right)\text{spt}(n) + \ell\,\text{spt}\left(\frac{n + s_\ell}{\ell^2}\right)$$

$$\equiv \chi_{12}(\ell)(1+\ell)\,\text{spt}(n) \pmod{3}, \tag{1.13}$$

where

$$s_\ell = \frac{1}{24}(\ell^2 - 1). \tag{1.14}$$

This result was observed earlier by the author. In this paper we prove a much stronger result.

Theorem 1.3. *(i) If $\ell \geq 5$ is prime, then*

$$\mathrm{spt}(\ell^2 n - s_\ell) + \chi_{12}(\ell)\left(\frac{1-24n}{\ell}\right)\mathrm{spt}(n) + \ell\,\mathrm{spt}\left(\frac{n+s_\ell}{\ell^2}\right) \equiv \chi_{12}(\ell)\,(1+\ell)\,\mathrm{spt}(n) \quad (\mathrm{mod}\ 72).$$
$$\tag{1.15}$$

(ii) If $\ell \geq 5$ is prime, $t = 5, 7,$ or 13, and $\ell \neq t$, then

$$\mathrm{spt}(\ell^2 n - s_\ell) + \chi_{12}(\ell)\left(\frac{1-24n}{\ell}\right)\mathrm{spt}(n) + \ell\,\mathrm{spt}\left(\frac{n+s_\ell}{\ell^2}\right) \equiv \chi_{12}(\ell)\,(1+\ell)\,\mathrm{spt}(n) \quad (\mathrm{mod}\ t).$$
$$\tag{1.16}$$

Remark 1.4. Theorem 1.3 implies Ramanujan-type congruences mod 8 and 9.

(i) If $\ell \geq 5$ is prime, $\ell \equiv 7 \pmod{8}$, $24n \equiv 1 \pmod{\ell}$, and $24n \not\equiv 1 \pmod{\ell^2}$, then

$$\mathrm{spt}(\ell^2 n - s_\ell) \equiv 0 \pmod{8}.$$

For example, with $\ell = 7$, we have

$$\mathrm{spt}(2401n + 243) \equiv \mathrm{spt}(2401n + 586) \equiv \mathrm{spt}(2401n + 929) \equiv \mathrm{spt}(2401n + 1272)$$
$$\equiv \mathrm{spt}(2401n + 1615) \equiv \mathrm{spt}(2401n + 1958) \equiv 0 \pmod{8}.$$

In particular,

$$\mathrm{spt}(243) = 1670037549780488 = 2^3 \cdot 7^2 \cdot 17 \cdot 489707 \cdot 5124131.$$

(ii) If $\ell \geq 5$ is prime, $\ell \equiv 7 \pmod{9}$, and $\left(\frac{1-24n}{\ell}\right) = -1$, then

$$\mathrm{spt}(\ell^2 n - s_\ell) \equiv 0 \pmod{9}.$$

For example, with $\ell = 7$, we have

$$\mathrm{spt}(343n + 47) \equiv \mathrm{spt}(343n + 145) \equiv \mathrm{spt}(343n + 194) \equiv 0 \pmod{9}.$$

(iii) If $\ell \geq 5$ is prime, $\ell \equiv 8 \pmod{9}$, $24n \equiv 1 \pmod{\ell}$, and $24n \not\equiv 1 \pmod{\ell^2}$, then

$$\mathrm{spt}(\ell^2 n - s_\ell) \equiv 0 \pmod{9}.$$

For example, with $\ell = 17$, we have

$$\mathrm{spt}(4913n + 1433) \equiv 0 \pmod{9}, \quad \text{if } n \not\equiv 16 \pmod{17}.$$

Of course Theorem 1.3 also implies the

Corollary 1.5. *If ℓ is prime and $\ell \notin \{2,3,5,7,13\}$, then*

$$\mathrm{spt}(\ell^2 n - s_\ell) + \chi_{12}(\ell)\left(\frac{1-24n}{\ell}\right)\mathrm{spt}(n) + \ell\,\mathrm{spt}\left(\frac{n+s_\ell}{\ell^2}\right)$$

$$\equiv \chi_{12}(\ell)(1+\ell)\,\mathrm{spt}(n) \pmod{32760}. \qquad (1.17)$$

This congruence modulo $32760 = 2^3 \cdot 3^2 \cdot 5 \cdot 7 \cdot 13$ is the congruence referred to in the title of this paper.

In 1966, Atkin [5] found a similar congruence for the partition function.

Theorem 1.6 (Atkin). *Let $t = 5$, 7, or 13, and $c = 6$, 4, or 2, respectively. Suppose $\ell \geq 5$ is prime and $\ell \neq t$. If $\left(\frac{1-24n}{t}\right) = -1$, then*

$$\ell^3 p(\ell^2 n - s_\ell) + \ell\chi_{12}(\ell)\left(\frac{1-24n}{\ell}\right)p(n) + p\left(\frac{n+s_\ell}{\ell^2}\right) \equiv \gamma_t\, p(n) \pmod{t^c},$$

$$(1.18)$$

where γ_t is an integral constant independent of n.

We find that there is a corresponding result for the function $\mathbf{a}(n)$ defined in (1.5).

Theorem 1.7. *Let $t = 5$, 7, or 13, and $c = 6$, 4, or 2, respectively. Suppose $\ell \geq 5$ is prime and $\ell \neq t$. If $\left(\frac{1-24n}{t}\right) = -1$, then*

$$\mathbf{a}(\ell^2 n - s_\ell) + \chi_{12}(\ell)\left(\frac{1-24n}{\ell}\right)\mathbf{a}(n) + \ell\,\mathbf{a}\left(\frac{n+s_\ell}{\ell^2}\right) \equiv \chi_{12}(\ell)(1+\ell)\mathbf{a}(n) \pmod{t^c}.$$

$$(1.19)$$

In Sect. 2 we prove Theorem 1.3. The method involves reviewing the action of weight $-\frac{1}{2}$ Hecke operators $T(\ell^2)$ on the function $\eta(z)^{-1}$ and doing a careful study of the action of weight $\frac{3}{2}$ Hecke operators on the function $\frac{d}{dz}\eta(z)^{-1}$ modulo 5, 7, 13, 27, and 32. In Sect. 3 we prove Theorem 1.7. This involves extending Atkin's method [5] of modular functions to weight two modular forms on $\Gamma_0(t)$ for $t = 5, 7$, and 13. The proofs of both Theorems 1.3 and 1.7 depend on Ono's Theorem 1.1.

2 Proof of Theorem 1.3

In this section we prove Theorem 1.3. Atkin [4] showed essentially that applying certain weight $-\frac{1}{2}$ Hecke operators $T(\ell^2)$ to the function $\eta(z)^{-1}$ produces a function with the same multiplier system as $\eta(z)^{-1}$, and thus $\eta(z)$ times this function is a certain polynomial (depending on ℓ) of Klein's modular invariant $j(z)$. We review Ono's [15] recent explicit form for these polynomials. Although our proof does not depend on it, Ono's result is quite useful for computational purposes. The action of the corresponding weight $\frac{3}{2}$ Hecke operators on $\frac{d}{dz}\eta(z)^{-1}$ can be given in terms of the same polynomials. See Theorem 2.3. To finish the proof of the theorem, we need to make a careful study of the action of these operators modulo 5, 7, 13, 27, and 32.

For $\ell \geq 5$ prime, we define

$$Z_\ell(z) = \sum_{n=-s_\ell}^{\infty} \left(\ell^3 \, p(\ell^2 n - s_\ell) + \ell \chi_{12}(\ell) \left(\frac{1 - 24n}{\ell} \right) p(n) + p \left(\frac{n + s_\ell}{\ell^2} \right) \right) q^{n - \frac{1}{24}}.$$

(2.1)

Proposition 2.1 (Atkin [5]). *The function $Z_\ell(z)\,\eta(z)$ is a modular function on the full modular group $\Gamma(1)$.*

It follows that $Z_\ell(z)\,\eta(z)$ is a polynomial in $j(z)$, where $j(z)$ is Klein's modular invariant

$$j(z) := \frac{E_4(z)^3}{\Delta(z)} = q^{-1} + 744 + 196884q + \cdots,$$

(2.2)

$E_2(z)$, $E_4(z)$, $E_6(z)$ are the usual Eisenstein series

$$E_2(z) := 1 - 24 \sum_{n=1}^{\infty} \sigma_1(n)q^n, \qquad E_4(z) := 1 + 240 \sum_{n=1}^{\infty} \sigma_3(n)q^n,$$
$$E_6(z) := 1 - 504 \sum_{n=1}^{\infty} \sigma_5(n)q^n,$$

(2.3)

$\sigma_k(n) = \sum_{d|n} q^k$, and $\Delta(z)$ is Ramanujan's function

$$\Delta(z) := \eta(z)^{24} = q \prod_{n=1}^{\infty} (1 - q^n)^{24}.$$

(2.4)

In a recent paper, Ono [15] has found a nice formula for this polynomial. We define

$$E(q) := \prod_{n=1}^{\infty} (1 - q^n) = q^{-\frac{1}{24}} \eta(z),$$

(2.5)

and a sequence of polynomials $A_m(x) \in \mathbb{Z}[x]$, by

$$\sum_{m=0}^{\infty} A_m(x)q^m = E(q) \frac{E_4(z)^2 E_6(z)}{\Delta(z)} \frac{1}{j(z) - x}$$
$$= 1 + (x - 745)q + (x^2 - 1489x + 160511)q^2 + \cdots.$$

(2.6)

Theorem 2.2 (Ono [15]). *For $\ell \geq 5$ prime,*

$$Z_\ell(z)\,\eta(z) = \ell \chi_{12}(\ell) + A_{s_\ell}(j(z)),$$

(2.7)

where $Z_\ell(z)$ is given in (2.1), and s_ℓ is given in (1.14).

We define a sequence of polynomials $C_\ell(x) \in \mathbb{Z}[x]$ by

$$C_\ell(x) := \ell\chi_{12}(\ell) + A_{s_\ell}(x), \tag{2.8}$$

$$= \sum_{n=0}^{s_\ell} c_{n,\ell}x^n,$$

so that

$$Z_\ell(z)\,\eta(z) = C_\ell(j(z)). \tag{2.9}$$

We define

$$d(n) := (24n - 1)\,p(n), \tag{2.10}$$

so that

$$\sum_{n=0}^{\infty} d(n)q^{24n-1} = q\frac{d}{dq}\frac{1}{\eta(24z)} = -\frac{E_2(24z)}{\eta(24z)}, \tag{2.11}$$

and

$$\mathbf{a}(n) = 12\mathrm{spt}(n) + d(n). \tag{2.12}$$

For $\ell \geq 5$ prime, we define

$$\Xi_\ell(z) = \sum_{n=-s_\ell}^{\infty} \left(d(\ell^2 n - s_\ell) + \chi_{12}(\ell)\left(\left(\frac{1-24n}{\ell}\right) - 1 - \ell\right)d(n) + \ell d\left(\frac{n+s_\ell}{\ell^2}\right)\right)q^{n-\frac{1}{24}}. \tag{2.13}$$

We then have the following analogue of Theorem 2.2.

Theorem 2.3. *For $\ell \geq 5$ prime, we have*

$$\ell\Xi_\ell(z)\,\eta(z)\,\Delta(z)^{s_\ell} = -\sum_{n=0}^{s_\ell} c_{n,\ell}E_4(z)^{3n-1}\Delta(z)^{s_\ell-n}\left(24nE_6(z) + E_4(z)E_2(z)\right) \tag{2.14}$$

$$+ \chi_{12}(\ell)\ell(1+\ell)E_2(z)\Delta(z)^{s_\ell},$$

where the coefficients $c_{n,\ell}$ are defined by (2.6) and (2.8).

Proof. Suppose $\ell \geq 5$ is prime. In Eq. (2.9) we replace z by $24z$, apply the operator $q\frac{d}{dq}$, and replace z by $\frac{1}{24}z$ to obtain

$$\ell\Xi_\ell(z)\,\eta(z) = 24C'_\ell(j(z))\,q\frac{d}{dq}(j(z)) + \left(\chi_{12}(\ell)\ell(1+\ell) - C_\ell(j(z))\right)E_2(z). \tag{2.15}$$

The result then follows easily from the identities

$$j(z)\Delta(z) = E_4(z)^3,\quad q\frac{d}{dq}(\Delta(z)) = \Delta(z)E_2(z),\text{ and } q\frac{d}{dq}(j(z))\Delta(z) = -E_4(z)^2E_6(z), \tag{2.16}$$

which we leave as an easy exercise. $\qquad\square$

We are now ready to prove Theorem 1.3. A standard calculation gives the following congruences:

$$E_4(z)^3 - 720\Delta(z) \equiv 1 \pmod{65520}, \text{ and } E_2(z) \equiv E_4(z)^2 E_6(z) \pmod{65520}.$$
(2.17)

We now use (2.17) to reduce (2.14) modulo 65520.

$$\ell \Xi_\ell(z) \eta(z) \Delta(z)^{s_\ell}$$

$$\equiv -\sum_{n=0}^{s_\ell} c_{n,\ell} E_4(z)^{3n-1} \Delta(z)^{s_\ell-n} \left(24n E_6(z)(E_4(z)^3 - 720\Delta(z)) + E_4(z)^3 E_6(z)\right)$$

$$+ \chi_{12}(\ell)\ell(1+\ell) E_4(z)^2 E_6(z) \Delta(z)^{s_\ell} \pmod{65520}$$

$$\equiv -\sum_{n=0}^{s_\ell} (24n+1)c_{n,\ell} E_4(z)^{3n+2} E_6(z) \Delta(z)^{s_\ell-n}$$

$$+ \sum_{n=0}^{s_\ell} 720 \cdot 24n c_{n,\ell} E_4(z)^{3n-1} E_6(z) \Delta(z)^{s_\ell-n+1}$$

$$+ \chi_{12}(\ell)\ell(1+\ell) E_4(z)^2 E_6(z) \Delta(z)^{s_\ell} \pmod{65520}$$

$$\equiv (720 \cdot 24 c_{1,\ell} - c_{0,\ell} + \chi_{12}(\ell)\ell(1+\ell)) E_4(z)^2 E_6(z) \Delta(z)^{s_\ell}$$

$$+ \sum_{n=1}^{s_\ell-1} \left(720 \cdot 24(n+1)c_{n+1,\ell} - (24n+1)c_{n,\ell}\right) E_4(z)^{3n+2} E_6(z) \Delta(z)^{s_\ell-n}$$

$$- (24s_\ell+1)c_{s_\ell,\ell} E_4(z)^{3s_\ell+2} E_6(z) \pmod{65520}.$$
(2.18)

We define

$$\mathcal{A}_\ell(z) := \sum_{n=-s_\ell}^{\infty} \left(\mathbf{a}(\ell^2 n - s_\ell) + \chi_{12}(\ell)\left(\left(\frac{1-24n}{\ell}\right) - 1 - \ell\right)\mathbf{a}(n) \right. $$

$$\left. + \ell \mathbf{a}\left(\frac{n+s_\ell}{\ell^2}\right)\right) q^{n-\frac{1}{24}}$$
(2.19)

and

$$\mathcal{S}_\ell(z) := \sum_{n=1}^{\infty} \left(\text{spt}(\ell^2 n - s_\ell) + \chi_{12}(\ell)\left(\left(\frac{1-24n}{\ell}\right) - 1 - \ell\right)\text{spt}(n) \right.$$

$$\left. + \ell \text{spt}\left(\frac{n+s_\ell}{\ell^2}\right)\right) q^{n-\frac{1}{24}},$$
(2.20)

so that

$$\mathcal{A}_\ell(z) = 12\mathcal{S}_\ell(z) + \Xi_\ell(z) = \mathcal{M}_\ell(z/24).$$
(2.21)

By Theorem 1.1 and Eq. (1.10) we see that the function

$$\ell \mathcal{A}_\ell(z)\, \eta(z)\Delta(z)^{s_\ell} \in M_{\frac{1}{2}(\ell^2+3)}(\Gamma(1)), \tag{2.22}$$

the space of entire modular forms of weight $\frac{1}{2}(\ell^2+3)$ on $\Gamma(1)$. Since $\frac{1}{2}(\ell^2+3) = 2+12s_\ell$, the set

$$\{E_4(z)^{3n-1}E_6(z)\Delta(z)^{s_\ell-n} : 1 \leq n \leq s_\ell\} \tag{2.23}$$

is a basis. Hence there are integers $b_{n,\ell}$ $(1 \leq n \leq s_\ell)$ such that

$$\mathcal{A}_\ell(z)\,\eta(z)\Delta(z)^{s_\ell} = \sum_{n=1}^{s_\ell} b_{n,\ell}E_4(z)^{3n-1}E_6(z)\Delta(z)^{s_\ell-n}. \tag{2.24}$$

Using (2.17) we find that

$$\mathcal{A}_\ell(z)\,\eta(z)\Delta(z)^{s_\ell} \equiv -720b_{1,\ell}E_4(z)^2 E_6(z)\Delta(z)^{s_\ell}$$

$$+ \sum_{n=1}^{s_\ell-1}(b_{n,\ell} - 720b_{n+1,\ell})E_4(z)^{3n+2}E_6(z)\Delta(z)^{s_\ell-n}$$

$$+ b_{s_\ell,\ell}E_4(z)^{3s_\ell+2}E_6(z) \pmod{65520}. \tag{2.25}$$

By (2.18), (2.21), and (2.25) we deduce that there are integers $a_{n,\ell}$ $(0 \leq n \leq s_\ell)$ such that

$$12\,\ell\, \mathcal{S}_\ell(z)\,\eta(z)\Delta(z)^{s_\ell} \equiv \sum_{n=0}^{s_\ell} a_{n,\ell}E_4(z)^{3n+2}E_6(z)\Delta(z)^{s_\ell-n} \pmod{65520}. \tag{2.26}$$

It follows that

$$12\,\ell\, \mathcal{S}_\ell(z) \equiv 0 \pmod{65520}, \tag{2.27}$$

since

$$\mathrm{ord}_{i\infty}\left(12\,\ell\, \mathcal{S}_\ell(z)\,\eta(z)\Delta(z)^{s_\ell}\right) = s_\ell + 1,$$

$$0 \leq \mathrm{ord}_{i\infty}\left(E_4(z)^{3n+2}E_6(z)\Delta(z)^{s_\ell-n}\right) \leq s_\ell,$$

$$E_4(z)^{3n+2}E_6(z)\Delta(z)^{s_\ell-n} = q^{s_\ell-n} + \cdots, \tag{2.28}$$

for $0 \leq n \leq s_\ell$ and all functions have integral coefficients. (2.27) implies Part (ii) of Theorem 1.3. To prove Part (i), we need to work a little harder. We note that the congruence (2.27) does imply

$$\mathcal{S}_\ell(z) \equiv 0 \pmod{12}. \tag{2.29}$$

We need to show this congruence actually holds modulo 72.

First, we show the congruence holds modulo 8 by studying $\Xi_\ell(z)$ modulo 32. We need the congruences,

$$E_2(z) \equiv E_4(z) E_6(z) + 16\Delta(z) \quad (\text{mod } 32) \text{ and } E_4(z)^2 \equiv 1 \quad (\text{mod } 32), \quad (2.30)$$

which are routine to prove. We proceed as in the proof of (2.18) to find that

$$\ell \Xi_\ell(z) \eta(z) \Delta(z)^{s_\ell}$$
$$\equiv \left(\chi_{12}(\ell)\ell(1+\ell) - c_{0,\ell} - 16c_{1,\ell}\right) E_2(z) \Delta(z)^{s_\ell}$$
$$- \sum_{n=1}^{s_\ell-1} \left((24n+1)c_{n,\ell} + 16c_{n+1,\ell}\right) E_4(z)^{3n-1} E_6(z) \Delta(z)^{s_\ell-n}$$
$$- (24s_\ell + 1)c_{s_\ell,\ell} E_4(z)^{3s_\ell-1} E_6(z) \quad (\text{mod } 32). \quad (2.31)$$

By (2.31), (2.21), and (2.24) we deduce that there are integers $a'_{n,\ell}$ ($0 \leq n \leq s_\ell$) such that

$$12\ell S_\ell(z) \eta(z) \Delta(z)^{s_\ell} \equiv \sum_{n=1}^{s_\ell} a'_{n,\ell} E_4(z)^{3n-1} E_6(z) \Delta(z)^{s_\ell-n}$$
$$+ a'_{0,\ell} E_2(z) \Delta(z)^{s_\ell} \quad (\text{mod } 32). \quad (2.32)$$

Arguing as before, it follows that

$$12S_\ell(z) \equiv 0 \quad (\text{mod } 32), \quad \text{and} \quad S_\ell(z) \equiv 0 \quad (\text{mod } 8). \quad (2.33)$$

To complete the proof, we need to study $\Xi_\ell(z)$ modulo 27. We need the congruences,

$$E_2(z) \equiv E_4(z)^5 + 18\Delta(z) \quad (\text{mod } 27) \quad \text{and} \quad E_6(z) \equiv E_4(z)^6 \quad (\text{mod } 27),$$
$$(2.34)$$

which are routine to prove. We proceed as in the proof of (2.18) and (2.31) to find that

$$\ell \Xi_\ell(z) \eta(z) \Delta(z)^{s_\ell}$$
$$\equiv \left(\chi_{12}(\ell)\ell(1+\ell) - c_{0,\ell} - 18c_{1,\ell}\right) E_2(z) \Delta(z)^{s_\ell}$$
$$- \sum_{n=1}^{s_\ell-1} \left((24n+1)c_{n,\ell} + 18c_{n+1,\ell}\right) E_4(z)^{3n-1} E_6(z) \Delta(z)^{s_\ell-n}$$
$$- (24s_\ell + 1)c_{s_\ell,\ell} E_4(z)^{3s_\ell-1} E_6(z) \quad (\text{mod } 27). \quad (2.35)$$

By (2.35), (2.21), and (2.24) we deduce that there are integers $a''_{n,\ell}$ ($0 \leq n \leq s_\ell$) such that

$$12\ell\,\mathcal{S}_\ell(z)\,\eta(z)\,\Delta(z)^{s_\ell} \equiv \sum_{n=1}^{s_\ell} a_{n,\ell}'' E_4(z)^{3n-1} E_6(z)\,\Delta(z)^{s_\ell-n}$$

$$+a_{0,\ell}'' E_2(z)\,\Delta(z)^{s_\ell} \pmod{27}. \tag{2.36}$$

Arguing as before, it follows that

$$12\mathcal{S}_\ell(z) \equiv 0 \pmod{27} \quad \text{and} \quad \mathcal{S}_\ell(z) \equiv 0 \pmod{9}. \tag{2.37}$$

The congruences (2.33) and (2.37) give (1.15) and this completes the proof of Theorem 1.3.

3   Proof of Theorem 1.7

In this section we prove Theorem 1.7. Atkin [5] proved Theorem 1.6 by constructing certain special modular functions on $\Gamma_0(t)$ and $\Gamma_0(t^2)$ for $t = 5, 7$, and 13. We attack the problem by extending Atkin's results to the corresponding weight 2 case.

Let $GL_2^+(\mathbb{R})$ denote the group of all real 2×2 matrices with positive determinant. $GL_2^+(\mathbb{R})$ acts on the complex upper half plane $\mathcal{H}$ by linear fractional transformations. We define the slash operator for modular forms of integer weight. Let $k \in \mathbb{Z}$.
For a function $f : \mathcal{H} \longrightarrow \mathbb{C}$ and $L = \begin{pmatrix} a & b \\ c & d \end{pmatrix} \in GL_2^+(\mathbb{R})$, we define

$$f(z) \mid_k L = f \mid_k L = f \mid L = (\det L)^{\frac{k}{2}} (cz+d)^{-k} f(Lz). \tag{3.1}$$

Let $\Gamma' \subset \Gamma(1)$ (a subgroup of finite index). We say $f(z)$ is a *weakly holomorphic modular form* of weight k on Γ' if $f(z)$ is holomorphic on the upper half plane $\mathcal{H}$, $f(z) \mid_k L = f(z)$ for all L in Γ', and $f(z)$ has at most polar singularities in the local variables at the cusps of the fundamental region of Γ'. We say $f(z)$ is a *weakly holomorphic modular function* if it is a weakly holomorphic modular form of weight 0. We say $f(z)$ is an *entire modular form* of weight k on Γ' if it is a *weakly holomorphic modular form* that is holomorphic at the cusps of the fundamental region of Γ'. We denote the space of entire modular forms of weight k on Γ' by $M_k(\Gamma')$.

Suppose that $t \geq 5$ is prime. We need

$$W_t = W = \begin{pmatrix} 0 & -1 \\ t & 0 \end{pmatrix}, \quad R = \begin{pmatrix} 1 & 0 \\ -1 & 1 \end{pmatrix}, \quad V_a = \begin{pmatrix} a & \lambda \\ t & a' \end{pmatrix}, \quad B_t = \begin{pmatrix} t & 0 \\ 0 & 1 \end{pmatrix},$$

$$T_{b,t} = \begin{pmatrix} 1 & b \\ 0 & 1 \end{pmatrix}, \quad Q_{b,t} = \begin{pmatrix} 1/t & b/t \\ 0 & 1 \end{pmatrix},$$

where for $1 \le a \le t-1$, a' is uniquely defined by $1 \le a' \le t-1$, and $a'a - \lambda t = 1$. We have

$$B_t R^{at} = -W_t V_a T_{-a'/t} \tag{3.2}$$

$$R^{at} W_t = W_{t^2} Q_{a,t}. \tag{3.3}$$

We define

$$\Phi_t(z) = \Phi(z) := \frac{\eta(z)}{\eta(t^2 z)}. \tag{3.4}$$

Then $\Phi_t(z)$ is a modular function on $\Gamma_0(t^2)$,

$$\Phi_t(z) \mid W_{t^2} = t\,\Phi_t(z)^{-1} \qquad ([5, (24)]), \tag{3.5}$$

and

$$\Phi_t(z) \mid R^{at} = \sqrt{t}\, e^{\pi i(t-1)/4}\, e^{-\pi i a't/12} \left(\frac{a'}{t}\right) \frac{\eta(z)}{\eta(z-a'/t)} \qquad ([5, (25)]). \tag{3.6}$$

Although $E_2(z)$ is not a modular form, it well-known that

$$\mathcal{E}_{2,t}(z) := \frac{1}{t-1}\,(t\,E_2(tz) - E_2(z)) \tag{3.7}$$

is an entire modular form of weight 2 on $\Gamma_0(t)$ and

$$\mathcal{E}_{2,t}(z) \mid W_t = -\mathcal{E}_{2,t}(z). \tag{3.8}$$

Proposition 3.1. *Suppose $t \ge 5$ is prime, $K(z)$ is a weakly holomorphic modular function on $\Gamma_0(t)$, and*

$$S(z) = \mathcal{E}_{2,t}(tz) K^*(tz) \frac{\eta(z)}{\eta(t^2 z)} - \chi_{12}(t)\,\eta(z) \sum_{n=m}^{\infty} \left(\frac{1-24n}{t}\right) \beta_t(n) q^{n-\frac{1}{24}}, \tag{3.9}$$

where

$$\mathcal{E}_{2,t}(z) \frac{K(z)}{\eta(z)} = \sum_{n=m}^{\infty} \beta_t(n) q^{n-\frac{1}{24}}, \tag{3.10}$$

and

$$K^*(z) = K(z) \mid W_t. \tag{3.11}$$

Then $S(z)$ is a weakly holomorphic modular form of weight 2 on $\Gamma_0(t)$.

Proof. Suppose $t \ge 5$ is prime and $K(z)$, $K^*(z)$, $S(z)$ are defined as in the statement of the proposition. The function

$$H(z) := \mathcal{E}_{2,t}(tz)\,\Phi_t(z) K^*(tz) \tag{3.12}$$

is a weakly holomorphic modular form of weight 2 on $\Gamma_0(t^2)$. As in [5, Lemma1] the function

$$S_1(z) := \sum_{a=0}^{t-1} H(z) \mid R^{at} \tag{3.13}$$

is a weakly holomorphic modular form of weight 2 on $\Gamma_0(t)$. Utilizing (3.2), (3.6), and (3.8) we find that

$$S_1(z) = \mathcal{E}_{2,t}(tz) K^*(tz) \frac{\eta(z)}{\eta(t^2 z)} - \frac{1}{\sqrt{t}} e^{\pi i (t-1)/4} \eta(z)$$

$$\times \sum_{a'=1}^{t-1} e^{-\pi i a' t/12} \left(\frac{a'}{t}\right) \mathcal{E}_{2,t}(z - a'/t) \frac{K(z - a'/t)}{\eta(z - a'/t)} \tag{3.14}$$

Here we have also used the fact that

$$\mathcal{E}_{2,t}(tz) \mid R^{at} = -\frac{1}{t} \mathcal{E}_{2,t}(z - a'/t), \tag{3.15}$$

where $aa' \equiv 1 \pmod{t}$. This fact follows from (3.2) and (3.8) After using the evaluation of a quadratic Gauss sum [6, (1.7)], we have

$$S_1(z) = S(z).$$

This gives the result. □

We illustrate Proposition 3.1 with two examples:

$$S(z) = \mathcal{E}_{2,5}(z) \left(\frac{\eta(z)}{\eta(5z)}\right)^6 \qquad (K(z) = 1 \text{ and } t = 5) \tag{3.16}$$

and

$$S(z) = \mathcal{E}_{2,7}(z) \left(\left(\frac{\eta(z)}{\eta(7z)}\right)^8 + 3\left(\frac{\eta(z)}{\eta(7z)}\right)^4\right) \qquad (K(z) = 1 \text{ and } t = 7). \tag{3.17}$$

Corollary 3.2. *Suppose $t \geq 5$ is prime and $S(z)$, $K(z)$ and the sequence $\beta_t(n)$ are defined as in Proposition 3.1. Then*

$$S(z) \mid W_t = -\eta(tz) \sum_{tn - s_t \geq m} \beta_t(tn - s_t) q^{n - \frac{t}{24}}. \tag{3.18}$$

Proof. The result follows easily from (3.3), (3.5), (3.8), and (3.13). □

We illustrate the corollary by applying W to both sides of the Eqs. (3.16)–(3.17):

$$\sum_{n=1}^{\infty} \beta_5(5n - 1) q^{n - \frac{5}{24}} = 5^3 \frac{\mathcal{E}_{2,5}(z)}{\eta(5z)} \left(\frac{\eta(5z)}{\eta(z)}\right)^6 \qquad (K(z) = 1 \text{ and } t = 5) \tag{3.19}$$

and

$$\sum_{n=1}^{\infty} \beta_7(7n-2)q^{n-\frac{7}{24}} = 7^2 \frac{\mathcal{E}_{2,7}(z)}{\eta(7z)} \left(3 \left(\frac{\eta(7z)}{\eta(z)}\right)^4 + 7^2 \left(\frac{\eta(7z)}{\eta(z)}\right)^8\right)$$

$$(K(z) = 1 \text{ and } t = 7). \tag{3.20}$$

For t and $K(z)$ as in Proposition 3.1, we define

$$\Psi_{t,K}(z) = \Psi_t(z) = \mathcal{E}_{2,t}(tz) \frac{K^*(tz)}{\eta(t^2z)} - \chi_{12}(t) \sum_{n=m}^{\infty} \left(\frac{1-24n}{t}\right) \beta_t(n) q^{n-\frac{1}{24}}$$

$$- \sum_{t^2n-s_t \geq m} \beta_t(t^2n - s_t) q^{n-\frac{1}{24}}, \tag{3.21}$$

where $K^*(z)$ and the sequence $\beta_t(n)$ are defined in (3.10)–(3.11). We have the following analogue of Proposition 2.1.

Corollary 3.3. *The function* $\Psi_{t,K}(z) \eta(z)$ *is a weakly holomorphic modular form of weight 2 on the full modular group* $\Gamma(1)$.

Proof. Let $S(z)$ be defined as in (3.9), so that $S(z)$ is a weakly holomorphic modular form of weight 2 on $\Gamma_0(t)$. By [16, Lemma 7], the function

$$S(z) + S(z) \mid W_t \mid U \tag{3.22}$$

is a modular form of weight 2 on $\Gamma(1)$. Here $U = U_t$ is the Atkin operator

$$g(z) \mid U_t = \frac{1}{t} \sum_{a=0}^{t-1} g\left(\frac{z+a}{t}\right). \tag{3.23}$$

The result then follows from applying the U-operator to Eq. (3.18). $\square$

We illustrate the $K(z) = 1$ case of Corollary 3.3 with two examples:

$$\Psi_5(z) = \frac{E_4(z)^2 E_6(z)}{\eta(z)^{25}} \tag{3.24}$$

and

$$\Psi_7(z) = \frac{1}{\eta(z)^{49}} \left(E_4(z)^5 E_6(z) - 745 E_4(z)^2 E_6(z) \Delta(z)\right). \tag{3.25}$$

We need a weight 2 analogue of [5, Lemma 3]. For $t = 5$, 7, or 13, the genus of $\Gamma_0(t)$ is zero, and a Hauptmodul is

$$G_t(z) := \left(\frac{\eta(z)}{\eta(tz)}\right)^{24/(t-1)}. \tag{3.26}$$

This function satisfies

$$G_t\left(\frac{-1}{tz}\right) = t^{12/(t-1)}G_t(z)^{-1}. \tag{3.27}$$

Proposition 3.4. *Suppose $t = 5$, 7, or 13, and let m be any negative integer such that $24m \not\equiv 1 \pmod{t}$. Suppose constants k_j $(1 \le j \le -m)$ are chosen so that*

$$\beta_t(n) = 0, \qquad for\ m+1 \le n \le -1, \tag{3.28}$$

where

$$K(z) = G_t(z)^{-m} + \sum_{j=1}^{-m-1} k_j G_t(z)^j \tag{3.29}$$

and

$$\sum_{n=m}^{\infty} \beta_t(n) q^{n-\frac{1}{24}} = \mathcal{E}_{2,t}(z)\frac{K(z)}{\eta(z)}. \tag{3.30}$$

Then

$$\beta_t(n) = 0, \qquad for \quad \left(\frac{1-24n}{t}\right) = -\left(\frac{1-24m}{t}\right). \tag{3.31}$$

Proof. Suppose $t = 5$, 7, or 13, and m is a negative integer such that $24m \not\equiv 1 \pmod{t}$. Suppose $K(z)$ is chosen so that (3.28) holds. Let $S(z)$ be defined as in (3.9), and define

$$B(z) := S(z) + \chi_{12}(t)\left(\frac{1-24m}{t}\right)\mathcal{E}_{2,t}(z)K(z), \tag{3.32}$$

so that

$$B(z) \mid W_t = S^*(z) - \chi_{12}(t)\left(\frac{1-24m}{t}\right)\mathcal{E}_{2,t}(z)K^*(z), \tag{3.33}$$

where

$$S^*(z) = S(z) \mid W_t. \tag{3.34}$$

Since $24m \not\equiv 1 \pmod{t}$, we see, using Corollary 3.2, that

$$\mathrm{ord}_0(S(z)) = \mathrm{ord}_{i\infty}(S^*(z)) > 0. \tag{3.35}$$

From (3.27) and (3.29) we see that

$$\mathrm{ord}_0(K(z)) = \mathrm{ord}_{i\infty}(K^*(z)) > 0, \tag{3.36}$$

and hence

$$\mathrm{ord}_0(B(z)) > 0. \tag{3.37}$$

Now

$$\mathrm{ord}_{i\infty}\left(\mathcal{E}_{2,t}(tz)\,K^*(tz)\frac{\eta(z)}{\eta(t^2z)}\right) \geq t - \frac{1}{24}(t^2-1) > 0, \qquad (3.38)$$

for $t = 5, 7, 13$. By construction the coefficient of q^m in $B(z)$ is zero and so (3.28), (3.38) imply that

$$\mathrm{ord}_{i\infty}(B(z)) \geq 0. \qquad (3.39)$$

Therefore, $B(z)$ is an entire modular form of weight 2 and hence a multiple of $\mathcal{E}_{2,t}(z)$ since there are no nontrivial cusp forms of weight 2 on $\Gamma_0(t)$ for $t = 5, 7,$ or 13 by [10]. This implies that $B(z)$ is identically zero by (3.37). Hence

$$\frac{B(z)}{E(q)} = q^{-s_t}\mathcal{E}_{2,t}(tz)\,K^*(tz)\frac{1}{E(q^{t^2})} - \chi_{12}(t)\sum_{n=m}^{\infty}\left(\left(\frac{1-24n}{t}\right)-\left(\frac{1-24m}{t}\right)\right)\beta_t(n)q^n = 0. \qquad (3.40)$$

Since $-24s_t - 1 \equiv 0 \pmod{t}$, this implies that $\beta_t(n) = 0$ whenever $\left(\frac{1-24n}{t}\right) = -\left(\frac{1-24m}{t}\right)$. □

We illustrate Proposition 3.4 with two examples:

$$\sum_{n=-2}^{\infty}\beta_5(n)q^n = \frac{\mathcal{E}_{2,5}(z)}{E(q)}\left(G_5(z)^2 + 5G_5(z)\right)$$

$$= q^{-2} + 1 - 379\,q^3 + 625\,q^4 + 869\,q^5 - 20125\,q^8 + 23125\,q^9$$

$$+ 25636\,q^{10} - 329236\,q^{13} + \cdots. \qquad (3.41)$$

In this example, $t = 5$ and $m = -2$, and we see that $\beta_5(n) = 0$ for $n \equiv 1, 2 \pmod 5$. In our second example, $t = 7$ and $m = -1$.

$$\sum_{n=-1}^{\infty}\beta_7(n)q^n = \frac{\mathcal{E}_{2,7}(z)}{E(q)}G_7(z) \qquad (3.42)$$

$$= q^{-1} + 1 - 15\,q^2 + 49\,q^5 - 24\,q^6 + 88\,q^7 - 311\,q^9 + 392\,q^{12}$$

$$- 182\,q^{13} + 811\,q^{14} - 1886\,q^{16} + \cdots. \qquad (3.43)$$

In this example we see that $\beta_7(n) = 0$ for $n \equiv 1, 3, 4 \pmod 7$.

The function

$$\frac{E_4(z)^2 E_6(z)}{\Delta(z)} = \frac{E_6(z)}{E_4(z)}j(z)$$

$$= q^{-1} - 196884\,q - 42987520\,q^2 - 2592899910\,q^3 - 80983425024\,q^4$$

$$- 1666013203000\,q^5 + \cdots \qquad (3.44)$$

is a modular form of weight 2 on $\Gamma(1)$. As a modular form on $\Gamma_0(t)$, it has a simple pole at $i\infty$ and a pole of order t at $z = 0$. When $t = 5$, 7, or 13, it is straightforward to show that there are integers $a_{j,t}$ $(-1 \leq j \leq t)$ such that

$$\frac{E_6(z)}{E_4(z)} j(z) = \mathcal{E}_{2,t}(z) \sum_{j=-1}^{t} a_{j,t} G_t(z)^j. \tag{3.45}$$

For example,

$$\frac{E_6(z)}{E_4(z)} j(z) = \mathcal{E}_{2,5}(z)\Big(G_5(z) - 3^2 \cdot 5^5 \cdot 7 G_5(z)^{-1} - 2^3 \cdot 5^8 \cdot 13 G_5(z)^{-2} - 3^3 \cdot 5^{10} \cdot 7 G_5(z)^{-3}$$
$$- 3 \cdot 2^3 \cdot 5^{13} G_5(z)^{-4} - 5^{16} G_5(z)^{-5}\Big). \tag{3.46}$$

Reducing (3.45) mod t^c, we obtain a weight 2 analogue of [5, Lemma 4].

Lemma 3.5. *We have*

$$\frac{E_6(z)}{E_4(z)} j(z) \equiv \mathcal{E}_{2,5}(z) \left(G_5(z) + 2 \cdot 31 \cdot 5^5 G_5(z)^{-1}\right) \pmod{5^8}, \tag{3.47}$$

$$\frac{E_6(z)}{E_4(z)} j(z) \equiv \mathcal{E}_{2,7}(z) G_7(z) \pmod{7^4}, \tag{3.48}$$

$$\frac{E_6(z)}{E_4(z)} j(z) \equiv \mathcal{E}_{2,13}(z) G_{13}(z) \pmod{13^2}. \tag{3.49}$$

We also need [5, Lemma 4].

Lemma 3.6 (Atkin [5]). *We have*

$$j(z) \equiv G_5(z) + 750 + 3^2 \cdot 7 \cdot 5^5 G_5(z)^{-1} \pmod{5^8}, \tag{3.50}$$

$$j(z) \equiv G_7(z) + 748 \pmod{7^4}, \tag{3.51}$$

$$j(z) \equiv G_{13}(z) + 70 \pmod{13^2}. \tag{3.52}$$

Remark 3.7. In equation (3.50) we have corrected a misprint in [5, Lemma 4].

To handle the $(t,c) = (5,6)$ case of Theorem 1.7, we will need

Lemma 3.8.

$$\frac{E_6(z)}{E_4(z)} j(z) \equiv \mathcal{E}_{2,5}(z) \left(G_5(z) + 2 \cdot 5^5 G_5(z)^{-1}\right) \pmod{5^6}, \tag{3.53}$$

$$\frac{E_6(z)}{E_4(z)} j(z)^2 \equiv \mathcal{E}_{2,5}(z) \left(2 \cdot 3 \cdot 5^3 G_5(z) + G_5(z)^2\right) \pmod{5^6}, \tag{3.54}$$

and

$$\frac{E_6(z)}{E_4(z)} j(z)^a \equiv \mathcal{E}_{2,5}(z) \left(\varepsilon_{1,a} G_5(z)^{a-2} + \varepsilon_{2,a} G_5(z)^{a-1} + G_5(z)^a \right) \quad (\mathrm{mod}\ 5^6),$$

(3.55)

for $a \geq 3$, where $\varepsilon_{1,a}$, $\varepsilon_{2,a}$ are integers satisfying

$$\varepsilon_{1,a} \equiv 0 \quad (\mathrm{mod}\ 5^5) \quad and \quad \varepsilon_{2,a} \equiv 0 \quad (\mathrm{mod}\ 5^3).$$

(3.56)

Proof. The result can be proved from Lemmas 3.5 and 3.6, some calculation and an easy induction argument. □

We need bases for $M_{2+12s_\ell}(\Gamma_0(t))$ for $t = 5$, 7, 13. The following result follows from [10], and by checking, the modular forms involved are holomorphic at the cusps $i\infty$ and 0.

Lemma 3.9. *Suppose $t = 5$, 7, or 13 and $\ell > 3$ is prime. Then*

$$\dim M_{2+12s_\ell}(\Gamma_0(t)) = 1 + (1+t)s_\ell,$$

(3.57)

and the set

$$\{ \mathcal{E}_{2,t}(z) \Delta(z)^{s_\ell} G_t(z)^a : -ts_\ell \leq a \leq s_\ell \}$$

(3.58)

is a basis for $M_{2+12s_\ell}(\Gamma_0(t))$.

We are now ready to prove Theorem 1.7. We have two cases:

Case 1. In the first case we assume that $(t,c) = (5,5)$, $(7,4)$, or $(13,2)$. Suppose $\ell > 3$ is prime and $\ell \neq t$. By (2.24) we have

$$\mathcal{A}_\ell(z) = \sum_{n=1}^{s_\ell} b_{n,\ell} \frac{E_6(z)}{E_4(z)} \frac{j(z)^n}{\eta(z)}.$$

(3.59)

By Theorem 1.1, Equation (2.21), and Lemma 3.9, we have

$$\mathcal{A}_\ell(z) = \sum_{n=-ts_\ell}^{s_\ell} d_{n,\ell} \frac{\mathcal{E}_{2,t}(z)}{\eta(z)} G_t(z)^n,$$

(3.60)

for some integers $d_{n,\ell}$ ($-ts_\ell \leq n \leq s_\ell$). Now let

$$K(z) = \sum_{n=1}^{s_\ell} d_{n,\ell} G_t(z)^n.$$

(3.61)

By using Lemmas 3.5 and 3.6 to reduce Equation (3.59) modulo t^c and comparing the result with (3.60) we deduce that

$$\mathcal{A}_\ell(z) \equiv \mathcal{E}_{2,t}(z) \frac{K(z)}{\eta(z)} \pmod{t^c} \tag{3.62}$$

and that

$$d_{n,\ell} \equiv 0 \pmod{t^c}, \tag{3.63}$$

for $-ts_\ell \le n \le 0$. By examining (3.60) we see that

$$\mathcal{A}_\ell(z) = \mathcal{E}_{2,t}(z) \frac{K(z)}{\eta(z)} + O(q^{-\frac{1}{24}}). \tag{3.64}$$

So if we let

$$\mathcal{E}_{2,t}(z) \frac{K(z)}{\eta(z)} = \sum_{n=-s_\ell}^{\infty} \beta_{t,\ell}(n) q^{n-\frac{1}{24}}, \tag{3.65}$$

then (3.62) may be rewritten as

$$\mathbf{a}(\ell^2 n - s_\ell) + \chi_{12}(\ell) \left(\left(\frac{1-24n}{\ell} \right) - 1 - \ell \right) \mathbf{a}(n) + \ell \mathbf{a} \left(\frac{n+s_\ell}{\ell^2} \right) \equiv \beta_{t,\ell}(n) \pmod{t^c} \tag{3.66}$$

and from (3.64), we have

$$\mathbf{a}(\ell^2 n - s_\ell) + \chi_{12}(\ell) \left(\left(\frac{1-24n}{\ell} \right) - 1 - \ell \right) \mathbf{a}(n) + \ell \mathbf{a} \left(\frac{n+s_\ell}{\ell^2} \right) = \beta_{t,\ell}(n) \tag{3.67}$$

for $-s_\ell \le n \le -1$. Equation (3.67) implies that

$$\beta_{t,\ell}(-s_\ell) = -\ell \tag{3.68}$$

and

$$\beta_{t,\ell}(n) = 0, \tag{3.69}$$

for $-s_\ell < n \le -1$. We can now apply Proposition 3.4 with $m = -s_\ell$ since $1 - 24m = \ell^2$ and $t \ne \ell$. Hence

$$\beta_{t,\ell}(n) = 0, \quad \text{provided} \quad \left(\frac{1-24n}{t} \right) = -1. \tag{3.70}$$

This gives Theorem 1.7 when $(t,c) = (5,5)$, $(7,4)$, or $(13,2)$ by (3.66).

Case 2. We consider the remaining case $(t,c) = (5,6)$ and assume $\ell > 5$ is prime. We proceed as in Case 1. This time, when we use Lemma 3.8 to reduce (2.24) modulo 5^6, we see that the only extra term occurs when $n = 1$. We find that with $K(z)$ as before we have

$$\mathcal{A}_\ell(z) \equiv \mathcal{E}_{2,t}(z) \frac{K(z)}{\eta(z)} + b_{1,\ell} \cdot 2 \cdot 5^5 \frac{\mathcal{E}_{2,5}(z)}{\eta(z)} G_5(z)^{-1} \pmod{5^6}. \tag{3.71}$$

All that remains is to show that

$$b_{1,\ell} \equiv 0 \pmod{5}, \tag{3.72}$$

since then (3.62) actually holds modulo 5^6 and the rest of the proof proceeds as in Case 1. Since $E_4(z) \equiv 1 \pmod 5$ we may reduce (2.18) modulo 5 to obtain

$$\ell \Xi_\ell \equiv (\chi_{12}(\ell)\ell(1+\ell) - c_{0,\ell}) \frac{E_6(z)}{E_4(z)} \frac{1}{\eta(z)} - \sum_{n=1}^{s_\ell} (24n+1)c_{n,\ell} \frac{E_6(z)}{E_4(z)} \frac{j(z)^n}{\eta(z)} \pmod 5. \tag{3.73}$$

But

$$\mathcal{S}_\ell(z) \equiv 0 \pmod 5, \tag{3.74}$$

by Theorem 1.3(ii). Hence

$$\ell \mathcal{A}_\ell(z) \equiv (\chi_{12}(\ell)\ell(1+\ell) - c_{0,\ell}) \frac{E_6(z)}{E_4(z)} \frac{1}{\eta(z)} - \sum_{n=1}^{s_\ell} (24n+1)c_{n,\ell} \frac{E_6(z)}{E_4(z)} \frac{j(z)^n}{\eta(z)} \pmod 5, \tag{3.75}$$

and we see that $b_{1,\ell}$ the coefficient of $\frac{E_6(z)}{E_4(z)} \frac{j(z)}{\eta(z)}$ is divisible by 5 as required. This completes the proof of Theorem 1.7.

We close by illustrating Theorem 1.7 when $t = 5$ and $\ell = 7$. In this case the theorem predicts that

$$\mathbf{a}(49n - 2) - \left(\frac{1 - 24n}{7}\right) \mathbf{a}(n) + 7\mathbf{a}\left(\frac{n+2}{49}\right) \equiv -8\mathbf{a}(n) \pmod{5^6},$$

when $n \equiv 1, 2 \pmod 5$. When $n = 1$, this says

$$149077845 \equiv -280 \pmod{5^6},$$

which is easy to check.

Acknowledgements I would like to thank Ken Ono for sending me preprints of his recent work [14, 15].

Note Added: Nickolas Andersen ("Hecke-type congruences for two smallest parts functions," preprint) has found Hecke-type congruences for certain powers of 2 for two spt-like functions $\overline{\text{spt1}}(n)$ and M2spt(n). These other spt-like functions were studied earlier by Ahlgren, Bringmann, Lovejoy, and Osburn [1, 8].

References

1. S. Ahlgren, K. Bringmann, J. Lovejoy, ℓ-Adic properties of smallest parts functions. Adv. Math. **228**, 629–645 (2011). URL: http://arxiv.org/abs/1011.6079
2. G.E. Andrews, The number of smallest parts in the partitions of n. J. Reine Angew. Math. **624**, 133–142 (2008). URL: http://dx.doi.org/10.1515/CRELLE.2008.083
3. A.O.L. Atkin, F.G. Garvan, Relations between the ranks and cranks of partitions. Ramanujan J. **7**, 343–366 (2003). URL: http://dx.doi.org/10.1023/A:1026219901284
4. A.O.L. Atkin, Ramanujan congruences for $p_{-k}(n)$. Can. J. Math. **20**, 67–78 (1968) corrigendum, ibid. **21**, 256 (1968). URL: http://cms.math.ca/cjm/v20/p67
5. A.O.L. Atkin, Multiplicative congruence properties and density problems for $p(n)$. Proc. Lond. Math. Soc. **18**(3), 563–576 (1968). URL: http://dx.doi.org/10.1112/plms/s3-18.3.563
6. B.C. Berndt, R.J. Evans, The determination of Gauss sums. Bull. Am. Math. Soc. (N.S.) **5**, 107–129 (1981). URL: http://dx.doi.org/10.1090/S0273-0979-1981-14930-2
7. K. Bringmann, On the explicit construction of higher deformations of partition statistics. Duke Math. J. **144**, 195–233 (2008). URL: http://dx.doi.org/10.1215/00127094-2008-035
8. K. Bringmann, J. Lovejoy, R. Osburn, Rank and crank moments for overpartitions. J. Number Theory **129**, 1758–1772 (2009). URL: http://dx.doi.org/10.1016/j.jnt.2008.10.017
9. K. Bringmann, J. Lovejoy, R. Osburn, Automorphic properties of generating functions for generalized rank moments and Durfee symbols. Int. Math. Res. Not. IMRN no. 2, 238–260 (2010). URL: http://dx.doi.org/10.1093/imrn/rnp131
10. H. Cohen, J. Oesterlé, Dimensions des espaces de formes modulaires. Springer Lect. Notes **627**, 69–78 (1977). URL: http://dx.doi.org/10.1007/BFb0065297
11. A. Folsom, K. Ono, The spt-function of Andrews. Proc. Natl. Acad. Sci. USA **105**, 20152–20156 (2008). URL: http://dx.doi.org/10.1073/pnas.0809431105
12. F.G. Garvan, Congruences for Andrews' smallest parts partition function and new congruences for Dyson's rank. Int. J. Number Theory **6**, 1–29 (2010). URL: http://dx.doi.org/10.1142/S179304211000296X
13. F.G. Garvan, Congruences for Andrews' spt-function modulo powers of 5, 7 and 13. Trans. Am. Math. Soc. **364**, 4847–4873 (2012). URL: http://www.math.ufl.edu/~fgarvan/papers/spt2.pdf
14. K. Ono, Congruences for the Andrews spt-function. Proc. Natl. Acad. Sci. USA **108**, 473–476 (2011). URL: http://mathcs.emory.edu/~ono/publications-cv/pdfs/131.pdf
15. K. Ono, The partition function and Hecke operators. Adv. Math. **228**, 527–534 (2011). URL: http://mathcs.emory.edu/~ono/publications-cv/pdfs/132.pdf
16. J.-P. Serre, Formes modulaires et fonctions zêta p-adiques, in *Modular Functions of One Variable, III* (*Proc. Internat. Summer School*, Univ. Antwerp, 1972). Lecture Notes in Math., vol 350 (Springer, Berlin, 1973), pp. 191–268. URL: http://dx.doi.org/10.1007/978-3-540-37802-0_4

Continued Fractions and Dedekind Sums
for Function Fields

Yoshinori Hamahata

Abstract For the classical Dedekind sum $d(a,c)$, Rademacher and Grosswald raised two questions: (1) Is $\{(a/c, d(a,c)) \mid a/c \in \mathbf{Q}^*\}$ dense in $\mathbf{R}^2$? (2) Is $\{d(a,c) \mid a/c \in \mathbf{Q}^*\}$ dense in $\mathbf{R}$? Using the theory of continued fractions, Hickerson answered these questions affirmatively. In function fields, there exists a Dedekind sum $s(a,c)$ (see Sect. 4) similar to $d(a,c)$. Using continued fractions, we answer the analogous problems for $s(a,c)$.

1 Introduction

Given relatively prime integers $c > 0$ and a, the classical Dedekind sum is defined as

$$d(a,c) = \frac{1}{4c} \sum_{k=1}^{c-1} \cot\left(\frac{\pi a k}{c}\right) \cot\left(\frac{\pi k}{c}\right).$$

It satisfies a famous relation called the *reciprocity law*

$$d(a,c) + d(c,a) = \frac{1}{12}\left(\frac{a}{c} + \frac{c}{a} + \frac{1}{ac} - 3\right)$$

Y. Hamahata
Institute for Teaching and Learning, Ritsumeikan University, 1-1-1 Noji-higashi,
Kusatsu, Shiga 525-8577, Japan
e-mail: hamahata@fc.ritsumei.ac.jp

J.M. Borwein et al. (eds.), *Number Theory and Related Fields: In Memory of Alf van der Poorten*, Springer Proceedings in Mathematics & Statistics 43,
DOI 10.1007/978-1-4614-6642-0_9, © Springer Science+Business Media New York 2013

for coprime $a, c > 0$ (see Rademacher–Grosswald [8] for details). A generalization of Dedekind sums to higher dimensions was presented by Zagier [10]. It is well known that $d(a, c)$ is a rational number. Rademacher and Grosswald [8, p. 28] raised two questions:

1. Is $\{(a/c, d(a, c)) \mid a/c \in \mathbf{Q}^*\}$ dense in $\mathbf{R}^2$?
2. Is $\{d(a, c) \mid a/c \in \mathbf{Q}^*\}$ dense in $\mathbf{R}$?

Using the theory of continued fractions, Hickerson [4] answered these questions affirmatively (see Sect. 3).

There is an analogy between number fields and function fields. For instance, $A := \mathbf{F}_q[T]$, $K := \mathbf{F}_q(T)$, and $K_\infty := \mathbf{F}_q((1/T))$ are analogous to $\mathbf{Z}$, $\mathbf{Q}$, and $\mathbf{R}$, respectively. Let C denote the completion of an algebraic closure of K_∞. Let Λ be an A-lattice in C, and ϕ be the corresponding Drinfeld module over C (see [2] for definitions). Let $e_\Lambda(z)$ be the exponential function for Λ defined by

$$e_\Lambda(z) = z \prod_{\lambda \in \Lambda \setminus \{0\}} (1 - z/\lambda)^{-1}.$$

The function $e_\Lambda(z)^{-1}$ is similar to $\pi \cot \pi z$ in that $\pi \cot \pi z$ is $\mathbf{Z}$-periodic and has the expression

$$\pi \cot \pi z = \frac{1}{z} + \sum_{n=1}^{\infty} \left(\frac{1}{z+n} + \frac{1}{z-n} \right);$$

$e_\Lambda(z)^{-1}$ is Λ-periodic and has the expression

$$e_\Lambda(z)^{-1} = \frac{1}{z} + \sum_{\lambda \in \Lambda \setminus \{0\}} \frac{1}{z - \lambda}.$$

Noting this observation, in our previous paper [1], we introduced a function field analog of the Dedekind sums and their higher-dimensional generalization. Above all, the Dedekind sum $s(a, c)$ for the A-lattice L, which corresponds to the Carlitz module, is given by

$$s(a, c) = \frac{1}{c} \sum_{0 \neq \ell \in L/cL} e_L \left(\frac{a\ell}{c} \right)^{-1} e_L \left(\frac{\ell}{c} \right)^{-1}$$

for relatively prime elements $a, c \in A \setminus \{0\}$ (see Sect. 4 for details). This Dedekind sum is very similar to $d(a, c)$. As shown in [1], $s(a, c)$ belongs to $\mathbf{F}_q(T)$. The objective of this study is to answer the problems analogous to 1, 2 stated above using continued fractions. More precisely, our main result is as follows:

Main Theorem (Theorems 6 and 7).

1. If $q > 3$, then

$$\{(a/c, s(a/c)) \mid a/c \in K^*\} = K^* \times \{0\}, \quad \{s(a/c) \mid a/c \in K^*\} = \{0\}.$$

2. If $q = 2$ or 3, then

  (a) $\{(a/c, s(a/c)) \mid a/c \in K^*\}$ is dense in K_∞^2.
  (b) $\{s(a/c) \mid a/c \in K^*\}$ is dense in K_∞.

The remainder of this paper is organized as follows. In Sect. 2, we review A-lattices and their exponential functions in function fields. We treat the A-lattice L corresponding to the Carlitz module. In Sect. 3, we recall continued fractions. In Sect. 4, we define the Dedekind sum $s(a, c)$, and we state the reciprocity law explicitly. Then, we describe the values of Dedekind sums by means of continued fractions. In Sect. 5, we establish the density results for our Dedekind sums.

Notation

  $\sum'$: sum over nonvanishing elements
  $\prod'$: product over nonvanishing elements
  $F^* = F \setminus \{0\}$ for a field F
  $\mathbf{F}_q$: finite field with q elements
  $A = \mathbf{F}_q[T]$: ring of polynomials in an indeterminate T
  $K = \mathbf{F}_q(T)$: quotient field of A
  $| \ |$: normalized absolute value on K such that $|T| = q$
  $K_\infty = \mathbf{F}_q((1/T))$: completion of K with respect to $| \ |$
  $\overline{K_\infty}$: fixed algebraic extension of K_∞
  C : completion of $\overline{K_\infty}$

2 Preliminaries

In this section, we give an overview of A-lattices and related periodic functions. For details, see Goss [2]. A rank r A-*lattice* Λ in C is a finitely generated A-submodule of rank r in C that is discrete in the topology of C. For such an A-lattice Λ, define the product

$$e_\Lambda(z) = z \prod_{\lambda \in \Lambda}' \left(1 - \frac{z}{\lambda}\right).$$

The product converges uniformly on bounded sets in C and defines a map $e_\Lambda : C \to C$. The map e_Λ has the following properties:

(E1) e_Λ is entire in the rigid analytic sense and surjective.
(E2) e_Λ is $\mathbf{F}_q$-linear and Λ-periodic.
(E3) e_Λ has simple zeros at the points of Λ and no other zeros.
(E4) $de_\Lambda(z)/dz = e_\Lambda'(z) = 1$. Hence, we have

$$\frac{1}{e_\Lambda(z)} = \frac{e_\Lambda'(z)}{e_\Lambda(z)} = \sum_{\lambda \in \Lambda} \frac{1}{z - \lambda}. \tag{1}$$

Let L be the rank one A-lattice in C that corresponds to the Carlitz module ρ : $A \to \mathrm{End}_C(\mathbf{G}_a)$ given by $\rho_T(z) = Tz + z^q$. For any $a \in A \setminus \{0\}$, let $\rho[a] := \{x \in C \mid \rho_a(x) = 0\}$ denote the A/aA-module of a-division points. Put

$$E_k(\rho[a]) := \sideset{}{'}\sum_{x \in \rho[a]} \frac{1}{x^k}$$

for each positive integer k. If $a \in \mathbf{F}_q \setminus \{0\}$, then $E_k(\rho[a]) = 0$ for any positive integer k.

3 Continued Fractions

3.1 The Classical Case

We recall continued fractions. The basic references are [3, 5].
Let $[a_0, a_1, a_2, \ldots]$ denote the continued fraction

$$a_0 + \cfrac{1}{a_1 + \cfrac{1}{a_2 + \cdots}}, \tag{2}$$

where a_n is called the nth *partial quotient*. Let

$$p_0 = a_0, \quad q_0 = 1, \quad p_1 = a_1 a_0 + 1, \quad q_1 = a_1,$$

$$p_n = a_n p_{n-1} + p_{n-2}, \quad q_n = a_n q_{n-1} + q_{n-2}.$$

Put

$$[a_0, a_1, \ldots, a_n] = a_0 + \cfrac{1}{a_1 + \cfrac{1}{a_2 + \cdots + \cfrac{1}{a_n}}}.$$

Then, we have some results:

$$p_n/q_n = [a_0, \ldots, a_n], \quad p_n q_{n-1} - p_{n-1} q_n = (-1)^{n+1},$$

$$p_n/p_{n-1} = [a_n, \ldots, a_0], \quad q_n/q_{n-1} = [a_n, \ldots, a_1],$$

$$\frac{1}{p_n q_n} = (-1)^{n+1} \left([a_n, \ldots, a_1] - [a_n, \ldots, a_0]\right).$$

The Dedekind sum $d(a, c)$ stated in the Introduction depends on a/c. Hence, we can write $d(a/c) = d(a, c)$. Hickerson proved the following result:

Theorem 1 (Hickerson [4]).

$$d([a_0,\ldots,a_r]) = \frac{-1+(-1)^r}{8} + \frac{1}{12}([0,a_1,\ldots,a_r]+(-1)^{r+1}[0,a_r,\ldots,a_1]$$

$$+a_1-a_2+\cdots+(-1)^{r+1}a_r).$$

Using this theorem, he proved the following density results:

Theorem 2 (Hickerson [4]).

1. The set $\{(a/c,d(a/c)) \mid a/c \in \mathbf{Q}^*\}$ is dense in $\mathbf{R}^2$.
2. The set $\{d(a/c) \mid a/c \in \mathbf{Q}^*\}$ is dense in $\mathbf{R}$.

We will establish results similar to these theorems in Sects. 4 and 5.

3.2 The Function Field Case

We refer to [6,7,9]. For a given $x \in K_\infty = \mathbf{F}_q((1/T))$, we define a sequence $(x_n)_{n\geq 0}$ by setting

$$x_0 = x, \quad x_{n+1} = \frac{1}{x_n - a_n}.$$

Here, $a_n = \sum_{i=0}^k A_i T^i$ is the polynomial part of the Laurent expansion $x_n = \sum_{i=-\infty}^k A_i T^i$. There is no sign condition on a_i in contrast to the positivity condition in the classical case. Instead, we force the condition $a_i \notin \mathbf{F}_q$. This sequence yields the continued fraction development (2) of x. It is easy to see that $(x_n)_{n\geq 0}$ terminates if and only if x belongs to $K = \mathbf{F}_q(T)$. There are convergence and uniqueness of $[a_0,a_1,a_2,\ldots]$. This continued fraction has properties similar to those in the classical case. The truncations $[a_0,a_1,\ldots,a_n]$ are rational functions p_n/q_n. Here, p_n,q_n are relatively prime polynomials in $A = \mathbf{F}_q[T]$ given by

$$\begin{pmatrix} a_0 & 1 \\ 1 & 0 \end{pmatrix}\begin{pmatrix} a_1 & 1 \\ 1 & 0 \end{pmatrix}\cdots\begin{pmatrix} a_n & 1 \\ 1 & 0 \end{pmatrix} = \begin{pmatrix} p_n & p_{n-1} \\ q_n & q_{n-1} \end{pmatrix}.$$

As $n \to \infty$, then we have $\deg p_n, \deg q_n \to \infty$, and $p_n/q_n \to x$.

Example 1. If $[a_0,a_1,\ldots,a_n] = p_n/q_n$, then we have the folding formula given by

$$\frac{p_n}{q_n} + \frac{(-1)^n}{yq_n^2} = [a_0,a_1,\ldots,a_n,y,-a_n,\ldots,-a_1].$$

(see [6,9]).

1. Put $y = -T$. Then we have

$$1+T^{-1} = [1,T],$$

$$1+T^{-1}+T^{-3} = [1,T,-T,-T],$$

$$1 + T^{-1} + T^{-3} + T^{-7} = [1, T, -T, -T, -T, T, T, -T],$$
$$1 + T^{-1} + T^{-3} + T^{-7} + T^{-15}$$
$$= [1, T, -T, -T, -T, T, T, -T, -T, T, -T, -T, T, T, T, -T].$$

2. Put $y = T$. Then we have

$$1 - T^{-2} = [1, -T^2],$$
$$1 - T^{-2} - T^{-5} = [1, -T^2, T, T^2],$$
$$1 - T^{-2} - T^{-5} - T^{-11} = [1, -T^2, T, -T^2, T, T^2, -T, T^2],$$
$$1 - T^{-2} - T^{-5} - T^{-11} - T^{-23}$$
$$= [1, -T^2, T, -T^2, T, T^2, -T, T^2, T, -T^2, T, -T^2, -T, T^2, -T, T^2].$$

4   Dedekind Sum

In this section, we recall our Dedekind sum and related results in [1]. Next, we prove a result related to continued fractions.

Definition 1. Let a, c be coprime elements of $A \setminus \{0\}$, and let L denote the A-lattice in C introduced in Sect. 2. The *Dedekind sum* $s(a,c)$ is defined as

$$s(a,c) = \frac{1}{c} \sum_{\ell \in L/cL} {}' e_L \left(\frac{a\ell}{c} \right)^{-1} e_L \left(\frac{\ell}{c} \right)^{-1}.$$

Here, if $L/cL = 0$, then $s(a,c)$ is defined to be zero.

Remark 1. If $q > 3$, then there exists an element $\zeta \in \mathbf{F}_q^*$ such that $\zeta^2 \neq 1$. Hence,

$$s(a,c) = \frac{1}{c} \sum_{\zeta \ell \in L/cL} {}' e_L \left(\frac{a\zeta\ell}{c} \right)^{-1} e_L \left(\frac{\zeta\ell}{c} \right)^{-1} = \zeta^{-2} s(a,c),$$

which gives $s(a,c) = 0$ for coprime $a, c \in A \setminus \{0\}$.

The Dedekind sum $s(a,c)$ has the following reciprocity law:

Theorem 3 (Reciprocity law [1]). *If a, c are coprime, then we have*

$$s(a,c) + s(c,a) = \frac{E_2(\rho[a]) + E_2(\rho[c]) - E_1(\rho[a])E_1(\rho[c])}{ac}.$$

This reciprocity law is expressed explicitly.

Theorem 4 (Reciprocity law). *If $a, c \in A$ are relatively prime, then*

$$
s(a,c) + s(c,a) =
\begin{cases}
0 & \text{if } q > 3, \\[2mm]
\dfrac{1}{T^3 - T}\left(\dfrac{a}{c} + \dfrac{c}{a} + \dfrac{1}{ac}\right) & \text{if } q = 3, \\[3mm]
\dfrac{1}{T^4 + T^2}\left(\dfrac{a}{c} + \dfrac{c}{a} + \dfrac{1}{a} + \dfrac{1}{c} + \dfrac{1}{ac} + 1\right) & \text{if } q = 2.
\end{cases}
$$

Proof. *Case $q > 3$:* This follows from Remark 1.

Case $q = 3$: See 4.3 in [1].

Case $q = 2$: For $a \in A \setminus \{0\}$, let $\rho_a(z) = \sum_{i=0}^m l_i(a) z^{q^i}$. As mentioned in Goss [2], $l_1(a) = (a^q - a)/(T^q - T)$. By Proposition 4.2 in [1], $E_1(\rho[a]) = l_1(a)/a = (a + 1)/(T^2 + T)$. By definition, we have $E_2(\rho[a]) = E_1(\rho[a])^2 = (a^2 + 1)/(T^4 + T^2)$. This completes the proof. □

Hereafter, we investigate the relation between $s(a, c)$ and continued fractions. It is easy to see that the value $s(a, c)$ depends on a/c. Hence, we write $s(a/c) = s(a, c)$. Note that

$$
s(a/c + b) = s(a/c) \tag{3}
$$

for $b \in A$. Indeed, it follows that $s(a/c + b) = s(a + bc, c) = s(a, c)$.

Here is a result that is an analog of Theorem 1.

Theorem 5. *(i)* *If $q > 3$, then $s([b_0, \ldots, b_r]) = 0$.*
(ii) *If $q = 3$, then $s([b_0, \ldots, b_r])$ is expressed as*

$$
\begin{cases}
\dfrac{1}{T^3 - T}([0, b_1, \ldots, b_r] + (-1)^{r+1}[0, b_r, \ldots, b_1] \\[2mm]
\qquad + b_1 - b_2 + \cdots + (-1)^{r+1} b_r) & \text{if } r \geq 1, \\[2mm]
\qquad\qquad 0 & \text{if } r = 0.
\end{cases}
$$

(iii) *If $q = 2$, then $s([b_0, \ldots, b_r])$ is expressed as*

$$
\begin{cases}
\dfrac{1}{T^4 + T^2}([0, b_1, \ldots, b_r] + (-1)^{r+1}[0, b_r, \ldots, b_1] + \prod_{i=1}^{r}[0, b_i, \ldots, b_r] \\[2mm]
\qquad + b_1 - b_2 + \cdots + (-1)^{r+1} b_r + r - 1) & \text{if } r \geq 1, \\[2mm]
\qquad\qquad 0 & \text{if } r = 0.
\end{cases}
$$

Proof. We prove the theorem using some formulas in 3.1. It should be noted that if $a/c = [0, b_1, \ldots, b_r]$, then $c/a = [b_1, \ldots, b_r]$. Since (i) is trivial by Remark 1, we prove (ii) and (iii).

(ii) We have $s([b_0]) = s([b_0], 1) = 0$. When $r \geq 1$, we prove the result by induction on r. When $r = 1$, using Theorem 4 and (3),

$$
s([b_0, b_1]) = s([0, b_1]) = s(1/b_1) = \frac{1}{T^3 - T}(1/b_1 + b_1 + 1/b_1) - s(b_1).
$$

Since $s(b_1) = 0$, the right-hand side of the last equality gives the case $r = 1$. Let $r > 1$, and assume that the case $r - 1$ holds. Denoting $[0, b_1, \ldots, b_r]$ by a/c,

$$
\begin{aligned}
s([b_0, \ldots, b_r]) &= s([0, b_1, \ldots, b_r]) \\
&= \frac{1}{T^3 - T}(a/c + c/a + 1/ac) - s([b_1, \ldots, b_r]) \\
&= \frac{1}{T^3 - T}([0, b_1, \ldots, b_r] + [b_1, \ldots, b_r] \\
&\qquad + (-1)^{r+1}[0, b_r, \ldots, b_1] - (-1)^{r+1}[0, b_r, \ldots, b_2]) \\
&\quad - \frac{1}{T^3 - T}([0, b_2, \ldots, b_r] + (-1)^r[0, b_r, \ldots, b_2] + b_2 - b_3 + \cdots + (-1)^r b_r),
\end{aligned}
$$

which yields the result for the case r.

(iii) Clearly, $s([b_0]) = 0$. Let $r \geq 1$. We prove the claim by induction on r. When $r = 1$, using Theorem 4 and (3),

$$
\begin{aligned}
s([b_0, b_1]) = s([0, b_1]) = s(1/b_1) &= \frac{1}{T^4 + T^2}(1/b_1 + b_1 + 1/b_1 + 1 + 1/b_1 + 1) - s(b_1) \\
&= \frac{1}{T^4 + T^2}([0, b_1] + [0, b_1] + [0, b_1] + b_1),
\end{aligned}
$$

which is the case $r = 1$. Let $r > 1$, and the case $r - 1$ is supposed to be true. Denoting $[0, b_1, \ldots, b_r]$ by a/c, we have

$$
\begin{aligned}
s([b_0, \ldots, b_r]) &= s([0, b_1, \ldots, b_r]) \\
&= \frac{1}{T^4 + T^2}([0, b_1, \ldots, b_r] + [b_1, \ldots, b_r] + 1/a + [0, b_1, \ldots, b_r]/a \\
&\quad + (-1)^{r+1}[0, b_r, \ldots, b_1] - (-1)^{r+1}[0, b_r, \ldots, b_2] \\
&\quad + 1) - s([b_1, \ldots, b_r]) \\
&= \frac{1}{T^4 + T^2}([0, b_1, \ldots, b_r] + b_1 + [0, b_2, \ldots, b_r] \\
&\quad + \prod_{i=2}^{r}[0, b_i, \ldots, b_r] + \prod_{i=1}^{r}[0, b_i, \ldots, b_r] \\
&\quad + (-1)^{r+1}[0, b_r, \ldots, b_1] - (-1)^{r+1}[0, b_r, \ldots, b_2] \\
&\quad + 1) - \frac{1}{T^4 + T^2}([0, b_2, \ldots, b_r] + (-1)^r[0, b_r, \ldots, b_2] + \prod_{i=2}^{r}[0, b_i, \ldots, b_r] \\
&\quad b_2 - b_3 + \cdots + (-1)^r b_r + r - 2).
\end{aligned}
$$

This yields the result for the case r. $\square$

5   Density Results

In this section, we consider the value distribution of our Dedekind sum $s(a/c) = s(a,c)$.

Let us define the norm on K_∞^2 by $|(x,y)| = \max\{|x|,|y|\}$. We need the following lemma later on.

Lemma 1. *The set* $(K_\infty \setminus K) \times K$ *is dense in* K_∞^2.

Proof. It suffices to show that $K_\infty \setminus K$ is dense in K_∞. We take any $x = \sum_{i=-\infty}^{k} A_i T^i \in K_\infty$. For any $\varepsilon > 0$, there exists $i = -n$ such that $|x - \sum_{i=-n}^{k} A_i T^i| < \varepsilon$. Select $y \in K_\infty \setminus K$. For fully large $m > 0$, $|T^{-m}y| < \varepsilon$. Then, if we set $z = \sum_{i=-n}^{k} A_i T^i + T^{-m}y$, then $z \in K_\infty \setminus K$ and $|x - z| < \varepsilon$. $\qquad\qquad\square$

We state a result for the values of the Dedekind sum $s(a/c)$.

Theorem 6.   (i) *If* $q > 3$, *then* $\{(a/c, s(a/c)) \mid a/c \in K^*\} = K^* \times \{0\}$.
(ii) *If* $q = 2$ *or* 3, *then* $\{(a/c, s(a/c)) \mid a/c \in K^*\}$ *is dense in* K_∞^2.

Proof.   (i) This follows from Remark 1.
(ii) *Case* $q = 3$: Let us take any element $(x,y) \in K_\infty^2$. We prove that for any $\varepsilon > 0$, there exists $a/c \in K$ such that $|x - a/c| < \varepsilon$ and $|y - s(a/c)| < 2\varepsilon$. By Lemma 1, we may assume that $x \in K_\infty \setminus K$ and $y \in K$. Write $x = [b_0, b_1, \ldots]$ and $p_i/q_i = [b_0, \ldots, b_i]$. For any $s \geq 1$ and $\alpha \in K_\infty^*$,

$$|x - [b_0, \ldots, b_{s-1}, \alpha]| \leq |x - p_{s-1}/q_{s-1}| + |p_{s-1}/q_{s-1} - [b_0, \ldots, b_{s-1}, \alpha]|$$

$$= |x - p_{s-1}/q_{s-1}| + \frac{1}{|q_{s-1}(\alpha q_{s-1} + q_{s-2})|}.$$

As $s \to \infty$, we have $p_{s-1}/q_{s-1} \to x$, $\deg p_{s-1} \to \infty$, and $\deg q_{s-1} \to \infty$. Take any $\varepsilon > 0$. If s is fully large, then $|x - [b_0, \ldots, b_{s-1}, \alpha]| < \varepsilon$ for any $\alpha \in K_\infty^*$. Since $y \in K$, $x - (T^3 - T)y \in K_\infty \setminus K$. Let $x - (T^3 - T)y = [d_0, d_1, \ldots]$. If t is fully large, then $|x - (T^3 - T)y - [d_0, \ldots, d_{t-1}, \alpha]| < \varepsilon$ for any $\alpha \in K_\infty^*$. We may assume that $s + t$ is even. Choose $m, n \in A \setminus \mathbf{F}_q$ such that

$$-b_0 + b_1 - b_2 + \cdots + (-1)^s b_{s-1} + (-1)^{t-1} d_{t-1} + \cdots - d_1 + d_0 = (-1)^s (m - n).$$

Put

$$a/c = [b_0, \ldots, b_{s-1}, m, n, d_{t-1}, \ldots, d_1], \qquad \alpha = [m, n, d_{t-1}, \ldots, d_1].$$

Then, $|x - a/c| < \varepsilon$ holds. By Theorem 5(ii), we have

$$s(a/c)$$

$$= \frac{1}{T^3 - T}([0, b_1, \ldots, b_{s-1}, m, n, d_{t-1}, \ldots, d_1] - [0, d_1, \ldots, d_{t-1}, n, m, b_{s-1}, \ldots, b_1]$$

$$+ b_1 - b_2 + \cdots + (-1)^s b_{s-1} + (-1)^{s+1} m + (-1)^{s+2} n + (-1)^{t-1} d_{t-1} + \cdots - d_1).$$

Putting $\delta_1 = x - a/c$ and $\delta_2 = x - (T^3 - T)y - [d_0, \ldots, d_{t-1}, \alpha]$, we get $|\delta_1| < \varepsilon$ and $|\delta_2| < \varepsilon$. Since $[0, b_1, \ldots, b_{s-1}, \alpha] = x - b_0 - \delta_1$ and $[0, d_1, \ldots, d_{t-1}, \alpha] = x - (T^3 - T)y - d_0 - \delta_2$,

$$s(a/c) = y + (\delta_2 - \delta_1)/(T^3 - T),$$

which gives $|y - s(a/c)| < 2\varepsilon$.

Case $q = 2$: We proceed along the same lines as the case $q = 3$. As seen above, we may assume that $x \in K_\infty \setminus K$ and $y \in K$. Write $x = [b_0, b_1, \ldots]$ and $p_i/q_i = [b_0, \ldots, b_i]$. For any $s \geq 1$ and $\alpha \in K_\infty^*$,

$$|x - [b_0, \ldots, b_{s-1}, \alpha]| \leq |x - p_{s-1}/q_{s-1}| + |p_{s-1}/q_{s-1} - [b_0, \ldots, b_{s-1}, \alpha]|$$

$$= |x - p_{s-1}/q_{s-1}| + \frac{1}{|q_{s-1}(\alpha q_{s-1} + q_{s-2})|}.$$

As $s \to \infty$, we have $p_{s-1}/q_{s-1} \to x$, $\deg p_{s-1} \to \infty$, and $\deg q_{s-1} \to \infty$. Take any $\varepsilon > 0$. If s is fully large, then $|x - [b_0, \ldots, b_{s-1}, \alpha]| < \varepsilon$ for any $\alpha \in K_\infty^*$. Since $y \in K$, $x - (T^4 + T^2)y \in K_\infty \setminus K$. Let $x - (T^4 + T^2)y = [d_0, d_1, \ldots]$. If t is fully large, then $|x - (T^4 + T^2)y - [d_0, \ldots, d_{t-1}, \alpha]| < \varepsilon$ for any $\alpha \in K_\infty^*$. We may assume that $s + t$ is even. Choose $m, n \in A \setminus \mathbf{F}_q$ such that

$$-b_0 + b_1 - b_2 + \cdots + (-1)^s b_{s-1} + (-1)^{t-1} d_{t-1} + \cdots - d_1 + d_0 = (-1)^s (m - n).$$

Put

$$a/c = [b_0, \ldots, b_{s-1}, m, n, d_{t-1}, \ldots, d_1], \qquad \alpha = [m, n, d_{t-1}, \ldots, d_1].$$

Then, $|x - a/c| < \varepsilon$ holds. We write $a/c = [h_0, h_1, \ldots, h_{s+t}]$. Taking fully large $s + t$, we may assume that $\prod_{i=1}^{s+t}[0, h_i, \ldots, h_{s+t}] < \varepsilon$. By Theorem 5(iii), we have

$$s(a/c)$$

$$= \frac{1}{T^4 + T^2}([0, b_1, \ldots, b_{s-1}, m, n, d_{t-1}, \ldots, d_1] - [0, d_1, \ldots, d_{t-1}, n, m, b_{s-1}, \ldots, b_1]$$

$$+ \prod_{i=1}^{s+t}[0, h_i, \ldots, h_{s+t}]$$

$$+ b_1 - b_2 + \cdots + (-1)^s b_{s-1} + (-1)^{s+1} m + (-1)^{s+2} n + (-1)^{t-1} d_{t-1} + \cdots - d_1).$$

Putting $\delta_1 = x - a/c$ and $\delta_2 = x - (T^4 + T^2)y - [d_0, \ldots, d_{t-1}, \alpha]$, we get $|\delta_1| < \varepsilon$ and $|\delta_2| < \varepsilon$. Since $[0, b_1, \ldots, b_{s-1}, \alpha] = x - b_0 - \delta_1$ and $[0, d_1, \ldots, d_{t-1}, \alpha] = x - (T^4 + T^2)y - d_0 - \delta_2$,

$$s(a/c) = y + (\delta_2 - \delta_1 + \prod_{i=1}^{s+t}[0, h_i, \ldots, h_{s+t}])/(T^4 + T^2),$$

which gives $|y - s(a/c)| < 3\varepsilon$. $\square$

From Theorem 6 and its proof, the following result is obtained:

Theorem 7. *(i) If $q > 3$, then $\{s(a/c) \mid a/c \in K^*\} = \{0\}$.*
(ii) If $q = 2$ or 3, then $\{s(a/c) \mid a/c \in K^*\}$ is dense in K_∞.

Theorems 6 and 7 are analogs of Theorem 2.

Acknowledgements The author greatly thanks the referee for helpful comments and suggestions which led to improvements of this paper. The author would like to dedicate this paper to the memory of A.J. van der Poorten.

References

1. A. Bayad, Y. Hamahata, Higher dimensional Dedekind sums in function fields. Acta Arith. **152**, 71–80 (2012)
2. D. Goss, *Basic Structures of Function Fields* (Springer, New York, 1998)
3. G.H. Hardy, E. Wright, *An Introduction to the Theory of Numbers*, 6th edn (Oxford Univ. Press, Oxford, 2008)
4. D. Hickerson, Continued fractions and density results for Dedekind sums. J. Reine Angew. Math. **290**, 113–116 (1977)
5. A.J. van der Poorten, An introduction to continued fractions, in *Diophantine Analysis*, ed. by J.H. Loxton, A.J. van der Poorten (Cambridge Univ. Press, Cambridge, 1986), pp. 99–138
6. A.J. van der Poorten, Continued fractions of formal power series, in *Advances in Number Theory* (*Proceedings of the 3rd Conference of the CNTA*, Kingston, 1991), ed. by F.Q. Gouvêa, N. Yui (The Clarendon Press, Oxford, 1993), pp. 453–466
7. A.J. van der Poorten, Formal power series and their continued fraction expansion, in *Algorithmic Number Theory* (*Proceedings of the 3rd International Symposium, ANTS-III*, Portland, Oregon, USA, 1998), ed. by J. Buhler (Lecture Notes in Computer Science, Springer), **1423**, 358–371 (1998)
8. H. Rademacher, E. Grosswald, *Dedekind Sums* (Math. Assoc. Amer., Washington, DC, 1972)
9. D. Thakur, Exponential and continued fractions. J. Number Theory **59**, 248–261 (1996)
10. D. Zagier, Higher dimensional Dedekind sums. Math. Ann. **202**, 149–172 (1973)

Burgess's Bounds for Character Sums

D.R. Heath-Brown

Abstract Let $S(N;H) = \sum_{N<n\leq N+H} \chi(n)$ be a character sum to modulus q. Then the standard Burgess bound takes the form $S(N;H) \ll_{\varepsilon,r} B_r$, where $B_r = H^{1-1/r}q^{(r+1)/4r^2+\varepsilon}$. We show that

$$\sum_{j=1}^{J} \max_{h\leq H} |S(N_j;h)|^{3r} \ll_{\varepsilon,r} B_r^{3r}$$

for any positive integers $N_j \leq q$ spaced at least H apart, so that even reducing to a single term of the sum recovers the Burgess estimate.

1 Introduction

Let $\chi(n)$ be a non-principal character to modulus q. Then the well-known estimates of Burgess [2, 4, 5] say that if

$$S(N;H) := \sum_{N<n\leq N+H} \chi(n),$$

then for any positive integer $r \geq 2$ and any $\varepsilon > 0$, we have

$$S(N;H) \ll_{\varepsilon,r} H^{1-1/r}q^{(r+1)/(4r^2)+\varepsilon} \tag{1}$$

uniformly in N, providing either that q is cube-free or that $r \leq 3$. Indeed one can make the dependence on r explicit, if one so wants. Similarly the q^{ε} factor may be

D.R. Heath-Brown (✉)
Mathematical Institute, 24-29 St. Giles', Oxford OX1 3LB, UK
e-mail: rhb@maths.ox.ac.uk

J.M. Borwein et al. (eds.), *Number Theory and Related Fields: In Memory of Alf van der Poorten*, Springer Proceedings in Mathematics & Statistics 43,
DOI 10.1007/978-1-4614-6642-0_10, © Springer Science+Business Media New York 2013

replaced by a power of $d(q)\log q$ if one wishes. The upper bound has been the best known for around 50 years. The purpose of this note is to establish the following estimate, which gives a mean-value estimate including the original Burgess bound as a special case:

Theorem. *Let $r \in \mathbb{N}$ and let $\varepsilon > 0$ be a real number. Suppose that $\chi(n)$ is a primitive character to modulus $q > 1$, and let a positive integer $H \leq q$ be given. Suppose that $0 \leq N_1 < N_2 < \ldots < N_J < q$ are integers such that*

$$N_{j+1} - N_j \geq H, \quad (1 \leq j < J). \tag{2}$$

Then

$$\sum_{j=1}^{J} \max_{h \leq H} |S(N_j; h)|^{3r} \ll_{\varepsilon,r} H^{3r-3} q^{3/4+3/(4r)+\varepsilon}$$

under any of the three conditions:

(i) $r = 1$.
(ii) $r \leq 3$ and $H \geq q^{1/(2r)+\varepsilon}$.
(iii) q is cube-free and $H \geq q^{1/(2r)+\varepsilon}$.

The case $J = 1$ reduces to the standard Burgess estimate (which would be trivial if one took $H \leq q^{1/2r}$). Moreover one can deduce that there are only $O_{\varepsilon,r}(q^{(3r+1)\varepsilon})$ points N_j for which

$$\max_{h \leq H} |S(N_j; h)| \geq H^{1-1/r} q^{(r+1)/(4r^2)-\varepsilon},$$

for example. It would be unreasonable to ask for such a result without the spacing condition (2), since if A and B are intervals that overlap, it is possible that the behaviour of both $\sum_{n \in A} \chi(n)$ and $\sum_{n \in B} \chi(n)$ is affected by $\sum_{n \in A \cap B} \chi(n)$.

There are other results in the literature with which this estimate should be compared. Friedlander and Iwaniec [8, Theorem 2′] establish a bound for

$$\sum_{j=1}^{J} S(N_j; h)$$

which can easily be used to obtain an estimate of the form

$$\sum_{j=1}^{J} |S(N_j; h)|^{2r} \ll_{\varepsilon,r} h^{2r-2} q^{1/2+1/(2r)+\varepsilon}.$$

This is superior to our result in that it involves a smaller exponent $2r$. However, they do not include a maximum over h and their result is subject to the condition that $h(N_J - N_1) \leq q^{1+1/(2r)}$.

We should also mention the work of Chang [6, Theorem 8]. The result here is not so readily compared with ours or indeed with the Burgess estimate (1). However, with a certain amount of effort one may show that our theorem gives a sharper bound at least when $JH^3 \leq q^2$.

It would have been nice to have established a result like our theorem but involving the $2r$-th moment. The present methods do not allow this in general. However, for the special case $r = 1$, one can indeed achieve this, in the following slightly more flexible form. Specifically, suppose that $\chi(n)$ is a primitive character to modulus q, and let $I_1, \ldots, I_J$ be disjoint subintervals of $(0, q]$. Then for any $\varepsilon > 0$, we have

$$\sum_{j=1}^{J} |\sum_{n \in I_j} \chi(n)|^2 \ll_\varepsilon q^{1+\varepsilon} \tag{3}$$

with an implied constant depending only on ε. This is a mild variant of Lemma 4 of Gallagher and Montgomery [9]. One can deduce the Pólya–Vinogradov as an immediate consequence of Lemma 4 (which is the same as Gallagher and Montgomery's Lemma 4). In fact there are variants of (3) for quite general character sums. For simplicity we suppose q is a prime p. Let $f(x)$ and $g(x)$ be rational functions on $\mathbb{F}_p$, possibly identically zero. Then (3) remains true if we replace $\chi(n)$ by $\chi(f(n))e_p(g(n))$, providing firstly that we exclude poles of f and g from the sum and secondly that we exclude the trivial case in which $\chi(f)$ is constant and g is constant or linear. (The implied constant will depend on the degrees of the numerators and denominators in f and g.) We leave the proof of this assertion to the reader.

For $r = 1$, the ideas of this paper are closely related to those in the article of Davenport and Erdős [7], which was a precursor of Burgess's work. For $r \geq 2$, the paper follows the route to Burgess's bounds developed in unpublished notes by Hugh Montgomery, written in the 1970s, which were later developed into the Gallagher and Montgomery article [9]. In particular the mean-value lemmas in Sect. 2 are essentially the same as in their paper, except that we have given the appropriate extension to general composite moduli q. We reproduce the arguments merely for the sake of completeness.

After the mean-value lemmas in Sect. 2 have been established, we begin the standard attack on the Burgess bounds in Sect. 3 but incorporating the sum over N_j in a nontrivial way in Sect. 4. It is this final step that involves the real novelty in the paper. This process will lead to the following key lemma:

Lemma 1. *Let a positive integer $r \geq 2$ and a real number $\varepsilon > 0$ be given. Let $0 \leq N_1 < N_2 < \ldots < N_J < q$ be integers such that (2) holds. Then for any primitive character χ to modulus q and any positive integer $H \in (q^{1/(2r)}, q]$, we have*

$$\sum_{j=1}^{J} \max_{h \leq H} |S(N_j; h)|^r$$

$$\ll_{\varepsilon,r} q^{1/4+1/(4r)+\varepsilon} H^{r-1}\{J^{2/3} + J(H^{-1}q^{1/(2r)} + Hq^{-1/2-1/(4r)})\}, \tag{4}$$

provided either that $r \leq 3$ or that q is cube-free.

Throughout the paper we shall assume that q is sufficiently large in terms of r and ε wherever it is convenient. The results are clearly trivial when $q \ll_{\varepsilon,r} 1$. We should also point out that we shall replace ε by a small multiple from time to time. This will not matter since all our results hold for all $\varepsilon > 0$. Using this convention we may write $q^\varepsilon \log q \ll_\varepsilon q^\varepsilon$, for example.

2   Preliminary Mean-Value Bounds

Our starting point, taken from previous treatments of Burgess's bounds, is the following pair of mean-value estimates.

Lemma 2. *Let r be a positive integer and let $\varepsilon > 0$. Then if χ is a primitive character to modulus q, we have*

$$\sum_{n=1}^{q} |S(n;h)|^2 \ll_\varepsilon q^{1+\varepsilon} h$$

for any q, while

$$\sum_{n=1}^{q} |S(n;h)|^{2r} \ll_{\varepsilon,r} q^\varepsilon (qh^r + q^{1/2}h^{2r})$$

under any of the three conditions:

  (i) q is cube-free.
 (ii) $r = 2$.
(iii) $r = 3$ and $h \leq q^{1/6}$.

The case $r = 1$ is given by Norton [11, (2.8)], though the proof is attributed to Gallagher. For $r \geq 2$ the validity of the lemma under the first two conditions follows from Burgess [4, Lemma 8], using the same method as in Burgess [3, Lemma 8]. The estimate under condition (iii) is given by Burgess [5, Theorem B].

We proceed to deduce a maximal version of Lemma 2, as in Gallagher and Montgomery [9, Lemma 3].

Lemma 3. *Let r be a positive integer and let $\varepsilon > 0$. Then if χ is a primitive character to modulus q and $H \in \mathbb{N}$, we have*

$$\sum_{n=1}^{q} \max_{h \leq H} |S(n;h)|^2 \ll_\varepsilon q^{1+\varepsilon} H$$

for all q, while

$$\sum_{n=1}^{q} \max_{h \leq H} |S(n;h)|^{2r} \ll_{\varepsilon,r} q^\varepsilon (qH^r + q^{1/2}H^{2r})$$

under either of the conditions:

(i) *q is cube-free.*
(ii) $2 \leq r \leq 3$.

The strategy for the proof goes back to independent work of Rademacher [12] and Menchov [10] from 1922 and 1923, respectively. It clearly suffices to consider the case in which $H = 2^t$ is a power of 2. We will first prove the result under the assumption that $H \leq q^{1/(2r)}$. We will assume that $r \geq 2$, the case $r = 1$ being similar. Suppose that $|S(n;h)|$ attains its maximum at a positive integer $h = h(n) \leq H$, say. We may write

$$h = \sum_{d \in \mathcal{D}} 2^{t-d}$$

for a certain set $\mathcal{D}$ of distinct nonnegative integers $d \leq t$. Then

$$S(n;h) = \sum_{d \in \mathcal{D}} S(n + v_{n,d} 2^{t-d}; 2^{t-d}),$$

where

$$v_{n,d} = \sum_{e \in \mathcal{D}, e < d} 2^{d-e} < 2^d.$$

By Hölder's inequality, we have

$$|S(n;h)|^{2r} \leq \{\#\mathcal{D}\}^{2r-1} \left\{ \sum_{d \in \mathcal{D}} |S(n + v_{n,d} 2^{t-d}; 2^{t-d})|^{2r} \right\}.$$

We now include all possible values of d and v to obtain

$$|S(n;h)|^{2r} \leq (t+1)^{2r-1} \sum_{0 \leq d \leq t} \sum_{0 \leq v < 2^d} |S(n + v2^{t-d}; 2^{t-d})|^{2r},$$

and hence

$$\max_{h \leq H} |S(n;h)|^{2r} \leq (t+1)^{2r-1} \sum_{0 \leq d \leq t} \sum_{0 \leq v < 2^d} |S(n + v2^{t-d}; 2^{t-d})|^{2r}.$$

We proceed to sum over n modulo q, using Lemma 2, and on recalling that $H = 2^t \leq q^{1/(2r)}$, we deduce that

$$\sum_{n=1}^{q} \max_{h \leq H} |S(n;h)|^{2r}$$

$$\ll_{\varepsilon, r} (t+1)^{2r-1} \sum_{0 \leq d \leq t} \sum_{0 \leq v < 2^d} q^{\varepsilon} (q 2^{r(t-d)} + q^{1/2} 2^{2r(t-d)}).$$

$$\ll_{\varepsilon,r} q^{\varepsilon}(t+1)^{2r-1} \sum_{0 \leq d \leq t} \sum_{0 \leq v < 2^d} (qH^r + q^{1/2}H^{2r})2^{-d}$$

$$= q^{\varepsilon}(t+1)^{2r}(qH^r + q^{1/2}H^{2r})$$

$$\ll_{\varepsilon,r} q^{2\varepsilon}(qH^r + q^{1/2}H^{2r}).$$

This establishes Lemma 3 when H is a power of 2 of size at most $q^{1/(2r)}$.

To extend this to the general case, write H_0 for the largest power of 2 of size at most $q^{1/(2r)}$. Then

$$\max_{h \leq H} |S(n;h)| \leq \sum_{0 \leq j \leq H/H_0} \max_{h \leq H_0} |S(n+jH_0;h)|$$

whence

$$\sum_{n=1}^{q} \max_{h \leq H} |S(n;h)|^{2r} \ll (H/H_0)^{2r-1} \sum_{n=1}^{q} \sum_{0 \leq j \leq H/H_0} \max_{h \leq H_0} |S(n+jH_0;h)|^{2r}$$

$$= (H/H_0)^{2r-1} \sum_{0 \leq j \leq H/H_0} \sum_{n=1}^{q} \max_{h \leq H_0} |S(n+jH_0;h)|^{2r}$$

$$= (H/H_0)^{2r-1} \sum_{0 \leq j \leq H/H_0} \sum_{n \pmod q} \max_{h \leq H_0} |S(n;h)|^{2r}$$

$$\ll_{\varepsilon,r} (H/H_0)^{2r-1} \sum_{0 \leq j \leq H/H_0} q^{\varepsilon}(qH_0^r + q^{1/2}H_0^{2r})$$

$$\ll_{\varepsilon,r} q^{\varepsilon}(H/H_0)^{2r}(qH_0^r + q^{1/2}H_0^{2r}).$$

However, our choice of H_0 ensures that $qH_0^r \ll_r q^{1/2}H_0^{2r}$ and the lemma follows.

A variant of Lemma 3 allows us to sum over well-spaced points.

Lemma 4. *Suppose that $\chi(n)$ is a primitive character to modulus $q > 1$, and let a positive integer $H \leq q$ be given. Suppose that $0 \leq N_1 < N_2 < \ldots < N_J < q$ are integers satisfying the spacing condition (2). Then*

$$\sum_{j=1}^{J} \max_{h \leq H} |S(N_j;h)|^2 \ll_{\varepsilon,r} q^{\varepsilon}(qH^{r-1} + q^{1/2}H^{2r-1})$$

for any $\varepsilon > 0$.

To prove this we follow the argument in Gallagher and Montgomery [9, Lemma 4]. We first observe that for any $n \leq N$, we have

$$S(N;h) = S(n;N-n+h) - S(n;N-n).$$

If $h \leq H$, it follows that

$$|S(N;h)| \leq 2 \max_{k \leq 2H} |S(n;k)|$$

whenever $N - H < n \leq N$. Then, summing over integers $n \in (N - H, N]$, we find that

$$H|S(N;h)| \leq 2 \sum_{n \in (N-H,N]} \max_{k \leq 2H} |S(n;k)| \tag{5}$$

whence Hölder's inequality yields

$$|S(N;h)|^{2r} \ll H^{-1} \sum_{n \in (N-H,N]} \max_{k \leq 2H} |S(n;k)|^{2r}.$$

Since the intervals $(N_j - H, N_j]$ are disjoint modulo q, we then deduce that

$$\sum_{j=1}^{J} \max_{h \leq H} |S(N_j;h)|^{2r} \ll H^{-1} \sum_{n=1}^{q} \max_{k \leq 2H} |S(n;k)|^{2r}$$

and the result follows from Lemma 3.

We can now deduce (3). By a dyadic subdivision, it will be enough to prove the result under the additional assumption that there is an integer H such that all the intervals I_j have length between $H/2$ and H. Thus we may write $I_j = (M_j, M_j + h_j]$ with $h_j \leq H$ for $1 \leq j \leq J$ and $M_{j+1} - M_j \geq H/2$ for $1 \leq j < J$. We may therefore apply the case $r = 1$ of Lemma 4 separately to the even-numbered intervals and the odd-numbered intervals to deduce (3).

3 Burgess's Method

In this section we will follow a mild variant of Burgess's method. Although there are small technical differences from previous works on the subject, there is no great novelty here.

For any prime $p < q$ which does not divide q, we will split the integers $n \in (N, N+h]$ into residue classes $n \equiv aq \pmod{p}$, for $0 \leq a < p$. Then we can write $n = aq + pm$ with $m \in (N', N' + h']$ say, where

$$N' = \frac{N - aq}{p}, \quad h' = \frac{h}{p}.$$

We then find that

$$S(N;h) = \chi(p) \sum_{0 \leq a < p} S(N';h')$$

and hence

$$|S(N;h)| \le \sum_{0 \le a < p} |S(N';h')|.$$

We now choose an integer parameter P in the range $(\log q)^2 \le P < q/2$ and sum the above estimate for all primes $p \in (P, 2P]$ not dividing q. Since the number of such primes is asymptotically $P/(\log P)$, we deduce that

$$P/(\log P)|S(N;h)| \ll \sum_{P < p \le 2P} \sum_{0 \le a < p} |S(N';h')|. \tag{6}$$

We now apply the inequality (5), with H replaced by H/P. Since we have $h' \le H/P$, we deduce that

$$HP^{-1}|S(N';h')| \ll \sum_{n \in (N'-H/P,N']} \max_{j \le 2H/P} |S(n;j)|.$$

Inserting this bound into (6), we find that

$$|S(N;h)| \ll (\log P)H^{-1}\sum_{n} A(n;N) \max_{j \le 2H/P} |S(n;j)|,$$

where

$$A(n,N) := \#\{(a,p) : P < p \le 2P, 0 \le a < p, n \le N' < n+H/P\},$$
$$= \#\{(a,p) : n \le (N-aq)/p < n+H/P\}.$$

Since

$$\sum_{n} A(n,N) = \sum_{a,p} \#\{n : n \le N' < n+H/P\} \le \sum_{a,p} \frac{H}{P} \ll PH,$$

we deduce from Hölder's inequality that

$$|S(N;h)|^r \ll (\log P)^r P^{r-1} H^{-1} \sum_{n} A(n;N) \max_{j \le 2H/P} |S(n;j)|^r,$$

for any $h \le H$. It should be noted that $A(n,N) = 0$ unless $|n| \le 2q$ so that the sum over n may be restricted to this range.

We proceed to sum over the values $N = N_j$ in Lemma 1, finding that

$$\sum_{j=1}^{J} \max_{h \le H} |S(N_j;h)|^r \ll (\log P)^r P^{r-1} H^{-1} \sum_{n} A(n) \max_{j \le 2H/P} |S(n;j)|^r,$$

where

$$A(n) := \#\{(a,p,N_j) : n \le (N_j - aq)/p < n+H/P\}.$$

From Cauchy's inequality, we then deduce that

$$\sum_{j=1}^{J} \max_{h \leq H} |S(N_j; h)|^r \ll (\log P)^r P^{r-1} H^{-1} \mathcal{N}^{1/2} \left\{ \sum_{|n| \leq 2q} \max_{j \leq 2H/P} |S(n; j)|^{2r} \right\}^{1/2},$$

where

$$\mathcal{N} := \sum_n A(n)^2 \leq H P^{-1} \mathcal{M},$$

with

$$\mathcal{M} := \#\{(a_1, a_2, p_1, p_2, N_j, N_k) : |(N_j - a_1 q)/p_1 - (N_k - a_2 q)/p_2| \leq H/P\}.$$

Thus

$$\sum_{j=1}^{J} \max_{h \leq H} |S(N_j; h)|^r$$

$$\ll (\log P)^r P^{r-3/2} H^{-1/2} \mathcal{M}^{1/2} \left\{ \sum_{|n| \leq 2q} \max_{j \leq 2H/P} |S(n; j)|^{2r} \right\}^{1/2}.$$

The second sum on the right may be bounded via Lemma 3, giving

$$\sum_{j=1}^{J} \max_{h \leq H} |S(N_j; h)|^r \ll_{\varepsilon, r} q^{\varepsilon} P^{r-3/2} H^{-1/2} (q^{1/2} (H/P)^{r/2} + q^{1/4} (H/P)^r) \mathcal{M}^{1/2},$$

on replacing ε by $\varepsilon/2$.

Naturally, in order to apply Lemma 3, we will need to have q cube-free, or $r \leq 3$. The natural choice for P is to take $2Hq^{-1/(2r)} \leq P \ll Hq^{-1/(2r)}$ so that $q^{1/2}(H/P)^{r/2}$ and $q^{1/4}(H/P)^r$ have the same order of magnitude. The conditions previously imposed on P are then satisfied provided that $H \geq q^{1/(2r)}$. With this choice for P, we deduce that

$$\sum_{j=1}^{J} \max_{h \leq H} |S(N_j; h)|^r \ll_{\varepsilon, r} q^{1/4 + 3/(4r) + \varepsilon} H^{r-2} \mathcal{M}^{1/2}. \tag{7}$$

4 Estimating $\mathcal{M}$

In this section we will estimate $\mathcal{M}$ and complete the proof of Lemma 1. It is the treatment of $\mathcal{M}$ which represents the most novel part of our argument.

We split $\mathcal{M}$ as $\mathcal{M}_1 + \mathcal{M}_2$ where $\mathcal{M}_1$ counts solutions with $p_1 = p_2$ and $\mathcal{M}_2$ corresponds to $p_1 \neq p_2$. When $p_1 = p_2$, the defining condition for $\mathcal{M}$ becomes

$$|(N_j - N_k) - q(a_1 - a_2)| \le p_1 H/P \le 2H.$$

Thus

$$|a_1 - a_2| \le q^{-1}(|N_j - N_k| + 2H) \le 3.$$

Moreover, given N_k and $a_1 - a_2$, there will be at most five choices for N_j, in view of the spacing condition (2). Thus we must allow for $O(P)$ choices for p_1, for $O(P)$ choices for a_1 and a_2, and $O(J)$ choices for N_j and N_k so that

$$\mathcal{M}_1 \ll P^2 J. \tag{8}$$

To handle $\mathcal{M}_2$, we begin by choosing a prime ℓ in the range

$$q/H < \ell \le 2q/H.$$

This is possible, by Bertrand's Postulate. We then set

$$M_j := \left[\frac{N_j \ell}{q}\right], \quad (1 \le j \le J)$$

so that the M_j are nonnegative integers in $[0, \ell)$. Moreover the spacing condition (2) implies that

$$M_{j+1} > \frac{N_{j+1}\ell}{q} - 1 \ge \frac{(N_j + H)\ell}{q} - 1 > \frac{N_j \ell}{q} \ge M_j$$

so that the integers M_j form a strictly increasing sequence. Since

$$|N_j - qM_j/\ell| \le q/\ell,$$

we now see that if $(a_1, a_2, p_1, p_2, N_j, N_k)$ is counted by $\mathcal{M}_2$, then

$$\left|\frac{qM_j/\ell - a_1 q}{p_1} - \frac{qM_k/\ell - a_2 q}{p_2}\right| \le \frac{H}{P} + \frac{q}{\ell p_1} + \frac{q}{\ell p_2}$$

whence

$$|p_2 M_j - p_1 M_k - \ell\delta| \le \frac{H\ell p_1 p_2}{Pq} + p_1 + p_2 \le 12P,$$

with $\delta = a_1 p_2 - a_2 p_1$. If p_1, p_2 and δ are given, there is at most one pair of integers a_1, a_2 with $0 \le a_1 < p_1$, $0 \le a_2 < p_2$ and $a_1 p_2 - a_2 p_1 = \delta$. Thus

$$\mathcal{M}_2 \le \sum_{M_j, M_k} \#\{(p_1, p_2, m) : |m| \le 12P, \ p_2 M_j - p_1 M_k \equiv m \pmod{\ell}\}.$$

We now consider how many pairs p_1, p_2 there may be for each choice of M_j, M_k. We define the set

$$\Lambda := \{(x, y, z) \in \mathbb{Z}^3 : xM_j - yM_k \equiv z \pmod{\ell}\},$$

which will be an integer lattice of determinant ℓ. Admissible pairs p_1, p_2 produce points $\mathbf{x} = (x, y, z) \in \Lambda$ with $x \neq y$ both prime and $|\mathbf{x}| \leq 12P$, where

$$|\mathbf{x}| := \max(|x|, |y|, |z|).$$

The lattice Λ has a $\mathbb{Z}$-basis $\mathbf{b}_1, \mathbf{b}_2, \mathbf{b}_3$ such that

$$|\mathbf{b}_1| \leq |\mathbf{b}_2| \leq |\mathbf{b}_3| \tag{9}$$

and

$$\det(\Lambda) \ll |\mathbf{b}_1| . |\mathbf{b}_2| . |\mathbf{b}_3| \ll \det(\Lambda) = \ell \tag{10}$$

and with the property that there is an absolute constant c_0 such that if $\mathbf{x} \in \Lambda$ is written as $\lambda_1 \mathbf{b}_1 + \lambda_2 \mathbf{b}_2 + \lambda_3 \mathbf{b}_3$, then

$$|\lambda_i| \leq c_0 |\mathbf{x}| / |\mathbf{b}_i|, \quad (1 \leq i \leq 3).$$

The existence of such a basis is a standard fact about lattices, see Browning and Heath-Brown [1, Lemma 1, (ii)], for example. When $|\mathbf{b}_3| \leq 12c_0 P$, we now see that the number of lattice elements of size at most $12P$ is

$$\leq \left(1 + \frac{24c_0 P}{|\mathbf{b}_1|}\right) \left(1 + \frac{24c_0 P}{|\mathbf{b}_2|}\right) \left(1 + \frac{24c_0 P}{|\mathbf{b}_3|}\right)$$

$$\ll \frac{P^3}{|\mathbf{b}_1| . |\mathbf{b}_2| . |\mathbf{b}_3|}$$

$$\ll \frac{P^3}{\det(\Lambda)}$$

$$\ll HP^3 q^{-1}$$

by (9) and (10). If $|\mathbf{b}_1| > 12c_0 P$, the only vector in Λ of norm at most $12P$ is the origin, while if $|\mathbf{b}_1| \leq 12c_0 P < |\mathbf{b}_2|$, the only possible vectors are of the form $\lambda_1 \mathbf{b}_1$. In this latter case $(p_2, p_1, m) = \lambda_1 \mathbf{b}_1$ so that λ_1 divides h.c.f.$(p_2, p_1) = 1$. Hence there is at most 1 solution in this case.

There remains the situation in which $|\mathbf{b}_2| \leq 12c_0 P < |\mathbf{b}_3|$ so that the admissible vectors are linear combinations $\lambda_1 \mathbf{b}_1 + \lambda_2 \mathbf{b}_2$. In this case we write $\mathbf{b}_i = (x_i, y_i, z_i)$ for $i = 1, 2$ and set $\Delta = x_1 y_2 - x_2 y_1$. If $\Delta = 0$, then (x_1, y_1) and (x_2, y_2) are proportional and hence are both integral scalar multiples of some primitive vector (x, y) say. However, we then see that if $(p_2, p_1, m) = \lambda_1 \mathbf{b}_1 + \lambda_2 \mathbf{b}_2$, then (p_2, p_1) is a scalar multiple of (x, y) so that $\mathbf{b}_1$ and $\mathbf{b}_2$ determine p_1 and p_2. Thus when $\Delta = 0$, the primes p_1 and p_2 are determined by M_j and M_k. In order to summarize our conclusions up to this point, we write $\mathcal{M}_3$ for the contribution to $\mathcal{M}_2$ corresponding to all cases except that in which $|\mathbf{b}_2| \leq 12c_0 P < |\mathbf{b}_3|$ and $\Delta \neq 0$. With this notation, we then have

$$\mathcal{M}_3 \ll (HP^3 q^{-1} + 1)J^2. \tag{11}$$

Suppose now that $|\mathbf{b}_2| \leq 12c_0 P < |\mathbf{b}_3|$ and $\Delta \neq 0$. We will write $\mathcal{M}_4$ for the corresponding contribution to $\mathcal{M}$. In this case we must have $\lambda_3 = 0$, and the number of choices for λ_1 and λ_2 will be

$$\leq \left(1 + \frac{24c_0 P}{|\mathbf{b}_1|}\right)\left(1 + \frac{24c_0 P}{|\mathbf{b}_2|}\right) \ll \frac{P^2}{|\mathbf{b}_1|.|\mathbf{b}_2|}.$$

Thus if $L < |\mathbf{b}_1|.|\mathbf{b}_2| \leq 2L$, say, the contribution to $\mathcal{M}_4$ will be $O(P^2 L^{-1})$ for each pair M_j, M_k.

To estimate the number of pairs of vectors $\mathbf{b}_1, \mathbf{b}_2$ with $L < |\mathbf{b}_1|.|\mathbf{b}_2| \leq 2L$, we observe that there are $O(B_1^3 B_2^3)$ possible choices with $B_1 < |\mathbf{b}_1| \leq 2B_1$ and $B_2 < |\mathbf{b}_2| \leq 2B_2$. A dyadic subdivision then shows that we will have to consider $O(L^3 \log L)$ pairs $\mathbf{b}_1, \mathbf{b}_2$. Writing $\mathbf{b}_i = (x_i, y_i, z_i)$ for $i = 1, 2$ as before, we will have

$$x_1 M_j - y_1 M_k \equiv z_1 \pmod{\ell}, \quad x_2 M_j - y_2 M_k \equiv z_2 \pmod{\ell}.$$

These congruences determine ΔM_j and ΔM_k modulo ℓ, and since ℓ is prime and $0 \leq M_j, M_k < \ell$, we see that $\mathbf{b}_1$ and $\mathbf{b}_2$ determine M_j, M_k precisely, providing that $\ell \nmid \Delta$. However,

$$|\Delta| \leq 2|\mathbf{b}_1||\mathbf{b}_2| \leq 2(|\mathbf{b}_1|.|\mathbf{b}_2|.|\mathbf{b}_3|)^{2/3} \ll \det(\Lambda)^{2/3} = \ell^{2/3}$$

by (9) and (10). Since $\Delta \neq 0$, we then see that $\ell \nmid \Delta$ providing that q/H, or equivalently ℓ, is sufficiently large. Under this assumption we therefore conclude that there are $O(L^3 \log L)$ pairs M_j, M_k for which $|\mathbf{b}_2| \leq 12c_0 P < |\mathbf{b}_3|$ and $\Delta \neq 0$ and for which $L < |\mathbf{b}_1|.|\mathbf{b}_2| \leq 2L$. Thus each dyadic range $(L, 2L]$ contributes $O(P^2 L^{-1} \min(J^2, L^3 \log L))$ to $\mathcal{M}_4$. Since

$$P^2 L^{-1} \min(J^2, L^3) \leq P^2 L^{-1}(J^2)^{2/3}(L^3)^{1/3} = P^2 J^{4/3},$$

we deduce that

$$\mathcal{M}_4 \ll P^2 J^{2/3} \log q,$$

and comparing this with the bounds (8) and (11), we then see that

$$\mathcal{M} \ll (HP^3 q^{-1} + 1)J^2 + P^2 J^{4/3} \log q.$$

We may now insert this bound into (7), recalling that P is of order $Hq^{-1/(2r)}$ to deduce, after replacing ε by $\varepsilon/2$ that

$$\sum_{j=1}^{J} \max_{h \leq H} |S(N_j;h)|^r$$

$$\ll_{\varepsilon,r} q^{1/4+1/(4r)+\varepsilon} H^{r-1} \{ J^{2/3} + J(H^{-1}q^{1/(2r)} + Hq^{-1/2-1/(4r)}) \},$$

as required for Lemma 1.

5   Deduction of the Theorem

We will prove the theorem by induction on r. The result for $r = 1$ is an immediate consequence of Lemma 4, together with the Pólya–Vinogradov inequality.

For $r \geq 2$, we will use a dyadic subdivision, classifying the N_j according to the value $V = 2^v$ for which

$$V/2 < \max_{h \leq H} |S(N_j;h)|^r \leq V. \tag{12}$$

Clearly numbers N_j for which the corresponding V is less than 1 make a satisfactory contribution in our theorem, and so it suffices to assume that (12) holds for all N_j.

We now give three separate arguments, depending on which of the three terms on the right of (4) dominates. If

$$\sum_{j=1}^{J} \max_{h \leq H} |S(N_j;h)|^r \ll_{\varepsilon,r} q^{1/4+1/(4r)+\varepsilon} H^{r-1} J^{2/3},$$

then

$$JV^r \ll_{\varepsilon,r} q^{1/4+1/(4r)+\varepsilon} H^{r-1} J^{2/3}$$

whence

$$JV^{3r} \ll_{\varepsilon,r} q^{3/4+3/(4r)+3\varepsilon} H^{3r-3},$$

which suffices for the theorem. If the second term dominates, we will have

$$\sum_{j=1}^{J} \max_{h \leq H} |S(N_j;h)|^r \ll_{\varepsilon,r} q^{1/4+1/(4r)+\varepsilon} H^{r-1} JH^{-1}q^{1/(2r)}$$

so that

$$JV^r \ll_{\varepsilon,r} q^{1/4+3/(4r)+\varepsilon} H^{r-2} J.$$

In this case, it follows that

$$V^r \ll_{\varepsilon,r} q^{1/4+3/(4r)+\varepsilon} H^{r-2}. \tag{13}$$

We now use Lemma 4, which implies that

$$JV^{2r} \ll_{\varepsilon,r} q^{\varepsilon}(qH^{r-1} + q^{1/2}H^{2r-1}) \ll_{\varepsilon,r} q^{1/2+\varepsilon}H^{2r-1} \tag{14}$$

since $H \geq q^{1/(2r)}$. Coupled with (13), this yields

$$JV^{3r} \ll_{\varepsilon,r} q^{3/4+3/(4r)+2\varepsilon}H^{3r-3}$$

which again suffices for the theorem. Finally, if the third term on the right of (4) dominates, we must have

$$JV^r \ll_{\varepsilon,r} q^{-1/4+\varepsilon}H^r J$$

whence $V \ll_{\varepsilon} Hq^{-1/(4r)+\varepsilon/r}$. Here we shall use the inductive hypothesis, which tells us that

$$JV^{3r-3} \ll_{\varepsilon,r} q^{3/4+3/(4r-4)+\varepsilon}H^{3r-6}$$

if either $r = 2$ or $H \geq q^{1/(2r-2)}$ and $r \geq 3$. Under this latter assumption we therefore deduce that

$$JV^{3r} \ll_{\varepsilon,r} q^{3/4+\phi+4\varepsilon}H^{3r-3}$$

with

$$\phi = \frac{3}{4r-4} - \frac{3}{4r} \leq \frac{3}{4r}$$

for $r \geq 2$. It therefore remains to consider the case in which $r \geq 3$ and $q^{1/(2r)} \leq H \leq q^{1/(2r-2)}$. However, for such H, we may again use the bound (14) whence

$$
\begin{aligned}
JV^{3r} &\ll_{\varepsilon,r} q^{1/2+\varepsilon}H^{2r-1} \cdot q^{-1/4+\varepsilon}H^r \\
&= q^{1/4+2\varepsilon}H^{3r-3}\{Hq^{-1/(2r-2)}\}^2 q^{1/(r-1)} \\
&\leq q^{1/4+1/(r-1)+2\varepsilon}H^{3r-3}.
\end{aligned}
$$

To complete the proof of this final case, it remains to observe that $1/4 + 1/(r-1) \leq 3/4 + 3/(4r)$ for $r \geq 3$.

References

1. T.D. Browning, D.R. Heath-Brown, Equal sums of three powers. Invent. Math. **157**, 553–573 (2004)
2. D.A. Burgess, On character sums and primitive roots. Proc. Lond. Math. Soc. **12**(3), 179–192 (1962)
3. D.A. Burgess, On character sums and L-series. Proc. Lond. Math. Soc. **12**(3), 193–206 (1962)
4. D.A. Burgess, On character sums and L-series. II. Proc. Lond. Math. Soc. **13**(3), 524–536 (1963)
5. D.A. Burgess, The character sum estimate with $r = 3$. J. Lond. Math. Soc. **33**(2), 219–226 (1986)

6. M.-C. Chang, On a question of Davenport and Lewis and new character sum bounds in finite fields. Duke Math. J. **145**, 409–442 (2008)
7. H. Davenport, P. Erdős, The distribution of quadratic and higher residues. Publ. Math. Debrecen **2**, 252–265 (1952)
8. J. Friedlander, H. Iwaniec, Estimates for character sums. Proc. Am. Math. Soc. **119**, 365–372 (1993)
9. P.X. Gallagher, H.L. Montgomery, A Note on Burgess's estimate. Math. Notes **88**, 321–329 (2010)
10. D. Menchov, Sur les séries de fonctions orthogonales. Fund. Math. **1**, 82–105 (1923)
11. K.K. Norton, On character sums and power residues. Trans. Am. Math. Soc. **167**, 203–226 (1972)
12. H. Rademacher, Einige Sätze über Reihen von allgemeinen Orthogonal-Funktionen. Math. Ann. **87**, 112–138 (1922)

Structured Hadamard Conjecture

Ilias S. Kotsireas

Étant donné un déterminant

$$\Delta = \begin{vmatrix} a_1 & b_1 & \dots & \ell_1 \\ a_2 & b_2 & \dots & \ell_2 \\ \dots & \dots & \dots & \dots \\ a_n & b_n & \dots & \ell_n \end{vmatrix}$$

dans lequel on sait que les éléments sont inférieurs en valeur absolue à une quantité déterminée A, il y a souvent lieu de chercher une limite que le module de Δ ne puisse dépasser.

– Jacques Hadamard, 1893

Abstract We present three different formalisms for a structured version of the Hadamard conjecture. Two of these formalisms are new, and we use them to provide independent verifications of some of the previously known computational results on this structured version of the Hadamard conjecture.

1 Introduction

Hadamard matrices are $n \times n$ matrices H with ± 1 elements such that $H \cdot H^t = nI_n$, where t denotes transposition and I_n denotes the $n \times n$ unit matrix. Aside from the trivial cases $n = 1$ and 2, a well-known necessary condition for the existence of a Hadamard matrix of order n is that $n \equiv 0 \pmod 4$; see [17]. The sufficiency of this condition is the celebrated *Hadamard conjecture* whose origins can be traced to a rather inauspicious beginning in [16] and which remains unsolved for more than a century. As it is often the case with other eponymous conjectures, the names of other pioneers (in this case Sylvester and Paley) are also associated with the conjecture.

I.S. Kotsireas (✉)
Wilfrid Laurier University, 75 University Avenue West, Waterloo ON, Canada N2L 3C5
e-mail: ikotsire@wlu.ca

J.M. Borwein et al. (eds.), *Number Theory and Related Fields: In Memory of Alf van der Poorten*, Springer Proceedings in Mathematics & Statistics 43,
DOI 10.1007/978-1-4614-6642-0_11, © Springer Science+Business Media New York 2013

A historical retrospective and recent progress on the Hadamard conjecture with an emphasis on asymptotic results are discussed by Jennifer Seberry in her article in these proceedings. Additional reliable and readable sources of information for Hadamard matrices in general include [18, 29].

There are several different kinds of Hadamard matrices; see [19, 32], and the progress update [20]. The book [19] exemplifies the group cohomological aspects of Hadamard matrix constructions. However, as it is explained in detail in Jennifer Seberry's article in these proceedings, most constructions fail to yield Hadamard matrices for every order which is a multiple of 4 and therefore are not suitable candidates for a proof of the Hadamard conjecture. Even the concerted use of all known constructions for Hadamard matrices is not enough to cover the entire range of multiples of 4, and in this context it turns out that the order $n = 668$ is currently the smallest order for which a Hadamard matrix is not known; see [4, 19]. Note that a Hadamard matrix of order 428 (the previously unknown smallest order) has been constructed in [21]. On the other hand, the sharpest asymptotic result on the density of the set of orders of Hadamard matrices known today is proved in [5].

A very valuable library of Hadamard matrices is maintained by Sloane [30]. Hadamard matrices often have unexpected uses, for example, Tao used a 12×12 Hadamard matrix (taken from Sloane's library, even though any Hadamard matrix of order 12 would have worked equally well, since they are all equivalent under the usual Hadamard equivalence) that is his disproof [35] of the Fuglede conjecture [14] and Barany [1] used Hadamard matrices to give a near-optimal algorithm for an NP-complete problem. One of the ways the importance of Jacques Hadamard in the Pantheon of Mathematicians has been recognized is the establishment of the *Fondation Mathématique Jacques Hadamard* in his native France [13]. The book [26] contains a systematic biography of Jacques Hadamard illustrated with several pictures from his life.

Hadamard and skew-Hadamard matrices of large orders have been constructed via a cyclotomy-based method (with group-theoretic underpinnings) by Djokovic; see [7–10]. There exist exponentially large lower bounds for the numbers of Hadamard matrices for various orders, as demonstrated by the powerful techniques of Lam, Lam, and Tonchev; see [23, 24].

There are two exact formulas in the literature that give the number of Hadamard matrices of order n. The first one is due to Eliahou in his landmark paper [11], and the second one is due to de Launey and Levin in their landmark paper [6]. The Eliahou formula uses coding theory, while the de Launey–Levin formula uses multidimensional integrals associated to lattice walks. Both formulas require a certain amount of technical definitions before they can be stated in a self-contained manner, so we omit their precise statements. Both formulas are difficult to evaluate for large n.

The Hadamard conjecture is related to the twin prime conjecture via Gruner's theorem; see [36]: if p and $p + 2$ are twin primes, then there exist Hadamard matrices of order $p(p+2)+1$. Therefore a proof of the twin prime conjecture would automatically furnish a new infinite class of Hadamard matrices, even though this class gives a quite sparse set of orders of Hadamard matrices, within the set of all multiples of 4. There is another strong number-theoretic connection of the Hadamard conjecture with Jacobi sums, via methods similar to the ones in [37].

The first aim of this paper is to point out that perhaps the Hadamard conjecture is so difficult to solve precisely because it is too general. This generality emanates from the fact that there are so many different constructions for Hadamard matrices and it also manifests itself upon careful examination of the Eliahou and de Launey–Levin enumeration formulas.

The second aim of this paper is to point out that there is one particular construction that seems to be one of the most prominent candidates to furnish a proof of the Hadamard conjecture, namely, the so-called *two-circulant core* construction; see [12, 22]. One of the important characteristics of this construction is that it introduces some *structure* into the more general Hadamard conjecture. This structure is described in terms of two circulant matrices with very specific properties that we shall detail in the sequel. The simplicity (and inherent elegance) of this structure along with the numerical evidence gathered in [22] led us to believe that this *structured Hadamard conjecture* is a worthwhile pursuit.

Another very promising candidate for a proof of the Hadamard conjecture seems to be the Ito Hadamard construction; see the numerical evidence in Table 6.2 on p.132 of [19]. This construction is based on a cocyclic technique using a single cohomology class for the dihedral group of the appropriate order. The matrix template for the Ito construction (see p. 130 of [19]) is arguably as simple as the two-circulant core one. The number of equivalence classes of two-circulant core Hadamard matrices has been calculated in [27], and it is shown that in the orders $n = 28, \ldots, 40$, there is always at least one equivalence class of such Hadamard matrices which is cocyclic but others which are not. In [28] the authors have also computed the numbers of equivalence classes of cocyclic Hadamard matrices in these orders; in each order there are more equivalence classes of cocyclic Hadamard matrices (over all index groups) than there are equivalence classes of two-circulant core Hadamard matrices.

The third aim of this paper is to demonstrate that the structured Hadamard conjecture has many different incarnations and therefore can be described via several equivalent formulations, and we attempted to make use of some of these equivalent formulations in order to verify and expand the numerical evidence presented [22].

The rest of this paper is organized as follows: In Sect. 2 we describe the two-circulant core construction for Hadamard matrices and survey some of its basic features. In the next three Sects. 3–5, we describe three alternative formulations of the ensuing structured Hadamard conjecture. In Sect. 6 we state our conclusions.

2 The Two-Circulant Core Construction for Hadamard Matrices

First, we introduce two concepts that are necessary in order to describe the two-circulant core construction for Hadamard matrices.

Definition 1. An $n \times n$ matrix $C(A)$ is called *circulant* if every row (except the first) is obtained by the previous row by a right cyclic shift by one. In particular,

$$C(A) = \begin{bmatrix} a_0 & a_1 & \cdots & a_{n-2} & a_{n-1} \\ a_{n-1} & a_0 & \cdots & a_{n-3} & a_{n-2} \\ \vdots & \vdots & \cdots & \vdots & \vdots \\ a_2 & a_3 & \cdots & a_0 & a_1 \\ a_1 & a_2 & \cdots & a_{n-1} & a_0 \end{bmatrix}.$$

Definition 2. The periodic autocorrelation function associated to a finite sequence $A = [a_0, \ldots, a_{n-1}]$ of length n is defined as

$$P_A(s) = \sum_{k=0}^{n-1} a_k a_{k+s}, \ s = 0, \ldots, n-1,$$

where $k+s$ is taken modulo n, when $k+s > n$.

Circulant matrices and periodic autocorrelation are linked by the following:

Property 1. Consider a finite sequence $A = [a_0, \ldots, a_{n-1}]$ of length n and the circulant matrix $C(A)$ whose first row is equal to A. Then $P_A(i)$ is the inner product of the first row of $C(A)$ and the $i+1$ row of $C(A)$.

Periodic autocorrelation values exhibit certain symmetries.

Property 2.
$$P_A(s) = P_A(n-s), s = 1, \ldots, n-1.$$

Periodic autocorrelation is also related with the second elementary symmetric function $e_2(a_0, \ldots, a_{n-1}) = \sum_{0 \leq i < j \leq n-1} a_i a_j$ in the n variables $a_0, \ldots, a_{n-1}$.

Property 3.
$$P_A(1) + \cdots + P_A(n-1) = 2 e_2(a_0, \ldots, a_{n-1}).$$

We are now ready to describe the two-circulant core construction for Hadamard matrices; see [12, 22].

Theorem 1. *Let ℓ be an odd integer, such that $\ell > 1$ and set $m = \frac{\ell-1}{2}$. If there exist two ± 1 sequences $A = [a_0, \ldots, a_{\ell-1}]$ and $B = [b_0, \ldots, b_{\ell-1}]$ of length ℓ each, such that*

$$P_A(s) + P_B(s) = -2, \text{ for } s = 1, \ldots, m, \tag{1}$$

then there exists a Hadamard matrix of order $2\ell + 2$ *with two circulant cores given by*

$$
H_{2\ell+2} = \begin{bmatrix}
-\ -\ & +\ \cdots\ +\ +\ \cdots\ + \\
-\ + & +\ \cdots\ +\ -\ \cdots\ - \\
+\ + & \\
\vdots\ \vdots & C(A) \qquad C(B) \\
+\ + & \\
\hline
+\ - & \\
\vdots\ \vdots & C(B)^t \qquad -C(A)^t \\
+\ - &
\end{bmatrix}. \tag{2}
$$

Note that in the expression for $H_{2\ell+2}$, we use the customary notation that $+$ stands for a $+1$ element and $-$ stands for a -1 element. Here is a 12×12 example of a Hadamard matrix with two circulant cores.

Example 1. Let $\ell = 5$, $m = 2$, and consider the two sequences $A = [1, 1, -1, -1, 1]$ and $B = [-1, 1, -1, 1, 1]$. Then we have that

s	$P_A(s)$	$P_B(s)$	$P_A(s) + P_B(s)$
1	1	-3	-2
2	-3	1	-2

and the corresponding matrix H_{12} with two circulant cores of the form (2) is a 12×12 Hadamard matrix.   □

It is easy to see that if we have a Hadamard matrix with two circulant cores, then the Diophantine equation

$$
(a_0 + \cdots + a_{\ell-1})^2 + (b_0 + \cdots + b_{\ell-1})^2 = 2 \tag{3}
$$

must hold. For technical reasons and without loss of generality, we restrict ourselves in the case where $a_0 + \cdots + a_{\ell-1} = 1$ and $b_0 + \cdots + b_{\ell-1} = 1$. Exhaustive searches for Hadamard matrices with two circulant cores of the form (2) have been undertaken in [22], and we reproduce here the relevant table. These exhaustive searches were performed by the author on the SHARcnet and WestGrid high-performance computing consortia in Canada, as well as on the AIST supercomputer in Japan, courtesy of Dr. Hiroshi Kai. In view of the numerical evidence of Table 1, together with the simple fact that the numbers $2\ell + 2$ for every odd ℓ cover the entire range of multiples of 4, it is reasonable to speculate that the two-circulant core construction for Hadamard matrices is one of the most promising candidates for a proof of the Hadamard conjecture and therefore it is worthy of being called *structured Hadamard conjecture*. The important realization here is that this structured form of the Hadamard conjecture is described solely via the two specific (and simple) concepts of circulant matrix and periodic autocorrelation function, and this can be used as a starting point to seek a proof for it.

Table 1 Exhaustive searches for Hadamard matrices with two circulant cores

ℓ	Order of $H_{2\ell+2}$	Total number of matrices	
3	8	9	$= 1 \times 3^2$
5	12	50	$= 2 \times 5^2$
7	16	196	$= 4 \times 7^2$
9	20	972	$= 12 \times 9^2$
11	24	2,904	$= 24 \times 11^2$
13	28	7,098	$= 42 \times 13^2$
15	32	38,700	$= 172 \times 15^2$
17	36	93,058	$= 322 \times 17^2$
19	40	161,728	$= 448 \times 19^2$
21	44	433,944	$= 984 \times 21^2$
23	48	1,235,744	$= 2336 \times 23^2$
25	52	2,075,000	$= 3320 \times 25^2$

In the next three sections of this paper, we state three different formulations of the structured Hadamard conjecture.

3 Algebraic Geometry and Computational Algebra Formulation

As it is pointed out in [22], one can view the structured Hadamard conjecture as a statement about certain ideals in a polynomial ring with 2ℓ variables $a_0, \ldots, a_{\ell-1}$, $b_0, \ldots, b_{\ell-1}$.

With ℓ, m defined as in (2), we reproduce the following definition from [22], using the definition of the periodic autocorrelation:

Definition 3. The ℓ**th Hadamard ideal** $\mathscr{H}_\ell$ (associated with the two-circulant cores construction) is defined by

$$\mathscr{H}_\ell = \langle P_A(1) + P_B(1) + 2, \ldots, P_A(m) + P_B(m) + 2,$$
$$a_0 + \ldots + a_{\ell-1} - 1, b_0 + \ldots + b_{\ell-1} - 1,$$
$$a_0^2 - 1, \ldots, a_{\ell-1}^2 - 1, b_0^2 - 1, \ldots, b_{\ell-1}^2 - 1 \rangle.$$

If one could use algebraic geometry and/or computational algebra methods to prove that the ideal $\mathscr{H}_\ell$ is nonempty, for all odd $\ell > 1$, then this would furnish an algebraic proof of the structured Hadamard conjecture. Such methods include Gröbner bases methods [2], possibly in conjunction with constructive/computational invariant theory methods [34]. Note that Gröbner bases can also be used to count (exactly) the number of solutions of systems of polynomial equations [33].

4 Inclusion–Exclusion Formulation

The principle of inclusion–exclusion is a classical combinatorial technique that can be used to compute the cardinality of the union of a finite number of sets; see, for instance, [15, 31].

Each of the m equations (1) can be thought of as a property of an element $[a_0, \ldots, a_{\ell-1}, b_0, \ldots, b_{\ell-1}]$ of the Boolean cube $\{-1, +1\}^{2\ell}$. These m properties $P_1, \ldots, P_m$ give rise to m corresponding subsets $A_1, \ldots, A_m$ of the Boolean cube $\{-1, +1\}^{2\ell}$. The intersection of these m subsets contains these (and only these) elements of the Boolean cube that satisfy equations (1) simultaneously, i.e., that give Hadamard matrices with two-circulant cores. In symbols we have

$$\#H_\ell = \frac{|A_1 \cap \ldots \cap A_m|}{4}. \tag{4}$$

Note that the factor 4 in the denominator of formula (4) comes from the fact that we have deliberately omitted the linear equations $a_0 + \cdots + a_{\ell-1} = 1$ and $b_0 + \cdots + b_{\ell-1} = 1$ from the defining equations of Hadamard matrices with two circulant cores, to simplify the computations. This omission can be made without loss of generality, due to the symmetry of the problem and the fact that the Diophantine equation $x^2 + y^2 = 2$ has the four solutions $(1, 1), (1, -1), (-1, 1), (-1, -1)$. Note that the Diophantine equation (3) is of the form $x^2 + y^2 = 2$.

We now present our computational results, for $\ell = 5, \ldots, 15$.

- Inclusion–exclusion formula for $\ell = 5$:

$$|A \cup B| = |A| + |B| - |A \cap B|$$

$$600 = 400 + 400 - 200$$

- Inclusion–exclusion formula $\ell = 7$:

$$|A \cup B \cup C| = |A| + |B| + |C| - |A \cap B| - |B \cap C| - |C \cap A| + |A \cap B \cap C|$$

$$12,544 = 6,076 + 6,076 + 6,076 - 2,156 - 2,156 - 2,156 + 784$$

- Inclusion–exclusion formula for $\ell = 9$:

$$|A_1 \cup A_2 \cup A_3 \cup A_4| = \sum_{i=1}^{4} |A_i| - \sum_{ij} |A_i \cap A_j| + \sum_{ijk} |A_i \cap A_j \cap A_k| - |A_1 \cap A_2 \cap A_3 \cap A_4|$$

$$204,984 = 355,104 - 188,568 + 42,336 - 3,888$$

- Inclusion–exclusion formula for $\ell = 11$:

$$|A_1 \cup A_2 \cup A_3 \cup A_4 \cup A_5| =$$

$$= \sum_{i=1}^{5} |A_i| - \sum_{ij} |A_i \cap A_j| + \sum_{ijk} |A_i \cap A_j \cap A_k| - \sum_{ijkl} |A_i \cap A_j \cap A_k \cap A_l| +$$

$$+ |A_1 \cap A_2 \cap A_3 \cap A_4 \cap A_5|$$

$$3{,}474{,}636 = 6{,}471{,}080 - 4{,}072{,}860 + 1{,}239{,}040 - 174{,}240 + 11{,}616$$

- Inclusion–exclusion formula for $\ell = 13$:

$$|A_1 \cup A_2 \cup A_3 \cup A_4 \cup A_5 \cup A_6| =$$

$$= \sum_{i=1}^{6} |A_i| - \sum_{ij} |A_i \cap A_j| + \sum_{ijk} |A_i \cap A_j \cap A_k| - \sum_{ijkl} |A_i \cap A_j \cap A_k \cap A_l| +$$

$$+ \sum_{ijklm} |A_i \cap A_j \cap A_k \cap A_l \cap A_m| - |A_1 \cap A_2 \cap A_3 \cap A_4 \cap A_5 \cap A_6|$$

$$57{,}404{,}568 =$$

$$= 115{,}871{,}808 - 85{,}431{,}528 + 33{,}413{,}328 - 7{,}166{,}952 + 746{,}304 - 28{,}392$$

- Inclusion–exclusion formula for $\ell = 15$:

$$|A_1 \cup A_2 \cup A_3 \cup A_4 \cup A_5 \cup A_6 \cup A_7| =$$

$$= \sum_{i=1}^{7} |A_i| - \sum_{ij} |A_i \cap A_j| + \sum_{ijk} |A_i \cap A_j \cap A_k| - \sum_{ijkl} |A_i \cap A_j \cap A_k \cap A_l| +$$

$$+ \sum_{ijklm} |A_i \cap A_j \cap A_k \cap A_l \cap A_m| - \sum_{ijklmn} |A_i \cap A_j \cap A_k \cap A_l \cap A_m \cap A_n| +$$

$$+ |A_1 \cap A_2 \cap A_3 \cap A_4 \cap A_5 \cap A_6 \cap A_7|$$

$$937{,}687{,}980 = 2{,}042{,}963{,}760 - 1{,}724{,}067{,}360 +$$

$$+ 812{,}185{,}560 - 228{,}187{,}440 + 38{,}177{,}460 - 3{,}538{,}800 + 154{,}800$$

It is important to point out that our computational results via the inclusion–exclusion formulation agree with (and verify independently) the computational results of Table 1 taking into account formula (4).

5 Eliahou Theory Formulation

The landmark paper [11] established a fundamental link between the number of
solutions of polynomial equations with positive integer coefficients and certain
associated binary linear codes and their weight enumerators. The weight enumerator
of a code is a generating function style codification of the numbers of codewords
of specific weights in the code. First, we introduce some notation from [11], in
order to summarize the relevant aspect of Eliahou theory which is important for our
purposes.

Let $f(x_1,\ldots,x_n)$ be a polynomial with nonnegative integer coefficients. Then
the enumeration of the values assumed by f on the Boolean cube $\{-1,+1\}^n$ is
equivalent to the enumeration of the weights in an associated binary linear code L_f.
Let M_n denote the set of square-free monomials in the n variables $x_1,\ldots,x_n$. Let
f be decomposed as $f = u_1 + \cdots + u_N$ with monomials $u_i \in M_n, i = 1,\ldots,N$. The
monomials u_i need not be distinct and they can be equal to 1. Eliahou associates
with f the $n \times N$ matrix $\Phi_f = (\Phi_{ij})$ over F_2 defined by

$$\Phi_{ij} = \begin{cases} 1, \text{if } x_i \text{ divides } u_j \\ 0, \text{otherwise} \end{cases}$$

Let L_f denote the binary linear code generated by the n rows of the matrix Φ_f. For
every $u \in Z$, let the binary fiber of u be defined as

$$f^{-1}(u) = \{p \in \{-1,+1\}^n | f(p) = u\}.$$

Eliahou theory furnishes a way to compute the cardinality of the binary fiber, for
every $u \in Z$. The following theorem is proved in [11].

Theorem 2. *With the notations described above, we have*

$$|f^{-1}(u)| = 2^{n-\dim L_f} \cdot \textit{coefficient of } (XY)^{(N-u)/2} \textit{ in the weight enumerator of } L_f$$

$$(5)$$

where $\dim L_f$ *denotes the dimension of the associated (to f) binary linear code L_f*
and we make use of the bivariate weight enumerator of L_f.

In the context of structured Hadamard conjecture, we are interested in the particular
case $u = 0$ in the above theorem, i.e., we are interested in computing the cardinality
of the binary fiber of 0, for the specific function of 2ℓ variables $a_1,\ldots,a_\ell, b_1,\ldots,b_\ell$,

$$f_H = \sum_{s=1}^m (P_A(s) + P_B(s) + 2)^2 \text{ and } m = \frac{\ell-1}{2}.$$ It is clear that for fixed ℓ, the cardinality

of the binary fiber of 0 for this function f_H is equal to four times the number of
Hadamard matrices $H_{2\ell+2}$ with two circulant cores. The reason of the appearance
of the factor of 4 is because we omitted the linear equations $a_1 + \cdots + a_\ell = 1$ and
$b_1 + \cdots + b_\ell = 1$ from the definition of the function f_H.

Example 2. We illustrate the construction of the code L_{f_H} with the case $\ell = 3$, i.e., $n = 2\ell = 6$. Since $m = 1$, we obtain $f_H = (P_A(1) + P_B(1) + 2)^2 = (a_1a_2 + a_2a_3 + a_3a_1 + b_1b_2 + b_2b_3 + b_3b_1 + 2)^2$, and upon expanding, using the simplifications $a_i^2 = b_i^2 = 1, i = 1, 2, 3$ and dividing by 2, we obtain

$$f_H = 5 + 3(P_A(1) + P_B(1)) + P_A(1)P_B(1)$$

which implies that f_H has $N = 5 + 3 \cdot 6 + 9 = 32$ and the associated 6×32 matrix Φ_{f_H} is

$$\Phi_{f_H} = \begin{bmatrix} 0,0,0,0,0,1,1,1,0,0,0,1,1,1,1,1,1,0,0,0,1,1,1,0,0,0,0,0,0,0,0,0 \\ 0,0,0,0,0,1,1,1,1,1,1,0,0,0,1,1,1,1,1,0,0,0,0,0,0,0,0,0,0,0,0,0 \\ 0,0,0,0,0,0,0,0,1,1,1,1,1,1,0,0,0,1,1,1,1,1,0,0,0,0,0,0,0,0,0,0 \\ 0,0,0,0,0,1,0,1,1,0,1,1,0,1,0,0,0,0,0,0,0,0,0,1,1,1,0,0,0,1,1,1 \\ 0,0,0,0,0,1,1,0,1,1,0,1,1,0,0,0,0,0,0,0,0,0,0,1,1,1,1,1,1,0,0,0 \\ 0,0,0,0,0,0,1,1,0,1,1,0,1,1,0,0,0,0,0,0,0,0,0,0,1,1,1,1,1,1,1,1 \end{bmatrix}$$

Let L_{f_H} denote the binary linear code generated by the six rows of the above matrix Φ_{f_H}. The bivariate weight enumerator of the binary linear code L_{f_H} can be computed with Magma and is equal to

$$X^{32} + 6X^{20}Y^{12} + 9X^{16}Y^{16}.$$

The dimension of the binary linear code L_{f_H} can also be computed in Magma and is equal to 4. Therefore Eq. (5) for $u = 0$ yields

$$|f^{-1}(0)| = 2^{n - \dim L_{f_H}} \cdot \text{coefficient of } (XY)^{N/2} \text{ in the weight enumerator of } L_{f_H} = $$
$$= 2^{6-4} \cdot 9 = 4 \cdot 9 = 36$$

which means that the equation $f_H = 0$ has 36 solutions in $\{-1, +1\}^6$. □

In order to apply Eliahou theory to the structured Hadamard conjecture, we constructed the codes L_{f_H} and computed their weight enumerators for $\ell = 3, 5, 7,$ 9, 11, 13, 15, 17, 19 using Magma. We summarize the results in Table 2. Since the factor $2^{n - \dim L_{f_H}}$ is equal to 4 for all these 9 values of ℓ, we see that the results obtained by Eliahou theory agree with (and independently verify) the results of Table 1. The Eliahou theory formulation of the structured Hadamard conjecture is the statement that the coefficient of $X^{N/2}Y^{N/2}$ in the weight enumerator of L_{f_H} is not equal to zero, for all odd integers $\ell (> 1)$.

Note that Eliahou theory can also be used to provide an independent verification of the exhaustive searches for Golay sequences undertaken by Borwein and Ferguson in [3]. In the case of Golay sequences, there are several values of the parameter (length of the sequence) for which Golay sequences do not exist, and this is consistently in all such cases detected by Eliahou theory simply by observing that the corresponding coefficient of the weight enumerator is equal to zero.

Table 2 Eliahou theory results for Hadamard matrices with two-circulant cores

ℓ	$n(=2\ell)$	N	$\dim L_{f_H}$	Coeff. of $X^{N/2}Y^{N/2}$ in the weight enum. of L_{f_H}
3	6	32	4	9
5	10	144	8	50
7	14	384	12	196
9	18	800	16	972
11	22	1,440	20	2,904
13	26	2,352	24	7,098
15	30	3,584	28	38,700
17	34	5,184	32	93,058
19	38	7,200	36	161,728

6 Conclusion

The Hadamard and the structured Hadamard conjectures are tempting flowers, in the terminology of [25], and the quest to find a proof for these will undoubtedly continue to attract generations of mathematicians for many years to come. Our humble hope is that we have contributed with our paper to draw the attention of number theorists, as well as mathematicians from other related disciplines, to the topic of the structured Hadamard conjecture. We firmly believe that it is possible to resolve the Hadamard conjecture by way of its structured alternative. We also believe that a synergistic approach will most likely be required to produce a proof.

Acknowledgements The author is grateful to the anonymous referees for their careful scrutiny of the original submission and their constructive and pertinent comments that led to a significantly improved version of this paper.

This work is supported by an NSERC grant.

References

1. I. Bárány, A vector-sum theorem and its application to improving flow shop guarantees. Math. Oper. Res. **6**(3), 445–452 (1981)
2. T. Becker, V. Weispfenning, *Gröbner Bases*. Graduate Texts in Mathematics, vol 141 (Springer, New York, 1993). A computational approach to commutative algebra, In cooperation with Heinz Kredel
3. P.B. Borwein, R.A. Ferguson, A complete description of Golay pairs for lengths up to 100. Math. Comp. **73**(246), 967–985 (2004) [electronic]
4. C.J. Colbourn, J.H. Dinitz (eds), *Handbook of Combinatorial Designs*, 2nd edn. Discrete Mathematics and its Applications (Boca Raton) (Chapman & Hall/CRC, Boca Raton, 2007)
5. W. de Launey, D.M. Gordon, On the density of the set of known Hadamard orders. Cryptogr. Commun. **2**(2), 233–246 (2010)
6. W. de Launey, D.A. Levin, A Fourier-analytic approach to counting partial Hadamard matrices. Cryptogr. Commun. **2**(2), 307–334 (2010)

7. D.Ž. Djoković, Two Hadamard matrices of order 956 of Goethals-Seidel type. Combinatorica **14**(3), 375–377 (1994)
8. D.Ž. Djoković, Hadamard matrices of order 764 exist. Combinatorica **28**(4), 487–489 (2008)
9. D.Ž. Djoković, Skew-Hadamard matrices of orders 188 and 388 exist. Int. Math. Forum **3**(21–24), 1063–1068 (2008)
10. D.Ž. Djoković, Skew-Hadamard matrices of orders 436, 580, and 988 exist. J. Combin. Des. **16**(6), 493–498 (2008)
11. S. Eliahou, Enumerative combinatorics and coding theory. Enseign. Math. (2) **40**(1–2), 171–185 (1994)
12. R.J. Fletcher, M. Gysin, J. Seberry, Application of the discrete Fourier transform to the search for generalised Legendre pairs and Hadamard matrices. Aust. J. Combin. **23**, 75–86 (2001)
13. Fondation Mathématique Jacques Hadamard (FMJH). http://www.fondation-hadamard.fr
14. B. Fuglede, Commuting self-adjoint partial differential operators and a group theoretic problem. J. Funct. Anal. **16**, 101–121 (1974)
15. I.P. Goulden, D.M. Jackson, *Combinatorial Enumeration* (Dover, Mineola, 2004). With a foreword by Gian-Carlo Rota, Reprint of the 1983 original.
16. J. Hadamard, Résolution d'une question relative aux déterminants. Bull. Sci. Mathé., **17**, 240–246 (1893)
17. M. Hall Jr., *Combinatorial Theory* (Blaisdell Publishing Co. Ginn and Co., Waltham/Toronto/London, 1967)
18. A. Hedayat, W.D. Wallis, Hadamard matrices and their applications. Ann. Statist. **6**(6), 1184–1238 (1978)
19. K.J. Horadam, *Hadamard Matrices and Their Applications* (Princeton University Press, Princeton, 2007)
20. K.J. Horadam, Hadamard matrices and their applications: progress 2007–2010. Cryptogr. Commun. **2**(2), 129–154 (2010)
21. H. Kharaghani, B. Tayfeh-Rezaie, A Hadamard matrix of order 428. J. Combin. Design **13**(6), 435–440 (2005)
22. I.S. Kotsireas, C. Koukouvinos, J. Seberry, Hadamard ideals and Hadamard matrices with two circulant cores. Eur. J. Combin. **27**(5), 658–668 (2006)
23. C. Lam, S. Lam, V.D. Tonchev, Bounds on the number of affine, symmetric, and Hadamard designs and matrices. J. Combin. Theory Ser. A **92**(2), 186–196 (2000)
24. C. Lam, S. Lam, V.D. Tonchev, Bounds on the number of Hadamard designs of even order. J. Combin. Design **9**(5), 363–378 (2001)
25. B. Mazur, Number theory as gadfly. Am. Math. Monthly **98**(7), 593–610 (1991)
26. V. Maz'ya, T. Shaposhnikova, Jacques Hadamard, a universal mathematician, in *History of Mathematics*, vol 14 (American Mathematical Society, Providence, 1998)
27. P. Ó Catháin, Group actions on Hadamard matrices. Master's Thesis, National University of Ireland, Galway, 2008
28. P. Ó Catháin and M. Röder, The cocyclic Hadamard matrices of order less than 40. Design Codes Cryptogr. **58**(1), 73–88 (2011)
29. J. Seberry, M. Yamada, Hadamard matrices, sequences, and block designs, in *Contemporary Design Theory*. Wiley-Intersci. Ser. Discrete Math. Optim. (Wiley, New York, 1992), pp. 431–560
30. N.J.A. Sloane, A library of Hadamard matrices. http://www2.research.att.com/~njas/hadamard/
31. R.P. Stanley, *Enumerative Combinatorics*, vol 1. Cambridge Studies in Advanced Mathematics, 2nd edn., vol 49 (Cambridge University Press, Cambridge, 2012)
32. D.R. Stinson, *Combinatorial Designs, Constructions and Analysis* (Springer, New York, 2004)
33. B. Sturmfels Solving systems of polynomial equations, in *CBMS Regional Conference Series in Mathematics*, vol 97 (Published for the Conference Board of the Mathematical Sciences, Washington, DC, 2002)

34. B. Sturmfels. Algorithms in invariant theory, in *Texts and Monographs in Symbolic Computation*, 2nd edn. (Springer, Wien/NewYork/Vienna, 2008)
35. T. Tao, Fuglede's conjecture is false in 5 and higher dimensions. Math. Res. Lett. **11**(2–3), 251–258 (2004)
36. A.L. Whiteman, A family of difference sets. Illinois J. Math. **6**, 107–121 (1962)
37. M. Yamada, Supplementary difference sets and Jacobi sums. Discrete Math. **103**(1), 75–90 (1992)

Families of Cubic Thue Equations with Effective Bounds for the Solutions

Claude Levesque and Michel Waldschmidt

Abstract To each nontotally real cubic extension K of $\mathbf{Q}$ and to each generator α of the cubic field K, we attach a family of cubic Thue equations, indexed by the units of K, and we prove that this family of cubic Thue equations has only a finite number of integer solutions, by giving an effective upper bound for these solutions.

1 Statements

Let us consider an irreducible binary cubic form having rational integers coefficients:

$$F(X,Y) = a_0 X^3 + a_1 X^2 Y + a_2 X Y^2 + a_3 Y^3 \in \mathbf{Z}[X,Y]$$

with the property that the polynomial $F(X,1)$ has exactly one real root α and two complex imaginary roots, namely, α' and $\overline{\alpha'}$. Hence, $\alpha \notin \mathbf{Q}$, $\alpha' \neq \overline{\alpha'}$ and

$$F(X,Y) = a_0(X - \alpha Y)(X - \alpha' Y)(X - \overline{\alpha'} Y).$$

Let K be the cubic number field $\mathbf{Q}(\alpha)$ which we view as a subfield of $\mathbf{R}$. Define $\sigma : K \to \mathbf{C}$ to be one of the two complex embeddings, the other one being the

C. Levesque
Département de mathématiques et de statistique, Université Laval,
Québec, QC, Canada G1V 0A6
e-mail: Claude.Levesque@mat.ulaval.ca

M. Waldschmidt (✉)
Institut de Mathématiques de Jussieu, Université Pierre et Marie Curie (Paris 6),
4 Place Jussieu, 75252 Paris Cedex 05, France
e-mail: miw@math.jussieu.fr

J.M. Borwein et al. (eds.), *Number Theory and Related Fields: In Memory of Alf van der Poorten*, Springer Proceedings in Mathematics & Statistics 43,
DOI 10.1007/978-1-4614-6642-0_12, © Springer Science+Business Media New York 2013

conjugate $\overline{\sigma}$. Hence, $\alpha' = \sigma(\alpha)$ and $\overline{\alpha'} = \overline{\sigma}(\alpha)$. If τ is defined to be the complex conjugation, we have $\overline{\sigma} = \tau \circ \sigma$ and $\sigma \circ \tau = \sigma$.

Let ε be a unit > 1 of the ring $\mathbf{Z}_K$ of algebraic integers of K and let $\varepsilon' = \sigma(\varepsilon)$ and $\overline{\varepsilon'} = \overline{\sigma}(\varepsilon)$ be the two other algebraic conjugates of ε. We have

$$|\varepsilon'| = |\overline{\varepsilon'}| = \frac{1}{\sqrt{\varepsilon}} < 1.$$

For $n \in \mathbf{Z}$, define

$$F_n(X,Y) = a_0\left(X - \varepsilon^n \alpha Y\right)\left(X - \varepsilon'^n \alpha' Y\right)\left(X - \overline{\varepsilon'}^n \overline{\alpha'} Y\right).$$

Let $k \in \mathbf{N}$, where $\mathbf{N} = \{1,2,\ldots\}$. We plan to study the family of Thue inequations

$$0 < |F_n(x,y)| \le k, \tag{1}$$

where the unknowns n,x,y take values in $\mathbf{Z}$.

Theorem 1. *There exist effectively computable positive constants κ_1 and κ_2, depending only on F, such that, for all $k \in \mathbf{Z}$ with $k \ge 1$ and for all $(n,x,y) \in \mathbf{Z} \times \mathbf{Z} \times \mathbf{Z}$ satisfying $\varepsilon^n \alpha \notin \mathbf{Q}$, $xy \neq 0$ and $|F_n(x,y)| \le k$, we have*

$$\max\left\{\varepsilon^{|n|}, |x|, |y|\right\} \le \kappa_1 k^{\kappa_2}.$$

From this theorem, we deduce the following corollary:

Corollary 1. *For $k \in \mathbf{Z}$, $k > 0$, the set*

$$\left\{(n,x,y) \in \mathbf{Z} \times \mathbf{Z} \times \mathbf{Z} \mid \varepsilon^n \alpha \notin \mathbf{Q}\,;\, xy \neq 0\,;\, |F_n(x,y)| \le k\right\}$$

is finite.

This corollary is a particular case of the main result of [2], but the proof in [2] is based on the Schmidt subspace theorem which does not allow to give an effective upper bound for the solutions (n,x,y).

Example. Let $D \in \mathbf{Z}$, $D \neq -1$. Let $\varepsilon := \left(\sqrt[3]{D^3 + 1} - D\right)^{-1}$. There exist two positive effectively computable absolute constants κ_3 and κ_4 with the following property. Define a sequence $(F_n)_{n \in \mathbf{Z}}$ of cubic forms in $\mathbf{Z}[X,Y]$ by

$$F_n(X,Y) = X^3 + a_n X^2 Y + b_n X Y^2 - Y^3,$$

where $(a_n)_{n \in \mathbf{Z}}$ is defined by the recurrence relation

$$a_{n+3} = 3D a_{n+2} + 3D^2 a_{n+1} + a_n$$

with the initial conditions $a_0 = 3D^2, a_{-1} = 3$ and $a_{-2} = -3D$ and where $(b_n)_{n\in\mathbf{Z}}$ is defined by $b_n = -a_{-n-2}$. Then, for x, y, n rational integers with $xy \neq 0$ and $n \neq -1$, we have

$$|F_n(x,y)| \geq \kappa_3 \max\{|x|, |y|, \varepsilon^{|n|}\}^{\kappa_4}.$$

This result follows from Theorem 1 with $\alpha = \varepsilon$ and

$$F(X,Y) = X^3 - 3DX^2Y - 3D^2XY^2 - Y^3.$$

Indeed, the irreducible polynomial of $\varepsilon^{-1} = \sqrt[3]{D^3+1} - D$ is

$$F_{-2}(X,1) = (X+D)^3 - D^3 - 1 = X^3 + 3DX^2 + 3D^2X - 1,$$

the irreducible polynomial of $\alpha = \varepsilon$ is

$$F(X,1) = F_0(X,1) = F_{-2}(1,X) = X^3 - 3D^2X^2 - 3DX - 1,$$

while

$$F_{-1}(X,Y) = (X-Y)^3 = X^3 - 3X^2Y + 3XY^2 - Y^3.$$

For $n \in \mathbf{Z}$, $n \neq -1$, $F_n(X,1)$ is the irreducible polynomial of $\alpha\varepsilon^n = \varepsilon^{n+1}$, while for any $n \in \mathbf{Z}$, $F_n(X,Y) = N_{\mathbf{Q}(\varepsilon)/\mathbf{Q}}(X - \varepsilon^{n+1}Y)$. The recurrence relation for

$$a_n = \varepsilon^{n+1} + \varepsilon'^{n+1} + \overline{\varepsilon'}^{n+1}$$

follows from

$$\varepsilon^{n+3} = 3D\varepsilon^{n+2} + 3D^2\varepsilon^{n+1} + \varepsilon^n$$

and for b_n, from $F_{-n}(X,Y) = -F_{n-2}(Y,X)$.

2 Elementary Estimates

For a given integer $k > 0$, we consider a solution (n,x,y) in $\mathbf{Z}^3$ of the Thue inequation (1) with $\varepsilon^n \alpha$ irrational and $xy \neq 0$. We will use $\kappa_5, \kappa_6, \ldots, \kappa_{55}$ to designate some constants depending only on α.

Let us firstly explain that in order to prove Theorem 1, we can assume $n \geq 0$ by eventually permuting x and y. Let us suppose that $n < 0$ and write

$$F(X,Y) = a_3(Y - \alpha^{-1}X)(Y - \alpha'^{-1}X)(Y - \overline{\alpha'}^{-1}X).$$

Then

$$F_n(X,Y) = a_3(Y - \varepsilon^{|n|}\alpha^{-1}X)(Y - \varepsilon'^{|n|}\alpha'^{-1}X)(Y - \overline{\varepsilon'}^{|n|}\overline{\alpha'}^{-1}X).$$

Now it is simply a matter of using the result for $|n|$ for the polynomial $G(X,Y) = F(Y,X)$.

Let us now check that, in order to prove the statements of Sect. 1, there is no restriction in assuming that α is an algebraic integer and that $a_0 = 1$. To achieve this goal, we define

$$\tilde{F}(T,Y) = T^3 + a_1 T^2 Y + a_0 a_2 T Y^2 + a_0^2 a_3 Y^3 \in \mathbf{Z}[T,Y],$$

so that $a_0^2 F(X,Y) = \tilde{F}(a_0 X, Y)$. If we define $\tilde{\alpha} = a_0 \alpha$ and $\tilde{\alpha}' = a_0 \alpha'$, then $\tilde{\alpha}$ is a nonzero algebraic integer, and we have

$$\tilde{F}(T,Y) = (T - \tilde{\alpha} Y)(T - \tilde{\alpha}' Y)(T - \overline{\tilde{\alpha}'} Y).$$

For $n \in \mathbf{Z}$, the binary form

$$\tilde{F}_n(T,Y) = (T - \varepsilon^n \tilde{\alpha} Y)(T - \varepsilon'^n \tilde{\alpha}' Y)(T - \overline{\varepsilon'}^n \overline{\tilde{\alpha}'} Y)$$

satisfies

$$a_0^2 F_n(X,Y) = \tilde{F}_n(a_0 X, Y).$$

The condition (1) implies $0 < |\tilde{F}_n(a_0 x, y)| \le a_0^2 k$. Therefore, it suffices to prove the statements for $\tilde{F}_n$ instead of F_n, with α and α' replaced by $\tilde{\alpha}$ and $\tilde{\alpha}'$. This allows us, from now on, to suppose $\alpha \in \mathbf{Z}_K$ and $a_0 = 1$.

As already explained, we can assume $n \ge 0$. There is no restriction in supposing $k \ge 2$ (if we prove the result for a value of $k \ge 2$, we deduce it right away for smaller values of k, since we consider Thue inequations and not Thue equations). If k were asumed to be ≥ 2, we would not need κ_1, as is easily seen, and the conclusion would read:

$$\max\{\varepsilon^{|n|}, |x|, |y|\} \le k^{\kappa_2}.$$

Without loss of generality we can assume that n is sufficiently large. As a matter of fact, if n is bounded, we are led to some given Thue equations, and Theorem 1 follows from Theorem 5.1 of [3].

Let us recall that for an algebraic number γ, the house of γ, denoted $\boxed{\gamma}$, is by definition the maximum of the absolute values of the conjugates of γ. Moreover, d is the degree of the algebraic number field K (viz., $d = 3$ here) and R is the regulator of K (viz. $R = \log \varepsilon$), where, from now on, ε is the fundamental unit > 1 of the non totally real cubic field K. The next statement is Lemma A.6 of [3].

Lemma 1. *Let γ be a nonzero element of $\mathbf{Z}_K$ of norm $\le M$. There exists a unit $\eta \in \mathbf{Z}_K^\times$ such that the house $\boxed{\eta \gamma}$ is bounded by an effectively computable constant which depends only on d, R, and M.*

We need to make explicit the dependence upon M, and for this, it suffices to apply Lemma A.15 of [3], which we want to state, under the assumption that the d embeddings of the algebraic number field K in $\mathbf{C}$ are noted $\sigma_1, \ldots, \sigma_d$.

Lemma 2. *Let K be an algebraic number field of degree d and let γ be a nonzero element of $\mathbf{Z}_K$ whose absolute value of the norm is m. Then there exists a unit $\eta \in \mathbf{Z}_K^\times$ such that,*

$$\frac{1}{R} \max_{1 \le j \le d} \left| \log(m^{-1/d} |\sigma_j(\eta\gamma)|) \right|$$

is bounded by an effectively computable constant which depends only on d.

Since $d = 3$, $K = \mathbf{Q}(\alpha)$ and the regulator R of K is an effectively computable constant (see for instance [1, §6.5]), the conclusion of Lemma 2 is

$$-\kappa_5 \le \log(|\sigma_j(\eta\gamma)|/\sqrt[3]{m}) \le \kappa_5,$$

which can also be written as

$$\kappa_6 \sqrt[3]{m} \le |\sigma_j(\eta\gamma)| \le \kappa_7 \sqrt[3]{m},$$

with two effectively computable positive constants κ_6 and κ_7. We will use only the upper bound[1]: under the hypotheses of Lemma 1 with $d = 3$, when γ is a nonzero element of $\mathbf{Z}_K$ of norm $\le M$, there exists a unit η of $\mathbf{Z}_K^\times$ such that

$$\overline{|\eta\gamma|} \le \kappa_7 \sqrt[3]{M}.$$

Since (n, x, y) satisfies (1), the element $\gamma = x - \varepsilon^n \alpha y$ of $\mathbf{Z}_K$ has a norm of absolute value $\le k$. It follows from Lemma 2 that γ can be written as

$$x - \varepsilon^n \alpha y = \varepsilon^\ell \xi_1 \tag{2}$$

with $\ell \in \mathbf{Z}$, $\xi_1 \in \mathbf{Z}_K$ and the house of ξ_1, $\overline{|\xi_1|} = \max\{|\xi_1|, |\xi_1'|\}$ satisfies

$$\overline{|\xi_1|} \le \kappa_8 \sqrt[3]{k}.$$

We will not use the full force of this upper bound, but only the consequence

$$\max\left\{ |\xi_1|^{-1}, |\xi_1'|^{-1}, \overline{|\xi_1|} \right\} \le k^{\kappa_9}. \tag{3}$$

Taking the conjugate of (2) by σ, we have

$$x - \varepsilon'^n \alpha' y = \varepsilon'^\ell \xi_1' \tag{4}$$

with $\xi_1' = \sigma(\xi_1)$.

---

[1] The lower bound follows from looking at the norm!

Our strategy is to prove that $|\ell|$ is bounded by a constant times $\log k$ and that $|n|$ is also bounded by a constant times $\log k$; then we will show that $|y|$ is bounded by a a constant power of k and deduce that $|x|$ is also bounded by a constant power of k.

Let us eliminate x in (2) and (4) to obtain

$$y = -\frac{\varepsilon^\ell \xi_1 - \varepsilon'^\ell \xi_1'}{\varepsilon^n \alpha - \varepsilon'^n \alpha'}; \tag{5}$$

since we supposed $\varepsilon^n \alpha$ irrational, we did not divide by 0. The complex conjugate of (4) is written as

$$x - \overline{\varepsilon'^n} \, \overline{\alpha'} y = \overline{\varepsilon'^\ell} \, \overline{\xi_1'}. \tag{6}$$

We eliminate x and y in the three equations (2), (4), and (6) to obtain a unit equation à la Siegel:

$$\varepsilon^\ell \xi_1 (\alpha' \varepsilon'^n - \overline{\alpha'} \, \overline{\varepsilon'^n}) + \varepsilon'^\ell \xi_1' (\overline{\alpha'} \, \overline{\varepsilon'^n} - \alpha \varepsilon^n) + \overline{\varepsilon'^\ell} \, \overline{\xi_1'} (\alpha \varepsilon^n - \alpha' \varepsilon'^n) = 0. \tag{7}$$

In the remaining part of this Sect. 2, we suppose

$$\varepsilon^n |\alpha| \geq 2|\varepsilon'^n \alpha'|. \tag{8}$$

Note that if this inequality is not satisfied, then we have

$$\varepsilon^{3n/2} < \frac{2|\alpha'|}{|\alpha|} < \kappa_{10},$$

and this leads to the inequality (18), and to the rest of the proof of Theorem 1 by using the argument following the inequality (18).

For $\ell > 0$, the absolute value of the numerator $\varepsilon^\ell \xi_1 - \varepsilon'^\ell \xi_1'$ in (5) is increasing like ε^ℓ and for $\ell < 0$ it is increasing like $\varepsilon^{|\ell|/2}$; for $n > 0$, the absolute value of the denominator $\varepsilon^n \alpha - \varepsilon'^n \alpha'$ is increasing like ε^n and for $n < 0$ it is increasing like $\varepsilon^{|n|/2}$. In order to extract some information from Equation (5), we write it in the form

$$y = \pm \frac{A - a}{B - b}$$

with

$$B = \varepsilon^n \alpha, \quad b = \varepsilon'^n \alpha', \quad \{A, a\} = \left\{ \varepsilon^\ell \xi_1 \,, \, \varepsilon'^\ell \xi_1' \right\},$$

the choice of A and a being dictated by

$$|A| = \max\{\varepsilon^\ell |\xi_1| \,, \, |\varepsilon'^\ell \xi_1'|\}, \quad |a| = \min\{\varepsilon^\ell |\xi_1| \,, \, |\varepsilon'^\ell \xi_1'|\}.$$

Since $|A - a| \leq 2|A|$ and since $|b| \leq |B|/2$ because of (8), we have $|B - b| \geq |B|/2$, so we get

$$|y| \leq 4\frac{|A|}{|B|}.$$

We will consider the two cases corresponding to the possible signs of ℓ (remember that n is positive).

First Case. Let $\ell \leq 0$. We have

$$|A| \leq \kappa_{11}\varepsilon^{|\ell|/2}k^{\kappa_9}.$$

We deduce from (5)

$$1 \leq |y| \leq 4\left|\frac{\xi_1'}{\alpha}\right|\varepsilon^{(|\ell|/2)-n} \leq \kappa_{12}\varepsilon^{(|\ell|/2)-n}k^{\kappa_9}. \tag{9}$$

Hence there exists κ_{13} such that

$$0 \leq \log|y| \leq \left(\frac{|\ell|}{2} - n\right)\log\varepsilon + \kappa_{13}\log k,$$

from which we deduce the inequality

$$n \leq \frac{|\ell|}{2} + \kappa_{14}\log k, \tag{10}$$

which will prove useful: n is roughly bounded by $|\ell|$. From (4) we deduce the existence of a constant κ_{15} such that

$$|x| \leq \varepsilon^{-n/2}|\alpha'y| + \kappa_{15}k^{\kappa_9}\varepsilon^{|\ell|/2}. \tag{11}$$

Second Case. Let $\ell > 0$. We have

$$|A| \leq \kappa_{16}\varepsilon^{\ell}k^{\kappa_9}.$$

We deduce from (5) the upper bound

$$1 \leq |y| \leq 4\left|\frac{\xi_1}{\alpha}\right|\varepsilon^{\ell-n} \leq \kappa_{17}k^{\kappa_9}\varepsilon^{\ell-n}; \tag{12}$$

hence there exists κ_{18} such that

$$0 \leq \log|y| \leq (\ell-n)\log\varepsilon + \kappa_{18}\log k.$$

Consequently,

$$n \leq \ell + \kappa_{19}\log k. \tag{13}$$

From the relation (4) we deduce the existence of a constant κ_{20} such that

$$1 \leq |x| \leq \varepsilon^{-n/2}|\alpha'y| + \kappa_{20}k^{\kappa_9}\varepsilon^{-\ell/2}. \tag{14}$$

By taking into account the inequalities (9)–(11) in the case $\ell \leq 0$, and the inequalities (12)–(14) in the case $\ell > 0$, let us show that the existence of a constant κ_{21} satisfying $|\ell| \leq \kappa_{21}\log k$ allows to conclude the proof of Theorem 1. As a matter of fact, suppose

$$|\ell| \leq \kappa_{21}\log k. \tag{15}$$

Then (10) and (13) imply $n \leq \kappa_{22}\log k$, whereupon $|\ell|$ and n are effectively bounded by a constant times $\log k$. This implies that the elements ε^t, with t being $(|\ell|/2) - n$, $\ell - n$, $-n/2$, $|\ell|/2$ or $-\ell/2$, appearing in (9), (12), (11), and (14) are bounded from above by $k^{\kappa_{23}}$ for some constant κ_{23}. Therefore, the upper bound of $|y|$ in the conclusion of Theorem 1 follows from (9) and (12), and the upper bound of $|x|$ is a consequence of (11) and (14). Our goal is to show that sooner or later, we end up with the inequality (15).

In the case $\ell > 0$, the lower bound $|x| \geq 1$ provides an extra piece of information. If the term $\varepsilon'^\ell\xi_1'$ on the right-hand side of (4) does not have an absolute value $< 1/2$, then the upper bound (15) holds true and this suffices to claim the proof of Theorem 1. Suppose now $|\varepsilon'^\ell\xi_1'| < 1/2$. Since the relation (12) implies

$$\varepsilon^{-n/2}|\alpha'y| \leq 4\left|\frac{\xi_1\alpha'}{\alpha}\right|\varepsilon^{\ell-(3n/2)},$$

we have

$$1 \leq |x| \leq 4\left|\frac{\xi_1\alpha'}{\alpha}\right|\varepsilon^{\ell-(3n/2)} + \frac{1}{2}$$

and

$$1 \leq 8\left|\frac{\xi_1\alpha'}{\alpha}\right|\varepsilon^{\ell-(3n/2)}.$$

We deduce

$$\frac{3}{2}n \leq \ell + \kappa_{24}\log k. \tag{16}$$

The upper bound in (16) is sharper than the one in (13), but, amazingly, we used (13) to establish (16).

When $\ell < 0$, we have $|\ell - n| = n + |\ell| \geq |\ell|$, while in the case $\ell \geq 0$ we have

$$|\ell - n| \geq \frac{1}{3}\ell + \frac{2}{3}\ell - n \geq \frac{1}{3}|\ell| - \kappa_{24}\log k,$$

because of (16). Therefore, if ℓ is positive (recall (16)), zero, or negative (recall (10)), we always have

$$n \leq \frac{2}{3}|\ell| + \kappa_{25}\log k \quad \text{and} \quad |\ell - n| \geq \frac{1}{3}|\ell| - \kappa_{24}\log k \tag{17}$$

with $\kappa_{24} > 0$ and $\kappa_{25} > 0$.

3 Diophantine Tool

Let us remind what we mean by the absolute logarithmic height $h(\alpha)$ of an algebraic number α (cf. [4, Chap. 3]). For L a number field and for $\alpha \in L$, we define

$$h(\alpha) = \frac{1}{[L:Q]}\log H_L(\alpha),$$

with

$$H_L(\alpha) = \prod_v \max\{1, |\alpha|_v\}^{d_v}$$

where v runs over the set of places of L, with d_v being the local degree of the place v if v is ultrametric, $d_v = 1$ if v is real, $d_v = 2$ if v is complex. When $f(X) \in \mathbf{Z}[X]$ is the minimal polynomial of α and $f(X) = a_0 \prod_{1 \leq j \leq d}(X - \alpha_j)$, with $\alpha_1 = \alpha$, it happens that

$$h(\alpha) = \frac{1}{d}\log M(f) \quad \text{with} \quad M(f) = |a_0| \prod_{1 \leq j \leq d} \max\{1, |\alpha_j|\}.$$

We will use two particular cases of Theorem 9.1 of [4]. The first one is a lower bound for the linear form of logarithms $b_0\lambda_0 + b_1\lambda_1 + b_2\lambda_2$, and the second one is a lower bound for $\gamma_1^{b_1}\gamma_2^{b_2} - 1$. Here is the first one.

Proposition 1. *There exists an explicit absolute constant $c_0 > 0$ with the following property: Let $\lambda_0, \lambda_1, \lambda_2$ be three logarithms of algebraic numbers and let b_0, b_1, b_2 be three rational integers such that $\Lambda = b_0\lambda_0 + b_1\lambda_1 + b_2\lambda_2$ be nonzero. Write*

$$\gamma_0 = e^{\lambda_0}, \quad \gamma_1 = e^{\lambda_1}, \quad \gamma_2 = e^{\lambda_2} \quad \text{and} \quad D = [\mathbf{Q}(\gamma_0, \gamma_1, \gamma_2) : \mathbf{Q}].$$

Let A_0, A_1, A_2 and B be real positive numbers satisfying

$$\log A_i \geq \max\left\{h(\gamma_i), \frac{|\lambda_i|}{D}, \frac{1}{D}\right\} \quad (i = 0, 1, 2)$$

and

$$B \geq \max\left\{e,\ D,\ \frac{|b_2|}{D\log A_0} + \frac{|b_0|}{D\log A_2},\ \frac{|b_2|}{D\log A_1} + \frac{|b_1|}{D\log A_2}\right\}.$$

Then

$$|\Lambda| \geq \exp\{-c_0 D^5(\log D)(\log A_0)(\log A_1)(\log A_2)(\log B)\}.$$

The second particular case of Theorem 9.1 in [4] that we will use is the next Proposition 2. It also follows from Corollary 9.22 of [4]. We could as well deduce it from Proposition 1.

Proposition 2. *Let D be a positive integer. There exists an explicit constant $c_1 > 0$, depending only on D with the following property: Let K be a number field of degree $\leq D$. Let γ_1, γ_2 be nonzero elements in K and let b_1, b_2 be rational integers. Assume $\gamma_1^{b_1}\gamma_2^{b_2} \neq 1$. Set*

$$B = \max\{2, |b_1|, |b_2|\} \quad and, for\ i = 1, 2, \quad A_i = \exp(\max\{e, h(\gamma_i)\}).$$

Then

$$|\gamma_1^{b_1}\gamma_2^{b_2} - 1| \geq \exp\{-c_1(\log B)(\log A_1)(\log A_2)\}.$$

Proposition 2 will come into play via its following consequence:

Corollary 2. *Let δ_1 and δ_2 be two real numbers in the interval $[0, 2\pi)$. Suppose that the numbers $e^{i\delta_1}$ and $e^{i\delta_2}$ are algebraic. There exists an explicit constant $c_2 > 0$, depending only upon δ_1 and δ_2, with the following property: for each $n \in \mathbf{Z}$ such that $\delta_1 + n\delta_2 \notin \mathbf{Z}\pi$, we have*

$$|\sin(\delta_1 + n\delta_2)| \geq (|n| + 2)^{-c_2}.$$

Proof. Write $\gamma_1 = e^{i\delta_1}$ and $\gamma_2 = e^{i\delta_2}$. By hypothesis, γ_1 and γ_2 are algebraic with $\gamma_1\gamma_2^n \neq 1$. Let us use Proposition 2 with $b_1 = 1$, $b_2 = n$. The parameters A_1 and A_2 depend only upon δ_1 and δ_2 and the number $B = \max\{2, |n|\}$ is bounded from above by $|n| + 2$. Hence

$$|\gamma_1\gamma_2^n - 1| \geq (|n| + 2)^{-c_3}$$

where c_3 depends only upon δ_1 and δ_2. Let ℓ be the nearest integer to $(\delta_1 + n\delta_2)/\pi$ (take the floor if there are two possible values) and let $t = \delta_1 + n\delta_2 - \ell\pi$. This real number t is in the interval $(-\pi/2, \pi/2]$. Now

$$|e^{it} + 1| = |1 + \cos(t) + i\,\sin(t)| = \sqrt{2(1 + \cos(t))} \geq \sqrt{2}.$$

Since $e^{it} = (-1)^{\ell}\gamma_1\gamma_2^n$, we deduce

$$|\sin(\delta_1 + n\delta_2)| = |\sin(t)| = \frac{1}{2}\left|(-1)^{2\ell}e^{2it} - 1\right|$$

$$= \frac{1}{2}\left|(-1)^{\ell}e^{it} + 1\right| \cdot \left|(-1)^{\ell}e^{it} - 1\right| \geq \frac{\sqrt{2}}{2}|\gamma_1\gamma_2^n - 1|.$$

This secures the proof of Corollary 2.   □

The following elementary lemma makes clear that $e^t \sim 1$ for $t \to 0$. The first (resp. second) part follows from Exercise 1.1.a (resp. 1.1.b or 1.1.c) of [4]. We will use only the second part; the first one shows that the number t in the proof of Corollary 2 is close to 0, but we did not need it.

Lemma 3.   (a) For $t \in \mathbf{C}$, we have

$$|e^t - 1| \leq |t|\max\{1, |e^t|\}.$$

(b) If a complex number z satisfies $|z - 1| < 1/2$, then there exists $t \in \mathbf{C}$ such that $e^t = z$ and $|t| \leq 2|z - 1|$. This t is unique and is the principal determination of the logarithm of z:

$$|\log z| \leq 2|z - 1|.$$

4   Proof of Theorem 1

Let us define some real numbers θ, δ and v in the interval $[0, 2\pi)$ by

$$\varepsilon' = \frac{1}{\varepsilon^{1/2}}e^{i\theta}, \quad \alpha' = |\alpha'|e^{i\delta}, \quad \xi_1' = |\xi_1'|e^{iv}.$$

By ordering the terms of (7), we can write this relation as

$$T_1 + T_2 + T_3 = 0,$$

and the three terms involved are

$$\begin{cases} T_1 := \varepsilon^{\ell}\xi_1(\alpha'\varepsilon'^n - \overline{\alpha'}\,\overline{\varepsilon'}^n) & = 2i\xi_1|\alpha'|\varepsilon^{\ell - n/2}\sin(\delta + n\theta), \\ T_2 := \alpha\varepsilon^n(\overline{\varepsilon'}^{\ell}\overline{\xi_1'} - \varepsilon'^{\ell}\xi_1') & = -2i|\xi_1'|\alpha\varepsilon^{n - \ell/2}\sin(v + \ell\theta), \\ T_3 := \xi_1'\varepsilon'^{\ell}\overline{\alpha'}\,\overline{\varepsilon'}^n - \overline{\xi_1'}\overline{\varepsilon'}^{\ell}\alpha'\varepsilon'^n = 2i|\xi_1'|\alpha'|\varepsilon^{-(n+\ell)/2}\sin(v - \delta + (\ell - n)\theta). \end{cases}$$

It turns out that these three terms are purely imaginary. We write this zero sum as

$$a + b + c = 0 \quad \text{with } |a| \geq |b| \geq |c|,$$

and we use the fact that this implies that $|a| \leq 2|b|$. Thanks to (17), Corollary 2 shows that a lower bound of the sinus terms is $|\ell|^{-\kappa_{26}}$ (and an obvious upper bound is 1!). Moreover,

- The T_1 term contains a constant factor and the factors:
  - $|\xi_1|$ with $k^{-\kappa_9} \leq |\xi_1| \leq k^{\kappa_9}$
  - $\varepsilon^{\ell-(n/2)}$ (which is the main term)
  - A sinus with a parameter n (a lower bound of the absolute value of that sinus being $n^{-\kappa_{27}}$)

- Similarly, T_2 contains a constant factor and the factors:
  - $|\xi_1'|$, with $k^{-\kappa_9} \leq |\xi_1'| \leq k^{\kappa_9}$
  - $\varepsilon^{n-(\ell/2)}$ (which the main term)
  - A sinus with a parameter ℓ (a lower bound of the absolute value of that sinus being $|\ell|^{-\kappa_{28}}$)

- Similarly, T_3 contains a constant factor and the factors:
  - $|\xi_1'|$, with $k^{-\kappa_9} \leq |\xi_1'| \leq k^{\kappa_9}$
  - $\varepsilon^{-(n+\ell)/2}$ (which the main term)
  - A sinus with a parameter $\ell - n$ (a lower bound of the absolute value of that sinus being $|\ell - n|^{-\kappa_{29}}$)

We will consider three cases, and we will use the inequalities (3) and (17). This will eventually allow us to conclude that there is an upper bound for $|\ell|$ and n by an effective constant times $\log k$.

First Case. If the two terms a and b with the largest absolute values are T_1 and T_2, from the inequalities $|T_1| \leq 2|T_2|$ and $|T_2| \leq 2|T_1|$ (which come from $|b| \leq |a| \leq 2|b|$), we deduce (thanks to (17))

$$k^{-\kappa_{30}} |\ell|^{-\kappa_{31}} \leq \varepsilon^{\frac{3}{2}(\ell-n)} \leq k^{\kappa_{32}} |\ell|^{\kappa_{33}},$$

whereupon, thanks again to (17), we have

$$-\kappa_{34} \log k + \frac{|\ell|}{3} \leq |\ell - n| \leq \kappa_{35} \log |\ell| + \kappa_{36} \log k,$$

which leads to $|\ell| \leq \kappa_{37}(\log k + \log |\ell|)$. This secures the upper bound (15) and ends the proof of Theorem 1.

Second Case. Suppose that the two terms a and b with the largest absolute values are T_1 and T_3. By writing $|T_1| \leq 2|T_3|$ and $|T_3| \leq 2|T_1|$, we obtain (thanks to (17))

$$k^{-1/3}|\ell|^{-\kappa_{38}} \leq \varepsilon^{3\ell/2} \leq k^{1/3}|\ell|^{\kappa_{39}};$$

hence

$$|\ell| \leq \kappa_{40}(\log k + \log|\ell|).$$

Once more, we have $\varepsilon^{|\ell|} \leq k^{\kappa_{41}}$, and we saw that the upper bound (17) allows to draw the conclusion.

Third Case. Let us consider the remaining case, namely, the two terms a and b with the largest absolute values being T_2 and T_3. Consequently, in the relation $T_1 + T_2 + T_3 = 0$, written in the form $a + b + c = 0$ with $|a| \geq |b| \geq |c|$, we have $c = T_1$. Writing $|T_2| \leq 2|T_3|$ and $|T_3| \leq 2|T_2|$, we obtain

$$k^{-1/3}|\ell|^{-\kappa_{42}} \leq \varepsilon^{3n/2} \leq k^{1/3}|\ell|^{\kappa_{43}}.$$

From the second of these inequalities, we deduce the existence of κ_{44} such that

$$n \leq \kappa_{44}(\log k + \log|\ell|). \tag{18}$$

Remark. The upper bound (18) allows to proceed as in the usual proof of the Thue theorem where n is fixed.

From the upper bound $|T_1| \leq |T_2|$, one deduces $n > \ell - \kappa_{45}\log k$, so that (18) leads right away to the conclusion if ℓ is positive.

Let us suppose now that ℓ is negative. Let us consider again Eq. (7) that we write in the form

$$\rho_n \varepsilon^\ell + \mu_n \varepsilon'^\ell - \overline{\mu}_n \overline{\varepsilon}'^\ell = 0 \tag{19}$$

with

$$\rho_n = \xi_1(\alpha' \varepsilon'^n - \overline{\alpha'} \, \overline{\varepsilon}'^n) \quad \text{and} \quad \mu_n = \xi_1'(\overline{\alpha'} \, \overline{\varepsilon}'^n - \alpha \varepsilon^n).$$

We check (cf. Property 3.3 of [4])

$$h(\mu_n) \leq \kappa_{46}(n + \log k).$$

Let us divide each side of (19) by $-\mu_n \varepsilon'^\ell$:

$$\frac{\overline{\mu}_n \overline{\varepsilon}'^\ell}{\mu_n \varepsilon'^\ell} - 1 = \frac{\rho_n \varepsilon^\ell}{\mu_n \varepsilon'^\ell}.$$

We have

$$|\alpha' \varepsilon'^n - \overline{\alpha'} \, \overline{\varepsilon}'^n| \leq |\alpha' \varepsilon'^n| + |\overline{\alpha'} \, \overline{\varepsilon}'^n| = 2|\varepsilon'^n \alpha'|$$

and, using (8),

$$\left| \overline{\alpha'} \, \overline{\varepsilon'}^n - \alpha \varepsilon^n \right| \geq \frac{1}{2} |\alpha| \varepsilon^n.$$

Since

$$\left| \overline{\xi_1} \right| \leq k^{\kappa_9} \quad \text{and} \quad |\xi_1'| > k^{-\kappa_9}$$

by (3), we come up with

$$|\rho_n| \leq \kappa_{47} k^{\kappa_9} \varepsilon^{n/2}, \quad |\mu_n| \geq \kappa_{48} \varepsilon^n k^{-\kappa_9}.$$

Therefore, since $|\varepsilon'|^{-1} = \varepsilon^{1/2}$, we have

$$\left| \frac{\overline{\mu_n} \, \overline{\varepsilon'}^\ell}{\mu_n \varepsilon'^\ell} - 1 \right| = \left| \frac{\rho_n \varepsilon^\ell}{\mu_n \varepsilon'^\ell} \right| \leq \kappa_{49} \varepsilon^{-(n+3|\ell|)/2} k^{\kappa_9}. \tag{20}$$

We denote by log the principal value of the logarithm, and we set

$$\lambda_1 = \log\left(\frac{\overline{\varepsilon'}}{\varepsilon'} \right), \quad \lambda_2 = \log\left(\frac{\overline{\mu_n}}{\mu_n} \right) \quad \text{and} \quad \Lambda = \log\left(\frac{\overline{\mu_n} \, \overline{\varepsilon'}^\ell}{\mu_n \varepsilon'^\ell} \right).$$

We have

$$\lambda_1 = 2i\pi v \quad \lambda_2 = 2i\pi \theta_n,$$

where v and θ_n are the real numbers in the interval $[0, 1)$ defined by

$$\frac{\overline{\varepsilon'}}{\varepsilon'} = e^{2i\pi v} \quad \text{and} \quad \frac{\overline{\mu_n}}{\mu_n} = e^{2i\pi \theta_n}.$$

From $e^\Lambda = e^{\ell \lambda_1 + \lambda_2}$, we deduce $\Lambda - \ell \lambda_1 - \lambda_2 = 2i\pi h$ with $h \in \mathbf{Z}$. From Lemma 3b we deduce $|\Lambda| \leq 2|e^\Lambda - 1|$. Using $|\Lambda| < 2\pi$ and writing

$$2i\pi h = \Lambda - 2i\pi \ell v - 2i\pi \theta_n,$$

we deduce $|h| \leq |\ell| + 2$.

In Proposition 1, let us take

$$b_0 = h, \quad b_1 = \ell, \quad b_2 = 1, \quad \gamma_0 = 1, \quad \lambda_0 = 2i\pi, \quad \gamma_1 = \frac{\overline{\varepsilon'}}{\varepsilon'}, \quad \gamma_2 = \frac{\overline{\mu_n}}{\mu_n},$$

$$A_0 = A_1 = \kappa_{50}, \quad A_2 = (k \, \varepsilon^n)^{\kappa_{51}}, \quad B = e + \frac{|\ell|}{\log A_2}.$$

Notice that the degree D of the field $\mathbf{Q}(\gamma_0, \gamma_1, \gamma_2)$ is ≤ 6. Then we obtain

$$\left| \frac{\overline{\mu}_n}{\mu_n} \left(\frac{\overline{\varepsilon}'}{\varepsilon'} \right)^{\ell} - 1 \right| = |e^{\Lambda} - 1| \geq \frac{1}{2}|\Lambda| \geq \exp\{-\kappa_{52}(\log A_2)(\log B)\}.$$

By combining this estimate with (20), we deduce

$$|\ell| \leq \kappa_{53}(n + \log k)\log B,$$

which can also be written as $B \leq \kappa_{54} \log B$; hence B is bounded. This allows to obtain

$$|\ell| \leq \kappa_{55}(n + \log k).$$

We use (18) to deduce $\varepsilon^{|\ell|} \leq k^{\kappa_{41}}$ and we saw that the upper bound (15) leads to the conclusion of the main Theorem 1.

References

1. H. Cohen, Advanced topics in computational number theory, in *Graduate Texts in Mathematics*, vol 193 (Springer, New York, 2000)
2. C. Levesque, M. Waldschmidt, Familles d'équations de Thue–Mahler n'ayant que des solutions triviales Acta Arith., **155**, 117–138 (2012)
3. T.N. Shorey, R. Tijdeman, Exponential Diophantine equations, in *Cambridge Tracts in Mathematics*, vol 87 (Cambridge University Press, Cambridge, 1986)
4. M. Waldschmidt, Diophantine approximation on linear algebraic groups, in *Grundlehren der Mathematischen Wissenschaften*, vol 326 (Springer, Berlin, 2000)

Consequences of a Factorization Theorem for Generalized Exponential Polynomials with Infinitely Many Integer Zeros

Ouamporn Phuksuwan and Vichian Laohakosol

Abstract Two consequences of our earlier result about factorization of generalized exponential polynomials are given. The first consequence shows that all but finitely many integer zeros of a generalized exponential polynomial form a finite union of arithmetic progressions. The second shows how to construct classes of transcendentally transcendental power series having the property that the index set of its zero coefficients is a finite union of arithmetic progressions plus a finite set.

1 Introduction

In our earlier paper [6] the following factorization theorem about generalized exponentials is proved:

Theorem 1. *If f is a nontrivial generalized exponential polynomial of the form*

$$f(t) = \sum_{i=1}^{k} P_i(t) A_i^{Q(t)}, \tag{1}$$

where the A_i's are distinct elements of $\mathbb{C} \setminus \{0\}$, $P_i(t) \in \mathbb{C}[t] \setminus \{0\}$ and $Q(t) \in \mathbb{Z}[t] \setminus \mathbb{Z}$, then there exist $T \in \mathbb{N}$ and a subset $E \subseteq \{0, 1, \ldots, T-1\}$ such that

O. Phuksuwan (✉)
Faculty of Science, Department of Mathematics and Computer Science, Chulalongkorn University, Bangkok 10330, Thailand
e-mail: ouamporn.p@chula.ac.th

V. Laohakosol
Faculty of Science, Department of Mathematics, Kasetsart University, Bangkok 10900, Thailand
e-mail: fscivil@ku.ac.th

J.M. Borwein et al. (eds.), *Number Theory and Related Fields: In Memory of Alf van der Poorten*, Springer Proceedings in Mathematics & Statistics 43, DOI 10.1007/978-1-4614-6642-0_13, © Springer Science+Business Media New York 2013

$$f(t) = \left(\prod_{r \in E} \left(\eta^{Q(t)} - \eta^{Q(r)} \right)^{m_r} \right) g(t), \qquad (2)$$

where η is a primitive T^{th} root of unity, $m_r \in \mathbb{N}$, and $g(t)$ is a generalized exponential polynomial of the same form (1) but with finitely many integer zeros.

Our objective is to derive consequences of our factorization theorem above. The first shows that such a generalized exponential polynomial possesses the so-called *Skolem–Mahler–Lech or SML property*, i.e., all but finitely many of its integer zeros form a finite union of arithmetic progressions, and the second shows how to construct classes of transcendentally transcendental power series (power series that satisfy no algebraic differential equations) with the SML property. In the last section, a correction to Example 2 on p. 218 of [6] is given.

2   The SML Property

In [8], Shapiro showed that an exponential polynomial of the form $E(x) = \sum_{i=1}^{m} P_i(x) A_i^x$ with infinitely many integer zeros can be factorized as $E(x) = \prod_d \left(\eta^x - \eta^d \right) E_0(x)$, where the product is taken over a subset of least positive residues modulo a specific number called the basic period of $E(x)$, η a primitive root of unity, and $E_0(x)$ an exponential polynomial having only finitely many integer zeros. Such a factorization enables us to deduce the classical Skolem–Mahler–Lech theorem [2–4, 9], which states that all but finitely many integer zeros of an exponential polynomial, if exist, form a finite union of arithmetic progressions.

Let $f(t)$ be a generalized exponential polynomial of the form (1). When $Q(t) = t$, we revert back to ordinary exponential polynomials. From the factorization (2), it is natural to ask whether the SML property remains valid for generalized exponential polynomials, namely, all but finitely many of integer zeros form a finite union of arithmetic progressions. We answer this question affirmatively in the next theorem.

Theorem 2. *Let $f(t)$ be a generalized exponential polynomial of the form as stated in (1). If $f(t)$ has infinitely many integer zeros, then all but finitely many of its integer zeros form a finite union of arithmetic progressions.*

We give three proofs of Theorem 2. The first one follows immediately from the proof of Lemma 3.2 in [6] which makes use of the p-adic method, and the remaining two proofs are derived from the factorization theorem above.

First Proof of Theorem 2. From lines 15–17 in the proof of [6, Lemma 3.2], we have that if $r_1, \ldots, r_m \in \{0, 1, \ldots, p-2\}$ is such that the function $f(x(p-1) + r_i)$ is zero, then $f(x(p-1) + r_i)$ is zero for all $x \in \mathbb{Z}$. Thus, the set of integer zeros of f is $\bigcup_{i=1}^{m} ((p-1)\mathbb{Z} + r_i)$ together with a finite set. Otherwise, if the function $f(x(p-1) + r)$ is not zero, then f has only a finite number of zeros in the closed unit disk of $\mathbb{C}_p$, and hence $f(x(p-1) + r)$ has only a finite numbers of zeros in $\mathbb{Z}$. $\square$

Second Proof of Theorem 2. To determine the integer zeros of $f(t)$ from the factorization (2), we must solve $\eta^{Q(t)} - \eta^{Q(d)} = 0$, i.e., we need to find all integers x such that

$$Q(x) - Q(d) \in T\mathbb{Z}, \tag{3}$$

where T is the order of η. Consider the set

$$A = \{x \in \mathbb{Z} : Q(x) \equiv Q(d) \bmod T\}.$$

We claim that A is a finite union of arithmetic progressions. From (3), since Q is periodic on $\mathbb{Z}$ mod T, it follows that if $0 \le x_0 < \cdots < x_m < T$ are the solutions of the congruence $Q(x) \equiv Q(d) \bmod T$ in $\{0, 1, \ldots, T-1\}$, then

$$B := \bigcup_{j=0}^{m} (x_j + T\mathbb{Z}) \subseteq A.$$

Conversely, let $x \in A$. Putting $x = Ty + z$ for some $z \in \{0, 1, \ldots, T-1\}$. By periodicity, we have $z \in A$, so $z = x_j$ for some j. This implies $x \in B$, and hence $B = A$. $\qquad\square$

The last proof of Theorem 2 is longer but gives a little more information. We need a simple fact about the intersection of two arithmetic progressions, which is of inidependent interest.

Lemma 1. *The intersection of two arithmetic progressions consisting only of integers, if nonempty, is again an arithmetic progression.*

Proof. Denote the two arithmetic progressions by

$$(ak+b)_k := \{ak+b \, ; \, k \in \mathbb{Z}\}, \quad (c\ell+d)_\ell := \{c\ell+d \, ; \, \ell \in \mathbb{Z}\},$$

where $a(\ne 0)$, b, $c(\ne 0)$, $d \in \mathbb{Z}$. Assume that their intersection is nonempty, i.e.,

$$\mathscr{I} := (ak+b)_k \cap (c\ell+d)_\ell \ne \emptyset.$$

Let $ak_0 + b = c\ell_0 + d$ be the least positive element of $\mathscr{I}$. The existence of k_0 and ℓ_0 indicates that $g := \gcd(a, c) \mid (d - b)$. Thus, $d - b = gN$ for some $N \in \mathbb{Z}$, and so $c\ell + d = c\ell + gN + b$. Consequently, $\mathscr{I}$ is an arithmetic progression:

$$\Longleftrightarrow (ak)_k \cap (c\ell + gN)_\ell \text{ is an arithmetic progression.}$$

$$\Longleftrightarrow \left(\frac{a}{g}k\right)_k \cap \left(\frac{c}{g}\ell + N\right)_\ell \text{ is an arithmetic progression.}$$

It is thus enough to consider the reduced arithmetic progressions $(Ak)_k$ and $(C\ell + N)_\ell$, where $A = a/g$ and $C = c/g$. Let $Ak_1 = C\ell_1 + N$ $(k_1, \ell_1 \in \mathbb{Z})$ be the least positive element of $(Ak)_k \cap (C\ell + N)_\ell$. Since the general solutions of the linear diophantine equation $Ax = Cy + N$ are of the form $x = k_1 - Cn$, $y = \ell_1 - An$ $(n \in \mathbb{Z})$, the elements in $(Ak)_k \cap (C\ell + N)_\ell$ are of the form $Ax = A(k_1 - Cn) = Ak_1 - ACn$ $(n \in \mathbb{Z})$ and correspondingly those in $\mathscr{I}$ are of the form

$$g(Ak_1 - ACn) + b = gAk_1 - gACn + b = ak_1 + b - aCn \ (n \in \mathbb{Z}),$$

showing that $\mathscr{I}$ is an arithmetic progression.   □

Third Proof of Theorem 2. To determine the integer zeros of $f(t)$ from the factorization (2), we must solve $\eta^{Q(t)} - \eta^{Q(d)} = 0$, i.e., we need to find all integral solutions (x, y) of the diophantine equation $Q(x) - Q(d) = yT$, where T is the order of η. Letting

$$Q(x) = a_n x^n + a_{n-1} x^{n-1} + \cdots + a_1 x + a_0 \in \mathbb{Z}[x] \setminus \mathbb{Z},$$

the equation is equivalent to

$$
\begin{aligned}
yT &= Q(x) - Q(d) \\
&= (x - d)\{ a_n (x^{n-1} + x^{n-2}d + \cdots + d^{n-1}) \\
&\quad + a_{n-1}(x^{n-2} + x^{n-3}d + \cdots + d^{n-2}) + \cdots + a_1 \}.
\end{aligned}
\tag{4}
$$

Thus, $x - d = kT_1$, where T_1 is a positive divisor of T and $k \in \mathbb{Z}$. Replacing $x = kT_1 + d$ in the right-hand expression of (4), we have

$$
\begin{aligned}
yT &= a_n((kT_1 + d)^n - d^n) + a_{n-1}((kT_1 + d)^{n-1} - d^{n-1}) \\
&\quad + \cdots + a_1((kT_1 + d) - d) =: H(k).
\end{aligned}
\tag{5}
$$

Let the unique prime representation of T be $T = p_1^{\alpha_1} p_2^{\alpha_2} \cdots p_s^{\alpha_s}$, where $p_1, p_2, \ldots, p_s$ are distinct primes and $\alpha_1, \alpha_2, \ldots, \alpha_s \in \mathbb{N}$. To solve (4) is thus equivalent to solving the system of congruences

$$H(k) \equiv 0 \pmod{p_i^{\alpha_i}} \ (i = 1, 2, \ldots, s).
\tag{6}$$

For fixed $i \in \{1, 2, \ldots, s\}$, consider

$$H(k) \equiv 0 \pmod{p_i}.
\tag{7}$$

If $\deg H = n \geq p_i$, by [7, Theorem 2.15], $H(k)$ can be reduced to a polynomial $g(k)$ of degree less than p_i, and the congruence $g(k) \equiv 0 \pmod{p_i}$ has the same integral solutions as those of (7). Thus, without loss of generality, assume that $\deg H = n < p_i$. Theorem 2.16 of [7] also tells us that the congruence (7) has at most n integral solutions mod p_i, and for each solution $k_0 \pmod{p_i}$ of (7), the number of integral solutions of (6) corresponding to k_0 is at most p_i. This means the solutions of (6) consist of a finite number of arithmetic progressions. Using Lemma 1, the intersection of the resulting arithmetic progressions derived from each of the solutions of (6) and the arithmetic progression $(kT_1 + d)_k$ is again an arithmetic progression. Thus, the x-values of the integral solutions of (4), which are identical with the integer zeros of the generalized exponential polynomial $f(t)$, form a finite union of arithmetic progressions as desired.   □

3 Transcendentally Transcendental Power Series with the SML Property

A formal power series

$$F(x) = \sum_{n=0}^{\infty} f(n)x^n, \tag{8}$$

alternatively, a sequence (f_n), is said to have the SML property if the index set of its zero coefficients $\{n \; ; \; f(n) = 0\}$ is a finite union of arithmetic progressions plus a finite set. An equivalent statement of the Skolem–Mahler–Lech theorem [2–4, 9] states that if $F(x)$ is a rational power series, then it has the SML property. It is well known that the coefficients $f(n)$ of a rational power series satisfy a linear recurrence with constant coefficients [4]. This in turn shows that a sequence satisfying a linear recurrence relation with constant coefficients necessarily possesses the SML property.

The Skolem–Mahler–Lech theorem has later been extended in various directions. Of particular interest to us here is the following proposition due to Bell et al. in [1].

Proposition 1. *Let K be a field of characteristic zero, let $d \in \mathbb{N}$, and let $P_1(z), \ldots,$ $P_d(z) \in K[z]$ be polynomials with $P_d(z)$ being a nonzero constant. Suppose that $f :$ $\mathbb{N} \to K$ is a sequence satisfying the polynomial-linear recurrence*

$$f(n) = \sum_{i=1}^{d} P_i(n)f(n-i)$$

for all n sufficiently large. Then $\{n \in \mathbb{N} \; ; \; f(n) = 0\}$ is a union of finitely many arithmetic progressions and a finite set.

Following Stanley, [10], a sequence (or its corresponding formal power series) is said to be *differentially finite*, or D-finite, if it satisfies a linear recurrence with polynomial coefficients. In [10], the following equivalences of D-finite power series are proved.

Proposition 2. *The following three conditions on a formal power series $y \in \mathbb{C}[[x]]$ are equivalent:*

(i) *y is D-finite.*
(ii) *There exist finitely many polynomials $q_0(x),\ldots,q_k(x)$, not all zero, and a polynomial $q(x)$ such that*

$$q_k(x)y^{(k)} + \cdots + q_1(x)y' + q_0(x)y = q(x). \qquad (9)$$

(iii) *There exist finitely many polynomials $p_0(x),\ldots,p_m(x)$, not all zero, such that*

$$p_m(x)y^{(m)} + \cdots + p_1(x)y' + p_0(x)y = 0. \qquad (10)$$

Combining Propositions 1 and 2, we see that a sequence which belongs to a special subclass of D-finite sequences, containing the ones corresponding to rational power series, has the SML property.

Our aim now is to construct classes of power series with the SML property that do not satisfy the differential equations of the form (9) or (10), which in turn are not contained in the class established by Bell, Burris, and Yeats. This shows that the SML and the D-finite properties are essentially independent.

Recall from [5] that a formal power series of the form (8) is said to be *transcendentally transcendental* if it does not satisfy any algebraic differential equation. Consider a power series $F(x)$ of the form (8), where the coefficients $f(n)$ are generalized exponential polynomials of the form (1). We now state and prove our second main result.

Theorem 3. *Let $F(x)$ be a power series of the form (8), where the coefficients $f(n)$ are nonzero generalized exponential polynomials of the form (1) and $f(n) = 0$ for infinitely many $n \in \mathbb{N}$. Assume that all the coefficients of P_i's and the A_i's are algebraic numbers. Using the notation and terminology of Theorem 1, we know that $f(t)$ has a factorization of the form*

$$f(t) = \left(\prod_{r \in E} \left(\eta^{\mathcal{Q}(t)} - \eta^{\mathcal{Q}(r)} \right)^{m_r} \right) g(t), \qquad (11)$$

where

$$g(t) := \sum_{i=i}^{k} \mathscr{P}_i(t) \mathscr{A}_i^{\mathscr{Q}(t)}, \qquad (12)$$

is a generalized exponential polynomial with only finitely many integer zeros. Assume that $\deg \mathscr{Q} \geq 2$, the leading coefficient of $\mathscr{Q}$ is positive, and

$$0 < |\mathscr{A}_1| \leq \cdots \leq |\mathscr{A}_k| < 1. \qquad (13)$$

Then $F(x)$ is a transcendentally transcendental power series with the SML property.

Proof. The power series $F(x)$ clearly has the SML property by Theorem 2, and we note in passing that its convergence can easily be deduced under the assumption

(13). Suppose to the contrary that $F(x)$ satisfies an algebraic differential equation over $\mathbb{C}$. By a theorem of Popken [5, Theorem 19, pp. 206–207], there exists $\Gamma \in \mathbb{N}$ such that

$$e^{-\Gamma n (\log n)^2} \le |f(n)|,$$

whenever the algebraic coefficient $f(n)$ is nonzero. On the other hand, estimating $|f(n)|$ from above using (11) and (12), we get

$$|f(n)| = C_f \left| \mathscr{P}_1(n) \mathscr{A}_1^{\mathscr{Q}(n)} + \cdots + \mathscr{P}_k(n) \mathscr{A}_k^{\mathscr{Q}(n)} \right|$$

$$\le C_f \left\{ |\mathscr{P}_1(n)| |\mathscr{A}_1|^{\mathscr{Q}(n)} + \cdots + |\mathscr{P}_k(n)| |\mathscr{A}_k|^{\mathscr{Q}(n)} \right\} \le C_f \, e^{-M \mathscr{Q}(n)} H(n),$$

where $C_f := \left| \prod_{r \in E} \left(\eta^{\mathscr{Q}(n)} - \eta^{\mathscr{Q}(r)} \right)^{m_r} \right| \le 2^{\Sigma m_r}$, $M = \min\{ -\ln|\mathscr{A}_1|, \ldots, -\ln|\mathscr{A}_k| \}$ > 0 and $H(n) := |\mathscr{P}_1(n)| + \cdots + |\mathscr{P}_k(n)|$. Thus, for those sufficiently large n with $f(n) \ne 0$, we have

$$e^{-\Gamma n (\log n)^2} \le C_f \, e^{-M \mathscr{Q}(n)} H(n). \tag{14}$$

Since $\deg \mathscr{Q} \ge 2$ and the leading coefficient of $\mathscr{Q}$ is positive, the inequality (14) is not tenable when n is large enough showing that $F(x)$ cannot satisfy any algebraic differential equation. $\qquad \square$

Theorem 3 is restrictive in the sense that the coefficients $f(n)$ need to be algebraic numbers in order to apply Popken's lower bound of $f(n)$. This restriction can be removed at the cost of strengthening the condition (13) and making use of an upper bound for $f(n)$ in a theorem of Maillet [5, Theorem 16, p. 200].

Theorem 4. *Let $F(x)$ be a power series of the form (8), whose coefficients $f(n)$ are nonzero generalized exponential polynomials of the form*

$$f(n) = \sum_{i=1}^{k} P_i(n) A_i^{Q(n)},$$

where the A_i's are distinct elements of $\mathbb{C} \setminus \{0\}$, $P_i(t) \in \mathbb{C}[t] \setminus \{0\}$ and $Q(t) \in \mathbb{Z}[t] \setminus \mathbb{Z}$. Suppose that $Q(t)$ has a positive leading coefficient, $\deg Q \ge 2$, and that

$$1 < |A_k|, \quad |A_i| < |A_k| \quad (i = 1, 2, \ldots, k-1). \tag{15}$$

If $f(n) \ne 0$, then $F(x)$ is a transcendentally transcendental power series with the SML property.

Proof. Suppose to the contrary that $F(x)$ satisfies an algebraic differential equation. Since $P_i(t) \in \mathbb{C}[t] \setminus \{0\}$, using (15), we easily deduce that there exist positive constants c_1 and c_2 such that

$$|f(n)| \ge c_1 n^{c_2} |A_k|^{Q(n)} \tag{16}$$

for all sufficiently large n. However, from [5, Theorem 16, p. 200], we know that there are positive constants Γ_1, Γ_2 for which

$$|f(n)| \leq \Gamma_1 (n!)^{\Gamma_2}. \tag{17}$$

Using $\deg Q \geq 2$ and (15), the two inequalities (16) and (17) are not compatible for large n, and we are done. $\square$

4   A Corrigendum

In Example 2, pp. 217–218, of [6], the parameters, expressions, and factorization of

$$F(x) = 1 - e^{\frac{\pi}{2}ix^2} - e^{\frac{\pi}{3}ix^2} + e^{\frac{5\pi}{6}ix^2}$$

given on p. 218, lines 7–8, should be replaced by

$$F(x) = \left(e^{\frac{\pi}{6}ix^2} - 1\right)^2 \left(e^{\frac{\pi}{6}ix^2} - e^{\frac{2\pi}{3}i}\right) \left(e^{\frac{\pi}{6}ix^2} + 1\right) \left(e^{\frac{\pi}{6}ix^2} + e^{\frac{\pi}{3}i}\right),$$

where $m_{d_1} = 2, m_{d_2} = 1$ and $G(x) = \left(e^{\frac{\pi}{6}ix^2} + 1\right) \left(e^{\frac{\pi}{6}ix^2} + e^{\frac{\pi}{3}i}\right)$.

References

1. J.P. Bell, S.N. Burris, K. Yeats, On the set of zero coefficients of a function satisfying a linear differential equation. Math. Proc. Camb. Phil. Soc. **153**, 235–247 (2012)
2. C. Lech, A note on recurring series. Ark. Mat. **2**, 417–421 (1953)
3. K. Mahler, Eine arithmetische Eigenschaft der Taylor-Koeffizienten rationaler Funktionen. Akad. Wetensch. Amsterdam Proc. **38**, 50–60 (1935)
4. K. Mahler, On the Taylor coefficients of rational functions. Proc. Camb. Phil. Soc. **52**, 39–48 (1956)
5. K. Mahler, Lectures on transcendental numbers, in ed. & completed by B. Divis, W.J. LeVeque. Lecture Notes in Math., vol 546 (Springer, Berlin, 1976)
6. O. Phuksuwan, V. Laohakosol, A factorization theorem for generalized exponential polynomials with infinitely many integer zeros. Acta Math. Acad. Paedagog. Nyházi. **27**(2), 211–221 (2011)
7. D. Redmond, *Number Theory—An Introduction* (Marcel Dekker, New York, 1996)
8. H.N. Shapiro, On a theorem concerning exponential polynomials, Comm. Pure Appl. Math. **12**, 487–500 (1959)
9. T. Skolem, Ein Verfahren zur Behandlung gewisser exponentialer Gleichungen und diophantischer Gleichungen, in *Comptes Rendus du Huitième Congrès des Mathématiciens Scandinaves*, Stockholm, 1934, pp. 163–188
10. R. Stanley, Differentiably finite power series. Eur. J. Combin. **1**, 175–188 (1980)

On Balanced Subgroups of the Multiplicative Group

Carl Pomerance and Douglas Ulmer

In memory of Alf van der Poorten

Abstract A subgroup H of $(\mathbb{Z}/d\mathbb{Z})^\times$ is called *balanced* if every coset of H is evenly distributed between the lower and upper halves of $(\mathbb{Z}/d\mathbb{Z})^\times$, i.e., has equal numbers of elements with representatives in $(0, d/2)$ and $(d/2, d)$. This notion has applications to ranks of elliptic curves. We give a simple criterion in terms of characters for a subgroup H to be balanced, and for a fixed integer p, we study the distribution of integers d such that the cyclic subgroup of $(\mathbb{Z}/d\mathbb{Z})^\times$ generated by p is balanced.

Mathematics Subject Classification (2010): Primary 11N37; Secondary 11G05

1 Introduction

Let $d > 2$ be an integer and consider $(\mathbb{Z}/d\mathbb{Z})^\times$, the group of units modulo d. Let A_d be the first half of $(\mathbb{Z}/d\mathbb{Z})^\times$, that is, A_d consists of residues with a representative in $(0, d/2)$. Let $B_d = (\mathbb{Z}/d\mathbb{Z})^\times \backslash A_d$ be the second half of $(\mathbb{Z}/d\mathbb{Z})^\times$. We say a subgroup H of $(\mathbb{Z}/d\mathbb{Z})^\times$ is *balanced* if for each $g \in (\mathbb{Z}/d\mathbb{Z})^\times$ we have $|gH \cap A_d| = |gH \cap B_d|$, that is, each coset of H has equally many members in the first half of $(\mathbb{Z}/d\mathbb{Z})^\times$ as in the second half.

C. Pomerance (✉)
Department of Mathematics, Dartmouth College, Hanover, NH 03755, USA
e-mail: carl.pomerance@dartmouth.edu

D. Ulmer
School of Mathematics, Georgia Institute of Technology, Atlanta, GA 30332, USA
e-mail: douglas.ulmer@math.gatech.edu

J.M. Borwein et al. (eds.), *Number Theory and Related Fields: In Memory of Alf van der Poorten*, Springer Proceedings in Mathematics & Statistics 43, DOI 10.1007/978-1-4614-6642-0_14, © Springer Science+Business Media New York 2013

Let φ denote Euler's function, so that $\varphi(d)$ is the cardinality of $(\mathbb{Z}/d\mathbb{Z})^\times$. If n and m are coprime integers with $m > 0$, let $l_n(m)$ denote the order of the cyclic subgroup $\langle n \bmod m \rangle$ generated by n in $(\mathbb{Z}/m\mathbb{Z})^\times$ (i.e., $l_n(m)$ is the multiplicative order of n modulo m).

Our interest in balanced subgroups stems from the following result:

Theorem 1.1 ([2]). *Let p be an odd prime number, let $\mathbb{F}_q$ be the finite field of cardinality $q = p^f$, and let $\mathbb{F}_q(u)$ be the rational function field over $\mathbb{F}_q$. Let d be a positive integer not divisible by p, and let E_d be the elliptic curve over $\mathbb{F}_q(u)$ defined by*

$$y^2 = x(x+1)(x+u^d).$$

Then we have

$$\mathrm{Rank}\, E_d(\mathbb{F}_q(u)) = \sum_{\substack{e \mid d,\ e > 2 \\ \langle p \bmod e \rangle \text{ balanced}}} \frac{\varphi(e)}{l_q(e)}.$$

A few simple observations are in order. It is easy to see that $\langle -1 \rangle$ is a balanced subgroup of $(\mathbb{Z}/d\mathbb{Z})^\times$. It is also easy to see that if $4 \mid d$, then $\langle \frac{1}{2}d + 1 \rangle$ is a balanced subgroup of $(\mathbb{Z}/d\mathbb{Z})^\times$. In addition, if H is a balanced subgroup of $(\mathbb{Z}/d\mathbb{Z})^\times$ and K is a subgroup of $(\mathbb{Z}/d\mathbb{Z})^\times$ containing H, then K is balanced as well. Indeed, K is a union of $[K : H]$ cosets of H, so for each $g \in (\mathbb{Z}/d\mathbb{Z})^\times$, gK is a union of $[K : H]$ cosets of H, each equally distributed between the first half of $(\mathbb{Z}/d\mathbb{Z})^\times$ and the second half. Thus, gK is also equally distributed between the first half and the second half.

It follows that if some power of p is congruent to -1 modulo d and if $q \equiv 1 \pmod d$, then the theorem implies that $\mathrm{Rank}\, E_d(\mathbb{F}_q(u)) = d - 2$ if d is even and $d - 1$ if d is odd. The rank of E_d when some power of p is -1 modulo d was first discussed in [12], and with hindsight it could have been expected to be large from considerations of "supersingularity." The results of [2] show, perhaps surprisingly, that there are many other classes of d for which high ranks occur. Our aim here is to make this observation more quantitative.

More precisely, the aim of this paper is to investigate various questions about balanced pairs (p, d), i.e., pairs such that $\langle p \bmod d \rangle$ is a balanced subgroup of $(\mathbb{Z}/d\mathbb{Z})^\times$. In particular, we give a simple criterion in terms of characters for a subgroup to be balanced (Theorem 2.1), and we use it to determine all balanced subgroups of order 2 (Theorem 3.2). We also investigate the distribution for a fixed p of the set of d's such that (p, d) is balanced (Theorems 4.1–4.3). We find that when p is odd, the divisors d of numbers of the form $p^n + 1$ are not the largest contributor to this set. Finally, we investigate the average rank and typical rank of the curves E_d in Theorem 1.1 for fixed q and varying d.

2   Balanced Subgroups and Characters

The goal of this section is to characterize balanced subgroups of $(\mathbb{Z}/d\mathbb{Z})^\times$ in terms of Dirichlet characters. If χ is a character modulo d, we define

$$c_\chi = \sum_{0 < a < d/2} \chi(a).$$

As usual, we say that χ is odd if $\chi(-1) = -1$ and χ is even if $\chi(-1) = 1$.

Theorem 2.1. *A subgroup $H \subset (\mathbb{Z}/d\mathbb{Z})^\times$ is balanced if and only if $c_\chi = 0$ for every odd character χ of $(\mathbb{Z}/d\mathbb{Z})^\times$ whose restriction to H is trivial.*

As an example, note that if $H = \langle -1 \rangle$, then there are no odd characters trivial on H and so the theorem implies that H is balanced.

Proof. Throughout the proof, we write G for $(\mathbb{Z}/d\mathbb{Z})^\times$. We also write A for A_d as above and similarly for B, so that G is the disjoint union $A \cup B$.

We write $\mathbf{1}_A$ for the characteristic function of $A \subset G$ and similarly for $\mathbf{1}_B$. Let $f : G \to \mathbb{C}$ be the sum over H of translates of $\mathbf{1}_A - \mathbf{1}_B$:

$$f(g) = \sum_{h \in H} (\mathbf{1}_A(gh) - \mathbf{1}_B(gh))$$

$$= \#(gH \cap A) - \#(gH \cap B).$$

By definition, H is balanced if and only if f is identically zero.

We write $\hat{G}$ for the set of complex characters of G, and we expand f in terms of these characters:

$$f = \sum_{\chi \in \hat{G}} \hat{f}(\chi) \chi$$

where

$$\hat{f}(\chi) = \frac{1}{\varphi(d)} \sum_{g \in G} f(g) \chi^{-1}(g).$$

Thus, H is balanced if and only if $\hat{f}(\chi) = 0$ for all $\chi \in \hat{G}$.

It is easy to see that $\hat{f}(\chi_{triv}) = 0$. Since $\mathbf{1}_A - \mathbf{1}_B = 2\mathbf{1}_A - \mathbf{1}_G$, for χ nontrivial, we find that

$$\hat{f}(\chi^{-1}) = \frac{2}{\varphi(d)} \left(\sum_{h \in H} \chi(h) \right) \left(\sum_{a \in A} \chi(a) \right).$$

Note that $\sum_{h \in H} \chi(h)$ is zero if and only if the restriction of χ to H is nontrivial. Note also that

$$c_\chi = \sum_{0 < a < d/2} \chi(a) = \sum_{a \in A} \chi(a)$$

since χ is a Dirichlet character. If χ is even and nontrivial, then

$$c_\chi = \frac{1}{2} \sum_{g \in G} \chi(g) = 0.$$

Thus, $\hat{f}(\chi) = 0$ for all $\chi \in \hat{G}$ if and only if $c_\chi = 0$ for all odd characters χ which are trivial on H. This completes the proof of the theorem.   □

We now give a non-vanishing criterion for c_χ.

Lemma 2.2. *If χ is a primitive, odd character of $(\mathbb{Z}/d\mathbb{Z})^\times$, then $c_\chi \neq 0$.*

Proof. Under the hypotheses on χ, the classical evaluation of $L(1,\chi)$ leads to the formula

$$L(1,\chi^{-1}) = \frac{\pi i \tau(\chi^{-1})}{d(\chi^{-1}(2) - 2)} c_\chi$$

where $\tau(\chi^{-1})$ is a Gauss sum. (See, e.g., [7, pp. 200–201] or [8, Theorem 9.21], though there is a small typo in the second reference.) By the theorem of Dirichlet, $L(1,\chi^{-1}) \neq 0$ and so $c_\chi \neq 0$.   □

In light of the lemma, we should consider imprimitive characters.

Lemma 2.3. *Suppose that ℓ is a prime number dividing d and set $d' = d/\ell$. Suppose also that χ is a nontrivial character modulo d induced by a character χ' modulo d'. If $\ell = 2$, then $c_\chi = -\chi'(2)c_{\chi'}$. If ℓ is odd, then $c_\chi = (1 - \chi'(\ell))c_{\chi'}$. Here, we employ the usual convention that $\chi'(\ell) = 0$ if $\ell \mid d'$.*

Proof. First suppose $\ell = 2$. We have

$$c_\chi = \sum_{\substack{a < d/2 \\ \gcd(a,d)=1}} \chi(a) = \sum_{\substack{a < d' \\ \gcd(a,2d')=1}} \chi'(a).$$

If $2 \mid d'$, this is a complete character sum and so vanishes. If $2 \nmid d'$, then

$$\sum_{\substack{a < d' \\ \gcd(a,2d')=1}} \chi'(a) = \sum_{\substack{a < d' \\ \gcd(a,d')=1}} \chi'(a) - \sum_{\substack{a < d'/2 \\ \gcd(a,d')=1}} \chi'(2a)$$

$$= - \sum_{\substack{a < d'/2 \\ \gcd(a,d')=1}} \chi'(2a)$$

$$= -\chi'(2)c_{\chi'}$$

as desired.

Now assume that ℓ is odd. We have

$$c_\chi = \sum_{\substack{a<d/2 \\ \gcd(a,d)=1}} \chi(a) = \sum_{\substack{a<\ell d'/2 \\ \gcd(a,\ell d')=1}} \chi'(a).$$

If $\ell \mid d'$, then

$$\sum_{\substack{a<\ell d'/2 \\ \gcd(a,\ell d')=1}} \chi'(a) = \sum_{\substack{a<d'/2 \\ \gcd(a,d')=1}} \chi'(a) = c_{\chi'}.$$

If $\ell \nmid d'$, then

$$\sum_{\substack{a<\ell d'/2 \\ \gcd(a,\ell d')=1}} \chi'(a) = \sum_{\substack{a<\ell d'/2 \\ \gcd(a,d')=1}} \chi'(a) - \sum_{\substack{a<d'/2 \\ \gcd(a,d')=1}} \chi'(\ell a)$$

$$= \sum_{\substack{a<d'/2 \\ \gcd(a,d')=1}} \chi'(a) - \chi'(\ell) \sum_{\substack{a<d'/2 \\ \gcd(a,d')=1}} \chi'(a)$$

$$= (1 - \chi'(\ell))c_{\chi'}$$

as desired.   □

Applying the lemma repeatedly, we arrive at the following non-vanishing criterion:

Proposition 2.4. *Suppose that χ is an odd character modulo d induced by a primitive character χ' modulo d'. Then $c_\chi \neq 0$ if and only if the following two conditions both hold: (i) $4 \nmid d$ or d/d' is odd, and (ii) for every odd prime ℓ which divides d and does not divide d', we have $\chi'(\ell) \neq 1$.*

As an example, suppose that $4 \mid d$ and $H = \langle \frac{1}{2}d + 1 \rangle$. Note that

$$(\mathbb{Z}/d\mathbb{Z})^\times / \langle \tfrac{1}{2}d + 1 \rangle \cong (\mathbb{Z}/\tfrac{1}{2}d\mathbb{Z})^\times.$$

Thus, if χ is an odd character modulo d and $\chi(\frac{1}{2}d+1) = 1$, then the conductor d' of χ divides $d/2$. This shows that d/d' is even and so condition (i) of the proposition fails and $c_\chi = 0$. Therefore, H is balanced.

3 Balanced Subgroups of Small Order

In this section, we discuss balanced subgroups of small order. We have already seen that a subgroup of $(\mathbb{Z}/d\mathbb{Z})^\times$ which contains -1 or $\frac{1}{2}d + 1$ is balanced. We will show that in a certain sense small balanced subgroups are controlled by these balanced subgroups of order 2.

Theorem 3.1. *For every positive integer n there is an integer $d(n)$ such that if $d > d(n)$ and H is a balanced subgroup of $(\mathbb{Z}/d\mathbb{Z})^\times$ of order n, then either $-1 \in H$ or $4 \mid d$ and $\frac{1}{2}d + 1 \in H$.*

We can make this much more explicit for subgroups of order 2.

Theorem 3.2. *A subgroup $H = \langle h \rangle$ of $(\mathbb{Z}/d\mathbb{Z})^\times$ of order 2 is balanced if and only if d and h satisfy one of the following conditions:*

1. $h \equiv -1 \pmod{d}$,
2. $d \equiv 0 \pmod 4$ and $h \equiv \frac{1}{2}d + 1 \pmod d$,
3. $d = 24$ and $h \equiv 17 \pmod d$ or $h \equiv 19 \pmod d$,
4. $d = 60$ and $h \equiv 41 \pmod d$ or $h \equiv 49 \pmod d$.

Proof of Theorem 3.1. Throughout this proof and the next, we write G for $(\mathbb{Z}/d\mathbb{Z})^\times$. Using Proposition 2.4, we will show that if d is sufficiently large with respect to n, then for any subgroup $H \subset G$ of order n which does not contain -1 or $\frac{1}{2}d + 1$, there is a character χ which is odd, trivial on H, and with $c_\chi \neq 0$. By Theorem 2.1, this implies that H is not balanced.

Note that a balanced subgroup obviously has even order, so there is no loss in assuming that n is even. We make this assumption for the rest of the proof.

Let H^+ be the subgroup of G generated by H, -1 and, if $4 \mid d$, by $\frac{1}{2}d + 1$. Fix a character χ_0 of G which is trivial on H, odd, and -1 on $\frac{1}{2}d + 1$ if $4 \mid d$. The set of all characters satisfying these restrictions is a homogeneous space for $\widehat{G/H^+} \subset \hat{G}$. We will argue that multiplying χ_0 by a suitable $\psi \in \widehat{G/H^+}$ yields a $\chi = \chi_0 \psi$ for which Proposition 2.4 implies that $c_\chi \neq 0$.

Note first that any character χ which is odd and, if $4 \mid d$, has $\chi(\frac{1}{2}d + 1) = -1$ automatically satisfies condition (i) in Proposition 2.4. Indeed, if $4 \mid d$, then the condition $\chi(\frac{1}{2}d + 1) = -1$ implies that χ is 2-primitive, i.e., the conductor d' of χ has d/d' odd. The rest of the argument relates to condition (ii) in Proposition 2.4.

Write $d = \prod_\ell \ell^{e_\ell}$ and write G_ℓ for $(\mathbb{Z}/\ell^{e_\ell}\mathbb{Z})^\times$ so that $G \cong \prod_\ell G_\ell$. Let $\chi = \prod_\ell \chi_\ell$. Note that ℓ divides the conductor of χ if and only if χ_ℓ is non-trivial.

We will sloppily write G_ℓ/H^+ for G_ℓ modulo the image of H^+ in G_ℓ. For odd ℓ, G_ℓ is cyclic and therefore so is G_ℓ/H^+; for $\ell = 2$, since $-1 \in H^+$, G_2/H^+ is also cyclic.

Note also that H^+ is the product of H and a group of exponent 2, namely, the subgroup of G generated by -1 or by -1 and $\frac{1}{2}d + 1$. Also, we have assumed that $n = |H|$ is even. If ℓ is odd, then G_ℓ is cyclic of even order, so it has a unique element of order 2. It follows that the order of the image of H^+ in G_ℓ divides n.

We define three sets of odd primes:

$$S_1 = \left\{ \text{odd } \ell : \ell \mid d, \ G_\ell/H^+ = \{1\} \right\},$$

$$S_2 = \left\{ \text{odd } \ell : \ell \mid d, \ \varphi(\ell^{e_\ell}) \mid n \right\},$$

and

$$S_3 = \{\text{odd } \ell : \varphi(\ell) \mid n\}.$$

Note that $S_1 \subset S_2 \subset S_3$ and S_3 depends only on n, not on d.

If ℓ is odd, $\ell \mid d$, and $\ell \notin S_1$, then G_ℓ/H^+ is nontrivial. Thus, choosing a suitable ψ, we may arrange that the conductor of $\chi_1 = \chi_0\psi$ is divisible by every prime dividing d which is not in S_1.

For the odd primes ℓ which divide d and do not divide the conductor of χ_1 (a subset of S_1, thus also a subset of S_3), we must arrange that $\chi'(\ell) \neq 1$ (where χ' is the primitive character inducing χ).

Recall that G_ℓ/H^+ is cyclic. We now remark that if C is a cyclic group and $a \in C$, then, for each $z \in \mathbb{C}$, the set of characters $\psi : C \to \mathbb{C}$ such that $\chi(a) \neq z$ has cardinality at least $|C|(1 - 1/|\langle a \rangle|)$ (where $|\langle a \rangle|$ is the order of a). If we have several elements $a_1, \ldots, a_n$ and several values $z_1, \ldots, z_n$ to avoid, then the number of characters ψ such that $\psi(a_i) \neq z_i$ is at least

$$|C| \left(1 - \frac{1}{|\langle a_1 \rangle|} - \cdots - \frac{1}{|\langle a_n \rangle|} \right).$$

Thus we can find such a character provided that each a_i has order $> n$.

Now we use that d is large to conclude that a large prime power ℓ^e divides d. (Note that ℓ might be 2 here.) Then G_ℓ/H^+ is a cyclic group in which the order of each prime in S_1 is large. (The primes in S_1 are also in S_3, so belong to a set fixed independently of d.) We want a character ψ of G_ℓ/H^+ which satisfies $\psi(r) \neq \chi_1^{-1}(r)$ for all $r \in S_1$. We also want $\psi\chi_1$ to have nontrivial ℓ component which, phrased in the language above, means that we want $\psi(a) \neq 1$ for some fixed generator of G_ℓ/H^+. Since the size of S_1 is bounded depending only on n, the discussion of the previous paragraph shows that these conditions can be met if ℓ^e is large enough.

Setting $\chi = \psi\chi_1$ with ψ as in the previous paragraph yields a character χ such that $c_\chi \neq 0$, and this completes the proof. $\qquad\square$

Proof of Theorem 3.2. We retain the concepts and notation of the proof of Theorem 3.1. We also say that a subgroup of order 2 is "exceptional" if it does not contain -1 or $\frac{1}{2}d + 1$.

Since $n = 2$, the set $S_3 = \{3\}$ and the set S_2 is either empty (if $3 \nmid d$ or $9 \mid d$) or $S_2 = \{3\}$ (if 3 exactly divides d). If S_2 is empty and H is an exceptional subgroup of order 2, then the first part of the proof of Theorem 3.1 provides a primitive odd character trivial on H, and so H is not balanced.

Suppose we are in the case where 3 exactly divides d. Following the first part of the proof of Theorem 3.1, we have a character χ_1 of G with conductor divisible by $d' = d/3$ which is odd, trivial on H, and, if $4 \mid d$, satisfies $\chi_1(\frac{1}{2}d + 1) = -1$. If the conductor of χ_1 is d or if the primitive character χ' inducing χ_1 has $\chi'(3) \neq 1$, then setting $\chi = \chi_1$ we have $c_\chi \neq 0$ and we see that H is not balanced.

If not, we will modify χ_1. Note that if $\ell = 2$ and $16 \mid d$, or $\ell = 5$ and $25 \mid d$, or ℓ is a prime ≥ 7 and $\ell \mid d$, then the order of 3 in G_ℓ/H^+ is at least 3. Thus, in these cases, there is a character ψ of G_ℓ/H^+ so that the ℓ part of $\chi = \chi_1 \psi$ is nontrivial and so that the primitive character χ' inducing χ satisfies $\chi'(3) \neq 1$. Then $c_\chi \neq 0$ and H is not balanced.

This leaves a small number of values of d to check for exceptional balanced subgroups of order 2. Namely, we just need to check divisors of $8 \cdot 3 \cdot 5 = 120$ which are divisible by 3. A quick computation which we leave to the reader finishes the proof.   □

4 Distribution of Numbers d with $\langle p \bmod d \rangle$ Balanced

For the rest of the paper, we write $\mathbb{U}_d$ for $(\mathbb{Z}/d\mathbb{Z})^\times$. Fix an integer p with $|p| > 1$. In our application to elliptic curves, p is an odd prime number, but it seems interesting to state our results on balanced subgroups in a more general context. Let $\mathscr{B}_p$ denote the set of integers $d > 2$ coprime to p for which $\langle p \bmod d \rangle$ is a balanced subgroup of $\mathbb{U}_d$. Further, define subsets of $\mathscr{B}_p$ as follows:

$$\mathscr{B}_{p,0} = \{d > 2 : (d,p) = 1,\ 4 \mid d,\ \tfrac{1}{2}d + 1 \in \langle p \bmod d \rangle\},$$

$$\mathscr{B}_{p,1} = \{d > 2 : (d,p) = 1,\ -1 \in \langle p \bmod d \rangle\}$$

$$\mathscr{B}_{p,*} = \mathscr{B}_p \setminus (\mathscr{B}_{p,0} \cup \mathscr{B}_{p,1}).$$

Note that if p is even then $\mathscr{B}_{p,0}$ is empty. For any set $\mathscr{A}$ of positive integers and x a real number at least 1, we let $\mathscr{A}(x) = |\mathscr{A} \cap [1,x]|$.

We state the principal results of this section, which show that when p is odd, most members of $\mathscr{B}_p$ lie in $\mathscr{B}_{p,0}$.

Theorem 4.1. *For each odd integer p with $|p| > 1$, there are positive numbers b_p, b'_p with*

$$b_p \frac{x}{\log\log x} \leq \mathscr{B}_{p,0}(x) \leq b'_p \frac{x}{\log\log x}$$

for all sufficiently large numbers x depending on the choice of p.

We remark that $\mathscr{B}_{p,1}$ has been studied by Moree. In particular we have the following result:

Theorem 4.2 ([9, Thm. 5]). *For each integer p with $|p| > 1$, there are positive numbers c_p, δ_p such that*

$$\mathscr{B}_{p,1}(x) \sim c_p \frac{x}{(\log x)^{\delta_p}} \quad as\ x \to \infty.$$

In particular, for p prime we have $\delta_p = \tfrac{2}{3}$.

Theorem 4.3. *For each integer p with $|p| > 1$, there is a number $\varepsilon_p > 0$ such that for all $x \geq 3$,*

$$\mathscr{B}_{p,*}(x) = O_p\left(\frac{x}{(\log x)^{\varepsilon_p}}\right).$$

Corollary 4.4. *If p is an odd integer with $|p| > 1$, then $\mathscr{B}_p(x) \sim \mathscr{B}_{p,0}(x)$ as $x \to \infty$.*

It is easy to see that $\mathscr{B}_{p,1} \cap \mathscr{B}_{p,0}$ has at most one element. Indeed, the cyclic group $\langle p \bmod d \rangle$ has at most one element of order exactly 2, so if $d \in \mathscr{B}_{p,1} \cap \mathscr{B}_{p,0}$, then for some f, we have $p^f \equiv -1 \equiv \frac{1}{2}d + 1 \pmod d$, and this can happen only when $d = 4$. This shows that for $x \geq 4$,

$$\mathscr{B}_p(x) \geq \mathscr{B}_{p,0}(x) + \mathscr{B}_{p,1}(x) - 1.$$

We believe that $\mathscr{B}_{p,0}$ and $\mathscr{B}_{p,1}$ comprise most of $\mathscr{B}_p$, and in fact we pose the following conjecture:

Conjecture 4.5. *For each integer p with $|p| > 1$ we have*

$$\mathscr{B}_p(x) = \mathscr{B}_{p,0}(x) + (1 + o(1))\mathscr{B}_{p,1}(x) \quad as \ x \to \infty,$$

that is, $\mathscr{B}_{p,*}(x) = o(\mathscr{B}_{p,1}(x))$ as $x \to \infty$.

We now begin a discussion leading to the proofs of Theorems 4.1 and 4.3. The following useful result comes from [4, Theorem 2.2]:

Proposition 4.6. *There is an absolute positive constant c such that for all numbers $x \geq 3$ and any set $\mathscr{R}$ of primes in $[1, x]$, the number of integers in $[1, x]$ not divisible by any member of $\mathscr{R}$ is at most*

$$cx \prod_{r \in \mathscr{R}} \left(1 - \frac{1}{r}\right) \leq cx \exp\left(-\sum_{r \in \mathscr{R}} \frac{1}{r}\right).$$

Note that the inequality in the display follows immediately from the inequality $1 - \theta < e^{-\theta}$ for every $\theta \in (0, 1)$.

For a positive integer m coprime to p, recall that $l_p(m)$ denotes the order of $\langle p \bmod m \rangle$. If r is a prime, we let $v_r(m)$ denote that integer v with $r^v \mid m$ and $r^{v+1} \nmid m$.

We would like to give a criterion for membership in $\mathscr{B}_{p,0}$, but before this, we establish an elementary lemma.

Lemma 4.7. *Let p be an odd integer with $|p| > 1$ and let k, i be positive integers. Then*

$$v_2\left(\frac{p^{2^i k} - 1}{p^{2k} - 1}\right) = i - 1.$$

Proof. The result is clear if $i = 1$. If $i > 1$, we see that

$$\frac{p^{2^i k} - 1}{p^{2k} - 1} = (p^{2k} + 1)(p^{4k} + 1) \ldots (p^{2^{i-1}k} + 1),$$

which is a product of $i - 1$ factors that are each 2 (mod 4). $\qquad \square$

The following result gives a criterion for membership in $\mathscr{B}_{p,0}$:

Proposition 4.8. *Let p be odd with $|p| > 1$ and let $m \geq 1$ be an odd integer coprime to p. If $l_p(m)$ is odd, then $2^j m \in \mathscr{B}_{p,0}$ if and only if $j = 1 + v_2(p - 1)$ or $j > v_2(p^2 - 1)$. If $l_p(m)$ is even, then $2^j m \in \mathscr{B}_{p,0}$ if and only if $j > v_2(p^{l_p(m)} - 1)$.*

Proof. We first prove the "only if" part. Assume that $d = 2^j m \in \mathscr{B}_{p,0}$ and let f be an integer with $p^f \equiv \frac{1}{2}d + 1 \pmod{d}$. Then $l_p(m) \mid f$ so that $j - 1 = v_2(p^f - 1) \geq v_2(p^{l_p(m)} - 1)$. This establishes the "only if" part if $l_p(m)$ is even, and it also shows that $j \geq 1 + v_2(p - 1)$ always, so in particular if $l_p(m)$ is odd. Suppose $l_p(m)$ is odd and $1 + v_2(p - 1) < j \leq v_2(p^2 - 1)$. Then, $j - 1 > v_2(p - 1)$, so that $l_p(2^{j-1}m)$ is even. Using $l_p(m)$ odd, this implies that $2l_p(m) \mid f$, so that $2^j \mid (p^2 - 1) \mid (p^f - 1)$, contradicting $p^f \equiv \frac{1}{2}d + 1 \pmod{d}$.

Towards showing the "if" part, let $v = v_2(p^{l_p(m)} - 1)$. We have $p^{l_p(m)} - 1 \equiv 2^v m$ (mod $2^{v+1}m$), so that $2^{v+1}m \in \mathscr{B}_{p,0}$. If $j > v + 1$ and $l_p(m)$ is even, then with $f = 2^{j-v-1}l_p(m)$, Lemma 4.7 implies that $p^f - 1 \equiv 2^{j-1}m \pmod{2^j m}$, so that $2^j m \in \mathscr{B}_{p,0}$. If $l_p(m)$ is odd, then $v = v_2(p - 1)$, so that $2^{v+1} \in \mathscr{B}_{p,0}$. Finally assume that $j > v_2(p^2 - 1)$ and $l_p(m)$ is odd. Then Lemma 4.7 implies that $p^{2^{j-v_2(p^2-1)}l_p(m)} - 1 \equiv 2^{j-1}m \pmod{2^j m}$, so that $2^j m \in \mathscr{B}_{p,0}$. This concludes the proof. $\qquad \square$

Proof of Theorem 4.1. For $m \geq 1$ coprime to $2p$, let

$$f_p(m) := v_2\left(p^{l_p(m)} - 1\right), \quad f_p'(m) := \max\left\{f_p(m), v_2\left(p^2 - 1\right)\right\}.$$

Proposition 4.8 implies that if $2^j m \in \mathscr{B}_{p,0}$ with m odd, then $j > f_p(m)$. Further, if $(m, 2p) = 1$ then $2^j m \in \mathscr{B}_{p,0}$ for all $j > f_p'(m)$.

Using this last property, we have $\mathscr{B}_{p,0}(x)$ at least as big as the number of choices for m coprime to $2p$ with $1 < m \leq x/2^{f_p'(m)+1}$. Thus, the lower bound in the theorem will follow if we show that there are at least $b_p x / \log\log x$ integers m coprime to $2p$ with $m \leq x/2^{f_p'(m)+1}$.

Let $\lambda(m)$ denote Carmichael's function at m, which is the order of the largest cyclic subgroup of $\mathbb{U}_m$. Then $l_p(m) \mid \lambda(m)$. Also, for $m > 2$, $\lambda(m)$ is even, so that

$$f_p'(m) \leq g_p(m) := v_2\left(p^{\lambda(m)} - 1\right).$$

Thus, the lower bound in the theorem will follow if we show that there are at least $b_p x / \log\log x$ integers m coprime to $2p$ with $m \leq x/2^{g_p(m)+1}$. Using Lemma 4.7, we have $g_p(m) + 1 = v_2(\lambda(m)) + v_2(p^2 - 1)$. Further, it is easy to see that $2^{v_2(p^2-1)} \leq 2(|p| + 1)$, with equality when $|p| + 1$ is a power of 2.

It follows from [10, Section 2, Remark 1] that uniformly for all $x \geq 3$ and all positive integers n,

$$\sum_{\substack{r \leq x \\ r \text{ prime} \\ n | r-1}} \frac{1}{r} = \frac{\log\log x}{\varphi(n)} + O\left(\frac{\log(2n)}{\varphi(n)}\right). \tag{4.9}$$

We apply this with $n = 2^{g_0+1}$, where g_0 is the first integer with $2^{g_0} \geq 4\log\log x$. Thus, if $\mathscr{R}$ is the set of primes $r \leq x$ with $v_2(r-1) > g_0$, we have for x sufficiently large,

$$\sum_{r \in \mathscr{R}} \frac{1}{r} < \frac{1}{3}.$$

Let $z = x/(25|p|\log\log x)$. In $[1, z]$ there are $(\varphi(|p|)/(2|p|))z + O_p(1)$ integers coprime to $2p$. And for a given value of $r \in \mathscr{R}$, there are at most $(\varphi(|p|)/(2|p|))z/r + O_p(1)$ numbers in $[1, z]$ coprime to $2p$ and divisible by r. It follows that for x sufficiently large depending on the choice of p, there are at least

$$\frac{\varphi(|p|)}{2|p|}z - \frac{\varphi(|p|)}{2|p|}z \sum_{r \in \mathscr{R}} \frac{1}{r} + O_p\left(\sum_{r \in \mathscr{R}} 1\right) > \frac{\varphi(|p|)}{4|p|}z$$

integers $m \leq z$ coprime to $2p$ and not divisible by any prime $r \in \mathscr{R}$. (We used that $|\mathscr{R}| = O(x/\log x)$ to estimate the O-term above.)

It remains to note that if $m \leq z$, m is coprime to $2p$, and m is not divisible by any prime in $\mathscr{R}$, then $v_2(\lambda(m)) \leq g_0$, so that

$$2^{g_p(m)+1} \leq 2^{v_2(\lambda(m))+v_2(p^2-1)} \leq 2^{g_0} 2^{v_2(p^2-1)}$$

$$\leq 2^{g_0} \cdot 2(|p| + 1) \leq 2^{g_0} \cdot 3|p| < 25|p|\log\log x.$$

Thus, $2^{g_p(m)+1}m \in \mathscr{B}_{p,0}$ and $2^{g_p(m)+1}m \leq x$, so that

$$\mathscr{B}_{p,0}(x) \geq \frac{\varphi(|p|)}{100p^2} \frac{x}{\log\log x},$$

for x sufficiently large depending on the choice of p. This completes our proof of the lower bound.

For the upper bound, it suffices to show that

$$N(x) := \mathscr{B}_{p,0}(x) - \mathscr{B}_{p,0}(x/2) = O_p\left(\frac{x}{\log\log x}\right).$$

(With this assumption, no two numbers d counted can have the same odd part.) We shall assume that p is not a square, the case when $p = p_0^{2^j}$ for some integer p_0 and $j \geq 1$ being only slightly more complicated. From Proposition 4.8, $N(x)$ is at most the number of odd numbers m coprime to p with $m \leq x/2^{f_p(m)+1}$. Let $N_k(x)$ be the number of odd numbers $m \leq x/2^{k+1}$ with m coprime to p and $f_p(m) = k$. Then

$$N(x) = \sum_k N_k(x) = \sum_{2^k \leq \log\log x} N_k(x) + O\left(\frac{x}{\log\log x}\right).$$

We now concentrate our attention on $N_k(x)$ with $2^k \leq \log\log x$. If $f_p(m) = k$, then m is not divisible by any prime r with $(p/r) = -1$ and $2^{k+1} \mid r - 1$. Then, using (4.9) and quadratic reciprocity,

$$\sum_{\substack{r \leq x \\ (p/r)=-1 \\ 2^{k+1}\mid r-1 \\ r \text{ prime}}} \frac{1}{r} = \frac{\log\log x}{2^{k+1}} + O_p\left(\frac{k}{2^k}\right).$$

By Proposition 4.6, the number of integers $m \leq x/2^{k+1}$ not divisible by any such prime r is at most

$$O\left(\frac{x}{2^{k+1}} \exp\left(-\sum_r \frac{1}{r}\right)\right) = O_p\left(\frac{x}{2^{k+1}} \exp\left(-\frac{\log\log x}{2^{k+1}}\right)\right).$$

Summing this expression for $2^k \leq \log\log x$ gives $O_p(x/\log\log x)$, which completes the proof of Theorem 4.1.

Remark 4.10. One might wonder if there is a positive constant β_p such that if p is odd with $|p| > 1$, then $\mathscr{B}_{p,0}(x) \sim \beta_p x/\log\log x$ as $x \to \infty$. Here we sketch an argument that no such β_p exists, that is,

$$0 < \liminf_{x\to\infty} \frac{\mathscr{B}_{p,0}(x)}{x/\log\log x} < \limsup_{x\to\infty} \frac{\mathscr{B}_{p,0}(x)}{x/\log\log x} < \infty.$$

First note that but for $O_p(x/(\log x)^{1/2})$ values of $d \leq x$, there is a prime $r \mid d$ with $(p/r) = -1$. (We are assuming here that p is not a square.) For such values of $d = 2^j m$, with m odd, we have $2 \mid l_p(m)$, so that in the notation above we have $f_p(m) = f_p'(m) \geq 3$. Thus it suffices to count numbers $2^j m \leq x$ with m odd and $j > f_p(m) \geq 3$. Note that

$$f_p(m) = v_2(l_p(m)) + v_2(p^2 - 1) - 1 = v_2(l_p(m)) + h_p - 1,$$

say. Further,

$$v_2(l_p(m)) = \max_{r\mid m} v_2(l_p(r)),$$

where r runs over the prime divisors of m. We have $\{r \text{ prime} : v_2(l_p(r)) = k\}$ equal to

$$\bigcup_{i \geq 0} \{r \text{ prime} : v_2(r-1) = k+i,$$

$$p \text{ is a } 2^i \text{ power } (\text{mod } r) \text{ and not a } 2^{i+1} \text{ power } (\text{mod } r)\}.$$

For $k > (\log\log\log x)^2$, the density of primes $r \equiv 1 \pmod{2^k}$ is so small that we may assume that no d is divisible by such a prime r. For k below this bound, the density of primes r with $v_2(l_p(r)) = k$ is $1/(3 \cdot 2^{k-1})$. Thus, there is a positive constant $c_{k,p}$ with $c_{k,p} \to 1$ as $k \to \infty$ such that the density of integers m coprime to $2p$ and with $f_p(m) < k + h_p$ is asymptotically equal to

$$c_p(\varphi(2|p|)/(2|p|)) \exp(-(\log\log x)/(3 \cdot 2^k)),$$

as $x \to \infty$. Thus, the number of $m \leq x/2^{k+h_p}$ coprime to $2p$ and with $f_p(m) = k + h_p - 1$ is asymptotically equal to

$$c_{k,p} \frac{\varphi(2|p|)}{2|p|} \frac{x}{2^{k+h_p}} \frac{\log\log x}{3 \cdot 2^k} \exp\left(-\frac{\log\log x}{3 \cdot 2^k}\right)$$

as $x \to \infty$. This expression then needs to be summed over k. For k small, the count is negligible because of the exp factor. For k larger, we can assume that the coefficients $c_{k,p}$ are all 1, and then the sum takes on the form

$$\frac{\varphi(2|p|)}{2|p|2^{h_p}} \frac{x}{\log\log x} \sum_k \frac{(\log\log x)^2}{3 \cdot 2^{2k}} \exp\left(-\frac{\log\log x}{3 \cdot 2^k}\right).$$

Letting this sum on k be denoted $g(x)$, it remains to note that $g(x)$ is bounded away from both 0 and ∞ yet does not tend to a limit, cf. [5, Theorem 3.25].

To prove Theorem 4.3, we first establish the following result.

Proposition 4.11. *Let p be an integer with $|p| > 1$. Let d be a positive integer coprime to p such that d is divisible by odd primes s, t with*

$$l_p(s) \equiv 2 \pmod 4, \quad l_p(t) \equiv 1 \pmod 2, \quad \langle p, -1 \bmod s \rangle \neq \mathbb{U}_s, \quad \langle p, -1 \bmod t \rangle \neq \mathbb{U}_t.$$

Assume that $4 \mid l_p(d)$. Then either $4 \mid d$ and $\frac{1}{2}d + 1 \in \langle p \bmod d \rangle$ or $\langle p \bmod d \rangle$ is not balanced.

Proof. Let $k = l_p(d)$. First assume that $4 \mid d$ and $\frac{1}{2}d + 1 \notin \langle p \bmod d \rangle$. Let 2^κ be the largest power of 2 in k. Write $d = 2^j m$ where m is odd, let 2^{κ_1} be the power of 2 in $l_p(m)$, and let $2^{\kappa_2} = l_p(2^j)$. Then $\kappa = \max\{\kappa_1, \kappa_2\}$. Suppose that $\kappa_2 > \kappa_1$. We have $p^{k/2} \equiv 1 \pmod m$ and $p^{k/2} \not\equiv 1 \pmod{2^j}$. Since $4 \mid k$, we have $p^{k/2} + 1 \equiv 2$

mod 4, and since $p^k - 1 = (p^{k/2} - 1)(p^{k/2} + 1)$, we have $p^{k/2} \equiv 1 \pmod{2^{e-1}}$. Thus, $p^{k/2} \equiv \frac{1}{2}d + 1 \pmod d$, contrary to our assumption. Hence, we may assume that $\kappa = \kappa_1 \geq \kappa_2$. Note that this inequality holds too in the case that $4 \nmid d$, since then $\kappa_2 = 0$.

We categorize the odd prime powers r^a coprime to p as follows:

- **Type 1**: $\langle p, -1 \bmod r^a \rangle = \mathbb{U}_{r^a}$.
- **Type 2**: $\langle p, -1 \bmod r^a \rangle \neq \mathbb{U}_{r^a}$.
- **Type 3**: It is Type 2 and also $l_p(r^a) \equiv 2 \pmod 4$.
- **Type 4**: It is Type 2 and also $l_p(r^a)$ is odd.

By assumption d has at least one Type 3 prime power component and at least one Type 4 prime power component. We will show that $\langle p \bmod d \rangle$ is not balanced in $\mathbb{U}_d$. By Proposition 2.4, it is sufficient to exhibit an odd character $\chi \pmod d$ that is trivial at p with conductor d' divisible by the same odd primes as are in d, and with either $d \equiv 2 \pmod 4$ or d/d' odd.

Let $r_1^{a_1} \| d$ where the power of 2 in $l_p(r_1^{a_1})$ is 2^{κ_1}. (Note that $r_1^{a_1}$ cannot be Type 3 nor Type 4, since we have $\kappa_1 = \kappa \geq 2$, so that $4 \mid l_p(r_1^{a_1})$.) Consider the Type 1 prime powers in d, other than possibly $r_1^{a_1}$ in case it is of Type 1. For each we take the quadratic character, and we multiply these together to get a character χ_1 whose conductor contains all of the primes involved in Type 1 prime powers, except possibly r_1.

If $j \leq 1$, we let ψ_{2^j} be the principal character mod 2^j. If $j \geq 2$, let ψ_{2^j} be a primitive character mod 2^j with $\psi_{2^j}(p) = \zeta$, a primitive 2^{κ_2}-th root of unity. Let $\chi_2 = \chi_1 \psi_{2^j}$.

We choose a character $\psi_{r_1^{a_1}}$ mod $r_1^{a_1}$ with $\psi_{r_1^{a_1}}(p) = \chi_2(p)^{-1}$ if $\chi_2(p) \neq 1$, and otherwise we choose it so that $\psi_{r_1^{a_1}}(p) = -1$. Thus, this character is non-principal. Let $\chi_3 = \psi_{r_1^{a_1}} \chi_2$. We now have $\chi_3(p) = \pm 1$.

If $\chi_3(p) = -1$ we use a Type 3 prime power $r_3^{a_3} \| d$ and choose a character $\psi_{r_3^{a_3}}$ $(\bmod\ r_3^{a_3})$ with $\psi_{r_3^{a_3}}(p) = -1$. Let $\chi_4 = \chi_3 \psi_{r_3^{a_3}}$. If $\chi_3(p) = 1$, we let $\chi_4 = \chi_3$. We now have $\chi_4(p) = 1$.

If $\chi_4(-1) = 1$, we use a Type 4 prime power $r_4^{a_4} \| d$ and choose a character $\psi_{r_4^{a_4}}$ $(\bmod\ r_4^{a_4})$ with $\psi_{r_4^{a_4}}(p) = 1$ and $\psi_{r_4^{a_4}}(-1) = -1$. Let $\chi_5 = \chi_4 \psi_{r^a}$. If $\chi_4(-1) = -1$, we let $\chi_5 = \chi_4$.

All remaining prime powers r^a in d are of Type 2. For these we take non-principal characters that are trivial on $\langle p, -1 \bmod r^a \rangle$ and multiply them in to χ_5 to form χ_6. This is the character we are looking for, and so $\langle p \bmod d \rangle$ is not balanced. This completes our proof. $\square$

Proof of Theorem 4.3. In the proof we shall assume that p is neither a square nor twice a square, showing in these cases that we may take $\varepsilon_p = 1/16$. The remaining cases are done with small adjustments to the basic argument but may require a smaller value for ε_p.

Let $d \leq x$ be coprime to p. The set of primes $r \nmid p$ with $r \equiv 1$ (mod 4) and for which p is a quadratic nonresidue has density $1/4$, and in fact, the sum of reciprocals of such primes $r \leq x$ is $\frac{1}{4}\log\log x + O_p(1)$. (This follows from either (4.9) and quadratic reciprocity or from the Chebotarev density theorem.) Thus by Proposition 4.6, the number of integers $d \leq x$ not divisible by any of these primes r is $O_p(x/(\log x)^{1/4})$. Thus, we may assume that d is divisible by such a prime r and so that $4 \mid l_p(d)$.

Note that if $r \equiv 5$ (mod 8) and that p is a quadratic residue modulo r, but not a fourth power, then any r^a is of Type 3. The density of these primes r is $1/16$, by the Chebotarev theorem; in fact, the sum of reciprocals of such primes $r \leq x$ is $\frac{1}{16}\log\log x + O_p(1)$. So the number of values of $d \in [3,x]$ not divisible by at least one of them is $O_p(x/(\log x)^{1/16})$, using Proposition 4.6. Also note that if $r \equiv 5$ (mod 8) and p is a nonzero fourth power modulo r, then any r^a is Type 4. The density of these primes r is also $1/16$, and again the number of $d \in [3,x]$ not divisible by at least one of them is $O_p(x/(\log x)^{1/16})$.

Thus, the number of values of $d \leq x$ coprime to p and not satisfying the hypotheses of Proposition 4.11 is $O(x/(\log x)^{1/16})$. This completes the proof of Theorem 4.3.

5   The Average and Normal Order of the Rank

In this section we consider the average and normal order of the rank of the curve E_d given in Theorem 1.1 as d varies.

It is clear from Theorem 1.1 that for q odd,

$$\operatorname{Rank} E_d(\mathbb{F}_q(u)) \leq \begin{cases} d-2 & \text{if } d \text{ is even} \\ d-1 & \text{if } d \text{ is odd} \end{cases}$$

with equality when $d \in \mathscr{B}_{p,1}$ and $q \equiv 1$ (mod d).

For all q and $d > 1$, it is known [1, Prop. 6.9] that

$$\operatorname{Rank} E_d(\mathbb{F}_q(u)) \leq \frac{d}{2\log_q d} + O\left(\frac{d}{(\log_q d)^2}\right).$$

(Here $\log_q d$ is the logarithm of d base q, i.e., $\log d/\log q$.) We do not include the details here, but this bound can be proved directly for the the curves in Theorem 1.1 using that theorem. In addition, for q odd, considering values of d of the form $q^f + 1$ for some positive integer f and using Theorem 1.1, we see that the main term in this inequality is sharp for this family of curves.

We show below that although the average rank of $E_d(\mathbb{F}_q(u))$ is large—its average for d up to x is at least $x^{1/2}$—for "most" values of d the rank is much smaller.

Theorem 5.1. *There is an absolute constant* $\alpha > \frac{1}{2}$ *with the following property: For each odd prime p and finite field* $\mathbb{F}_q$ *of characteristic p, with* $\mathbb{F}_q(u)$ *and* E_d *as in Theorem 1.1, we have*

$$x^{\alpha} \leq \frac{1}{x} \sum_{d \leq x} \mathrm{Rank}\, E_d(\mathbb{F}_q(u)) \leq x^{1-\log\log\log x/(2\log\log x)}$$

for all sufficiently large x depending on the choice of p.

Proof. This result follows almost immediately from [11, Theorem 1]. A result is proved there for the average value of the rank of curves in a different family also parametrized by a positive integer d. Using the notation from this chapter, if $d \in \mathscr{B}_{p,1}$, the rank of the curve considered in [11] is within 4 of

$$\sum_{\substack{e|d \\ e>2}} \frac{\varphi(e)}{l_q(e)}. \tag{5.2}$$

We have $d \in \mathscr{B}_{p,1}$ implies that $e \in \mathscr{B}_{p,1}$ for all $e \mid d$ with $e > 2$. By Theorem 1.1, formula (5.2) is exactly the rank of $E_d(\mathbb{F}_q(u))$ for $d \in \mathscr{B}_{p,1}$. Since the proof of the lower bound x^{α} in [11] uses only values of $d \in \mathscr{B}_{p,1}$, we have the lower bound x^{α} in the present theorem.

Since the rank of $E_d(\mathbb{F}_q(u))$ is bounded above by the formula (5.2) whether or not d is in $\mathscr{B}_{p,1}$, and in fact whether or not $\langle p \bmod d \rangle$ is balanced, the argument given in [11] for the upper bound gives our upper bound here. □

Theorem 5.3. *For each odd prime p and finite field* $\mathbb{F}_q$ *of characteristic p, with* $\mathbb{F}_q(u)$ *and* E_d *as in Theorem 1.1, we have but for* $o(x/\log\log x)$ *values of* $d \leq x$ *with* $d \in \mathscr{B}_p$ *that*

$$\mathrm{Rank}\, E_d(\mathbb{F}_q(u)) \geq (\log d)^{(1+o(1))\log\log\log d}$$

as $x \to \infty$. *Further, assuming the GRH, we have but for* $o(x/\log\log x)$ *values of* $d \leq x$ *with* $d \in \mathscr{B}_p$ *that*

$$\mathrm{Rank}\, E_d(\mathbb{F}_q(u)) \leq (\log d)^{(1+o(1))\log\log\log d}$$

as $x \to \infty$. *Assuming the GRH, this upper bound holds but for* $o(x)$ *values of* $d \leq x$ *coprime to p as* $x \to \infty$, *regardless of whether* $d \in \mathscr{B}_p$.

Proof. For $d \in \mathscr{B}_p$, Theorem 1.1 implies that the rank of $E_d(\mathbb{F}_q(u))$ is at least $\varphi(d)/l_q(d) \geq \varphi(d)/\lambda(d)$, where λ was defined in the previous section as the order of the largest cyclic subgroup of $\mathbb{U}_d$. It is shown in the proof of Theorem 2 in [3] that on a set of asymptotic density 1, we have $\varphi(d)/\lambda(d) = (\log d)^{(1+o(1))\log\log\log d}$. We would like to show this holds for almost all $d \in \mathscr{B}_p$. Note that we have $\varphi(m)/\lambda(m) = (\log m)^{(1+o(1))\log\log\log m}$ for almost all odd numbers m. We have for all odd m and every integer $j \geq 0$ that

$$\frac{\varphi(m)}{\lambda(m)} \le \frac{\varphi(2^j m)}{\lambda(2^j m)} \le 2^j \frac{\varphi(m)}{\lambda(m)}. \tag{5.4}$$

Thus, for almost all odd numbers m, we have for all nonnegative integers j with $2^j \le \log m$ that $\varphi(2^j m)/\lambda(2^j m) = (\log(2^j m))^{(1+o(1))\log\log\log(2^j m)}$. Further, it follows from (4.9) that but for a set of odd numbers m of asymptotic density 0, we have $v_2(\lambda(m)) \le 2\log\log\log m$. It thus follows from Proposition 4.8 that for almost all odd numbers m, there is some nonnegative j with $2^j m \in \mathscr{B}_p$ and $2^j \le \log m$. By Theorems 4.1 and 4.3 almost all members of $\mathscr{B}_p$ are of this form, and so we have the lower bound in the theorem.

For the upper bound, we use an argument in [6]. There, Corollary 2 and the following remark imply that under the assumption of the GRH, for almost all numbers d coprime to p, we have $\varphi(d)/l_q(d) = (\log d)^{(1+o(1))\log\log\log d}$. We use that $\varphi(e)/l_q(e) \mid \varphi(d)/l_q(d)$ for $e \mid d$ and from the normal order of the number-of-divisors function $\tau(d)$, that most numbers d have $\tau(d) \le \log d$. It thus follows from Theorem 1.1 and the GRH that for almost all numbers d coprime to p that

$$\mathrm{Rank}\, E_d(\mathbb{F}_q(u)) \le \tau(d)\frac{\varphi(d)}{l_q(d)} \le (\log d)\frac{\varphi(d)}{l_q(d)} = (\log d)^{(1+o(1))\log\log\log d}.$$

We would like to show as well that this inequality continues to hold for almost all d that are in $\mathscr{B}_p$. As above, the GRH implies that for almost all odd numbers m coprime to p, we have $\varphi(m)/l_q(m) = (\log m)^{(1+o(1))\log\log\log m}$. Since (5.4) continues to hold with l_q in place of λ, it follows that for almost all odd m and for all j with $1 \le 2^j \le \log m$, that $\varphi(2^j m)/l_q(2^j m) = (\log(2^j m))^{(1+o(1))\log\log\log(2^j m)}$. Again using the normal order of the number-of-divisors function τ, we have that for almost all odd m and all j with $1 \le 2^j \le \log m$, that $\tau(2^j m) \le \log m$. Further, as we noted above, from Theorems 4.1 and 4.3, it follows that almost all members d of $\mathscr{B}_p$ are of the form $2^j m$ with m odd and $2^j \le \log m$. The rank formula in Theorem 1.1 implies that the rank of $E_d(\mathbb{F}_q(u))$ is bounded above by $\tau(d)\varphi(d)/l_q(d)$. Thus, for almost all $d \in \mathscr{B}_p$, we have the rank at most $(\log d)^{(1+o(1))\log\log\log d}$. This completes the proof. □

Acknowledgements The authors gratefully thank the referee for some useful suggestions. In addition, CP acknowledges partial support from NSF grant DMS-1001180.

References

1. A. Brumer, The average rank of elliptic curves. I. Invent. Math. **109**, 445–472 (1992)
2. R. Conceição, C. Hall, D. Ulmer, Explicit points on the Legendre curve II (2013, in preparation)
3. P. Erdős, C. Pomerance, E. Schmutz, Carmichael's lambda function. Acta Arith. **58**, 363–385 (1991)
4. H. Halberstam, H.-E. Richert, *Sieve Methods* (Academic Press, London, 1974)

5. S. Li, On Artin's conjecture for composite moduli. Ph.D. Thesis, U. Georgia, 1998
6. S. Li, C. Pomerance, On generalizing Artin's conjecture on primitive roots to composite moduli. J. Reine Angew. Math. **556**, 205–224 (2003)
7. D. Marcus, *Number Fields* (Springer, New York, 1977)
8. H.L. Montgomery, R.C. Vaughan, *Multiplicative Number Theory. I. Classical Theory* (Cambridge U. Press, Cambridge, 2007)
9. P. Moree, On the divisors of $a^k + b^k$. Acta Arith. **80**, 197–212 (1997)
10. C. Pomerance, On the distribution of amicable numbers. J. Reine Angew. Math. **293/294**, 217–222 (1977)
11. C. Pomerance, I.E. Shparlinski, Rank statistics for a family of elliptic curves over a function field. Pure Appl. Math. Q. **6**, 21–40 (2010)
12. D. Ulmer, Explicit points on the Legendre curve. Preprint available at arXiv:1002.3313

Some Extensions of the Lucas Functions

E.L. Roettger, H.C. Williams, and R.K. Guy

Abstract From 1876 to 1880 Lucas developed his theory of the functions v_n and u_n which now bear his name. Today these functions find use in primality testing and integer factorization, among other computational techniques. The functions v_n and u_n can be expressed in terms of the nth powers of the zeroes of a quadratic polynomial, and throughout his writings Lucas speculated about the possible extension of these functions to those which could be expressed in terms of the nth powers of the zeroes of a cubic polynomial or of a quartic polynomial. Indeed, at the end of his life he stated that by searching for the addition formulas of the numerical functions which originate from recurrence sequences of the third or fourth degree and by studying in a general way the laws of the residues of these functions for prime moduli, we would arrive at important new properties of prime numbers. We only have scattered hints concerning what functions Lucas had in mind because he provided so little information about them in his published and unpublished work. In this paper we discuss two pairs of functions that are easily expressed as certain combinations of the n th powers of the zeroes of a quartic polynomial and of a sextic polynomial, respectively. We also present several new results, which illustrate the striking similarity between these functions and those of Lucas. The methods that we use to obtain these results are for the most part elementary and would likely have been known to Lucas.

E.L. Roettger (✉)
Mount Royal University, Calgary, AB, Canada
e-mail: eroettger@mtroyal.ca

H.C. Williams • R.K. Guy
University of Calgary, Calgary, AB, Canada
e-mail: williams@math.ucalgary.ca; rkg@cpsc.ucalgary.ca

J.M. Borwein et al. (eds.), *Number Theory and Related Fields: In Memory of Alf van der Poorten*, Springer Proceedings in Mathematics & Statistics 43,
DOI 10.1007/978-1-4614-6642-0_15, © Springer Science+Business Media New York 2013

1   Introduction

Let p and q be coprime integers and α, β the zeroes of the polynomial $f(x) = x^2 - px + q$; the Lucas functions u_n and v_n are defined by:

$$u_n = (\alpha^n - \beta^n)/(\alpha - \beta), \qquad v_n = \alpha^n + \beta^n.$$

When $p = \sqrt{r}$, where r and q are coprime integers, we call these functions the Lehmer functions (see [22, §8.4]).

Since both u_n and v_n are symmetric functions of the zeroes of a polynomial with integer coefficients, they must be integers for all nonnegative integral values of n. Furthermore, they must both satisfy the simple second-order linear recurrence: $X_{n+1} = pX_n - qX_{n-1}$. (This is mentioned as formulas (10) in Lucas [9].) Since $u_0 = 0$, $u_1 = 1$, $v_0 = 2$, $v_1 = p$, this recurrence can be used to compute u_n and v_n for any integer value of n.

From 1876 until about 1880, Édouard Lucas discovered many properties of these functions. In particular, he was interested in how these functions could be used to produce new results concerning prime numbers. Indeed, it was during this period that he used these properties to develop tests for the primality of large integers, including what is now called the Lucas-Lehmer test for the primality of Mersenne numbers (see §5.4 of Williams [22]). These tests were usually sufficiency tests, which could be used to prove whether a number N of a certain special form is a prime. As Lucas well realized, these tests were quite novel for their time, because instead of having to trial divide N by a large number of integers, for example, all the primes less than $\sqrt{N}$, it was only necessary to compute some integer S and test whether $N \mid S$.

It is important to recognize, however, that Lucas found many other applications of his functions. He was particularly struck by the similarity of his functions with the sine and cosine functions; in fact, he noticed that if i is used to denote a zero of $x^2 + 1$, then

$$u_n = (2q^{n/2}/\sqrt{-\Delta})\sin[(ni/2)\log(\alpha/\beta)] \quad \text{and} \quad v_n = 2q^{n/2}\cos[(ni/2)\log(\alpha/\beta)]$$

$$(1.1)$$

where $\Delta = (\alpha - \beta)^2 = p^2 - 4q$. As both sine and cosine are singly periodic functions with period 2π, Lucas (see Section XXVI of [9]) regarded u_n and v_n as simply periodic numerical functions, where for any particular m, the (numerical) period in this case is the least positive integer p such that both $u_{n+p} \equiv u_n \pmod{m}$ and $v_{n+p} \equiv v_n \pmod{m}$ hold.

Throughout his several papers on u_n and v_n, Lucas alluded to the problem of extending or generalizing these functions and offered various suggestions by which this might be done. However, in spite of having these ideas, he seems never to have produced any consistent theory concerning this that was analogous to his work on the Lucas functions. The purpose of this paper is to provide an extension of the

Lucas functions which makes use of the zeroes of a polynomial and to develop a theory of these functions which is very much analogous to that of the Lucas functions. In doing this we will make use of elementary techniques that Lucas would likely have known (see §5.1 of Roettger [18]).

We begin by reviewing some of the fundamental properties of these functions. It is from these basic results that Lucas was able to develop his theory. In doing this, we will make use of his extensive memoir [9] concerning his functions, a work which he regarded as one of his most important [10, p. 42].

In Section IX of [9], Lucas derived what he called his *addition formulas* for u_n and v_n. He called them addition formulas because they correspond to the formulas for addition of angles in the sine and cosine functions of trigonometry. They are

$$2u_{n+m} = u_n v_m + u_m v_n, \qquad 2v_{n+m} = v_n v_m + \Delta u_n u_m. \qquad (1.2)$$

In Sections XII and XIII he developed his *multiplication formulas* for u_n and v_n. He stated that these formulas express the functions u_{mn} and v_{mn} as functions of u_n and v_n, and he appeared to use mathematical induction to obtain three formulas. The first of these can be produced by combining his formulas (74) and (76):

$$u_{mn} = u_n \sum_{k=0}^{m/2-1} (-1)^k \binom{m-k-1}{k} q^{nk} v_n^{m-2k-1} \qquad (m \text{ even}). \qquad (1.3)$$

The next two are his formulas (79) and (86):

$$u_{mn} = u_n \sum_{k=0}^{\lfloor m/2 \rfloor} \frac{m}{k} \binom{m-k-1}{k-1} q^{nk} \Delta^{\lfloor m/2 \rfloor - k} u_n^{m-2k-1} \qquad (m \text{ odd}), \qquad (1.4)$$

$$v_{mn} = \sum_{k=0}^{\lfloor m/2 \rfloor} (-1)^k \frac{m}{k} \binom{m-k-1}{k-1} q^{nk} v_n^{m-2k}. \qquad (1.5)$$

That he called these his multiplication formulas is mentioned more explicitly in the headings of Sections XIII and XIV of [8], a somewhat earlier version of [9]. Furthermore, he could have proved the second and third of these formulas by using Waring's formula, something he seems to have realized in [11, pp. 274–275].

Lucas was also very interested in the divisibility properties of his functions. In Section VII of [9], he noted that u_m is always divisible by u_n whenever n divides m. In today's language we would say that the sequence $\{u_n\}$ is a divisibility sequence. This means that (with the exception of some trivial examples) u_n can be a prime only when n is a prime; he also knew that the converse of this statement is not true. The observation that $\{u_n\}$ is a divisibility sequence was made by Lucas in his earliest papers on his functions. Indeed, on commenting on his first paper [6] in [10, p. 14], he stated that he discovered some arithmetic properties of the Fibonacci numbers after factoring the first few by trial. His table of the factorization of the first 60 terms of the Fibonacci sequence appears in [7, p. 169]. (The first chapter of [7] seems to be a version of the thesis mentioned as XXIV in [10, p. 15].) This table was subsequently published in [9, p. 299]. In view of his use of the terms proper and improper divisors, it is very clear that he made use of the divisibility sequence property of the Fibonacci sequence to assist in the production of this table. Many of his subsequent discoveries of the arithmetical properties of u_n made use of this divisibility property of the $\{u_n\}$ sequence. In the introduction to [11], Lucas gives us some indication of his thoughts and methods:

> The theory of recurrent sequences is an inexhaustible mine which contains all the properties of numbers; by calculating the successive terms of such sequences, decomposing them into their prime factors and seeking out by experimentation the laws of appearance and reproduction of the prime numbers, one can advance in a systematic manner the study of the properties of numbers and their application to all branches of mathematics.

After developing the above results, Lucas was able to provide the most important results of his research into his sequences. For example, in Section XIII of [9], he used formula (1.4) to derive his *law of repetition*.

If r is a prime and $r^\lambda \| u_n$, then

$$r^{\lambda+\mu} \| u_{nr^\mu} \quad \text{if} \quad r^\lambda \neq 2,$$

$$r^{\lambda+\mu} | u_{nr^\mu} \quad \text{if} \quad r^\lambda = 2.$$

He subsequently went on to prove his *law of apparition*:

> Suppose r is any odd prime and r does not divide q. If ε is the value of the Legendre symbol (Δ/r), then $r \mid u_{r-\varepsilon}$.

He continued to develop further results, and the work culminates in what he called his Fundamental Theorem. It was this theorem which he applied to the problem of developing primality tests for numbers of special forms, such as the Mersenne numbers.

Theorem 1.1. *Suppose N is an odd integer and let $T = N - 1$ or $N + 1$. If $N \mid u_T$, but $N \nmid u_{T/d}$ for all d such that d is a divisor of T and $d < T$, then N is a prime.*

Later, Lehmer [5] realized that this could be expressed in a more efficient manner.

Theorem 1.2. *Suppose N is an odd integer and let $T = N - 1$ or $N + 1$. If $N \mid u_T$, but $N \nmid u_{T/q}$ for each distinct prime divisor q of T, then N is a prime.*

Another version of this result, requiring only that N divides certain numbers, is provided below. (See [18, Corollary 2.23.1].)

Theorem 1.3. *Suppose N is an odd integer. Let $T = N - 1$ or $N + 1$. If $N \mid u_T$ and $N \mid u_T/u_{T/q}$ for each distinct prime divisor of T, then N is a prime.*

In much of his published work on the u_n and v_n functions, Lucas mentioned the problem of extending or generalizing them. In what follows, we will attempt to summarize Lucas's ideas concerning this. The most important source is [9]. In Section IX he stated:

We will complete this section with the proof of formulas of extreme importance, because these will serve later as a basis for the theory of the numerical functions of double period, derived from the consideration of the symmetric functions of the roots of third and fourth degree equations with rational coefficients.

He then derived his formulas (A) and (A$'$). In fact, Bell [2] noted that (A$'$) was incorrect. We supply the correct version here:

(A) $\qquad u_n^2 u_{m-1} u_{m+1} - u_m^2 u_{n-1} u_{n+1} = q^{n-1} u_{m-n} u_{m+n},$

(A$'$) $\qquad v_n^2 v_{m-1} v_{m+1} - v_m^2 v_{n-1} v_{n+1} = -\Delta^2 q^{n-1} u_{m-n} u_{m+n}.$

He concluded by mentioning that these formulas belong to the theory of elliptic functions and in particular to the Jacobi Θ and H functions. Thus, it is clear from what little he did write on this matter that he considered an attempt at generalizing his functions should begin by looking at doubly periodic functions, such as elliptic functions. It seems that Lucas believed (probably by analogy) that the numerical functions that would be derived through this analysis should exhibit the property of being doubly (numerically) periodic and likely be related to some doubly periodic functions in a manner similar to the relationship (1.1) between his u_n and v_n functions and the sine and cosine functions. However, it is not clear what this property would have been. Lucas's intuition that elliptic functions would be helpful was moving him in a very productive direction. Unfortunately, he did not possess the mathematical knowledge, nor did such knowledge exist until the twentieth century, to take advantage of his rather vague ideas. This is explained in some detail in Chapter 17 of [22] and needs no further elaboration here. What is important to note is that his belief that linear recurring sequences would play a role in this approach led him to a dead end. However, what is useful for us to know is that he was looking to generalize his theory by continuing to make use of two functions.

Later, in Section XXI he briefly discussed the function S_n, the sum of the nth powers of the zeroes of a polynomial in $\mathbb{Z}[x]$, and obtained a few results, at least one of dubious validity, concerning it when the polynomial is the cubic $x^3 - x - 1$. In Section XXIX, he stated:

We have further indicated (Sections IX and XXI) a first generalization of the principal idea of this memoir in the study of recurrence sequences which arise from the symmetric functions of the roots of algebraic equations of the third and fourth degree and, more generally, of the roots of equations of any degree with rational coefficients. One finds, in particular, in the study of the function

$$U_n = \Delta(a^n, b^n, c^n, \ldots)/\Delta(a, b, c, \ldots)$$

in which a, b, c,... designate the roots of the equation, and $\Delta(a,b,c,\ldots)$, the *alternating function* of the roots, or the square root of the discriminant of the equation, the generalization of the principal formulas contained in the first part of this work.

This function U_n, in the case where the equation is a cubic, was discussed in great detail in Roettger [18] and Müller et al. [13]. Note that $\{U_n\}$ is a divisibility sequence.

In writing later concerning his memoir, Lucas [10] remarked:

> Since the publication of this work, the author has added to it twenty other sections, as yet unpublished, which altogether form the arithmetical theory of the symmetric functions of the roots of equations of the second degree. The author hopes to find the time to write up in a similar manner the theory of doubly periodic functions, in their connection to symmetric functions of the roots of equations of the third and fourth degree, and with elliptic functions.

Finally, in the year of his death, Lucas [12] wrote at much greater length concerning the problem of generalizing his functions. We will only quote part of it here; the rest can be found in Chapter 1 of [18]:

> We think that, by developing these new methods [concerning higher order linear recurrences], by searching for the addition and multiplication formulas of the numerical functions which originate from recurrence sequences of the third and of the fourth degree, and by studying in a general way the laws of the residues of these functions for prime moduli ... that we would arrive at important new properties of prime numbers.

On examining this material, we note several aspects of Lucas's investigation into his functions and those that he might have considered as proper generalizations. We certainly see that he was interested in functions satisfying linear recurring sequences; these functions should be functions of the zeroes of a defining polynomial with rational (in practice, usually integral) coefficients, and there is more than one function to be considered and likely only two. He seems to have been particularly interested in defining polynomials of degree three or four. He indicated the need to find addition and multiplication formulas involving these functions; this is certainly what he did in order to prove the many properties of his own functions. His method of approach was to use empirical methods to attempt to elucidate what the laws of apparition and repetition for these functions would be, and from this material he should be able, as he did in the case of u_n and v_n, to derive primality testing algorithms. Further information on this can be found in Chapter 1 of [18] and in Chapter 4 of [22].

2 A Possible Generalization of Lucas Functions

In view of the preceding discussion, we note that there are five fundamental properties which characterize the Lucas functions:

P1. There are two functions of an integer parameter n (u_n and v_n).
P2. Both functions are integer valued for $n \geq 0$ and satisfy the same linear recurrence (of order 2).
P3. One of the functions (u_n) produces a divisibility sequence ($\{u_n\}$).
P4. There are addition formulas for the functions.
P5. There are multiplication formulas for the functions.

It seems that any generalization of the Lucas functions should satisfy these conditions. We will now attempt to find a more general pair of functions U_n, V_n which possess these properties. We begin with a discussion of linear divisibility sequences.

Definition 2.1. A sequence $\{A_n\} \subseteq \mathbb{Z}$ is a *linear recurrence of order t* if, for all integers n,

$$A_{n+t} = c_1 A_{n+t-1} + c_2 A_{n+t-2} + \cdots + c_t A_n$$

where $c_1, c_2, \ldots c_t$ are fixed integers ($c_t \neq 0$) and $A_0, A_1, \ldots, A_{t-1}$ are chosen.

The polynomial

$$x^t - c_1 x^{t-1} - c^2 x^{t-2} - \cdots - c_t$$

is called the *characteristic polynomial* of the sequence $\{A_n\}$.

Definition 2.2. If $\{A_n\}$ is a linear recurrence of order t and $A_m \mid A_n$ whenever $m \mid n$ and $A_m \neq 0$, then we say that $\{A_n\}$ is a *linear divisibility sequence of order t*.

We mention that linear divisibility sequences (of order t) were termed "Lucasian" by Ward [20].

Hall [4] showed that if $A_0 \neq 0$, then all of the primes which divide the terms of a linear divisibility sequence (of order t) $\{A_n\}$ are contained in the set of primes which divide A_0 and c_t, but if $A_0 = 0$, then the totality of primes dividing the terms of $\{A_n\}$ includes all of the primes which do not divide c_t. We will therefore assume that $A_0 = 0$ and since A_n is a divisibility sequence, we may assume with no loss of generality that $A_1 = 1$.

Lucas believed that his functions could be extended to linear recurrences of order 3 or 4, but the following longstanding conjecture of Hall [4] suggests that this is not possible in the case of order 3:

Conjecture 2.3. The only linear divisibility sequences of order 3 are given by

$$A_n = n^2 a^{n-1}, \quad A_n = n u_n, \quad A_n = u_n^2,$$

where a is any integer and u_n is either a Lucas or a Lehmer function.

In fact, when $t = 3$ and $(c_2, c_3) = 1$, Hall showed that there is no linear divisibility sequence whose characteristic polynomial is irreducible over $\mathbb{Q}$. (The restriction that $(c_2, c_3) = 1$ was later removed by Ward [21].) Hall [4] also reported that Ward showed that this is also the case if the characteristic polynomial factors into a linear and irreducible quadratic over $\mathbb{Q}$.

In an attempt to characterize all linear divisibility sequences, Bézivin et al. [1] proved the following result:

Theorem 2.4. *If $\{A_n\}$ is a linear divisibility sequence, then there is a linear recurring sequence $\{B_n\}$ and a nonnegative integer r such that*

$$B_n = n^r \prod_{i=1}^{k} (\alpha_i^n - \beta_i^n)/(\alpha_i - \beta_i)$$

and $A_n \mid B_n$ for $n = 1, 2, 3, \ldots$. Here α_i and β_i represent nonzero algebraic numbers for $i = 1, 2, \ldots, k$.

If we put $r = 0$, $\gamma_i = \alpha_i/\beta_i$ $(i = 1,2,\ldots,k)$ and $\lambda = \beta_1\beta_2\cdots\beta_k$, then

$$B_n = \lambda^{k-1}\prod_{i=1}^{k}(\gamma_i^n - 1)/(\gamma_i - 1),$$

where λ, γ_i $(i = 1,2,\ldots,k)$ are algebraic numbers.

We say that a linear recurring sequence is *degenerate* if the quotient of any two of the zeroes of its characteristic polynomial is a root of unity. Recently Oosterhout [14] has made the following conjecture concerning nondegenerate linear divisibility sequences.

Conjecture 2.5. Let $\{A_n\}$ be a nondegenerate linear divisibility sequence and let θ_1, $\theta_2, \ldots, \theta_m$ denote the zeroes of its characteristic polynomial. Let Γ be the subgroup of $\overline{\mathbb{Q}}^*$ generated by θ_1, θ_2, $\ldots$, θ_m. Then there exists a c in $\overline{\mathbb{Q}}$, elements λ, γ_1, γ_2, $\ldots$, $\gamma_k \in \Gamma$, and Lucas polynomials $p_1, p_2, \ldots, p_k$ such that

$$A_n = c\lambda^n p_1(\gamma_1)p_2(\gamma_2)\cdots p_k(\gamma_k)$$

for all $n \geq 0$.

Lucas polynomials are polynomials $p(x) \in \overline{\mathbb{Q}}[x]$ such that $p(x) \mid p(x^k)$ for every $k \in \mathbb{Z}^{>0}$. For example, consider $p(x) = x^d - 1$ for any $d \in \mathbb{Z}^{>0}$. Of course, if we consider $p_i(x) = x - 1$ $(i = 1,2,\ldots,k)$ and insist that $A_1 = 1$, we get

$$A_n = \lambda^{n-1}\prod_{i=1}^{k}(\gamma_i^n - 1)/(\gamma_i - 1) \tag{2.1}$$

where λ, γ_1, γ_2, $\ldots$, $\gamma_k \in \overline{\mathbb{Q}}^*$.

In 1916 Pierce [15] attempted to find functions with properties similar to those of u_n and v_n. He considered

$$\Delta_n = \prod_{i=1}^{k}(1 - \gamma_i^n), \qquad S_n = \prod_{i=1}^{k}(1 + \gamma_i^n),$$

where the products are taken over the zeroes γ_i $(i = 1,2,\ldots,k)$ of a monic polynomial of degree k with integer coefficients. He was able to produce a number of results concerning these functions, many of which are analogous to properties exhibited by the Lucas functions u_n and v_n, but neither Δ_n nor S_n is a generalization of u_n in that there are no k, γ_1, γ_2, $\ldots$, γ_k such that $\Delta_n = u_n$. However, we can combine the definition of Pierce's functions, Δ_n and S_n, with the idea of (2.1) to produce the functions

$$U_n = \lambda^{n-1}\prod_{i=1}^{k}(1 - \gamma_i^n)/(1 - \gamma_i), \qquad V_n = \lambda^n\prod_{i=1}^{k}(1 + \gamma_i^n), \tag{2.2}$$

where $\lambda, \gamma_1, \gamma_2, \ldots, \gamma_k \in \overline{\mathbb{Q}}^*$. For the purposes of this paper we shall insist that $\{U_n\}$ and $\{V_n\}$ be linear recurrences and that $V_1 \neq 0$.

In the case of $k = 1$ we see that

$$U_n = \lambda^{n-1}(\gamma_1^n - 1)/(\gamma_1 - 1) \quad \text{and} \quad V_n = \lambda^n(\gamma_1^n + 1).$$

If we put $p = \lambda(\gamma_1 + 1), q = \lambda^2 \gamma_1$, then both $\{U_n\}$ and $\{V_n\}$ satisfy

$$X_{n+1} = pX_n - qX_{n-1}.$$

If $p, q \in \mathbb{Z}$, then $U_n = u_n(p.q)$, $V_n = v_n(p,q)$. Also, $\alpha = \lambda$, $\beta = \lambda \gamma_1$. Thus, (2.2) represents a generalization of the Lucas functions.

3   Is $\{U_n\}$ a Linear Divisibility Sequence?

We notice that if we put $\delta = \lambda \prod_{i=1}^{k}(1 - \gamma_i)$, then

$$\begin{aligned}
\delta U_n &= \lambda^n - (\lambda \gamma_1)^n - (\lambda \gamma_2)^n - \cdots - (\lambda \gamma_k)^n \\
&\quad + (\lambda \gamma_1 \gamma_2)^n + (\lambda \gamma_1 \gamma_3)^n + \cdots + (\lambda \gamma_{k-1} \gamma_k)^n \\
&\quad - (\lambda \gamma_1 \gamma_2 \gamma_3)^n + \cdots \pm (\lambda \gamma_1 \gamma_2 \gamma_3 \cdots \gamma_k)^n \\
V_n &= \lambda^n + (\lambda \gamma_1)^n + (\lambda \gamma_2)^n + \cdots + (\lambda \gamma_1 \gamma_2 \cdots \gamma_k)^n.
\end{aligned}$$

Thus, the characteristic polynomial for V_n and U_n is given by

$$F(x) = (x - \lambda)(x - \lambda \gamma_1)(x - \lambda \gamma_2) \cdots (x - \lambda \gamma_1 \gamma_2 \cdots \gamma_k),$$

a polynomial of degree $r = 2^k$. Now if $\{U_n\}$ and $\{V_n\}$ are to be linear recurrences, we must have $F(x) \in \mathbb{Z}[x]$. We must also have $V_i, U_i \in \mathbb{Z}$ $(i = 0, 1, 2, \ldots, r - 1)$ and $\{U_n\}$ must be a divisibility sequence.

We do not know what the necessary and sufficient conditions on $\lambda, \gamma_1, \gamma_2, \ldots, \gamma_k$ are in order to guarantee that $\{U_n\}$ will be a divisibility sequence, but we prove that for any k, there will always exist some values of $\lambda, \gamma_1, \gamma_2, \ldots, \gamma_k$ such that $\{U_n\}$ is a divisibility sequence. Of course, if $\lambda = 1$ and $\gamma_1, \gamma_2, \ldots, \gamma_k$ are the zeroes of a monic polynomial of degree k in $\mathbb{Z}[x]$, then $\{U_n\}$ is the divisibility sequence $\{\Delta_n\}$, but we can extend this simple observation somewhat further.

We let

$$F(x) = \sum_{i=0}^{r}(-1)^i T_i x^{r-i}. \tag{3.1}$$

We have $T_0 = 1$,

$$T_1 = \lambda(1 + \gamma_1 + \gamma_2 + \cdots + \gamma_1\gamma_2 + \cdots + \gamma_1\gamma_2\cdots\gamma_k)$$

$$= \lambda\prod_{i=1}^{k}(1 + \gamma_i) = V_1 \neq 0.$$

Also

$$T_r = \lambda(\lambda\gamma_1\lambda\gamma_2\cdots\lambda\gamma_k)(\lambda\gamma_1\gamma_2\lambda\gamma_1\gamma_3\cdots\lambda\gamma_{k-1}\gamma_k)$$

$$\times\cdots\times(\lambda\gamma_1\gamma_2\cdots\gamma_k)$$

$$= \lambda^{p_1}G^{p_2},$$

where $p_1 = 1 + k + \binom{k}{2} + \binom{k}{3} + \cdots + \binom{k}{k} = 2^k$, $G = \prod_{i=1}^{k}\gamma_i$. Now since the number of ways to select m objects from k when one of the objects is specified is $\binom{k-1}{m-1}$, we see that $p_2 = 1 + \binom{k-1}{1} + \binom{k-1}{2} + \cdots + \binom{k-1}{k-1} = 2^{k-1}$. Putting $Q = \lambda^2 G$, we get $T_r = Q^{r/2}$. Also,

$$T_{r/2} = \lambda^{r-2}G^{r/2-1}\lambda\prod_{i=1}^{k}(1 + \gamma_i)$$

$$= Q^{r/2-1}T_1.$$

We see that if $T_1, T_{r/2}, T_r \in \mathbb{Z}$, then $Q^{r/2} \in \mathbb{Z}$, $Q^{r/2-1} \in \mathbb{Q}$, which means that $Q \in \mathbb{Q}$, and since $Q^{r/2} \in \mathbb{Z}$, we get $Q \in \mathbb{Z}$.

We let $\mu_i = \gamma_i + 1/\gamma_i$ $(1 = 1, 2, \ldots, k)$ and let e_i denote the ith elementary symmetric function on k variables. Put

$$\kappa_i = e_i(\mu_1, \mu_2, \ldots, \mu_k).$$

We are now able to establish the following result:

Theorem 3.1. *If $T_1, Q, Q\kappa_i \in \mathbb{Z}$ $(i = 1, 2, \ldots, k)$, then $V_n, U_n \in \mathbb{Z}$ $(n \geq 0)$.*

Proof. We put $\mu_i^{(n)} = \gamma_i^n + \gamma_i^{-n}$. It is well known (see, e.g., (1.5), or §4.2 of [22]) that

$$\mu_i^{(n)} = \sum_{j=0}^{\lfloor n/2 \rfloor}\frac{n}{j}\binom{n-j-1}{j-1}(-1)^j\mu_i^{n-2j};$$

thus, we can write

$$\mu_i^{(n)} = \sum_{j=0}^{n}c_j\mu_i^j, \tag{3.2}$$

where $c_j \in \mathbb{Z}$ and the value of c_j is independent of the value of i. Now

$$V_{2n} = \lambda^{2n} \prod_{i=1}^{k}(1 + \gamma_i^{2n}) = Q^n \prod_{i=1}^{k} \mu_i^{(n)}.$$

It follows from (3.2) that $\prod_{i=1}^{k} \mu_i^{(n)}$ is an integral symmetric polynomial in μ_1, μ_2, ..., μ_k. Indeed, it can be written as a polynomial of degree n in κ_i ($i = 1, 2, \ldots, k$). Hence V_{2n} is an integral symmetric polynomial in $Q\mu_i$ ($i = 1, 2, \ldots, k$), and this means that $V_{2n} \in \mathbb{Z}$ by the symmetric function theorem.

Next, consider

$$V_{2n+1} = \lambda^{2n+1} \prod_{i=1}^{k}(1 + \gamma_i^{2n+1})$$

$$= T_1 \lambda^{2n} \prod \frac{1 + \gamma_i^{2n+1}}{1 + \gamma_i}.$$

We have

$$\frac{1 + \gamma_i^{2n+1}}{1 + \gamma_i} = \gamma_i^n \sum_{j=1}^{n} \mu_i^{(j)}(-1)^{j-1} + (-1)^n \gamma_i^n;$$

hence, by (3.2) there exist integers b_j ($j = 0, 1, 2, \ldots, n$) such that

$$\frac{1 + \gamma_i^{2n+1}}{1 + \gamma_i} = \gamma_i^n \sum_{j=0}^{n} b_j \mu_i^j,$$

where b_j is independent of i. By our previous reasoning we see that $V_{2n+1} \in \mathbb{Z}$ ($n \geq 0$). By similar reasoning we also can show that $U_{2n+1} \in \mathbb{Z}$ ($n \geq 0$). Since $U_{2n} = U_n V_n$, we see that $U_{2n} \in \mathbb{Z}$ ($n \geq 0$). $\qquad \square$

Theorem 3.2. *Under the same conditions as Theorem 3.1, $\{U_n\}$ is a divisibility sequence.*

Proof. Put $\kappa_i^{(n)} = e_i(\mu_1^{(n)}, \mu_2^{(n)}), \ldots, \mu_k^{(n)})$. By the same reasoning as that employed in the proof of Theorem 3.1, $\kappa_i^{(n)}$ is an integral symmetric polynomial in $\mu_1, \mu_2, \ldots, \mu_k$; indeed, it can be written as a polynomial of degree n in $\kappa_1, \kappa_2, \ldots \kappa_k$. Hence $Q^n \kappa_i^{(n)} \in \mathbb{Z}$. Now consider

$$U_{mn}/U_n = \lambda^{mn-n} \prod_{i=1}^{k} \frac{\gamma_i^{mn} - 1}{\gamma_i^n - 1}.$$

If m is odd, say $m = 2s + 1$, then

$$\frac{\gamma_i^{mn} - 1}{\gamma_i^n - 1} = \gamma_i^{ns}\left[\gamma_i^{ns} + \gamma_i^{-ns} + \gamma_i^{n(s-1)} + \gamma_i^{-n(s-1)} + \cdots + 1\right]$$

$$= \gamma_i^{ns} \sum_{j=0}^{s} a_j \mu_i^{(n)j},$$

where $a_j \in \mathbb{Z}$ and a_j is independent of i. It follows that

$$U_{mn}/U_n = Q^{ns} \prod_{i=1}^{k} \sum_{j=0}^{s} a_j \mu_i^{(n)j},$$

is an integral symmetric polynomial in $\mu_i^{(n)}$ $(i = 1, 2, \ldots, k)$ of degree s. Thus, by our previous reasoning, $U_{mn}/U_n \in \mathbb{Z}$ whenever m is odd.

If m is even, then $m = 2^t s$ and s is odd. Since

$$U_{mn}/U_n = U_{2^t sn}/U_n = V_{2^{t-1}sn} V_{2^{t-2}sn} \cdots V_{sn} U_{sn}/U_n$$

and $V_p \in \mathbb{Z}$ for any positive integer p, we see that $U_{mn}/U_n \in \mathbb{Z}$.   □

Thus, we now know that, subject to the conditions of Theorem 3.1, we have $\{U_n\}$ a divisibility sequence. However, it is not clear that it is a linear divisibility sequence. Of course, if $T_1, T_2, \ldots, T_r \in \mathbb{Z}$, we are done, but it is not clear that this will be the case. By Newton's identities we do know that

$$V_n = T_1 V_{n-1} - T_2 V_{n-2} + T_3 V_{n-3} - \cdots + (-1)^{n-1} n T_n$$

for $1 \leq n \leq r$. It follows by induction on n that since $V_n \in \mathbb{Z}$ for $n \geq 0$, we have $T_1, T_2, \ldots, T_r \in \mathbb{Q}$. Thus, we can put $T_i = S_i/G$ where[1] $G \in \mathbb{Z}$, $G \neq 0$, $S_i \in \mathbb{Z}$ and $(G, S_1, S_2, \ldots, S_r) = 1$. Since

$$GU_{n+r} = S_1 U_{n+r-1} - S_2 U_{n+r-2} + \cdots + S_r U_n,\qquad(3.3)$$

we can multiply (3.3) by G^{n+r-2}, and we find that $\{G^{m-1}U_m\}$ is a linear recurring sequence of order r. Also, $G^{n-1}U_n \mid G^{nm-1}U_{mn}$. Thus, if we replace λ by $G\lambda$ in (2.2), we see that under the conditions of Theorem 3.1, the corresponding $\{U_n\}$ is a linear divisibility sequence of order r.

---

[1] It seems that G is always 1, but we have not been able to prove this in general.

4 The Case of $k = 2$

We now deal with some special cases of (2.2). We have already mentioned that when $k = 1$, we get the Lucas functions. When $k = 2$ we have

$$T_1 = \lambda(1 + \gamma_1 + \gamma_2 + \gamma_1\gamma_2)$$
$$T_2 = \lambda^2(2\gamma_1\gamma_2 + \gamma_1 + \gamma_2 + \gamma_1^2\gamma_2 + \gamma_1\gamma_2^2)$$
$$= \lambda^2(\gamma_1 + \gamma_2)(1 + \gamma_1\gamma_2) + 2\lambda^2\gamma_1\gamma_2$$
$$T_3 = \lambda^3\gamma_1\gamma_2(1 + \gamma_1 + \gamma_2 + \gamma_1\gamma_2)$$
$$T_4 = (\lambda^2\gamma_1\gamma_2)^2 = Q^2.$$

Thus, if we put $P_1 = T_1 \in \mathbb{Z}$, $P_2 + 2Q = T_2 \in \mathbb{Z}$, we find that in order for $\{U_n\}$ to be a linear recurring sequence, we must have $P_1, P_2, Q \in \mathbb{Z}$, and $\{U_n\}, \{V_n\}$ will satisfy the fourth order linear recurrence

$$X_{n+4} = P_1 X_{n+3} - (P_2 + 2Q)X_{n+2} + P_1 Q X_{n+1} - Q^2 X_n$$

with $U_0 = 0$, $U_1 = 1$, $U_2 = P_1$, $U_3 = P_1^2 - 2P_2 - 3Q$, $V_0 = 4$, $V_1 = P_1$, $V_2 = P_1^2 - 2P_2 - 4Q$, $V_3 = P_1(P_1^2 - 3P_2 - 3Q)$.

If $\rho_1 = \lambda(\gamma_1 + \gamma_2)$, $\rho_2 = \lambda(1 + \gamma_1\gamma_2)$, then $\rho_1 + \rho_2 = P_1$, $\rho_1\rho_2 = P_2$, and the characteristic polynomial for $\{U_n\}$ or $\{V_n\}$ can be written as

$$(x^2 - \rho_1 x + Q)(x^2 - \rho_2 x + Q).$$

Putting $\alpha_1 = \lambda\gamma_1$, $\beta_1 = \lambda\gamma_2$, $\alpha_2 = \lambda$, $\beta_2 = \lambda\gamma_1\gamma_2$, we get

$$U_n = (\alpha_1^n + \beta_1^n - \alpha_2^n - \beta_2^n)/(\alpha_1 + \beta_1 - \alpha_2 - \beta_2)$$
$$V_n = \alpha_1^n + \beta_1^n + \alpha_2^n + \beta_2^n.$$

Also, if $\delta = \alpha_1 + \beta_1 - \alpha_2 - \beta_2$, we find that

$$\Delta = \delta^2 = (\rho_1 - \rho_2)^2 = P_1^2 - 4P_2.$$

In this case we have $Q\kappa_1 = P_2$, $Q\kappa_2 = V_2$; thus $\{U_n\}$ is a linear divisibility sequence of order four by Theorem 3.2.

Of course, if we select any integers P_1, P_2, Q, then there certainly exist algebraic numbers $\lambda, \gamma_1, \gamma_2$ such that

$$\lambda(\gamma_1 + \gamma_2 + 1 + \gamma_1\gamma_2) = P_1, \quad \lambda^2(\gamma_1 + \gamma_2)(1 + \gamma_1 + \gamma_2) = P_2, \quad \lambda^2\gamma_1\gamma_2 = Q.$$

These sequences $\{U_n\}$ and $\{V_n\}$ have been discussed in some detail by Williams and Guy [23]. They possess the five basic properties of the Lucas functions mentioned at the beginning of Sect. 2. Also, when $\gcd(P_1, P_2, Q) = 1$, there is a law of apparition, a law of repetition, and a Fundamental Theorem analogous to those for the Lucas functions.

In addition to the examples mentioned in [23], we point out that arithmetical properties of the families of sequences, defined analogously to U_n, with characteristic polynomial $f_a = x^4 - ax^3 + bx^2 + ax + 1$ ($P_1 = a, P_2 = b - 2, Q = 1$), $b = -1$, -3 were used in [16] to prove that a certain set of units is a set of fundamental units in the ring $\mathbb{Z}[\alpha]$, where α is a zero of f_a. In [17] properties of U_n and V_n were investigated for the more general sequences with characteristic polynomial $x^4 - ax^3 + bx^2 + \delta ax + 1$, $a, b \in \mathbb{Z}$, $\delta \in \{-1, 1\}$ ($P_1 = a, P_2 = b - 2\delta, Q = \delta$) and such that $a^2 - 4(b - 2\delta) \neq 0$, $b\delta \neq 2$ and if $\delta = 1$, then $b \neq 2a - 2$. We are grateful to an anonymous referee for bringing these references to our attention.

We remark here that the addition formulas for these functions are

$$2V_{n+m} = V_n V_m + \Delta U_n U_m - 2Q^m V_{n-m}$$

$$2U_{n+m} = U_n V_m + U_m V_n - 2Q^m U_{n-m}.$$

These are very similar to the addition formulas for the Lucas functions except that we now need V_{n-m} and U_{n-m} as well as V_n, V_m, U_n, U_m to compute V_{n+m}, U_{n+m}. If we put $n = m$, we get the duplication formulas

$$2V_{2n} = V_n^2 + \Delta U_n^2 - 8Q^n$$

$$U_{2n} = U_n V_n.s$$

Thus, when $k = 2$, we get a very close analog to the Lucas functions which involves linear recurrences of order 4. We next examine the case of $k = 3$.

5   The Case of $k = 3$

When $k = 3$, put $\rho_1 = \lambda(\gamma_1 + \gamma_2\gamma_3)$, $\rho_2 = \lambda(\gamma_2 + \gamma_3\gamma_1)$, $\rho_3 = \lambda(\gamma_3 + \gamma_1\gamma_2)$, $\rho_4 = \lambda(1 + \gamma_1\gamma_2\gamma_3)$, $Q = \lambda^2\gamma_1\gamma_2\gamma_3$. We have already seen that we must have $Q \in \mathbb{Z}$. Let $\rho_1, \rho_2, \rho_3, \rho_4$ be the zeroes of

$$f(x) = x^4 - P_1 x^3 + P_2 x^2 - P_3 x + P_4.$$

We have $T_1 = P_1$, $T_2 = P_2 + 4Q$, $T_3 = P_3 + 3QP_1$, $T_4 = P_4 + 2QP_2 + 6Q^2$, $T_5 = QT_3$, $T_6 = Q^2 T_2$, $T_7 = Q^3 T_1$, and $T_8 = Q^4$ in (3.1). Thus if $\{U_n\}$ is to satisfy a linear recurrence, we must have $T_1, T_2, T_3, T_4 \in \mathbb{Z}$ which, since $Q \in \mathbb{Z}$, is equivalent to $P_1, P_2, P_3, P_4 \in \mathbb{Z}$. In this case the characteristic polynomial of $\{U_n\}$ and $\{V_n\}$ is given by

$$\prod_{i=1}^{4}(x^2 - \rho_i x + Q).$$

Also, if α_i, β_i are the zeroes of $x^2 - p_i x + Q$ $(i = 1, 2, 3, 4)$, then

$$U_n = \frac{\alpha_1^n - \beta_1^n + \alpha_2^n - \beta_2^n + \alpha_3^n - \beta_3^n + \alpha_4^n - \beta_4^n}{\alpha_1 - \beta_1 + \alpha_2 - \beta_2 + \alpha_3 - \beta_3 + \alpha_4 - \beta_4}$$

$$V_n = \alpha_1^n + \beta_1^n + \alpha_2^n + \beta_2^n + \alpha_3^n + \beta_3^n + \alpha_4^n + \beta_4^n.$$

This is similar to the $k = 2$ case; however, unlike that case, simply selecting any P_1, P_2, P_3, P_4, Q will not necessarily guarantee that $\{U_n\}$, $\{V_n\} \subseteq \mathbb{Z}$ for all $n \geq 0$. We will now determine what conditions must be satisfied by P_1, P_2, P_3, P_4 and Q in order that that will be the case.

We need to have V_n, $U_n \in \mathbb{Z}$ whenever $n \geq 0$. We know that $U_0 = 0$, $U_1 = 1$. Also, $U_2 = \lambda(\gamma_1 + 1)(\gamma_2 + 1)(\gamma_3 + 1) = P_1(\neq 0) \in \mathbb{Z}$. Consider

$$U_3 = \lambda^2(\gamma_1^2 + \gamma_1 + 1)(\gamma_2^2 + \gamma_2 + 1)(\gamma_3^2 + \gamma_3 + 1)$$
$$= Q(\mu_1 + 1)(\mu_2 + 1)(\mu_3 + 1).$$

Thus, in order for $U_3 \in \mathbb{Z}$, we require $Q(\kappa_3 + \kappa_2 + \kappa_1 + 1) \in \mathbb{Z}$. Now $Q\kappa_3 = Q\mu_1\mu_2\mu_3 = V_2 \in \mathbb{Z}$ and

$$U_2^2 = Q(\mu_1 + 2)(\mu_2 + 2)(\mu_3 + 2) = Q(\kappa_3 + 2\kappa_2 + 4\kappa_1 + 8) \in \mathbb{Z}.$$

Hence we must have $Q\kappa_2 + Q\kappa_1 \in \mathbb{Z}$ and $Q\kappa_2 + 2Q\kappa_1 \in \mathbb{Z}$. It follows that $Q\kappa_1 \in \mathbb{Z}$ and $Q\kappa_2 \in \mathbb{Z}$. By Theorem 3.1 this means that U_n, $V_n \in \mathbb{Z}$ for all $n \geq 0$. Also by Theorem 3.2 we see that $\{U_n\}$ is a linear divisibility sequence of order eight.

We have seen that if $Q \in \mathbb{Z}$ and $\{U_n\}$, $\{V_n\} \in \mathbb{Z}$ $(n \geq 0)$, then $Q\kappa_1$, $Q\kappa_2$, $Q\kappa_3 \in \mathbb{Z}$. We next prove the following result:

Theorem 5.1. If Q, P_1, $Q\kappa_1$, $Q\kappa_2$, $Q\kappa_3 \in \mathbb{Z}$, then P_2, P_3, $P_4 \in \mathbb{Z}$, $P_1 \mid P_3$ and $P_4 = (P_3/P_1)^2 + 8(P_3/P_1)Q - 4QP_2 + QP_1^2$.

Proof. It is not difficult to establish that

$$P_1^2 = Q(\kappa_3 + 2\kappa_2 + 4\kappa_1 + 8)$$
$$P_2 = Q(\kappa_2 + 2\kappa_1)$$
$$P_3 = QP_1(\kappa_1 - 2)$$
$$P_4 = Q^2(\kappa_1^2 - 2\kappa_2 + \kappa_3 - 4).$$

Thus we must have P_2, P_3, $P_4 \in \mathbb{Z}$, and $P_1 \mid P_3$. Furthermore, since

$$\left.\begin{array}{l} Q\kappa_1 = P_3/P_1 + 2Q \\ Q\kappa_2 = P_2 - 2P_3/P_1 - 4Q \\ Q\kappa_3 = P_1^2 - 2P_2 - 8Q \end{array}\right\} \tag{5.1}$$

we get

$$P_4 = (P_3/P_1 + 2Q)^2 - 2Q(P_2 - 2P_3/P_1 - 4Q) + Q(P_1^2 - 2P_2 - 8Q) - 4Q^2$$
$$= (P_3/P_1)^2 + 8(P_3/P_1)Q - 4QP_2 + QP_1^2. \qquad \qquad \square$$

Notice that we have shown that in the case of $k = 3$, we have $G = 1$ in Sect. 3. We have also proved the following theorem:

Theorem 5.2. *Put* $R_i = Q\kappa_i$ $(i = 1, 2, 3)$. *If* $k = 3$ *and* $\{U_n\}$ *and* $\{V_n\}$ *in (2.2) are to be linear recurring sequences, it is necessary and sufficient that the following three conditions hold:*

(1) $Q = \lambda^2 \gamma_1 \gamma_2 \gamma_3 \in \mathbb{Z}.$

(2) $P_1 = \lambda(1 + \gamma_1)(1 + \gamma_2)(1 + \gamma_3) \in \mathbb{Z}.$

(3) $R_1, R_2, R_3 \in \mathbb{Z}.$

We can now characterize those values of Q, P_1, P_2, P_3, P_4 *such that both* $\{U_n\}$ *and* $\{V_n\}$ *will be sequences of integers for* $n \geq 0$.

Theorem 5.3. *Let* Q, P_1, P_2, P_3, $P_4 \in \mathbb{Z}$ *be such that* $P_1 \neq 0$, $P_1 \mid P_3$ *and*

$$P_4 = (P_3/P_1)^2 + 8Q(P_3/P_1) + QP_1^2 - 4QP_2. \qquad (5.2)$$

If $\{U_n\}$ *and* $\{V_n\}$ *are linear recurring sequences defined by the characteristic polynomial,*

$$F(x) = x^8 - P_1 x^7 + (P_2 + 4Q)x^6 - (P_3 + 3QP_1)x^5 + (P_4 + 2QP_2 + 6Q^2)x^4$$
$$- Q(P_3 + 3QP_1)x^3 + Q^2(P_2 + 4Q)x^2 - Q^3 P_1 x + Q^4, \qquad (5.3)$$

and initial conditions

$$U_0 = 0, U_1 = 1, U_2 = P_1, U_3 = P_1^2 - P_2 - P_3/P_1 - 9Q,$$

$$U_4 = P_1(P_1^2 - 2P_2 - 8Q), U_{-n} = -U_n/Q^n (n = 1, 2, 3, 4);$$

$$V_0 = 8, V_1 = P_1, V_2 = P_1^2 - 2P_2 - 8Q, V_3 = P_1(P_1^2 + 3(P_3/P_1) - 3P_2 - 3Q)$$

$$V_4 = P_1^4 + 2P_2^2 - 4P_1^2 P_2 - 8QP_1^2 - 4(P_3/P_1)^2 + 4P_3 P_1 - 8QP_2 + 24Q^2,$$

$$V_{-n} = V_n/Q^n (n = 1, 2, 3, 4);$$

then when $k = 3$ *there exist* λ, γ_1, γ_2, $\gamma_3 \in \overline{\mathbb{Q}}$, *such that* U_n *and* V_n *are given by (2.2).*

Proof. Put $R_1 = P_3/P_1 + 2Q$, $R_2 = P_2 - 2P_3/P_1 - 4Q$, $R_3 = P_1^2 - 2P_2 - 8Q$, and let v_1, v_2, v_3 be the three zeroes of

$$g(x) = x^3 - R_1 x^2 + QR_2 x - Q^2 R_3. \qquad (5.4)$$

Solve for γ_1, γ_2, γ_3 such that

$$\gamma_i + 1/\gamma_i = v_i/Q \quad (i = 1, 2, 3),$$

and put

$$\lambda^2 = Q/(\gamma_1\gamma_2\gamma_3). \tag{5.5}$$

Note that γ_1, γ_2, γ_3, $\lambda \in \overline{\mathbb{Q}}$.

Now consider

$$U_n^* = \lambda^{n-1}(1 - \gamma_1^n)(1 - \gamma_2^n)(1 - \gamma_3^n)/(1 - \gamma_1)(1 - \gamma_2)(1 - \gamma_3)$$
$$V_n^* = \lambda^n(1 + \gamma_1^n)(1 + \gamma_2^n)(1 + \gamma_3^n).$$

We have

$$P_1^2 = R_3 + 2R_2 + 4R_1 + 8Q,$$

$$P_2 = R_2 + 2R_1, \tag{5.6}$$

$$P_3 = P_1(R_1 - 2Q), \tag{5.7}$$

$$P_4 = R_1^2 - 2QR_1 + QR_3 - 4Q^2, \tag{5.8}$$

by (5.2).

Since

$$R_i = Qe_i(\gamma_1 + 1/\gamma_1, \gamma_2 + 1/\gamma_2, \gamma_3 + 1/\gamma_3), \tag{5.9}$$

we get

$$P_1^2 = \lambda^2(1 + \gamma_1)^2(1 + \gamma_2)^2(1 + \gamma_3)^2.$$

Select that value of λ in (5.5) such that

$$P_1 = \lambda(1 + \gamma_1)(1 + \gamma_2)(1 + \gamma_3). \tag{5.10}$$

Put $\rho_1 = \lambda(\gamma_1 + \gamma_2\gamma_3)$, $\rho_2 = \lambda(\gamma_2 + \gamma_3\gamma_1)$, $\rho_3 = \lambda(\gamma_3 + \gamma_1\gamma_2)$, $\rho_4 = \lambda(1 + \gamma_1\gamma_2\gamma_3)$. By (5.10), (5.6), (5.7), (5.8), and (5.9), we have $P_i = e_i(\rho_1, \rho_2, \rho_3, \rho_4)$ $(i = 1, 2, 3, 4)$. Thus the zeroes of

$$x^4 - P_1 x^3 + P_2 x^2 - P_3 x + P_4$$

are ρ_1, ρ_2, ρ_3, ρ_4. It follows that the characteristic polynomial of $\{U_n^*\}$ and $\{V_n^*\}$ is given by $F(x)$. Since

$$U_i^* = U_i, V_i^* = V_i (i = -4, -3, -2, \ldots, 4),$$

we must have $V_n^* = V_n$, $U_n^* = U_n$ for all $n \in \mathbb{Z}$. $\qquad\square$

We also notice that the conditions of Theorem 5.3 also imply that $\{U_n\}$, $\{V_n\} \in \mathbb{Z}$ for all $n \geq 0$ and that $\{U_n\}$ is a divisibility sequence. As an example consider $Q = 1$, $P_1 = 56$, $P_2 = 668$, $P_3 = 2464$ $(P_3/P_1 = 44)$ and $P_4 = 44^2 + 8(44) + 56^2 - 4(668) = 2752$. We get

$$\{U_n\} = 0, 1, 56, 2415, 100352, 4140081, 170537640, 7022359583, \ldots.$$

This is A003696 in the Online Encyclopedia of Integer Sequences [19], and gives the number of spanning trees in $P_4 \times P_n$.

Another example is the Pierce function $\Delta_n = (1 - \gamma_1^n)(1 - \gamma_2^n)(1 - \gamma_3^n)$ where γ_1, γ_2, γ_3 are the zeroes of the cubic $x^3 - c_1 x^2 + c_2 x - c_3 \in \mathbb{Z}[x]$.

Here we have $\Delta_n = \delta U_n$, where

$$\delta = (1 - \gamma_1)(1 - \gamma_2)(1 - \gamma_3) = 1 - c_1 + c_2 - c_3 \in \mathbb{Z}.$$

Also, $\lambda = 1$, $Q = c_3$, $P_1 = 1 + c_1 + c_2 + c_3 \in \mathbb{Z}$, $P_2 = c_1 + 2c_2 - 3c_3 + 2c_1 c_3 + c_1 c_2 + c_2 c_3 \in \mathbb{Z}$, $P_3 = P_1(c_1 c_3 + c_2 - 2c_3)$,

$$P_4 = (1 + c_3)(c_3^2 + c_3 + c_2^2 - 2c_1 c_3 + c_1^2 c_3 - 2c_2 c_3)$$
$$= (P_3/P_1)^2 + 8Q(P_3/P_1) - 4QP_2 + QP_1^2.$$

6 Addition and Multiplication Formulas When $k = 3$

We have seen the conditions under which the sequences given by (2.2) satisfy properties P1, P2, and P3. We must now attempt to produce addition and multiplication formulas for this case. We begin by putting

$$\delta = \alpha_1 - \beta_1 + \alpha_2 - \beta_2 + \alpha_3 - \beta_3 + \alpha_4 - \beta_4 = \lambda(1 - \gamma_1)(1 - \gamma_2)(1 - \gamma_3).$$

Define

$$\Delta = \delta^2 = Q(\mu_1 - 2)(\mu_2 - 2)(\mu_3 - 2) = Q\kappa_3 - 2Q\kappa_2 + 4Q\kappa_1 - 8Q.$$

If $R_i = Q\kappa_i$ $(i = 1, 2, 3)$, we see by (5.1) that

$$\Delta = R_3 - 2R_2 + 4R_1 - 8Q = P_1^2 - 4P_2 + 8P_3/P_1.$$

Let

$$\mu_i^{(n)} = \gamma_i^n + \gamma_i^{-n} \quad (i = 1, 2, 3),$$

and put

$$\kappa_i^{(n)} = e_i(\mu_1^{(n)}, \mu_2^{(n)}, \mu_3^{(n)}) \quad (i = 1, 2, 3).$$

Since $Q\kappa_i \in \mathbb{Z}$ $(i = 1, 2, 3)$, we see by (3.2) that

$$Q^n \kappa_i^{(n)} \in \mathbb{Z} \quad (i = 1, 2, 3).$$

Now

$$
\begin{aligned}
V_n^2 &= \lambda^{2n}(1 + \gamma_1^n)^2(1 + \gamma_2^n)^2(1 + \gamma_3^n)^2 \\
&= Q^n(\kappa_3^{(n)} + 2\kappa_2^{(n)} + 4\kappa_1^{(n)} + 8) \\
\Delta U_n^2 &= \lambda^{2n}(1 - \gamma_1^n)^2(1 - \gamma_2^n)^2(1 - \gamma_3^n)^2 \\
&= Q^n(\kappa_3^{(n)} - 2\kappa_2^{(n)} + 4\kappa_1^{(n)} - 8).
\end{aligned}
$$

Also, $V_{2n} = Q^n \kappa_3^{(n)}$. It follows that

$$
\begin{aligned}
2V_{2n} &= V_n^2 + \Delta U_n^2 - 8Q^n \kappa_1^{(n)} \\
4Q^n \kappa_2^{(n)} &= V_n^2 - \Delta U_n^2 - 16Q^n.
\end{aligned}
$$

Notice that although we have $U_{2n} = U_n V_n$, there is no simple formula that relates V_{2n} to U_n and V_n as is the case for $k = 1$ and $k = 2$.

In order to obtain our duplication formulas in the case of $k = 3$, we introduce

$$Z_n = Q^n \kappa_1^{(n)} = Q^n(\gamma_1^n + \gamma_1^{-n} + \gamma_2^n + \gamma_2^{-n} + \gamma_3^n + \gamma_3^{-n}).$$

We have

$$
\begin{aligned}
Z_{2n} &= Q^{2n}(\gamma_1^n + \gamma_1^{-n} + \gamma_2^n + \gamma_2^{-n} + \gamma_3^n + \gamma_3^{-n})^2 - 2Q^{2n}\kappa_2^{(n)} - 6Q^n \\
&= Z_n^2 - Q^n(V_n^2 - \Delta U_n^2 - 16Q^n)/2 - 6Q^{2n}.
\end{aligned}
$$

Hence

$$
\begin{aligned}
2Z_{2n} &= 2Z_n^2 - Q^n V_n^2 + Q^n \Delta U_n^2 + 4Q^{2n} \\
2V_{2n} &= \Delta U_n^2 + V_n^2 - 8Z_n \\
U_{2n} &= U_n V_n
\end{aligned}
$$

are the duplication formulas in this case.

If we put

$$
\begin{aligned}
A_n &= \alpha_1^n + \alpha_2^n + \alpha_3^n + \alpha_4^n, \\
B_n &= \beta_1^n + \beta_2^n + \beta_3^n + \beta_4^n,
\end{aligned}
$$

we have

$$U_n = (A_n - B_n)/\delta$$
$$V_n = A_n + B_n;$$

hence

$$\left.\begin{array}{l} A_n = (V_n + \delta U_n)/2 \\ B_n = (V_n - \delta U_n)/2. \end{array}\right\}$$ (6.1)

We next put

$$\sigma_i^{(n)} = e_i(\alpha_1^n, \alpha_2^n, \alpha_3^n, \alpha_4^n), \quad \tau_i^{(n)} = e_i(\beta_1^n, \beta_2^n, \beta_3^n, \beta_4^n),$$

where $\alpha_1 = \lambda\gamma_1$, $\alpha_2 = \lambda\gamma_2$, $\alpha_3 = \lambda\gamma_3$, $\alpha_1 = \lambda\gamma_1\gamma_2\gamma_3$, $\beta_i = Q/\alpha_i$ ($i = 1,2,3,4$). Note that $\sigma_1^{(n)} = A_n$, $\tau_1^{(n)} = B_n$, $\sigma_4^{(n)} = \tau_4^{(n)} = Q^{2n}$, $\sigma_2^{(n)} = \tau_2^{(n)} = Z_n$, $\sigma_3^{(n)} = Q^n\tau_1^{(n)} = Q^n B_n$, $\tau_3^{(n)} = Q^n\sigma_1^{(n)} = Q^n A_n$. Thus

$$\alpha_i^{4n} = A_n\alpha_i^{3n} - Z_n\alpha_i^{2n} + Q^n B_n\alpha_i^n - Q^{2n}$$

and

$$\alpha_i^{4n+m} = A_n\alpha_i^{3n+m} - Z_n\alpha_i^{2n+m} + Q^n B_n\alpha_i^{n+m} - Q^{2n}\alpha_i^m \quad (i = 1,2,3,4).$$

It follows that

$$A_{4n+m} = A_n A_{3n+m} - Z_n A_{2n+m} + Q^n B_n A_{n+m} - Q^{2n}A_m$$

and similarly

$$B_{4n+m} = B_n B_{3n+m} - Z_n B_{2n+m} + Q^n A_n B_{n+m} - Q^{2n}B_m.$$

If we substitute for the A's and B's using (6.1), we find that

$$\frac{V_{4n+m}}{2} = \frac{V_n V_{3n+m} + \Delta U_n U_{3n+m}}{4} - \frac{V_{2n+m}Z_n}{2} + Q^n\frac{V_n V_{n+m} - \Delta U_n U_{n+m}}{4} - Q^{2n}\frac{V_m}{2}$$

$$\frac{U_{4n+m}}{2} = \frac{V_n U_{3n+m} + V_{3n+m}U_n}{4} - \frac{U_{2n+m}Z_n}{2} + Q^n\frac{V_n U_{n+m} - U_n V_{n+m}}{4} - Q^{2n}\frac{U_m}{2}.$$

If we replace m by $m - 2n$, we get

$$2V_{m+2n} = V_n V_{m+n} + \Delta U_n U_{m+n} - V_m Z_n + Q^n(V_n V_{m-n} - \Delta U_n U_{m-n}) - 2Q^{2n}V_{m-2n}$$

$$2U_{m+2n} = V_n U_{m+n} + U_n V_{m+n} - U_m Z_n + Q^n(V_n U_{m-n} - U_n V_{m-n}) - 2Q^{2n}U_{m-2n},$$

the addition formulas for $\{V_n\}$ and $\{U_n\}$. Unfortunately, there does not seem to be an analogous formula for Z_{m+2n}.

We now look at the problem of developing multiplication formulas for U_n and V_n. In view of the fact that we could not find simple duplication formulas without introducing Z_n, it is not likely that we can find multiplication formulas without introducing a third function. From Waring's formula we have

$$A_{mn} = \sum (-1)^{\lambda_0} \frac{m(m - \lambda_0 - 1)!}{\lambda_1! \lambda_2! \lambda_3! \lambda_4!} \sigma_1^{(n)\lambda_1} \sigma_2^{(n)\lambda_2} \sigma_3^{(n)\lambda_3} \sigma_4^{(n)\lambda_4},$$

where the sum is taken over all non-negative integers $\lambda_0, \lambda_1, \lambda_2, \lambda_3, \lambda_4$ such that

$$\sum_{i=0}^{4} \lambda_i = \sum_{i=0}^{4} i\lambda_i = m.$$

Similarly

$$B_{mn} = \sum (-1)^{\lambda_0} \frac{m(m - \lambda_0 - 1)!}{\lambda_1! \lambda_2! \lambda_3! \lambda_4!} \tau_1^{(n)\lambda_1} \tau_2^{(n)\lambda_2} \tau_3^{(n)\lambda_3} \tau_4^{(n)\lambda_4}.$$

Now $\sigma_1^{(n)\lambda_1} \sigma_3^{(n)\lambda_3} = Q^{n\lambda_3} A_n^{\lambda_1} B_n^{\lambda_3}$, $\tau_1^{(n)\lambda_1} \tau_3^{(n)\lambda_3} = Q^{n\lambda_3} A_n^{\lambda_3} B_n^{\lambda_1}$. Also,

$$A_n^s = ((V_n + \delta U_n)/2)^s = (v_s(\tilde{p}_n, \tilde{q}_n) + \tilde{\delta}_n u_s(\tilde{p}_n, \tilde{q}_n))/2$$
$$B_n^s = ((V_n - \delta U_n)/2)^s = (v_s(\tilde{p}_n, \tilde{q}_n) - \tilde{\delta}_n u_s(\tilde{p}_n, \tilde{q}_n))/2,$$

where u_n and v_n are the Lucas functions and $\tilde{p}_n = V_n$, $\tilde{q}_n = (V_n^2 - \Delta U_n^2)/4$, $\tilde{\delta}_n = \delta U_n$, $\tilde{\Delta}_n = \Delta U_n^2$. Hence,

$$\sigma_1^{(n)\lambda_1} \sigma_3^{(n)\lambda_3} = Q^{n\lambda_3}(v_{\lambda_1} v_{\lambda_3} - \tilde{\Delta}_n u_{\lambda_1} u_{\lambda_3} + \tilde{\delta}_n v_{\lambda_3} u_{\lambda_1} - \tilde{\delta}_n v_{\lambda_1} u_{\lambda_3})/4$$
$$\tau_1^{(n)\lambda_1} \tau_3^{(n)\lambda_3} = Q^{n\lambda_3}(v_{\lambda_1} v_{\lambda_3} - \tilde{\Delta}_n u_{\lambda_1} u_{\lambda_3} - \tilde{\delta}_n v_{\lambda_3} u_{\lambda_1} + \tilde{\delta}_n v_{\lambda_1} u_{\lambda_3})/4$$

and

$$\sigma_1^{(n)\lambda_1} \sigma_3^{(n)\lambda_3} + \tau_1^{(n)\lambda_1} \tau_3^{(n)\lambda_3} = Q^{n\lambda_3}(v_{\lambda_1} v_{\lambda_3} - \tilde{\Delta}_n u_{\lambda_1} u_{\lambda_3})/2$$
$$\sigma_1^{(n)\lambda_1} \sigma_3^{(n)\lambda_3} - \tau_1^{(n)\lambda_1} \tau_3^{(n)\lambda_3} = Q^{n\lambda_3}(u_{\lambda_1} v_{\lambda_3} - v_{\lambda_1} u_{\lambda_3})/2.$$

Since

$$\frac{v_{\lambda_1} v_{\lambda_3} - \tilde{\Delta}_n u_{\lambda_1} u_{\lambda_3}}{2} = \tilde{q}_n^{\lambda_3} v_{\lambda_1 - \lambda_3}$$

$$\frac{u_{\lambda_1} v_{\lambda_3} - v_{\lambda_1} u_{\lambda_3}}{2} = \tilde{q}_n^{\lambda_3} u_{\lambda_1 - \lambda_3},$$

we get the multiplication formulas

$$V_{mn} = \sum (-1)^{\lambda_0} \frac{m(m-\lambda_0-1)!}{\lambda_1!\lambda_2!\lambda_3!\lambda_4!} Z_n^{\lambda_2} Q^{n(\lambda_3+2\lambda_4)} \tilde{q}_n^{\lambda_3} v_{\lambda_1-\lambda_3}(\tilde{p}_n, \tilde{q}_n)$$

$$U_{mn}/U_n = \sum (-1)^{\lambda_0} \frac{m(m-\lambda_0-1)!}{\lambda_1!\lambda_2!\lambda_3!\lambda_4!} Z_n^{\lambda_2} Q^{n(\lambda_3+2\lambda_4)} \tilde{q}_n^{\lambda_3} u_{\lambda_1-\lambda_3}(\tilde{p}_n, \tilde{q}_n),$$

where the sums are taken over all non-negative integers λ_0, λ_1, λ_2, λ_3, λ_4 such that

$$\sum_{i=0}^{4} \lambda_i = \sum_{i=0}^{4} i\lambda_i = m.$$

We observe, however, that both of these formulas involve Z_n; consequently, we must also produce a multiplication formula for Z_n.

Put $e_i = e_i(\gamma_1^n, \gamma_2^n, \gamma_3^n, \gamma_1^{-n}, \gamma_2^{-n}, \gamma_3^{-n})$ $(i = 1, 2, \ldots, 6)$. It is a simple matter to verify that

$$Q^n e_2 = 3Q^n + Q^n \kappa_2^{(n)}, \quad Q^n e_3 = Q^n \kappa_3^{(n)} + 2Q^n \kappa_1^{(n)},$$

$$Q^n e_4 = 3Q^n + Q^n \kappa_2^{(n)}, \quad Q^n e_5 = Z_n, \quad Q^n e_6 = Q^n.$$

Hence

$$Q^n e_1 = Z_n = Q^n e_5$$

$$Q^n e_2 = \frac{V_n^2 - \Delta U_n^2}{4} - Q^n = \tilde{q}_n - Q^n = Q^n e_4$$

$$Q^n e_3 = \frac{V_n^2 + \Delta U_n^2}{4} - 2Z_n = V_n^2 - 2\tilde{q}_n - 2Z_n.$$

Since

$$Z_{mn} = Q^{mn}(\gamma_1^{mn} + \gamma_1^{-mn} + \gamma_2^{mn} + \gamma_2^{-mn} + \gamma_3^{mn} + \gamma_3^{-mn}),$$

we find from Waring's formula that

$$Z_{mn} = \sum (-1)^{\lambda_0} \frac{m(m-\lambda_0-1)!}{\lambda_1!\lambda_2!\cdots\lambda_6!} Q^{\lambda_0 n} Z_n^{\lambda_1+\lambda_5} (\tilde{q}_n - Q^n)^{\lambda_2+\lambda_4} (V_n^2 - 2\tilde{q}_n - 2Z_n)^{\lambda_3},$$

where the sum is taken over all non-negative integers λ_0, λ_1, $\ldots$, λ_6 such that

$$\sum_{i=0}^{6} \lambda_i = \sum_{i=0}^{6} i\lambda_i = m.$$

Thus, there are multiplication formulas in the case of $k = 3$, but they involve three functions, not just two.

7 A Subcase

Suppose that the U_n and V_n given by (2.2) with $k = 3$ are linear recurring sequences and that $\gamma_1 \gamma_2 \gamma_3 = 1$. By Theorem 5.2 we have

$$Q = \lambda^2 \gamma_1 \gamma_2 \gamma_3 = \lambda^2 \in \mathbb{Z}, \quad P_1 \in \mathbb{Z}, \quad Q\kappa_1 \in \mathbb{Z}.$$

Since in this case $\gamma_1 \gamma_2 \gamma_3 = 1$, we have $P_1 = \lambda \kappa_1 + 2\lambda$ and $\lambda P_1 = Q\kappa_1 + 2Q \in \mathbb{Z}$; hence $\lambda \in \mathbb{Q}$, which, since $\lambda^2 \in \mathbb{Z}$, means that $\lambda \in \mathbb{Z}$. Put $R = \lambda$. We have $\rho_i = R(\gamma_i + \gamma_i^{-1}) = R\mu_i$ $(i = 1,2,3)$ and $\rho_4 = 2R$. We find that

$$P_1 = S_1 + 2R, \qquad P_2 = S_2 + 2RS_1, \qquad P_3 = S_3 + 2RS_2, \qquad P_4 = 2RS_3, \quad (7.1)$$

where $S_i = R^i e_i$ and $e_i = e_i(\mu_1, \mu_2, \mu_3)$ $(i = 1,2,3)$. Since $R, P_1, P_2, P_3, P_4 \in \mathbb{Z}$, we must have $S_1, S_2, S_3 \in \mathbb{Z}$. Also,

$$S_3 = R^3(\gamma_1 + \gamma_1^{-1})(\gamma_2 + \gamma_2^{-1})(\gamma_1 \gamma_2 + (\gamma_1 \gamma_2)^{-1}) = R^3(e_1^2 - 2e_2 - 4).$$

Hence

$$S_3 = RS_1^2 - 2RS_2 - 4R^3. \tag{7.2}$$

If we put $R = \lambda$, $\gamma_3 = (\gamma_1 \gamma_2)^{-1}$ in (2.2), we find that

$$G(x) = (x - R\gamma_1)(x - R\gamma_2)(x - R(\gamma_1 \gamma_2)^{-1})(x - R\gamma_1^{-1})(x - R\gamma_2^{-1})(x - R\gamma_1 \gamma_2)$$

is the characteristic polynomial for $\{U_n\}$ and $\{V_n - 2R^n\}$.

Since ρ_i $(i = 1,2,3)$ are the zeroes of

$$g(x) = x^3 - S_1 x^2 + S_2 x - S_3, \tag{7.3}$$

we find that

$$G(x) = x^6 - S_1 x^5 + (S_2 + 3R^2)x^4 - (S_3 + 2R^2 S_1)x^3 + R^2(S_2 + 3R^2)x^2 - R^4 S_1 x + R^6. \tag{7.4}$$

Thus, if we define

$$W_n = V_n - 2R^n,$$

we see that both $\{W_n\}$ and $\{U_n\}$ satisfy the sixth order recurrence

$$X_{n+6} = S_1 X_{n+5} - (S_2 + 3R^2)X_{n+4} + (S_3 + 2R^2 S_1)X_{n+3} - R^2(S_2 + 3R^2)X_{n+2}$$
$$+ R^4 S_1 X_{n+1} - R^6 X_n.$$

In addition we have the initial conditions (from (7.1) and Theorem 5.3) $U_0 = 0$, $U_1 = 1$, $U_2 = S_1 + 2R$, $U_3 = S_1^2 + RS_1 - S_2 - 3R^2$, $U_{-n} = -U_n/R^{2n}$ $(n = 1,2,3)$, $W_0 = 6$, $W_1 = S_1$, $W_2 = S_1^2 - 2S_2 - 6R^2$, $W_3 = S_1^3 - 3S_1 S_2 + 3RS_1^2 - 6RS_2 - 3R^2 S_1 - 12R^3$, $W_{-n} = W_n/R^{2n}$ $(n = 1,2,3)$.

We now examine the converse of the above situation. We first require a simple lemma.

Lemma 7.1. *Let R, S_1, $S_2 \in \mathbb{Z}$ and suppose S_3 is given by (7.2). Let σ_1, σ_2, σ_3 be the zeroes of $g(x)$ in (7.3), and let τ_1, τ_2, $\tau_3 \in \overline{\mathbb{Q}}$ be such that*

$$\sigma_i = R(\tau_i + \tau_i^{-1}) \quad (i = 1, 2, 3).$$

We must have $\tau_3 \in \{\tau_1 \tau_2, 1/(\tau_1 \tau_2), \tau_1/\tau_2, \tau_2/\tau_1\}$.

Proof. We have

$$S_3 = \sigma_1 \sigma_2 \sigma_3 = R(\sigma_1 + \sigma_2 + \sigma_3)^2 - 2R(\sigma_1 \sigma_2 + \sigma_2 \sigma_3 + \sigma_3 \sigma_1) - 4R^3$$
$$= R(\sigma_1^2 + \sigma_2^2 + \sigma_3^2) - 4R^3.$$

Solving for σ_3, we get

$$\sigma_3 = \frac{\sigma_1 \sigma_2 \pm \sqrt{(\sigma_1 \sigma_2)^2 - 4R^2(\sigma_1^2 + \sigma_2^2) + 16R^4}}{2R}.$$

Now

$$(\sigma_1 \sigma_2)^2 - 4R^2(\sigma_1^2 + \sigma_2^2) + 16R^4 = (\sigma_1^2 - 4R^2)(\sigma_2^2 - 4R^2)$$
$$= R^4(\tau_1 - 1/\tau_1)^2(\tau_2 - 1/\tau_2)^2.$$

Hence,

$$\sigma_3 = \frac{R^2(\tau_1 + 1/\tau_1)(\tau_2 + 1/\tau_2) \pm R^2(\tau_1 - 1/\tau_1)(\tau_2 - 1/\tau_2)}{2R}$$
$$= \begin{cases} R(\tau_1 \tau_2 + (\tau_1 \tau_2)^{-1}) \\ R(\tau_1/\tau_2 + \tau_2/\tau_1). \end{cases}$$

It follows that $\tau_3 \in \{\tau_1 \tau_2, 1/(\tau_1 \tau_2), \tau_1/\tau_2, \tau_2/\tau_1\}$. $\square$

From this result we are now able to show that if $\{W_n\}$ and $\{U_n\}$ have characteristic function $G(x)$ and the above initial conditions, then we can express them in terms of $\{V_n\}$ and $\{U_n\}$ in (2.2) (k = 3).

Theorem 7.2. *Select R, S_1 ($\neq -2R$), $S_2 \in \mathbb{Z}$ and find S_3 using (7.2). If $\{U_n\}$ and $\{W_n\}$ have characteristic polynomial $G(x)$ in (7.4) with the above initial conditions, then there exist γ_1, γ_2, $\gamma_3 \in \mathbb{Q}$ such that $\gamma_1 \gamma_2 \gamma_3 = 1$ and U_n, $V_n (= W_n + 2R^n)$ are given by (2.2) with $k = 3$ and $\lambda = R$.*

Proof. According to Lemma 7.1, we have

$$\tau_1 \tau_2 \tau_3 = 1, \quad \tau_1^{-1} \tau_2 \tau_3 = 1, \quad \tau_1 \tau_2^{-1} \tau_3 = 1 \quad \text{or} \quad \tau_1 \tau_2 \tau_3^{-1} = 1.$$

With no loss of generality, suppose that $\tau_1^{-1} \tau_2 \tau_3 = 1$. Put $\gamma_1 = \tau_1^{-1}$, $\gamma_2 = \tau_2$, $\gamma_3 = \tau_3$. We have $\gamma_1 \gamma_2 \gamma_3 = 1$ and $\gamma_1, \gamma_1^{-1}, \gamma_2, \gamma_2^{-1}, \gamma_3, \gamma_3^{-1}$ are the six zeroes of $G(x)$. If we put

$$U_n = R^{n-1} \frac{(\gamma_1^n - 1)}{(\gamma_1 - 1)} \frac{(\gamma_2^n - 1)}{(\gamma_2 - 1)} \frac{(\gamma_3^n - 1)}{(\gamma_3 - 1)},$$

$$V_n = R^n (\gamma_1^n + 1)(\gamma_2^n + 1)(\gamma_3^n + 1),$$

where $\gamma_1 \gamma_2 \gamma_3 = 1$, we see that $G(x)$ is the characteristic function of $\{U_n\}$ and $\{V_n - 2R^n\}$. Since the initial values of U_n and W_n are those given above, the theorem follows. □

Corollary 7.3. *Under the conditions of the theorem, $\{U_n\}$ is a divisibility sequence.*

Proof. We have $Q = R^2 \gamma_1 \gamma_2 \gamma_3 = R^2 \in \mathbb{Z}$, $P_1 = 2R + S_1 \in \mathbb{Z}$ and $\kappa_i = e_i(\mu_1, \mu_2, \mu_3)$. Since $R\mu_i$ are the zeroes of $g(x)$, we have $S_1 = R\kappa_1$, $S_2 = R^2 \kappa_2$, $S_3 = R^3 \kappa_3$. Now $R \mid S_3$ by (7.2); hence $Q\kappa_1, Q\kappa_2, Q\kappa_3 \in \mathbb{Z}$. The result follows from Theorem 3.2. □

Notice that if we put $\alpha_i = R\gamma_i$ $(i = 1, 2, 3)$, $\beta_i = R^2/\alpha_i$ $(i = 1, 2, 3)$, we get

$$U_n = \frac{\alpha_1^n - \beta_1^n + \alpha_2^n - \beta_2^n + \alpha_3^n - \beta_3^n}{\alpha_1 - \beta_1 + \alpha_2 - \beta_2 + \alpha_3 - \beta_3}$$

$$W_n = \alpha_1^n + \beta_1^n + \alpha_2^n + \beta_2^n + \alpha_3^n + \beta_3^n.$$

If we put $\delta = \alpha_1 - \beta_1 + \alpha_2 - \beta_2 + \alpha_3 - \beta_3$, then

$$\Delta = \delta^2 = \lambda^2 (1 - \gamma_1)^2 (1 - \gamma_2)^2 (1 - \gamma_3)^2 = P_1^2 - 4P_2 + 8P_3/P_1$$

$$= (S_1 + 2R)^2 - 4(S_2 + 2RS_1) + 8R(S_1 - 2R).$$

Hence

$$\Delta = S_1^2 - 4S_2 + 4RS_1 - 12R^2. \tag{7.5}$$

One example of this sequence $\{U_n\}$ occurs in Hall [3], where $U_0 = 0$, $U_1 = 1$, $U_2 = 1$, $U_3 = 1$, $U_4 = 5$, $U_5 = 1$, $U_6 = 7$, $U_7 = 8$, $U_8 = 5, \ldots$, and

$$U_{n+6} = -U_{n+5} + U_{n+4} + 3U_{n+3} + U_{n+2} - U_{n+1} - U_n.$$

Here we have $S_1 = -1$, $S_2 = -4$, $S_3 = 5$, $R = 1$. Another example is an unpublished sequence of Elkies (personal communication). For this sequence we have $U_0 = 0$, $U_1 = 1$, $U_2 = 1$, $U_3 = 2$, $U_4 = 7$, $U_5 = 5$, $U_6 = 20$, $U_7 = 27$, $U_8 = 49, \ldots$, and

$$U_{n+6} = -U_{n+5} + 2U_{n+4} + 5U_{n+3} + 2U_{n+2} - U_{n+1} - U_n.$$

Here we have $S_1 = -1, S_2 = -5, S_3 = 7, R = 1$.

If we let P', Q', R' be arbitrary integers and put $R = R'$,

$$S_1 = P'Q' - 3R', S_2 = P'^3R' + Q'^3 - 5P'Q'R' + 3R'^2, \left.\vphantom{\begin{array}{c}a\\b\end{array}}\right\}$$
$$S_3 = R'(P'^2Q'^2 - 2Q'^3 - 2P'^3R' + 4P'Q'R' - R'^2), \tag{7.6}$$

then

$$U_n = \frac{(\alpha^n - \beta^n)}{(\alpha - \beta)} \frac{(\beta^n - \gamma^n)}{(\beta - \gamma)} \frac{(\gamma^n - \alpha^n)}{(\gamma - \alpha)}$$
$$V_n = (\alpha^n + \beta^n)(\beta^n + \gamma^n)(\gamma^n + \alpha^n),$$

where α, β, γ are the zeroes of $x^3 - P'x^2 + Q'x - R'$. In this case we have $\gamma_1 = \alpha/\beta$, $\gamma_2 = \beta/\gamma$, $\gamma_3 = \gamma/\alpha$, $\lambda = R = R' = \alpha\beta\gamma$. This sequence and its companion sequence $W_n(= V_n - 2R^n)$ were the objects of a lengthy study by Roettger [18] (see also Müller et al. [13]). In [18] it is argued that the U_n sequence and some of its properties may have been known to Lucas.

We next put

$$C_n = R^n(\gamma_1^n + \gamma_2^n + (\gamma_1\gamma_2)^{-n}), \quad D_n = R^n(\gamma_1^{-n} + \gamma_2^{-n} + (\gamma_1\gamma_2)^n).$$

We have

$$R^n e_1(\gamma_1^n, \gamma_2^n, (\gamma_1\gamma_2)^{-n}) = C_n, \quad R^{2n} e_2(\gamma_1^n, \gamma_2^n, (\gamma_1\gamma_2)^{-n}) = R^n D_n,$$
$$R^{3n} e_3(\gamma_1^n, \gamma_2^n, (\gamma_1\gamma_2)^{-n}) = R^{3n}.$$

Similarly,

$$R^n e_1(\gamma_1^{-n}, \gamma_2^{-n}, (\gamma_1\gamma_2)^n) = D_n, \quad R^{2n} e_2(\gamma_1^{-n}, \gamma_2^{-n}, (\gamma_1\gamma_2)^n) = R^n C_n,$$
$$R^{3n} e_3(\gamma_1^{-n}, \gamma_2^{-n}, (\gamma_1\gamma_2)^n) = R^{3n}.$$

We have

$$C_{2n+m} = C_n C_{n+m} - R^n D_n C_m + R^{3n} C_{m-n},$$
$$D_{2n+m} = D_n D_{n+m} - R^n C_n D_m + R^{3n} D_{m-n}.$$

Since, for any n, $(W_n + \delta U_n)/2 = C_n$, $(W_n - \delta U_n) = D_n$, we get

$$2W_{2n+m} = W_{n+m}W_n + \Delta U_{n+m}U_n - R^n W_n W_m + R^n \Delta U_n U_m + 2R^{3n}W_{m-n},$$
$$2U_{2n+m} = U_{n+m}W_n + U_n W_{n+m} - R^n W_n U_m + R^n U_n W_m + 2R^{3n}U_{m-n}.$$

Since $W_{m-n} = W_{n-m}/R^{2(n-m)}$, $U_{m-n} = -U_{n-m}/R^{2(n-m)}$, we get the addition formulas

$$2W_{2n+m} = W_n W_{n+m} + \Delta U_{n+m} U_n - R^n W_n W_m + R^n \Delta U_n U_m + 2R^{n+2m} W_{n-m},$$

$$2U_{2n+m} = W_n U_{n+m} + U_n W_{n+m} - R^n W_n U_m + R^n U_n W_m - 2R^{n+2m} U_{n-m}.$$

On putting $m = 0$, we get the duplication formulas

$$2W_{2n} = W_n^2 + \Delta U_n^2 - 4R^n W_n,$$

$$U_{2n} = U_n(W_n + 2R^n).$$

Notice that there is no need for the extra function Z_n here. This is because in this case, $Z_n = R^n W_n$. We also see that if $2 \mid U_n$, then $2 \mid W_n$. If we put $m = n$ and use the duplication formulas, we get the triplication formulas

$$4W_{3n} = 3\Delta U_n^2(W_n + 2R^n) + W_n^2(W_n - 6R^n) + 24R^{3n},$$

$$4U_{3n} = U_n(3W_n^2 + \Delta U_n^2).$$

One can also use Waring's formula to compute C_{mn} and D_{mn} and the methods of §6 of [18] to obtain the multiplication formulas:

$$W_{mn} = \sum (-1)^{\lambda_0} \frac{m(m-\lambda_0-1)!}{\lambda_1!\lambda_2!\lambda_3!} R^{n(\lambda_0+\lambda_3)} \tilde{Q}_n^{\lambda_2} v_{\lambda_1-\lambda_2}, \tag{7.7}$$

$$U_{mn} = U_n \sum (-1)^{\lambda_0} \frac{m(m-\lambda_0-1)!}{\lambda_1!\lambda_2!\lambda_3!} R^{n(\lambda_0+\lambda_3)} \tilde{Q}_n^{\lambda_2} u_{\lambda_1-\lambda_2}. \tag{7.8}$$

Here the sums are taken over all nonzero integers $\lambda_0, \lambda_1, \lambda_2, \lambda_3$ such that

$$\sum_{i=0}^{3} \lambda_i = \sum_{i=0}^{3} i\lambda_i = m$$

and $u_n = u_n(\tilde{P}_n, \tilde{Q}_n)$, $v_n = v_n(\tilde{P}_n, \tilde{Q}_n)$, where $\tilde{P}_n = W_n$, $\tilde{Q}_n = (W_n^2 - \Delta U_n^2)/4$. Note that $\tilde{P}_1 = S_1$, $\tilde{Q}_1 = S_2 - S_1 R + 3R^2$. Thus, for a general m no extra functions (besides W_n and U_n) are needed in the multiplication formulas.

8 The Law of Repetition

We have seen that the sequences $\{U_n\}$ and $\{W_n\}$ mentioned in Sect. 7 satisfy the five basic properties of the Lucas functions. We will now investigate some properties of these sequences. By analogy to the constraints $\gcd(p,q) = 1$ for the Lucas sequences and $\gcd(P_1, P_2, Q) = 1$ for the sequences with $k = 2$, we will insist that $\gcd(S_1, S_2, S_3, R) = 1$. By (7.2) this means that $\gcd(S_1, S_2, R) = 1$. We can now prove the following simple lemma:

Lemma 8.1. *If* $(R, S_1, S_2) = 1$, *then* $(U_m, W_m, R) \mid 2$ *for all* $m \geq 0$.

Proof. From the multiplication formulas we observe that $U_m \equiv u_m(S_1, S_2)$, $W_m \equiv v_m(S_1, S_2) \pmod{R}$. Since $(R, S_1, S_2) = 1$, we may assume with no loss of generality that $S_2 \equiv S_2' \pmod{R}$ and $(S_2', S_1) = 1$. By the properties of Lucas functions (see, e.g., [22, p. 86]), we have $(u_m(S_1, S_2'), v_m(S_1, S_2')) \mid 2$; thus if p is a prime and $p \mid (U_m, W_m, R)$, then $p = 2$. Also, if $4 \mid R$, then $(U_m, W_m, R) \mid 2$. $\qquad\square$

This brings us to the problem of how $\{U_n\}$ and $\{W_n\}$ behave modulo 2. We begin with the case R even. From the multiplication formulas, we see that when $2 \mid R$ we have $\tilde{Q}_1 \equiv S_2 - RS_1$ and $\tilde{P}_1 \equiv S_1 \pmod{2R}$. Also,

$$U_m \equiv u_m(S_1, S_2 - RS_1) + (S_2 - RS_1)Rmu_{m-3}(S_1, S_2 - RS_1) \quad (\mathrm{mod}\ 2R)$$

$$W_m \equiv v_m(S_1, S_2 - RS_1) + (S_2 - RS_1)Rmv_{m-3}(S_1, S_2 - RS_1) \quad (\mathrm{mod}\ 2R).$$

Now if $2 \mid R$ and $2 \mid S_1$, then $2 \nmid S_2$ and $2 \mid u_m(S_1, S_2 - RS_1)$ if and only if $2 \mid m$; also, $v_m(S_1, S_2 - RS_1) \equiv 2 \pmod{4}$ when $m \geq 2$. It follows that $W_m \equiv 2 \pmod{4}$ when $m \geq 0$.

If $2 \mid R$ and $2 \nmid S_1$ and $2 \mid S_2$, then $2 \nmid U_m$. In the case where $2 \nmid S_1$ and $2 \nmid S_2$, we find that $2 \mid U_m$ if and only if $3 \mid m$. Also, if $3 \mid m$, then $4 \nmid (W_m, U_m)$.

If $2 \nmid R$, we have four possibilities and the results below can be proved by using the techniques of [18, §4.1]:

$$\text{If } 2 \mid S_1,\ 2 \nmid S_2, \text{ then } 2 \mid U_m \iff (m, 6) > 1;$$
$$\text{if } 2 \mid S_1,\ 2 \mid S_2, \text{ then } \quad 2 \mid U_m \iff 2 \mid m;$$
$$\text{if } 2 \nmid S_1,\ 2 \mid S_2, \text{ then } \quad 2 \mid U_m \iff 7 \mid m;$$
$$\text{if } 2 \nmid S_1,\ 2 \nmid S_2, \text{ then } \quad 2 \mid U_m \iff 3 \mid m.$$

We will now use this information to establish the law of repetition for $\{U_n\}$ when the prime $p = 2$. We first point out that we have already proved that when $2 \mid R$, we have $4 \nmid (W_m, U_m)$. We consider two cases:

Case 1. $2 \mid R$. We recall that $U_{2n} = U_n(W_n + 2R^n)$.

If $2^\lambda \| U_n$ and $\lambda \geq 2$, then $2 \| W_n$ and $2^{\lambda+1} \| U_{2n}$. By induction, $2^{\lambda+\mu} \| U_{2^\mu n}$.

If $2^\lambda \| U_n$ and $\lambda = 1$, then $2^{\lambda+\nu} \| U_{2n}$, where $2^\nu \| W_n + 2R^n$. Since $\nu \geq 1$, we have $2^{\lambda+\mu+\nu-1} \| U_{2^\mu n}$.

In the second case, where $2 \nmid R$, we note the following lemmas which are easy to establish by making use of the duplication formulas:

Lemma 8.2. *If* $2 \nmid R$, $8 \mid W_n + 2R^n$ *and* $8 \mid U_n$, *then* $8 \| W_{2n} + 2R^{2n}$.

Lemma 8.3. *If $2 \nmid R$, $2^\lambda \| U_n$ ($\lambda \geq 1$) and $2^\nu \| W_{2n} + 2R^{2n}$ ($\nu > 1$), then $8 \mid W_{4n} + 2R^{4n}$ and $8 \mid U_{4n}$.*

Lemma 8.4. *If $2 \nmid R$, $2^\lambda \| U_n$ ($\lambda \geq 2$), $2 \| W_n + 2R^n$, then $2^{\lambda+1} \| U_{2n}$ and $2 \| W_{2n} + 2R^{2n}$.*

Lemma 8.5. *If $2 \nmid R$, $2^\lambda \| U_n$, $2^\nu \| W_n + 2R^n$, where $\lambda \geq 2$, $\nu \geq 2$, then $2^{\lambda+\nu} \| U_{2n}$ and $8 \mid W_{2n} + 2R^{2n}$.*

If we now define ν_i by $2^{\nu_i} \| W_{2^i n} + 2R^{2^i n}$ ($i = 0, 1, 2$), we can use the above lemmas and induction to find the law of repetition for $p = 2$ in Case 2.

Case 2. $2 \nmid R$. If $2^\lambda \| U_n$, we have

$$2^{\lambda+\mu} \| U_{2^\mu n} \text{ when } \lambda \geq 2, \nu_0 = 1, \mu \geq 1;$$

$$2^{\lambda+\mu+\nu_0-1} \| U_{2^\mu n} \text{ when } \lambda = 1, \nu_1 = 1, \mu \geq 1;$$

$$2^{\lambda+3\mu+\nu_0-3} \| U_{2^\mu n} \text{ when } \lambda \geq 3, \nu_0 \geq 2, \mu \geq 1;$$

$$2^{\lambda+3\mu+\nu_0+\nu_1-6} \| U_{2^\mu n} \text{ when } \lambda \geq 2, \nu_0 \geq 2, \mu \geq 2;$$

$$2^{\lambda+3\mu+\nu_0+\nu_1+\nu_2-9} \| U_{2^\mu n} \text{ when } \lambda = 1, \nu_1 \geq 2, \mu \geq 3.$$

Thus, the law of repetition for the prime $p = 2$ is quite detailed. Fortunately, this is not the case for the odd primes, but we must first develop some theory before we can prove this law.

Let d be the discriminant of $g(x)$. Since $\rho_i = R(\gamma_i + \gamma_i^{-1})$ ($i = 1, 2, 3$) are the zeroes of $g(x)$, we have

$$(\rho_2 - \rho_3)(\rho_3 - \rho_1)(\rho_1 - \rho_2)$$

$$= R^3(\gamma_2 - \gamma_3)(\gamma_3 - \gamma_1)(\gamma_1 - \gamma_2)(1 - \frac{1}{\gamma_2\gamma_3})(1 - \frac{1}{\gamma_3\gamma_1})(1 - \frac{1}{\gamma_1\gamma_2}).$$

Since $\gamma_1\gamma_2\gamma_3 = 1$, we have

$$(1 - \frac{1}{\gamma_2\gamma_3})(1 - \frac{1}{\gamma_3\gamma_1})(1 - \frac{1}{\gamma_1\gamma_2}) = (1 - \gamma_1)(1 - \gamma_2)(1 - \gamma_3).$$

Hence $d = R^4 G\Delta$, where

$$G = (\gamma_2 - \gamma_3)^2(\gamma_3 - \gamma_1)^2(\gamma_1 - \gamma_2)^2$$

and Δ is given by (7.5).

If we put $X = R^2(\gamma_1/\gamma_2 + \gamma_1\gamma_2^2 + 1/\gamma_1^2\gamma_2)$, $Y = R^2(\gamma_2/\gamma_1 + \gamma_1^2\gamma_2 + 1/\gamma_1\gamma_2^2)$, then

$$R^4(\gamma_2 - \gamma_3)^2(\gamma_3 - \gamma_1)^2(\gamma_1 - \gamma_2)^2 = (X - Y)^2$$

and $d = \Delta(X - Y)^2$.

Now $X + Y = S_2 - RS_1$ and $XY = 3R^4 + RW_3$; thus,

$$(X - Y)^2 = (X + Y)^2 - 4XY = (S_2 - RS_1)^2 - 4(3R^4 + RW_3).$$

Since
$$W_3 = S_1^3 - 3S_1 S_2 + 3RS_1^2 - 6RS_2 - 3R^2 S_1 - 12R^3,$$

we get $d = \Delta \Gamma$, where

$$\Gamma = S_2^2 + 10RS_1 S_2 - 4RS_1^3 - 11R^2 S_1^2 + 12R^3 S_1 + 24R^2 S_2 + 36R^4.$$

Notice that if S_1, S_2, S_3, R are given by (7.6), then $\Gamma = (R'P'^3 - Q'^3)^2$ and Δ is the discriminant of $x^3 - P'x^2 + Q'x - R'$.

If D is the discriminant of $G(x)$, we have (see [22, §10.1])

$$D = Ed^2 R^{12},$$

where
$$E = \prod_{i=1}^{3} (\rho_i^2 - 4R^2).$$

Also, since $\rho_1^2 - 4R^2 = R^2(\gamma_i - \gamma_i^{-1})$, we get

$$E = \gamma_1^2 \gamma_2^2 \gamma_3^2 E = R^6 \prod_{i=1}^{3} (\gamma_i^2 - 1)^2 = R^6 (\gamma_1 + 1)^2 (\gamma_2 + 1)^2 (\gamma_3 + 1)^2 (\gamma_1 - 1)^2 (\gamma_2 - 1)^2 (\gamma_3 - 1)^2.$$

Hence
$$E = R^2 \Delta (S_1 + 2R)^2 \tag{8.1}$$

and
$$D = R^{14} (S_1 + 2R)^2 \Delta^3 \Gamma^2. \tag{8.2}$$

We can use the identity

$$W_n - 6R^n = \frac{2}{R^n \gamma_2^n} (R^n - R^n \gamma_2^n)^2 - (R^n - R^n \gamma_1^n)(1 - \gamma_3^n)(1 + \gamma_2^n) \tag{8.3}$$

to prove the following result:

Theorem 8.6. *If p is any prime such that $p \nmid 6D$ and $p \mid U_n$, $p \mid W_n - 6R^n$, then $p^3 \mid U_n$ and $p^2 \mid W_n - 6R^n$.*

Proof. This can be shown by using essentially the same argument as that used in the proof of Theorem 4.18 of [18].   □

In order to produce a complete law of repetition for $\{U_n\}$, we need to establish a few more results.

Lemma 8.7. *If p is a prime and $p \mid \Gamma$, $p \nmid 6R\Delta$ and $p \nmid S_1^2 - 3S_2$, and then if $\mathfrak{p}$ is any prime ideal lying over p in $\mathbb{K} = \mathbb{Q}(\gamma_1, \gamma_2)$, we have $\mathfrak{p}^3 \nmid p$.*

Proof. We note that if $p \mid \Gamma$, then $p \mid d$. Thus if ρ_1, ρ_2, ρ_3 are the zeroes of $g(x) \in \mathbb{F}_p[x]$, then $\rho_1, \rho_2, \rho_3 \in \mathbb{F}_p$ and without loss of generality, $\rho_1 = \rho_2$. It follows that

$$G(x) = (x^2 - \rho_1 x + R^2)^2 (x^2 - \rho_3 x + R^2)$$

in $\mathbb{F}_p[x]$. Since $\rho_1 = \rho_2$, we have

$$S_1^2 - 3S_2 = (\rho_1 + \rho_2 + \rho_3)^2 - 3(\rho_2\rho_3 + \rho_3\rho_1 + \rho_1\rho_2) = \rho_1^2 - 2\rho_1\rho_3 + \rho_3^2 = (\rho_1 - \rho_3)^2.$$

Since $p \nmid S_1^2 - 3S_2$, we cannot have $\rho_1 = \rho_3$. Now if $x^2 - \rho_1 x + R^2$ and $x^2 - \rho_3 x + R^2$ have a common zero in $\mathbb{F}_p$, then $\rho_1 = \rho_3$, which we have seen is impossible. Furthermore, if $x^2 - \rho_1 x + R^2$ is reducible over $\mathbb{F}_p$, then $R\gamma_1, R/\gamma_1 \in \mathbb{F}_p$. Since $\rho_3 = 1/\rho_1\rho_2 = 1/\rho_1^2$, we have $x^2 - \rho_3 x + R^2$ reducible over $\mathbb{F}_p$. Thus if both $x^2 - \rho_1 x + R^2$ and $x^2 - \rho_2 x + R^2$ are irreducible over $\mathbb{F}_p$, then by Dedekind's theorem, we have $p = \mathfrak{p}^2 \mathfrak{q}$, where $\mathfrak{p}$ and $\mathfrak{q}$ are distinct ideals in $\mathbb{K}$. If $x^2 - \rho_1 x + R^2$ is irreducible and $x^2 - \rho_3 x + R^2$ is reducible over $\mathbb{F}_p$, then $p = \mathfrak{p}^2 \mathfrak{q}_1 \mathfrak{q}_2$, where $\mathfrak{p}, \mathfrak{q}_1, \mathfrak{q}_2$ are prime ideals and $\mathfrak{p}$ is distinct from $\mathfrak{q}_1$ and $\mathfrak{q}_2$. Finally, if $x^2 - \rho_1 x + R^2$ is reducible, then $p = \mathfrak{p}_1^2 \mathfrak{p}_2^2 \mathfrak{q}_1 \mathfrak{q}_2$, where the prime ideals $\mathfrak{p}_1$ and $\mathfrak{p}_2$ are distinct from $\mathfrak{q}_1$ and $\mathfrak{q}_2$. It remains to show that $\mathfrak{p}_1 \neq \mathfrak{p}_2$. If $x^2 - \rho_1 x + R^2$ has a multiple zero in $\mathbb{F}_p$, then $R\gamma_1 = R/\gamma_1$ in $\mathbb{F}_p$ and $\gamma_1 = \pm 1$. If $\gamma_1 = 1$, then $p \mid \Delta$; if $\gamma_1 = -1$, then $\gamma_3 = 1/\gamma_1^2 = 1$ and $p \mid \Delta$. Since $p \nmid \Delta$, we must have $\mathfrak{p}_1 \neq \mathfrak{p}_2$. $\square$

Lemma 8.8. *If p is a prime, $p \nmid 6\Delta R$ and $p \mid S_1 + 2R$, then if $\mathfrak{p}$ is any prime lying over p in $\mathbb{K}$, we have $\mathfrak{p}^3 \nmid p$.*

Proof. This is proved in much the same manner as the previous result. We note that since $S_1 \equiv -2R \pmod{p}$, we have $S_3 \equiv -2RS_2 \pmod{p}$ and

$$G(x) = (x+R)^2 (x^4 + (S_2 + 2R^2)x^2 + R^4)$$

in $\mathbb{F}_p[x]$. Also, $x + R$ cannot be a factor of $x^4 + (S_2 + 2R^2)x^2 + R^4$ in $\mathbb{F}_p[x]$ unless $p \mid R^2(S_2 + 4R^2)$, which means that $p \mid \Delta R$, an impossibility.

If $x^4 + (S_2 + 2R^2)x^2 + R^4$ has multiple zeroes in $\mathbb{F}_p$, then it shares a zero with its derivative $4x^3 + 2(S_2 + 2R^2)x$. This means that $p \mid S_2(S_2 + 4R^2)$. We have already seen that $p \nmid S_2 + 4R^2$; hence $p \mid S_2$. In this case,

$$x^4 + (S_2 + 2R^2)x^2 + R^4 = (x^2 + R^2)^2$$

in $\mathbb{F}_p$, and $x^2 + R^2$ is either irreducible or the product of distinct linear factors in $\mathbb{F}_p[x]$. $\square$

Theorem 8.9. *Suppose p is a prime such that $p \nmid 6\Delta R$. If $p \mid S_1 + 2R$ or if $p \mid \Gamma$ and $p \nmid S_1^2 - 3S_2$, then if $p \mid U_n$ and $p \mid W_n - 6R^n$, we get $p^2 \mid U_n$.*

Proof. Since $p \mid U_n$ and $p \nmid \Delta$, we may assume without loss of generality that $\mathfrak{p} \mid \gamma_1^n - 1$ in $\mathbb{K}$, where $\mathfrak{p}$ is a prime ideal lying over p in $\mathbb{K}$. Since $\mathfrak{p} \mid W_n - 6R^n$, we can use (8.3) to get $\mathfrak{p} \mid \gamma_2^n - 1$. Since $\gamma_3 = 1/\gamma_1\gamma_2$, we also get $\mathfrak{p} \mid \gamma_3^n - 1$. It follows that $\mathfrak{p}^3 \mid U_n$ or $(\mathfrak{p}^2)^3 \mid U_n^2$. Hence, by Lemma 8.8 or Lemma 8.4, $p^3 \mid U_n^2$ and $p^2 \mid U_n$. $\square$

Theorem 8.10. *Suppose that p is a prime such that $p \nmid 6\Delta R$, $p \mid \Gamma$ and $p \mid S_1^2 - 3S_2$. If $p \mid U_n$, then $p \mid W_n - 6R^n$.*

Proof. If $p \mid S_1^2 - 3S_2$, then

$$9\Gamma \equiv S_1^4 - 6RS_1^3 - 27R^2S_1^2 + 108R^3S_1 + 224R^4$$
$$\equiv (S_1 + 3R)^2(S_1 - 6R)^2 \pmod{p}.$$

Since $p \nmid \Delta$, we cannot have $S_1 \equiv 6R \pmod{p}$; hence $S_1 \equiv -3R \pmod{p}$, $S_2 \equiv 3R^2$ and $S_3 \equiv -R^3 \pmod{p}$. Thus $g(x) \equiv (x - R)^3 \pmod{p}$. It follows that if ρ_1, ρ_2, ρ_3 are the zeroes of $g(x)$ in $\mathbb{F}_p$, then $\rho_1 \equiv \rho_2 \equiv \rho_3 \equiv R \pmod{p}$. Hence $\gamma_1 = \gamma_2 = \gamma_3$ in the splitting field of $G(x)$. (Since $p \nmid \Delta$, we cannot have γ_1, γ_2 or $\gamma_3 = 1$.) Also, since $\gamma_3 = 1/\gamma_1\gamma_2$, we have $\gamma_3^3 = 1$ and therefore $\gamma_1^2 + \gamma_1 + 1 = 0$, $\gamma_2^2 + \gamma_2 + 1 = 0$ and $\gamma_3^2 + \gamma_3 + 1 = 0$. Thus, $\gamma_1^n - 1 = 0$ if and only if $3 \mid n$. If $\mathfrak{p}$ is any prime ideal lying over p in $\mathbb{K}$, and $\mathfrak{p} \mid U_n$, then $\mathfrak{p} \mid \gamma_1^n - 1$, say, and we must have $3 \mid n$. Hence $\mathfrak{p} \mid \gamma_2^n - 1$ and $\mathfrak{p} \mid \gamma_3^n - 1$. Hence, $\mathfrak{p} \mid W_n - 6R^n$ by (8.1) and therefore $p \mid W_n - 6R^n$. $\square$

We are now able to complete our law of repetition for $\{U_n\}$. We first point out, however, that if p is a prime and $p \mid U_n$, $p \mid W_n - 6R^n$ and $p \mid R$, then p can only be 2 by Lemma 8.1.

Definition 8.11. We say that a prime $p \,(> 2)$ is *special* if $p = 3$ or if $p \mid R$ or if $p \nmid 6R\Delta$ and $p \mid \Gamma$.

By using the methods of Chapter 4 of [18], we can prove the law of repetition for all primes $p > 2$.

Law of Repetition. Let $p^\lambda \| U_n \,(\lambda \geq 1, p > 2)$.
   If p is a special prime, then

$$p^{\mu+\lambda} \| U_{p^\mu n} \text{ when } p \nmid W_n - 6R^n;$$

$$p^{3\mu+\lambda} \| U_{p^\mu n} \text{ when } p \mid W_n - 6R^n \text{ and } \lambda \geq 2;$$

$$p^{3\mu+\nu-2} \| U_{p^\mu n} \text{ when } p \mid W_n - 6R^n \text{ and } p^\nu \| U_{pn}/U_n.$$

If p is not a special prime, then

$$p^{\mu+\lambda} \| U_{p^\mu n} \text{ when } p \nmid W_n - 6R^n;$$

$$p^{3\mu+\lambda} \| U_{p^\mu n} \text{ when } p \mid W_n - 6R^n.$$

9 The Law of Apparition

We will now begin the process of establishing the law of apparition for the sequence $\{U_n\}$ of Sect. 7. We begin with some preliminary observations. If m_1 and m_2 are coprime integers and $m_1 \mid U_{n_1}$, $m_2 \mid U_{n_2}$, then since $\{U_n\}$ is a divisibility sequence, we know that $m_1 m_2 \mid U_{n_3}$, where $n_3 = [n_1, n_2]$, the least common multiple of n_1 and n_2. Thus, in order to discover the terms of the sequence $\{U_n\}$ where m appears as a factor, we need only to concern ourselves with where p^α occurs as a factor of terms of $\{U_n\}$, when p is a prime and $p^\alpha \| m$. The law of repetition tells us that we need first to find those ranks n where $p \mid U_n$.

Definition 9.1. Let p be a fixed prime and $\{U_n\}$ be a fixed sequence given in Sect. 7. Let r_i (if it exists) be the least positive integer such that $p \mid U_{r_i}$. For $i = 1, 2, \ldots$ define r_{i+1} (if it exists) to be the least positive integer such that $p \mid U_{r_{i+1}}$, $r_{i+1} > r_i$ and $r_j \nmid r_{i+1}$ for any $1 \leq j \leq i$. We define the $r_1, r_2, r_3, \ldots$ to be the *ranks of apparition* of p in $\{U_n\}$.

The following result can be easily established:

Theorem 9.2. *For any prime p such that $p \nmid dR$, there can be at most three ranks of apparition of p in $\{U_n\}$, and these values are the least positive integers r_i such that $\gamma_i^{r_i} = 1$ in the splitting field of $G(x) \in \mathbb{F}_p[x]$.*

Indeed, as shown in [18, Theorem 4.27], there exist sequences $\{U_n\}$ and primes p for which p has three ranks of apparition. In the previous section we showed that if $p = 2$, then p has no more than two ranks of apparition in $\{U_n\}$.

The situations for $p = 2$ and 3 are summarized in Tables 1 and 2.

We now turn our attention to those primes $p > 3$ such that $p \mid dR$. We note by the multiplication formulas in Sect. 7 that if $p \mid R$, then

$$U_m \equiv u_m(S_1, S_2) \pmod{p}.$$

Thus, in this case the theory of where p divides m reduces to that of where p divides the Lucas sequence $u_m(S_1, S_2)$. This means that there can be at most one rank of apparition of p. In what follows, we will now assume that $p \nmid R$.

Table 1 Ranks of apparition for $p = 2$

R	S_1	S_2	S_3	$U_m \equiv 0 \pmod 2$ iff	$d \pmod 2$
0	0	1	0	$m \equiv 0 \pmod 2$	0
0	1	0	0	only for $m = 0$	0
0	1	1	0	$m \equiv 0 \pmod 3$	1
1	0	0	0	$m \equiv 0 \pmod 2$	0
1	0	1	0	$m \equiv 0 \pmod 2$ or $m \equiv 0 \pmod 3$	0
1	1	0	1	$m \equiv 0 \pmod 7$	1
1	1	1	1	$m \equiv 0 \pmod 3$	0

Table 2 Ranks of apparition for $p = 3$

R	S_1	S_2	S_3	$U_m \equiv 0 \pmod 3$ iff	$d \pmod 3$
0	0	1	0	$m \equiv 0 \pmod 2$	2
0	0	2	0	$m \equiv 0 \pmod 2$	1
0	1	0	0	$m = 0$	0
0	1	1	0	$m \equiv 0 \pmod 3$	0
0	1	2	0	$m \equiv 0 \pmod 4$	2
0	2	0	0	$m = 0$	0
0	2	1	0	$m \equiv 0 \pmod 3$	0
0	2	2	0	$m \equiv 0 \pmod 4$	2
1	0	0	2	$m \equiv 0 \pmod 3$	0
1	0	1	0	$m \equiv 0 \pmod 4$	2
1	0	2	1	$m \equiv 0 \pmod{13}$	1
1	1	0	0	$m \equiv 0 \pmod 2$	0
1	1	1	1	$m \equiv 0 \pmod 2$	2
1	1	2	2	$m \equiv 0 \pmod 2$ or $m \equiv 0 \pmod 3$	0
1	2	0	0	$m \equiv 0 \pmod 3$ or $m \equiv 0 \pmod 4$	0
1	2	1	1	$m \equiv 0 \pmod 7$	1
1	2	2	2	$m \equiv 0 \pmod{13}$	1
2	0	0	1	$m \equiv 0 \pmod 3$	0
2	0	1	0	$m \equiv 0 \pmod 4$	2
2	0	2	2	$m \equiv 0 \pmod{13}$	1
2	1	0	0	$m \equiv 0 \pmod 3$ or $m \equiv 0 \pmod 4$	0
2	1	1	2	$m \equiv 0 \pmod 7$	1
2	1	2	1	$m \equiv 0 \pmod{13}$	1
2	2	0	0	$m \equiv 0 \pmod 2$	0
2	2	1	2	$m \equiv 0 \pmod 2$	2
2	2	2	1	$m \equiv 0 \pmod 2$ or $m \equiv 0 \pmod 3$	0

If $p \mid d$, we know that the splitting field of $g(x) \pmod p$ is $\mathbb{F}_p$ and there are two possibilities.

Case 1. $p \nmid R$, $p \mid d$, $p \mid S_1^2 - 3S_2$. In this case we get $\rho_1 = \rho_2 = \rho_3$ in $\mathbb{F}_p$. Since $\rho_i = R(\gamma_i + \gamma_i^{-1})$, $(i = 1, 2, 3)$, and $\gamma_1 \gamma_2 \gamma_3 = 1$, this means that either $\rho_1 = \rho_2 = \rho_3 = 0$ or $\rho_1 = \rho_2 = \rho_3 = 2R$. If $\rho_1 = \rho_2 = \rho_3 = 0$, we must have $\gamma_1 = \gamma_2 = \gamma_3 = \zeta$, where $\zeta^2 + \zeta + 1 = 0$ in the splitting field of $G(x) \pmod p$. This means that $p \mid S_1$, $p \mid S_2$ and

$$ U_n \equiv \begin{cases} R^n & n \equiv 1 \pmod 3 \\ 0 & n \equiv 0 \pmod 3 \\ -R^n & n \equiv -1 \pmod 3. \end{cases} $$

It follows that p has a single rank of apparition, r_1, in $\{U_n\}$ and $r_1 = 3$. If $\rho_1 = \rho_2 = \rho_3 = 2R$, we get $\gamma_1 = \gamma_2 = \gamma_3 = 1$ and

$$ U_n \equiv n^3 R^n \pmod p. $$

Hence p has a single rank of apparition, r_1, in $\{U_n\}$ and $r_1 = p$.

Case 2. $p \nmid R$, $p \mid d$, $p \nmid S_1^2 - 3S_2$. In this case we may assume that $\rho_1 = \rho_2 \neq \rho_3 \in \mathbb{F}_p$. Hence we get two possibilities: $\gamma_3 = 1$ or $\gamma_1 = \gamma_2$. Now $\gamma_3 = 1$ only when $2R$ is a zero of $g(x)$ and this also means that $p \mid \Delta$. Thus, if $p \mid \Delta$, we have $\gamma_2 = 1/\gamma_1$, and

$$U_n = R^{n-1}\gamma_1^{1-n}n\left(\frac{\gamma_1^n - 1}{\gamma_1 - 1}\right)^2$$

in $\mathbb{F}_p$, $\gamma_1 \in \mathbb{F}_{p^2}$ and $\gamma_1 \neq 1$. In this case we have two ranks of apparition, r_1 and r_2, of U_n. Here $r_1 = p$ and r_2 is some divisor of $p \pm 1$. If $p \nmid \Delta$, then $\gamma_1 = \gamma_2$, $\gamma_3 = 1/\gamma_1^2$, $\gamma_1 \neq \pm 1$ and

$$U_n = R^{n-1}\gamma_1^{2-2n}\left(\frac{\gamma_1^n - 1}{\gamma_1 - 1}\right)^2\left(\frac{\gamma_1^{2n} - 1}{\gamma_1^2 - 1}\right)$$

in $\mathbb{F}_p$. Thus, in this case we see that U_n has a single rank of apparition, r_1, in $\{U_n\}$ and $r_1 \mid p \pm 1$.

We next deal with the case where $p \nmid 6Rd$. We say that p is an S-, Q-, or I-prime with respect to $g(x)$ if the splitting field of $g(x) \in \mathbb{F}_p[x]$ is, respectively, $\mathbb{F}_p$, $\mathbb{F}_{p^2}$, or $\mathbb{F}_{p^3}$ (more information on this can be found in [18, §4.6]).

Lemma 9.3. *If p is an S-prime and $p \nmid (S_1 + 2R)$, then the splitting field of $G(x)$ is $\mathbb{F}_{p^2}$ and in $\mathbb{F}_{p^2}$*

$$(\rho_i^2 - 4R^2)^{\frac{p-1}{2}} = (\gamma_i - \gamma_i^{-1})^{p-1} = \varepsilon \quad (i = 1, 2, 3),$$

where $\varepsilon = (\Delta/p)$.

Proof. Since p is an S-prime, the zeroes ρ_1, ρ_2, ρ_3 of $g(x) \in \mathbb{F}_p[x]$ are in $\mathbb{F}_p$. Since $\rho_i = R(\gamma_i + \gamma_i^{-1})$ $(i = 1, 2, 3)$, we get

$$(\gamma_i + \gamma_i^{-1})^p = \gamma_i + \gamma_i^{-1} \quad (i = 1, 2, 3).$$

Now $E^{\frac{p-1}{2}} = (R^2(S_1 + 2R)^2\Delta)^{\frac{p-1}{2}} = \varepsilon$, and therefore

$$((\rho_1^2 - 4R^2)(\rho_2^2 - 4R^2)(\rho_3^2 - 4R^2))^{\frac{p-1}{2}} = \varepsilon.$$

Also, since $\rho_i^2 - 4R^2 \in \mathbb{F}_p$, we have

$$(\rho_i^2 - 4R^2)^{\frac{p-1}{2}} = \pm 1 \quad (i = 1, 2, 3).$$

$\square$

Case 1. $\varepsilon = 1$. Suppose with no loss of generality that $(\rho_1^2 - 4R^2)^{\frac{p-1}{2}} = 1$. If

$$(\rho_2^2 - 4R^2)^{\frac{p-1}{2}} = (\rho_3^2 - 4R^2)^{\frac{p-1}{2}} = -1,$$

then

$$(\gamma_2 - \gamma_2^{-1})^{p-1} = (\gamma_3 - \gamma_3^{-1})^{p-1} = -1.$$

We have

$$\gamma_1^p - \gamma_1^{-p} = \gamma_1 - \gamma_1^{-1}$$
$$\gamma_2^p - \gamma_2^{-p} = \gamma_2^{-1} - \gamma_2$$
$$\gamma_3^p - \gamma_3^{-p} = \gamma_3^{-1} - \gamma_3$$

and $\gamma_i^p + \gamma_i^{-p} = \gamma_i + \gamma_i^{-1}$ ($i = 1,2,3$). It follows that $\gamma_1^p = \gamma_1$, $\gamma_2^p = \gamma_2^{-1}$, $\gamma_3^p = \gamma_3^{-1}$, and $(\gamma_2\gamma_3)^p = (\gamma_2\gamma_3)^{-1}$; hence, $\gamma_1^p = \gamma_1^{-1}$ or $\gamma_1^{p+1} = 1$. Since $\gamma_1^{p-1} = 1$, we see that $\gamma_1^2 = 1$ and $p \mid U_2$. Since $U_2 = S_1 + 2R$, this is impossible. Thus, if $\varepsilon = 1$ and $(\rho_1^2 - 4R^2)^{\frac{p-1}{2}} = 1$, then $(\rho_2^2 - 4R^2)^{\frac{p-1}{2}} = (\rho_3^2 - 4R^2)^{\frac{p-1}{2}} = 1$.

Case 2. $\varepsilon = -1$. By reasoning similar to that employed in Case 1, we get a contradiction if $(\rho_1^2 - 4R^2)^{\frac{p-1}{2}} = -1$ and $(\rho_2^2 - 4R^2)^{\frac{p-1}{2}} = (\rho_3^2 - 4R^2)^{\frac{p-1}{2}} = 1$. $\square$

We are now able to state the law of apparition for S-primes.

Theorem 9.4. *If p is an S-prime, then $p \mid U_{p-\varepsilon}$ where $\varepsilon = (\Delta/p)$.*

Proof. If $p \nmid S_1 + 2R$, we have $(\gamma_i - \gamma_i^{-1})^{p-1} = \varepsilon$ and since $\gamma_i + \gamma_i^{-1}$ is in $\mathbb{F}_p$, we also have $(\gamma_i + \gamma_i^{-1})^p = \gamma_i + \gamma_i^{-1}$ and $(\gamma_i + \gamma_i^{-1})^{p-1} = 1$; hence $\gamma_i^{p-\varepsilon} = 1$ ($i = 1,2,3$) and $p \mid U_{p-\varepsilon}$. If $p \mid S_1 + 2R$, then $p \mid U_2$ and $2 \mid p - \varepsilon$. $\square$

Corollary 9.5. *If r is any rank of apparition of an S-prime in $\{U_n\}$, then $r \mid p - \varepsilon$.*

Proof. If $p \mid U_r$, then $\gamma_j^r = 1$ for some $j \in \{1,2,3\}$. Also, we know that $\gamma_i^{p-\varepsilon} = 1$ ($i = 1,2,3$). We may assume by Theorem 9.2 that r is the least positive integer such that $\gamma_j^r = 1$. If $r \nmid p - \varepsilon$, then $\gamma_j^s = 1$ and $p \mid U_s$ with $s < r$, an impossibility; thus $r \mid p - \varepsilon$. $\square$

We now consider the case where p is a Q-prime.

Lemma 9.6. *Suppose p is a Q-prime and $p \nmid S_1 + 2R$. Let $s \in \mathbb{F}_p$ be such that $g(s) = 0$ in $\mathbb{F}_p$. If $\eta = (s^2 - 4R^2/p)$, we have $\eta = (\Delta/p)$.*

Proof. Let $\rho_1 = s$ and ρ_2, ρ_3 be the zeroes of $g(x)$ in $\mathbb{F}_{p^2}$, where $\rho_2, \rho_3 \in \mathbb{F}_{p^2} \backslash \mathbb{F}_p$. Then $\rho_2^p = \rho_3$ and $\rho_3^p = \rho_2$ in $\mathbb{F}_{p^2}$. Now

$$(\rho_2^2 - 4R^2)^p = \rho_3^2 - 4R^2$$

means that

$$(\gamma_2 - \gamma_2^{-1})^{2p} = (\gamma_3 - \gamma_3^{-1})^2$$

in the splitting field of $G(x)$ in $\mathbb{F}_p[x]$. Thus, we have

$$(\gamma_2 - \gamma_2^{-1})^p = \sigma_1(\gamma_3 - \gamma_3^{-1})$$

and

$$(\gamma_3 - \gamma_3^{-1})^p = \sigma_2(\gamma_2 - \gamma_2^{-1})$$

where $\sigma_1, \sigma_2 \in \{-1, 1\}$. Since

$$\gamma_3 + \gamma_3^{-1} = (\gamma_2 + \gamma_2^{-1})^p = \gamma_2^p + \gamma_2^{-p},$$

we get $\gamma_2^p = \gamma_3^{\sigma_1}$ and also $\gamma_3^p = \gamma_2^{\sigma_2}$, $\gamma_1^p = \gamma_1^\eta$. By using the techniques employed in the proof of Lemma 9.3, we find that it is impossible for $\sigma_1 \neq \sigma_2$. Thus $\sigma_1 = \sigma_2$ and $\eta = \sigma_1 = \sigma_2$. Also, $\varepsilon = \eta \sigma_1 \sigma_2 = \eta$. $\square$

We can now prove the law of apparition for Q-primes.

Theorem 9.7. *If p is a Q-prime, then p has a single rank of apparition, r, in $\{U_n\}$ and $r \mid p - \varepsilon$, where $\varepsilon = (\Delta/p)$.*

Proof. We have $\gamma_1^{p-\eta} = 1$; hence, $p \mid U_{p-\eta}$ and since $\varepsilon = \eta$ by Lemma 9.6, we have $p \mid U_{p-\varepsilon}$ when $p \nmid S_1 + 2R$. Now let r be the least positive integer such that $p \mid U_r$. Since $p \nmid d$, we have either $\gamma_1 = 1$, $\gamma_2 = 1$ or $\gamma_3 = 1$ in the splitting field of $G(x) \in \mathbb{F}[x]$. Now suppose $\gamma_1^m = 1$ or $\gamma_2^m = 1$ or $\gamma_3^m = 1$. If $\varepsilon = 1$, then $\gamma_2^p = \gamma_3$, $\gamma_3^p = \gamma_2$. If $\gamma_2^m = 1$, then $\gamma_2^{pm} = \gamma_3^m = 1$. Since $\gamma_1 \gamma_2 \gamma_3 = 1$, we see that $\gamma_1^m = 1$. Similarly, if $\gamma_3^m = 1$, then $\gamma_2^m = \gamma_1^m = 1$. By a similar argument we can show that if $\varepsilon = -1$, and $\gamma_2^m = 1$ or $\gamma_3^m = 1$, then $\gamma_1^m = \gamma_2^m = \gamma_3^m = 1$. It follows from the definition of r that if $p \mid U_m$, then $\gamma_1^m = 1$, $r \mid m$, and there can be only one rank of apparition of p in $\{U_n\}$. Since $p \mid U_{p-\varepsilon}$, we have $r \mid p - \varepsilon$ when $p \nmid S_1 + 2R$. If $p \mid S_1 + 2R$, then $p \mid U_2$ and it is easy to show that $p \nmid U_n$ whenever n is odd. $\square$

We now consider the possibility that p is an I-prime. In this case we have the following results:

Theorem 9.8. *If p is an I-prime and $\varepsilon = (\Delta/p)$, then $p \mid U_{p^2+\varepsilon p+1}$.*

Proof. In the splitting field of $G(x) \in \mathbb{F}_p[x]$, we have $\rho_2 = \rho_1^p$, $\rho_3 = \rho_2^p$, $\rho_1 = \rho_3^p$. Hence

$$(\gamma_1 + \gamma_1^{-1})^p = \gamma_2 + \gamma_2^{-1}, \quad (\gamma_2 + \gamma_2^{-1})^p = \gamma_3 + \gamma_3^{-1}, \quad (\gamma_3 + \gamma_3^{-1})^p = \gamma_1 + \gamma_1^{-1}.$$

Also

$$(\gamma_1 - \gamma_1^{-1})^p = \sigma_1(\gamma_2 - \gamma_2^{-1}), \quad (\gamma_2 - \gamma_2^{-1})^p = \sigma_2(\gamma_3 - \gamma_3^{-1}),$$
$$(\gamma_3 - \gamma_3^{-1})^p = \sigma_3(\gamma_1 - \gamma_1^{-1})$$

where $\sigma_1, \sigma_2, \sigma_3 \in \{-1, 1\}$. Hence $\gamma_1^p = \gamma_2^{\sigma_1}$, $\gamma_2^p = \gamma_3^{\sigma_2}$, $\gamma_3^p = \gamma_1^{\sigma_3}$, and $\varepsilon = \sigma_1 \sigma_2 \sigma_3$. We have $\gamma_1^{\sigma_3} \gamma_2^{\sigma_1} \gamma_3^{\sigma_2} = 1$; hence $\gamma_1^{\sigma_3 - \sigma_2} \gamma_2^{\sigma_1 - \sigma_2} = 1$. This holds if $\sigma_1 = \sigma_2 = \sigma_3$; otherwise, we get $\gamma_1^2 = 1$, $\gamma_2^2 = 1$ or $\gamma_3^2 = 1$. However, in these latter cases, we find that $p \mid U_2$ or $p \mid S_1 + 2R$. If $p \mid S_1 + 2R$, then $g(x) \equiv (x + 2R)(x^2 + S_2) \pmod{p}$, which means that $g(x)$ is not irreducible in $\mathbb{F}_p[x]$, a contradiction. Thus we must have $\sigma_1 = \sigma_2 = \sigma_3$. If $\varepsilon = 1$, then $\sigma_1 = \sigma_2 = \sigma_3 = 1$ and $\gamma_1^{p^2 + p + 1} = \gamma_2^p \gamma_2 \gamma_1 = \gamma_3 \gamma_2 \gamma_1 = 1$; If $\varepsilon = -1$, then $\sigma_1 = \sigma_2 = \sigma_3 = -1$ and $\gamma_1^{p^2 - p + 1} = \gamma_2^{-p} \gamma_2 \gamma_1 = \gamma_3 \gamma_2 \gamma_1 = 1$. Hence, $p \mid U_{p^2 + \varepsilon p + 1}$.   □

Theorem 9.9. *If p is an I-prime, then p has only one rank of apparition, r, in $\{U_n\}$ and $r \mid p^2 + \varepsilon p + 1$.*

Proof. Suppose $p \mid U_m$. Then either $\gamma_1^m = 1$, $\gamma_2^m = 1$, or $\gamma_3^m = 1$ in the splitting field of $G(x) \in \mathbb{F}_p[x]$. We may assume with no loss of generality that $\gamma_1^m = 1$. We have already seen that $\gamma_1^p = \gamma_2$ or $\gamma_1^p = 1/\gamma_2$. In the first case, we get $1 = \gamma_1^{mp} = \gamma_2^m$, and since $\gamma_1 \gamma_2 \gamma_3 = 1$, we get $\gamma_3^m = 1$. It is also true in the second case that $\gamma_1^m = \gamma_2^m = \gamma_3^m = 1$. If r is the least possible integer such that $p \mid U_r$, then by the same argument as that used in the proof of Theorem 9.7, we have $r \mid m$. Thus r is the rank of apparition of p in $\{U_n\}$ and $r \mid p^2 + \varepsilon p + 1$ by Theorem 9.8.   □

Corollary 9.10. *If p is an I-prime and $p \mid U_m$, then $p \mid W_m - 6R^m$.*

Thus, if p is an I-prime and $p \mid U_m$, then $p^3 \mid U_m$ by Theorem 8.6.

There are many further number-theoretic properties of $\{U_n\}$ and $\{V_n\}$ which we do not have the space to include here. Some idea of the range and variety of this theory can be seen by looking at [18]; most of the properties of the functions discussed in that work can be extended to the more general functions discussed in this section. Indeed, if we define D_n to be the gcd of $W_n - 6R^n$ and U_n (this should not be confused with the D_n defined in Sect. 7) and E_n to be the gcd of W_n and U_n, then it can be shown that these have many arithmetic properties in common with u_n and v_n, respectively. For example, it can be shown that $\{D_n\}$ is a divisibility sequence, that $E_n \mid D_{3n}$, that I-primes do not divide E_n, and that if p is any prime such that $p > 3$ and p divides E_n, then $p \equiv (\Gamma/3) \pmod 3$. In view of this it should not be surprising that we are able to produce results analogous to Lucas's Fundamental Theorem. For the sake of brevity we will not supply proofs of these results, but as mentioned above the methods employed in [18], especially Chapter 7, can be modified to produce these proofs.

We require a simple lemma.

Lemma 9.11. *If $m \geq 2$, $X = \prod_{i=1}^{m} r_i$ and $r_i \geq 5$ $(i = 1, 2, \ldots, m)$, then*

$$X^2 - X + 1 > 2 \prod_{i=1}^{m} \frac{r_i^2 + r_i + 1}{2}.$$

With this we can produce an analog to Theorem 1.2.

Theorem 9.12. *Let N be an integer such that $(N,6) = 1$. Let $T = N^2 + N + 1$ or $T = N^2 - N + 1$. If $N \mid D_T$, $N \nmid D_{T/q}$ for each distinct prime divisor, q, of T and $(D_{T/q'}, N) = 1$ for at least one prime divisor, q', of T, then N is a prime.*

We remark that for technical reasons it is necessary to introduce the prime q' here, something that was not necessary in Theorem 1.2. We can eliminate this in the following result, which is analogous to Theorem 1.3:

Theorem 9.13. *Let N be an integer such that $(N,6) = 1$ and let $T = N^2 + N + 1$ or $N^2 - N + 1$. If $W_T \equiv 6R^T$ (mod N), $N \mid U_T$ and $N \mid U_T/U_{T/q}$ for each distinct prime divisor, q, of T, then N is a prime.*

If $3 \mid N^2 \pm N + 1$, we can produce another analog of Theorem 1.2 where q' is not needed. (Actually $q' = 3$ in this result.) We first require an elementary lemma.

Lemma 9.14. *If $(S_1, S_2, R) = 1$, then $(W_n, \Delta, R) \mid 4$.*

Proof. Suppose p is some prime such that $p \mid W_n$, $p \mid \Delta$ and $p \mid R$. Since $p \mid \Delta$ and $p \mid R$, we have $p \mid S_1^2 - 4S_2$. Also,

$$W_n \equiv v_n(S_1, S_2) \quad (\text{mod } p)$$

and from the theory of Lucas functions,

$$v_n^2(S_1, S_2) - (S_1^2 - 4S_2)u_n^2(S_1, S_2) = 4S_2^n.$$

Thus, if $p \mid W_n$, we have $p \mid 4S_2$. If p is odd, then $p \mid S_2$, $p \mid S_1$ and $p \mid R$, a contradiction. If $8 \mid (W_n, \Delta, R)$ and then since $W_n \equiv v_n(S_1, S_2)$ (mod 8) and $8 \mid S_1^2 - 4S_2$, we get $2 \mid S_2$ and $2 \mid S_1$, which is also a contradiction. $\qquad\square$

By means of Lemma 9.14 and the triplication formulas, we can prove the next lemma.

Lemma 9.15. *Suppose that N is an integer and $(N,6) = 1$. If $\Delta U_n^2 \equiv -27R^{2n}$ and $W_n \equiv -3R^n$ (mod N), then $N \mid D_{3n}$ and $(N, \Delta R U_n) = 1$.*

With this information we can now prove the following version of Theorem 9.12.

Theorem 9.16. *Suppose N is an integer and $(N,6) = 1$. Put $t = (N^2 \pm N + 1)/3 \in \mathbb{Z}$. If $W_t \equiv -3R^t$, $\Delta U_t^2 \equiv -27R^{2t}$ (mod N) and N does not divide $U_{3t/q}$ for each distinct prime q which divides t, then N is prime.*

We conclude this section by mentioning that many other results like those of Chapter 7 of [18] can also be developed, but those given here should suffice to indicate the extent of this theory.

10 Conclusion

We have seen that Lucas was correct about the existence of fourth-order analogs of his functions. He was wrong about the existence of third-order analogs, but we have shown that there do exist sixth-order analogs, and the zeroes of a cubic polynomial $g(x)$ play an important role in this theory. Certainly, a subcase (see [13, 18]) of the sixth-order functions does involve a symmetric function of the zeroes of a cubic polynomial and is a particular instance of a function explicitly mentioned by Lucas as being a possible extension of his u_n function.

There may exist other functions for $k \geq 3$, but it is likely that more than two functions will be necessary in order to advance the theory. We have only just begun the process of examining the properties of U_n and V_n when $k \geq 3$. It would be most interesting to discover what further results might be achieved when $k = 4$ or $k > 4$.

Acknowledgements H.C. Williams: Research supported in part by NSERC of Canada.

References

1. J.P. Bézivin, A. Pethö, A.J. van der Poorten, A full characterization of divisibility sequences. Am. J. Math. **112**, 985–1001 (1990)
2. E.T. Bell, Analogies between the u_n, v_n of Lucas and elliptic functions. Bull. Am. Math. Soc. **29**, 401–406 (1923)
3. M. Hall, Slowly increasing arithmetic series. J. Lond. Math. Soc. **8**, 162–166 (1933)
4. M. Hall, Divisibility sequences of the third order. Am. J. Math. **58**, 577–584 (1936)
5. D.H. Lehmer, Tests for primality by the converse of Fermat's theorem. Bull. Am. Math. Soc. **33**, 327–340 (1927)
6. É. Lucas, Note sur l'application des séries récurrentes à la recherche de la loi de distribution des nombres premiers. Comp. Rend. Acad. Sci. Paris **82**, 165–167 (1876)
7. É. Lucas, Recherches sur plusiers ouvrages de Leonard de Pise et sur diverse questions d'arithmétique supérieure. Boll. Bibliogr. Storia Sci. Matemat. Fisiche **10**, 129–193, 239–293 (1877)
8. É. Lucas, Sur la théorie des fonctions numériques simplement périodiques. Nouv. Corresp. Math. 3 369–376, 401–407 (1877); **4**, 1–8, 33–40, 65–71, 97–102, 129–134, 225–228 (1878)
9. É. Lucas, Théorie des fonctions numériques simplement périodiques. Am. J. Math. **1**, 184–240, 289–321 (1878)
10. É. Lucas, *Notice sur les Titres et Travaux Scientifiques de M. Édouard Lucas* (D. Jouaust, Paris, 1880)
11. É. Lucas, *Théorie des Nombres* (Gauthier-Villars, Paris, 1891)
12. É. Lucas, Questions proposées à la discussion des 1re et 2e sections, 1° Questions d'arithmétique supérieure. Assoc. Française l'Avancement Sci. Comp. Rend. Sessions **20**, 149–151 (1891)
13. S. Müller, E. Roettger & H.C. Williams, A cubic extension of the Lucas functions. Ann. Sci. Math. Québec **33**, 185–224 (2009)
14. A.D. Oosterhout, Characterization of divisibility sequences. Master's Thesis, Utrecht University, June 2011
15. T.A. Pierce, The numerical factors of the arithmetic forms $\prod_{i=1}^{n}(1 \pm \alpha_i^m)$. Ann. Math. (2nd Ser.) **18**, 53–64 (1916)

16. A. Pethö, Complete solutions to a family of quartic diophantine equations. Math. Comp. **57**, 777–798 (1991)
17. A. Pethö, Egy negyedrendü rekurzív sorozatcsaládról. Acta Acad. Paed. Agriensis Sect. Math. **30**, 115–122 (2003). MR2054721 (2005g:11018)
18. E. Roettger, A cubic extension of the Lucas functions. PhD Thesis, Univ. of Calgary, 2009. http://math.ucalgary.ca/~hwilliam/files/A Cubic Extension of the Lucas Functions.pdf
19. N. Sloane, Online encyclopedia of integer sequences. http://oeis.org/wiki/
20. M. Ward, The law of apparition of primes in a Lucasian sequence. Trans. Am. Math. Soc. **44**, 68–86 (1938)
21. M. Ward, The laws of apparition and repetition of primes in a cubic recurrence. Trans. Am. Math. Soc. **79**, 72–90 (1955)
22. H.C. Williams, *Édouard Lucas and Primality Testing*. Canadian Mathematical Society Series of Monographs and Advanced Texts, vol 22 (Wiley, New York, 1998)
23. H.C. Williams, R.K. Guy, Some fourth order linear divisibility sequences. Int. J. Number Theory **7**(5), 1255–1277 (2011)

The Impact of Number Theory and Computer-Aided Mathematics on Solving the Hadamard Matrix Conjecture

Jennifer Seberry

In memory of Alf van der Poorten

Abstract The Hadamard conjecture has been studied since the pioneering paper of Sylvester, "Thoughts on inverse orthogonal matrices, simultaneous sign successions, tessellated pavements in two or more colours, with applications to Newtons rule, ornamental tile work and the theory of numbers," *Phil Mag*, 34 (1867), 461–475, first appeared.

We review the importance of primes on those occasions that the conjecture, that for every odd number, t there exists an Hadamard matrix of order $4t$ – is confirmed. Although substantial advances have been made into the question of the density of this odd number, t it has still not been shown to have positive density.

We survey the results of some computer-aided construction algorithms for Hadamard matrices.

Key words Hadamard matrix • Skew-Hadamard matrix • Symmetric Hadamard matrix • Asymptotic Hadamard matrices • Prime number theorem • Extended Riemann hypothesis • Computer-aided construction

AMS Subject Classification: 20B20

J. Seberry (✉)
Centre for Computer and Information Security Research,
SCSSE, University of Wollongong, Wollongong, NSW 2522, Australia
e-mail: j.seberry@uow.edu.au

J.M. Borwein et al. (eds.), *Number Theory and Related Fields: In Memory of Alf van der Poorten*, Springer Proceedings in Mathematics & Statistics 43, DOI 10.1007/978-1-4614-6642-0_16, © Springer Science+Business Media New York 2013

1   Introduction

The Hadamard conjecture, attributed to Paley [24] in 1933, is that "an Hadamard matrix exists for every order 1, 2 and $4t$, t a non-negative integer." This question has attracted research for over 110 years. In the early 1970s, orthogonal designs were introduced to further their study. We do not cover orthogonal designs but refer the reader to Geramita and Seberry [16]. Hadamard matrices themselves have many applications: They give codes to correct the maximum number of errors in electromagnetic signals, they allow for the best possible weights to be obtained for tiny objects (e.g., jewels, chemicals, drugs), they allow the most exact spectral analysis of the content of solutions, and they give the binary functions with highest non-linearity for cryptographic primitives.

Amicable orthogonal designs derived from Hadamard matrices have led to the best CDMA (code division multiple access) codes for mobile communications.

Number theory has been extensively used by Yamada and Yamamoto [37, 38] of Japan, Ming Yuan Xia [32, 34] and his team in China, Whiteman [31] and Warwick de Launey [9]. The constructions use Legendre characters, Gaussian sums and relative Gauss sums, Jacobi sums, the theory of cyclotomy, quadratic forms, to name a few.

However, Hadamard matrices also come with a very curious property, while most of the constructions that give infinite families of orders which satisfy the Hadamard conjecture are based on primes, other constructions seem to hit a solid wall concerning upper sizes when computer-aided searches are undertaken to look for them.

We give a small taste of the problems with computer-assisted research.

2   Definitions and Basics

An Hadamard matrix, H, of order h is a square matrix with entries ± 1, in which any pair of distinct rows (or columns) are orthogonal, that is, $HH^T = hI_h$, where I is the identity matrix.

We first consider some results about non-prime orders, then prime orders, and then computer-aided construction.

We believe that orders of Hadamard matrices $4p$, $p \equiv 3 \pmod{4}$ are the toughest to find.

3 Why Are the Primes So Important for the Hadamard Conjecture?

3.1 Multiplying Hadamard Matrices

Hadamard matrices can be multiplied together to give new Hadamard matrices using the Kronecker product. However this increases the power of two. Summarizing

Matrix sizes	Multiplication size	Reference
$4a$, $4b$	$2^4 ab$	Classical result due to Sylvester [27]
$4a$, $4b$	$2^3 ab$	Agayan [1]
$4a$, $4b$, $4c$, $4d$	$2^5 abcd$	Agayan [1]
$4a$, $4b$, $4c$, $4d$	$2^4 abcd$	Craigen et al. [4]
$4a_1, \cdots, 4a_{12}$	$2^{10} \prod_{i=1}^{12} a_i$	Previous methods
$4a_1, \cdots, 4a_{12}$	$2^9 \prod_{i=1}^{12} a_i$	de Launey [6]

3.2 Asymptotic Existence of Hadamard Matrices

The Asymptotic Form of the Hadamard Conjecture

In 1975 Seberry [25] and in 1995 Craigen [2] had found some asymptotic type existence theorems for Hadamard matrices but were unable to establish that their results had positive density in the set of natural numbers $4t$.

Let $S(x)$ be the number $n \leq x$ for which an Hadamard matrix of order n exists. Then the Hadamard conjecture states that $S(x)$ is about $\frac{x}{4}$.

Then in a groundbreaking paper, de Launey and Gordon [10] increased the bound for the density from Paley's constructions $(O(\pi(x))$ by a factor of $\exp((\log\log\log x)^2)$, but this bound is still $o(x)$. The question of whether $S(x)$ has positive density is still open. Their lovely theorem is

Theorem 3.1 (de Launey and Gordon). *For all $\varepsilon > 0$, there is a natural number x_ε such that, for all $x \geq x_\varepsilon$,*

$$S(x) \geq \frac{x}{\log x} \exp((C+\varepsilon)(\log\log\log x)^2)$$

for $C = 0.8178\ldots$.

de Launey and Gordon's original insights [7, 9] were derived under the extended Riemann hypothesis, using the multiplication theorem of the type in Sect. 3.1 and the Paley theorems of Sect. 4. Later results by a number of authors including Kevin

Ford, David Levin, Iwaniec, Sidney Graham, and Igor Shparlinski [18] meant that the extended Riemann hypothesis was not required for the proof.

Looking at these results in a slightly different way, de Launey and Gordon [11] say: "Let a be any positive real number. Then for some absolute positive constant c_2 there is a Hadamard matrix of order 2^{tr} whenever $2^t \geq c_2 r^a$".

In 1975, Seberry [25] had proved this fact for the first time for $c_2 = 1$ and $a = 2$. In 1995 Craigen [2] improved Seberry's result to $c_2 = 2^{\frac{13}{8}}$ and $a = \frac{3}{8}$. de Launey and Gordon, now had the $a < 1$ they needed to remove the assumption of the extended Riemann hypothesis for their result in [9]. Further results appear in [13] and [22].

These were summarized by Gordon [11]. "For g odd, the fraction of orders $2^t g$ with orders $2^t > c_2 g^a$ is at least

Author	a	c_2	Fraction
Seberry	2	1	1
Craigen	$\frac{3}{8}$	$2^{13/8}$	1
Livinski	$\frac{1}{5}$		1
de Launey and Gordon	$\varepsilon > 0$	1	$c(\varepsilon) > 0$

"

Craigen, de Launey, Holzmann, Kharaghani, and Smith have also been involved in other wonderful asymptotic results [3, 5, 8, 12, 14].

3.3 Multiplication and Asymptotics

Since we can always multiply orders of Hadamard matrices together even though we get higher powers of two and the asymptotic results mean that we can always get an Hadamard matrix of order $2^t n$ for any odd n provided, we have all the factors of n.

This means we need to concentrate on looking for those orders $4p$ where p is a prime, even though the primes are not dense in the positive integers. Thus we first confine ourselves to the primes.

4 Constructions Using Primes

The earliest construction were based on primes or ad hoc constructions. We discuss computer-aided construction of ad hoc orders in the next section.

Yamada [36] has given a taste of the importance of primes to resolving the Hadamard conjecture:

Author	Year	Constructions
Sylvester	1867	Orders 2^t
Hadamard	1893	Orders 12, 20
Paley	1933	Orders $p^r + 1$, $p \equiv 3 \pmod 4$ a prime
		Orders $2(q^s + 1)$, $q \equiv 1 \pmod 4$ a prime

Theorem 4.2 (Yamada 1). *If $q \equiv 1 \pmod 8$ is a prime power and there exists an Hadamard matrix of order $(q-1)/2$, then we can construct an Hadamard matrix of order $4q$.*

Theorem 4.3 (Yamada 2). *If $q \equiv 5 \pmod 8$ is a prime power and there exists a skew-Hadamard matrix of order $(q+3)/2$, then we can construct an Hadamard matrix of order $4(q+2)$.*

Theorem 4.4 (Yamada 3). *If $q \equiv 1 \pmod 8$ is a prime power and there exists a conference matrix of order $(q+3)/2$, then we can construct an Hadamard matrix of order $4(q+2)$.*

Zuo and Xia introduced a special class of T-matrices in Zuo and Xia [39] which relies on primes. Xia and Colleagues also showed that

Theorem 4.5 (Xia et al. [33]). *There exists an infinite family of T-matrices of order q^2 with q prime power $q \equiv 3 \pmod 8$.*

5 By Hand and Computer-Aided Construction of Hadamard Matrices

5.1 Williamson Matrices (or Zero Periodic Autocorrelation Function)

In 1944, Williamson [30] gave a beautiful method to construct most orders ≤ 200 by hand. The only infinite families which suits this construction method were found by Turyn [28] and Yamada [35], but they give no new Hadamard matrix orders beyond those given by Paley.

However this led many of us to conjecture that "Williamson's method can be used to construct all Hadamard matrices of order $4n$ for n odd."

Alas, this is not true. In 1993, Djokovic [15] showed, by complete search, that there were matrices of this kind for order $4n$, $n = 33, 39$ and none of order 35. Thirty-five being a composite number, some thought that it may be true that such matrices exist of order $4n$, n a prime number. A further search appears in [23]. Then in 2008, Holzmann et al. [20] dashed hopes by using an exhaustive computer search to show none of this kind exist for $n = 47, 53$, or 59.

Some researchers now conjecture (verbally) that "Williamson's method can be used to construct Hadamard matrices of order $4n$ for n odd $n \leq 45$, $n \neq 35$, and orders $4n$, for some primes $n \equiv 1 \pmod 4$ and no other primes". Multiples of known orders have not been settled.

5.2 Constructions Using Other Sequences with Zero Autocorrelations Function

We refer our patient reader to Geramita and Seberry [16] and Seberry and Yamada [26] for any terms we do not now explain and for examples of the types of constructions we now describe.

Williamson's method inspired many authors to look for 1, 2, 3, 4, 6, and 8 matrices (or sequences which are their first rows) with elements $\{0, \pm 1\}$ and zero autocorrelation functions that might be used to find Hadamard matrices. The limit of 8 is related to the algebra of the quaternions. The methods found have been beautiful, clever, and very inventive. They include Golay sequences, T-matrices, base sequences, Yang sequences, Turyn sequences, good matrices, and Turyn 6-sequences. Often they give many new Hadamard matrices, but in the cases where computer searches have been undertaken, we have usually found that after a certain length n (the length seems to depend on the method) no more of that kind exist, see for example [17]. Many construction results have the proviso that no prime p, $p \equiv 3 \pmod 4$ divides the length n.

6 Equivalence of Hadamard Matrices by Computer

Two Hadamard matrices are said to be equivalent to each other if one can be obtained from the other by a series of operations of the type (a) interchange any pair of rows (or columns) and/or of the type (b) multiply any row or column by -1, see [19, 21, 29]. The results in the following table were found by hand upto order 20 and with computer assistance for higher orders:

Order	Number inequivalent	Authors
4, 8, 12	Unique	
16	5	Marshall Hall Jr
20	3	Marshall Hall Jr
24	60	Ito
28	487	Kimura
32	13,710,027	Many[a]
36	>30,000,000	Many[a]

[a] Among the authors are Holtzman, Kharaghani, Tayfeh-Rezaie, Koukouvinos, and Orrick.

7 The Future

Thanks to de Launey we now know more regarding the known orders for Hadamard matrices in the natural number $4t$: but this set has not yet been shown to have positive density. However we have merely scratched the surface of how important number theory is to solving the Hadamard conjecture which remains unresolved. A search (unpublished) for existence orders up to 40,000 showed that some kind of construction, including the multiplicative constructions, gave some result for 80–90 % of the odd numbers, but in the cases where the asymptotic results had to be invoked, it was usually a prime $p > 3$, $p \equiv 3 \pmod 4$ that was involved.

Acknowledgements The author sincerely thanks Professor Igor Shparlinski and Dr Daniel Gordon for alerting her to her confusion in Sect. 3.2 and pointing out pertinent references.

References

1. S.S. Agayan, *Hadamard Matrices and Their Applications*. Lecture Notes in Mathematics, vol 1168 (Springer, New York, 1985)
2. R. Craigen, Signed groups, sequences and the asymptotic existence of Hadamard matrices. J. Combin. Theory **71**, 241–254 (1995)
3. R. Craigen, W.H. Holzmann, H. Kharaghani, On the asymptotic existence of complex Hadamard matrices. J. Combin. Des. **5**, 319–327 (1997)
4. R. Craigen, J. Seberry, X.-M. Zhang, Product of four Hadamard matrices. J. Combin. Theory (Ser. A) **59**, 318–320 (1992)
5. W. de Launey, On the asymptotic existence of partial complex Hadamard matrices and related combinatorial objects, in *Conference on Coding, Cryptography and Computer Security*, Lethbridge, AB, 1998. Discrete Appl. Math. **102**, 37–45 (2000)
6. W. de Launey, A product for twelve Hadamard matrices. Aust. J. Combin. **7**, 123–127 (1993)
7. W. de Launey, On the asymptotic existence of Hadamard matrices. J. Combin. Theory (Ser. A) **116**, 1002–1008 (2009)
8. W. de Launey, J. Dawson, An asymptotic result on the existence of generalised Hadamard matrices. J. Combin. Theory (Ser. A) **65**, 158–163 (1994)
9. W. de Launey, D.M. Gordon, A comment on the Hadamard conjecture. J. Combin. Theory (Ser. A) **95**, 180–184 (2001)
10. W. de Launey, D.M. Gordon, On the density of the set of known Hadamard orders. Cryptogr. Commun. **2**, 233–246 (2010)
11. W. de Launey, D.M. Gordon, Should we believe the Hadamard conjecture?, in *Conference in Honor of Warwick de Launey, IDA/CCR*, La Jolla, CA, 16 May 2011
12. W. de Launey, H. Kharaghani, On the asymptotic existence of cocyclic Hadamard matrices. J. Combin. Theory (Ser. A) **116**, 1140–1153 (2009)
13. W. de Launey, D.A Levin, A Fourier analytic approach to counting partial Hadamard matrices (21 March 2010). arXiv:1003.4003v1 [math.CO]
14. W. de Launey, M. Smith, Cocyclic orthogonal designs and the asymptotic existence of cocyclic Hadamard matrices and maximal size relative difference sets with forbidden subgroup of size 2. J. Combin. Theory (Ser. A) **93**, 37–92 (2001)
15. D.Z Djokovic, Williamson matrices of order $4n$ of order $n = 33, 35, 39$. Discrete Math. **115**, 267–271 (1993)

16. A.V. Geramita, J. Seberry, *Orthogonal Designs: Quadratic Forms and Hadamard Matrices* (Marcel Dekker, Boston, 1969)
17. G.M'. Edmonson, J. Seberry, M. Anderson, On the existence of Turyn sequences of length less than 43. Math. Comput. **62**, 351–362 (1994)
18. S.W. Graham, I.E. Shparlinski, On RSA moduli with almost half of the bits prescribed. Discrete Appl. Math. **156**, 3150–3154 (2008)
19. M. Hall Jr., *Combinatorial Theory*, 2nd edn. (Wiley, New York, 1986)
20. W.H. Holzmann, H. Kharaghani, B. Tayfeh-Rezaie, Williamson matrices up to order 59. Designs Codes Cryptogr. **46**, 343–352 (2008)
21. K.J. Horadam, *Hadamard Matrices and Their Applications* (Princeton University Press, Princeton, 2007)
22. I. Livinski, Asymptotic existence of hadamard matrices. Master of Science Thesis, University of Manitoba, 2012
23. J. Horton, C. Koukouvinos, J. Seberry, A search for Hadamard matrices constructed from Williamson matrices. Bull. Inst. Combin. Appl. **35**, 75–88 (2002)
24. R.E.A.C. Paley, On orthogonal matrices. J. Math. Phys. **12**, 311–320 (1933)
25. J. Seberry Wallis, On Hadamard matrices. J. Combin. Theory (Ser. A) **18**, 149–164 (1975)
26. J. Seberry, M. Yamada, Hadamard matrices, sequences, and block designs, in *Contemporary Design Theory: A Collection of Surveys*, ed. by J.H. Dinitz, D.R. Stinson (Wiley, New York, 1992), pp. 431–560
27. J.J. Sylvester, Thoughts on inverse orthogonal matrices, simultaneous sign successions, tessellated pavements in two or more colours, with applications to Newton's rule, ornamental tile work and the theory of numbers. Phil. Mag. **34**, 461–475 (1867)
28. R. Turyn, An infinite class of Hadamard matrices. J. Combin. Theory (Ser. A) **12**, 319–321 (1972)
29. J. Seberry Wallis, Hadamard matrices, in *Combinatorics: Room Squares, Sum-Free Sets and Hadamard Matrices*. Lecture Notes in Mathematics, ed. by W.D. Wallis, A. Penfold Street, J. Seberry Wallis (Springer, Berlin, 1972)
30. J. Williamson, Hadamard's determinant theorem and the sum of four squares. Duke Math. J. **11**, 65–81 (1944)
31. A.L. Whiteman, An infinite family of Hadamard matrices of Williamson type. J. Combin. Theory (Ser. A) **14**, 334–340 (1972)
32. M.Y. Xia, Some infinite families of Williamson matrices and difference sets. J. Combin. Theory (Ser. A) **61**, 230–242 (1992)
33. M.Y. Xia, T.B. Xia, J. Seberry, G.X. Zuo, A new method for constructing T-matrices. Aust. J. Combin. **32**, 61–78 (2005)
34. M.Y. Xia, T. Xia, J. Seberry, A new method for constructing Williamson matrices. Designs Codes Cryptogr. **35**, 191–209 (2005)
35. M. Yamada, On the Williamson type j matrices of orders $4 \cdot 29$, $4 \cdot 41$ and $4 \cdot 37$. J. Combin. Theory (Ser. A) **27**, 378–381 (1979)
36. M. Yamada, Some new series of Hadamard matrices. Proc. Japan Acad. (Ser. A) **63**, 86–89 (1987)
37. K. Yamamoto, M. Yamada, Williamson Hadamard matrices and Gauss sums. J. Math. Soc. Japan **37**, 703–717 (1985)
38. K. Yamamoto, On congruences arising from relative Gauss sums. *Number Theory and Combinatorics* Japan 1984 (Tokyo, Okayama and Kyoto, 1984). World Scientific Publ., Singapore, pp. 423–446 (1985)
39. G.X. Zuo, M.Y. Xia, A special class of T-matrices. Designs Codes Cryptogr. **54**, 21–28 (2010)

Description of Generalized Continued Fractions by Finite Automata

Jeffrey Shallit

In memory of Alf van der Poorten, the sorcerer of continued fractions

Abstract A generalized continued fraction algorithm associates every real number x with a sequence of integers; x is rational iff the sequence is finite. For a fixed algorithm A, call a sequence of integers *valid* if it is the result of A on some input x_0. We show that, if the algorithm is sufficiently well behaved, then the set of all valid sequences is accepted by a finite automaton.

1 Introduction

Simple continued fractions are finite expressions of the form

$$a_0 + \cfrac{1}{a_1 + \cfrac{1}{a_2 + \cdots + \cfrac{1}{a_n}}},$$

usually abbreviated $[a_0, a_1, \ldots, a_n]$, or infinite expressions of the form

$$a_0 + \cfrac{1}{a_1 + \cfrac{1}{a_2 + \cdots}},$$

J. Shallit (✉)
School of Computer Science, University of Waterloo, Waterloo, ON, Canada N2L 3G1
e-mail: shallit@cs.uwaterloo.ca

J.M. Borwein et al. (eds.), *Number Theory and Related Fields: In Memory of Alf van der Poorten*, Springer Proceedings in Mathematics & Statistics 43, DOI 10.1007/978-1-4614-6642-0_17, © Springer Science+Business Media New York 2013

usually abbreviated $[a_0, a_1, \ldots]$. The latter expression is defined as the limit as $n \to \infty$, if it exists, of the corresponding finite expression ending in a_n. The a_i are called the *partial quotients* of the continued fraction.

The standard bracket notation above for continued fractions conflicts with the standard notation $[x, y]$ for closed intervals of the real line. We abuse notation by using both and trust that the reader can disambiguate if necessary.

The *simple continued fraction algorithm*, on the other hand, is the following algorithm that, given a real number x, produces a finite sequence of partial quotients $a_0, a_1, \ldots, a_n$ or infinite sequence of partial quotients $a_0, a_1, \ldots$, such that $x = [a_0, a_1, \ldots, a_n]$ or $x = [a_0, a_1, \ldots]$.

Algorithm SCF(x); outputs $(a_0, a_1, \ldots)$:
SCF1. Set $x_0 \leftarrow x$; set $i \leftarrow 0$.
SCF2. Set $a_i \leftarrow \lfloor x_i \rfloor$.
SCF3. If $a_i = x_i$ then stop. Otherwise set $x_{i+1} \leftarrow 1/(x_i - a_i)$; set $i \leftarrow i+1$ and
      go to step SCF2.

For example,

$$\mathrm{SCF}(\frac{52}{43}) = (1, 4, 1, 3, 2)$$

$$\mathrm{SCF}(\pi) = (3, 7, 15, 1, 292, \ldots)$$

$$\mathrm{SCF}(\sqrt{2}) = (1, 2, 2, 2, \ldots)$$

$$\mathrm{SCF}(e) = (2, 1, 2, 1, 1, 4, 1, 1, 6, 1, 1, 8, \ldots)$$

In the literature, these two different concepts

- The *function* mapping $(a_0, \ldots, a_n)$ to the rational number $[a_0, \ldots, a_n]$, or $(a_0, a_1, \ldots)$ to the real number $[a_0, a_1, \ldots,]$; and
- The *algorithm* taking x as input and producing the a_i

have, unfortunately, often been confused. This is probably due to two reasons: the fact that the birth of continued fractions [4] long predates the appreciation of algorithms as mathematical objects [10] and the fortunate happenstance that by imposing some simple rules on the partial quotients, we can ensure that that there is exactly one valid expansion for each real number.

The concept of "rules" that describe the set of possible outputs of a continued fraction expansion has appeared before in many places. For example, Hurwitz [7] used them to describe a nearest integer continued fraction algorithm in $\mathbb{Z}[i]$. It is our goal to formalize the concept and explain it in terms of automata theory.

As has long been known, if x is a real irrational number, then the set of outputs produced by the SCF algorithm is exactly

$$\{(a_0, a_1, \ldots) : \forall i \; a_i \in \mathbb{Z} \text{ and } a_i \geq 1 \text{ for } i \geq 1 \}.$$

On the other hand, if x is a rational number, then the set of outputs produced by the SCF algorithm is exactly

$$\{(a_0, a_1, \ldots, a_n) : \forall i \; a_i \in \mathbb{Z} \text{ and } a_i \geq 1 \text{ for } 1 \leq i \leq n \text{ and } a_n \geq 2 \text{ if } n \geq 1 \}.$$

Hence, if we insist that in any expression $[a_0, a_1, \ldots]$, we must have

- $\forall i \; a_i \in \mathbb{Z}$;
- $a_i \geq 1$ for $i \geq 1$;
- If the expansion terminates with a_n, then $a_n \geq 2$;

then the ambiguity between the function and the algorithm disappears. The question remains about how we could *discover* rules like this. This becomes important because there exist many other versions of the continued fraction algorithm, and we would like to have a similar characterization of the outputs.

For example, the *ceiling algorithm* (CCF) replaces the use of the floor function with the ceiling; that is, it replaces step SCF2 with

SCF2′. Set $a_i \leftarrow \lceil x_i \rceil$.

For example,

$$\mathrm{CCF}\left(\frac{52}{43}\right) = [2, -1, -3, -1, -3, -2]$$

$$\mathrm{CCF}(\pi) = [4, -1, -6, -15, -1, -292, \ldots]$$

$$\mathrm{CCF}(\sqrt{2}) = [2, -1, -1, -2, -2 - 2, -2, -2, \ldots]$$

$$\mathrm{CCF}(e) = [3, -3, -1, -1, -4, -1, -1, -6, -1, -1, -8, \ldots]$$

The expansions produced by CCF include negative partial quotients, and obey the following rules:

- $\forall i \; a_i \in \mathbb{Z}$.
- $a_i \leq -1$ for $i \geq 1$.
- If the expansion ends with a_n and $n \geq 1$, then $a_n \neq -1$.

Indeed, it is easy to see that if

$$\mathrm{SCF}(-x) = [a_0, a_1, a_2, \ldots],$$

then

$$\mathrm{CCF}(x) = [-a_0, -a_1, -a_2, \ldots].$$

Yet another expansion is the so-called *nearest integer continued fraction* (NICF). It is generated by an algorithm similar to SCF above, except that step SCF2 is replaced by

SCF2″. Set $a_i \leftarrow \lfloor x_i + \frac{1}{2} \rfloor$.

For example,

$$\text{NICF}(\frac{52}{43}) = (1, 5, -4, -2)$$

$$\text{NICF}(\pi) = (3, 7, 16, -294, \ldots)$$

$$\text{NICF}(\sqrt{2}) = (1, 2, 2, 2, 2, \ldots)$$

$$\text{NICF}(e) = (3, -4, 2, 5, -2, -7, 2, 9, -2, -11, \ldots)$$

The partial quotients generated by NICF satisfy the following rules:

- $\forall i \; a_i \in \mathbb{Z}$;
- $a_i \leq -2$ or $a_i \geq 2$ for $i \geq 1$;
- If $a_i = -2$ then $a_{i+1} \leq -2$;
- If $a_i = 2$ then $a_{i+1} \geq 2$; and
- If the expansion terminates with a_n, then $a_n \neq 2$.

(Actually, the NICF is usually described slightly differently in the literature, but our formulation is essentially the same. See [8].)

In this paper, we are concerned with the following questions:

1. Which functions f are suitable replacements for the floor function in algorithm SCF (i.e., yield generalized continued fraction algorithms)?
2. Which of these functions correspond to generalized continued fraction algorithms that have "easily describable" outputs (i.e., accepted by a finite automaton)?

In this paper, we will answer question (1) by *fiat*, and then examine the consequences for question (2). Before we do, however, we mention a useful connection with another famous type of continued fraction.

2 Semiregular Continued Fractions

There is a close relationship between the continued fractions we study here and what is called *semiregular continued fractions* in the literature. A semiregular continued fraction is a finite or infinite expression of the form

$$b_0 + \cfrac{\varepsilon_1}{b_1 + \cfrac{\varepsilon_2}{b_2 + \cfrac{\varepsilon_3}{b_3 + \cdots}}} \tag{1}$$

where

- $b_i \in \mathbb{Z}$;
- $b_i \geq 1$ for $i \geq 1$;
- $\varepsilon_i = \pm 1$ for $i \geq 1$;
- $b_i + \varepsilon_{i+1} \geq 1$ for $i \geq 1$.

It is easily seen that (1) is equivalent to

$$a_0 + \cfrac{1}{a_1 + \cfrac{1}{a_2 + \cfrac{1}{a_3 + \cdots}}}, \tag{2}$$

where $a_i = \varepsilon_1 \varepsilon_2 \cdots \varepsilon_i b_i$ for $i \geq 0$. This expresses a semiregular continued fraction in the form that we study.

This connection extends to the convergents. Setting, as usual, $p_{-1} = 1$, $q_{-1} = 0$, $p_0 = a_0$, and $q_0 = 1$, and $p_n = a_n p_{n-1} + p_{n-2}$, $q_n = a_n q_{n-1} + q_{n-2}$, the theory of continued fractions (or an easy induction) gives $p_n/q_n = [a_0, \ldots, a_n]$. The convergents to a semiregular continued fraction are defined similarly: $p'_{-1} = 1$, $q'_{-1} = 0$, $p'_0 = b_0$, $q'_0 = 1$, $p'_n = b_n p'_{n-1} + \varepsilon_n p'_{n-2}$, and $q'_n = b_n q'_{n-1} + \varepsilon_n q'_{n-2}$. If the a_i and b_i are related as above, an easy induction gives

$$p'_{2n} = \varepsilon_2 \varepsilon_4 \cdots \varepsilon_{2n} p_{2n}$$

$$q'_{2n} = \varepsilon_2 \varepsilon_4 \cdots \varepsilon_{2n} q_{2n}$$

and

$$p'_{2n-1} = \varepsilon_1 \varepsilon_3 \cdots \varepsilon_{2n-1} p_{2n-1}$$

$$q'_{2n-1} = \varepsilon_1 \varepsilon_3 \cdots \varepsilon_{2n-1} q_{2n-1}$$

for $n \geq 1$.

We will need the following classical results on semiregular continued fractions (see, e.g., [13, 18, §37, 38, pp. 135–143]):

Theorem 2.1. *Let*

$$b_0 + \cfrac{\varepsilon_1}{b_1 + \cfrac{\varepsilon_2}{b_2 + \cfrac{\varepsilon_3}{b_3 + \cdots}}}$$

be a semiregular continued fraction obeying the rules above. Then

(a) The sequence p'_n/q'_n converges;
(b) $\lim_{n \to \infty} |q'_n| = +\infty$;
(c) For a given infinite sequence of signs $(\varepsilon_i)_{i \geq 1}$, the expansion of a given irrational real number exists and is unique, provided $b_i + \varepsilon_{i+1} \geq 2$ infinitely often.

3 Real Integer Functions and Finite Automata

Let $f : \mathbb{R} \to \mathbb{Z}$. We say f is a *real integer function* if

(a) $|f(x) - x| < 1$ for all $x \in \mathbb{R}$;
(b) $f(x + j) = f(x) + j$ for all $x \in \mathbb{R}$, $j \in \mathbb{Z}$.

Examples include the floor function $f(x) = \lfloor x \rfloor$, the ceiling function $f(x) = \lceil x \rceil$, and the round function $f(x) = \lfloor x + \frac{1}{2} \rfloor$.

Real integer functions induce generalized continued fraction algorithms by imitating algorithm SCF above:

Algorithm $\mathrm{CF}_f(x)$; outputs $(a_0, a_1, \ldots)$:
CF1. Set $x_0 \leftarrow x$; set $i \leftarrow 0$.
CF2. Set $a_i \leftarrow f(x_i)$.
CF3. If $a_i = x_i$ then stop. Otherwise set $x_{i+1} \leftarrow 1/(x_i - a_i)$, $i \leftarrow i + 1$
      and go to step CF2.

For each such expansion, we have an associated sequence of *convergents* p_n and q_n, defined as in Sect. 2. The theory of continued fractions (or an easy induction) gives $p_n/q_n = [a_0, \ldots, a_n]$ and furthermore $x = [a_0, \ldots, a_n, x_{n+1}]$, for all $n \geq 0$.

Now we examine the properties of the expansion of rational numbers.

Theorem 3.2. *Let f be an integer function and let x be a real number. The algorithm $\mathrm{CF}_f(x)$ terminates iff x is rational. Furthermore, if $\mathrm{CF}_f(x)$ terminates, with $(a_0, a_1, \ldots a_n)$ as output, then $x = [a_0, a_1, \ldots, a_n]$.*

Proof. Suppose x is rational. The algorithm successively replaces x_i by x_{i+1}. Suppose $i \geq 1$ and $x_i = p/q$ for $p, q \in \mathbb{Z}$ with $q \geq 1$. If $x_i \in \mathbb{Z}$, that is, if $q \mid p$, then $x_i = f(x_i)$ and the algorithm terminates immediately. Then either $a_i = \lfloor x_i \rfloor$ or $a_i = \lceil x_i \rceil$. Since $x_{i+1} = (x_i - a_i)^{-1}$, we have either $x_{i+1} = q/(p \bmod q)$ or $x_{i+1} = -q/((-p) \bmod q)$. In both cases we have have replaced a denominator of q with a number strictly less than q. Thus after at most $q - 1$ steps, we will reach a denominator of 1, and the algorithm terminates.

For the other direction, an easy induction gives $x = [a_0, a_1, \ldots, a_{n-1}, x_n]$. If the algorithm terminates, then $x_n = a_n$ and we have $x = [a_0, a_1, \ldots, a_n]$, a rational function of the integers $a_0, \ldots, a_n$. $\qquad\square$

Next, we prove two useful lemmas. The first concerns occurrences of partial quotients ± 1, and the second concerns convergents.

Lemma 3.3. *Suppose f is an integer function and let a_i and x_i be defined as in the algorithm CF_f. Then*

(a) *If $a_i = 1$ for $i \geq 1$, then $x_{i+1} > 1$ and $a_{i+1} \geq 1$.*
(b) *If $a_i = -1$ for $i \geq 1$, then $x_{i+1} < -1$ and $a_{i+1} \leq -1$.*
(c) *There exists no i such that $a_{i+t} = (-1)^t 2$ for $t \geq 0$.*

Proof. (a) Suppose $a_i = 1$. Then there is a corresponding x_i from the algorithm with $f(x_i) = a_i$. Since $i \geq 1$, we have $x_i > 1$ or $x_i < -1$. Since $|x_i - f(x_i)| < 1$ by the definition of integer function, we have $|x_i - 1| < 1$. It follows that $1 < x_i < 2$, so $x_{i+1} = 1/(x_i - 1)$ satisfies $x_{i+1} > 1$. Since $|x_{i+1} - a_{i+1}| < 1$, we have $a_{i+1} \geq 1$.

(b) Analogous to (a).

(c) It is easy to see that $1 = [2, -2, 2, -2, \ldots]$. If there exists i such that $a_{i+t} = (-1)^t 2$ for $t \geq 0$, then in the algorithm we have $x_i = a_i - 1$, a contradiction. $\qquad\square$

Lemma 3.4. *If* $|q_n| \leq |q_{n-1}|$ *then* $|q_{n-2}| < |q_{n-1}|$ *and either*

(a) $\operatorname{sgn} q_{n-1} \neq \operatorname{sgn} q_{n-2}$ *and* $a_n = 1$; *or*

(b) $\operatorname{sgn} q_{n-1} = \operatorname{sgn} q_{n-2}$ *and* $a_n = -1$.

In both cases we have $|q_{n+1}| > |q_n|$.

Proof. We verify this by induction. Consider the smallest index n such that $|q_n| \leq |q_{n-1}|$. Then necessarily $|q_{n-2}| < |q_{n-1}|$. Suppose $\operatorname{sgn} q_{n-1} \neq \operatorname{sgn} q_{n-2}$. Since $q_n = a_n q_{n-1} + q_{n-2}$, if $a_n \leq -1$ or $a_n \geq 2$ then $|q_n| = |a_n q_{n-1} + q_{n-2}| > |q_{n-1}|$, a contradiction. So $a_n = 1$, and furthermore $\operatorname{sgn} q_{n-1} = \operatorname{sgn} q_n$.

Now it follows from Lemma 3.3 that $a_{n+1} \geq 1$, so $|q_{n+1}| = |a_{n+1} q_n + q_{n-1}| \geq |q_n + q_{n-1}| > |q_n|$.

The analogous analysis works if $\operatorname{sgn} q_{n-1} = \operatorname{sgn} q_{n-2}$.

Since $|q_{n+1}| \geq |q_n|$, if we let n' be the smallest index $n' > n$ such that $|q_{n'}| \leq |q_{n'-1}|$, then $n' > n + 1$ and $|q_{n'-2}| < |q_{n'-1}|$. Now we are in exactly the same situation as above, with n' replacing n, and the induction step proceeds in the same way. $\qquad\square$

Next, we discuss the properties of the expansion of irrational numbers. We start by characterizing those real numbers whose expansion has a given prefix.

For a list $\mathbf{a} = (a_0, a_1, a_2, \ldots, a_n, \ldots)$ containing at least $n + 1$ elements, we let $\operatorname{pref}_n(\mathbf{a}) = (a_0, \ldots, a_n)$ be the prefix consisting of the first $n + 1$ elements.

Theorem 3.5. *Let f be an integer function, and $a_0, a_1, \ldots$ a sequence of integers. Define $S_f(0) = f^{-1}[0]$ and*

$$S_f(n) = (S_f(n-1)^{-1} \cap f^{-1}[a_n]) - a_n = (S_f(n-1)^{-1} - a_n) \cap f^{-1}[0]$$

for $n \geq 1$. Then

$$S_f(n) = \{\xi \in \mathbb{R} : \operatorname{pref}_n(\operatorname{CF}_f([a_0, a_1, \ldots, a_{n-1}, a_n + \xi])) = (a_0, a_1, \ldots, a_n)\}$$

for $n \geq 0$.

Proof. By induction on n. The base case is $n = 0$. Here $\operatorname{pref}_0(\operatorname{CF}_f([a_0 + \xi])) = (a_0)$ iff $f(a_0 + \xi) = a_0$ iff $f(\xi) = 0$, and so $\xi \in f^{-1}[0]$, as required.

Now assume the result is true for $n' \leq n$; we prove it for $n + 1$. By definition $S_f(n+1) = (S_f(n)^{-1} - a_{n+1}) \cap f^{-1}[a_{n+1}]$. Then

$$\xi \in S_f(n+1) \iff \xi \in S_f(n)^{-1} - a_{n+1} \text{ and } \xi \in f^{-1}[0]$$

$$\iff a_{n+1} + \xi \in S_f(n)^{-1} \text{ and } \xi \in f^{-1}[0]$$

$$\iff (a_{n+1} + \xi)^{-1} \in S_f(n) \text{ and } \xi \in f^{-1}[0]$$

$$\iff \operatorname{pref}_n(\operatorname{CF}_f([a_0, \ldots, a_{n-1}, a_n + (a_{n+1} + \xi)^{-1}]))$$

$$= (a_0, \ldots, a_n) \text{ and } \xi \in f^{-1}[0]$$

$$\iff \operatorname{pref}_n(\operatorname{CF}_f([a_0, \ldots, a_{n-1}, a_n, a_{n+1} + \xi]))$$

$$= (a_0, \ldots, a_n) \text{ and } \xi \in f^{-1}[0]$$

$$\iff \operatorname{pref}_{n+1}(\operatorname{CF}_f([a_0, \ldots, a_{n-1}, a_n, a_{n+1} + \xi])) = (a_0, \ldots, a_{n+1}),$$

which completes the proof. $\qquad\qquad\qquad\qquad\qquad\qquad\qquad\qquad\qquad\qquad\qquad\qquad\square$

Corollary 3.6. *Let f be an integer function, and let the sets $S_f(n)$ be defined as in the previous theorem. Then for $n \geq 0$ we have*

$$\{x \in \mathbb{R} : \operatorname{pref}_n(\operatorname{CF}_f(x)) = (a_0, a_1, \ldots, a_n)\} = \left\{ \frac{p_n + \xi p_{n-1}}{q_n + \xi q_{n-1}} : \xi \in S_f(n) \right\}.$$

Next, we prove an approximation theorem.

Theorem 3.7. *Suppose $\operatorname{pref}_n(\operatorname{CF}_f(x)) = (a_0, a_1, \ldots, a_n)$. Then*

$$\left| x - \frac{p_n}{q_n} \right| \leq \left| \frac{1}{q_n} \right|.$$

Proof. From Corollary 3.6 we know that $x = \frac{p_n + \xi p_{n-1}}{q_n + \xi q_{n-1}}$ for some $\xi \in S_f(n)$. But each $S_f(n)$ is a subset of $f^{-1}[0]$, so $-1 < \xi < 1$. Hence

$$\left| x - \frac{p_n}{q_n} \right| = \left| \frac{p_n + \xi p_{n-1}}{q_n + \xi q_{n-1}} - \frac{p_n}{q_n} \right|$$

$$= \left| \frac{\xi}{q_n(q_n + \xi q_{n-1})} \right|$$

$$< \left| \frac{1}{q_n(q_n + \xi q_{n-1})} \right|.$$

Now suppose $|q_n| > |q_{n-1}|$. Then since $|\xi| < 1$, and the q_i are integers, we have $|q_n + \xi q_{n-1}| \geq 1$.

Otherwise $|q_n| \leq |q_{n-1}|$. Then from Lemma 3.4, we know that $a_n = \pm 1$. Suppose $a_n = 1$ (the case $a_n = -1$ is analogous). Then also from Lemma 3.4, we know that

$\operatorname{sgn} q_{n-1} = \operatorname{sgn} q_n$. Also from Lemma 3.3, we know that $x_{n+1} > 1$ and so $\xi > 0$. Thus $|q_n + \xi q_{n-1}| > |q_n|$. It follows that $|q_n(q_n + \xi q_{n-1})| \geq |q_n|$, which completes the proof. □

Theorem 3.8. *If* $\operatorname{CF}_f(x) = (a_0, a_1, \ldots)$ *then* $\lim_{n \to \infty} [a_0, a_1, \ldots, a_n]$ *exists and equals* x.

Proof. Suppose $\operatorname{CF}_f(x) = (a_0, a_1, \ldots)$. Consider $p_n/q_n = [a_0, \ldots, a_n]$. From Theorem 3.7, we know that $|x - \frac{p_n}{q_n}| < |\frac{1}{q_n}|$. From Lemma 3.3 we know that the partial quotients fulfill the rules corresponding to a semiregular continued fraction, and hence from Theorem 2.1 we know that the sequence $[a_0, \ldots, a_n]$ converges to some limit α. Since from Theorem 2.1 we also know that $|q_n| \to +\infty$, it follows that $x = \alpha$. □

The main result of this paper is that the outputs of CF_f are easily describable in most of the interesting cases, including the examples SCF, CCF, and NICF mentioned previously. Let us define more rigorously what we mean by "easily describable."

Call a finite sequence of integers *valid* if it is the result of $\operatorname{CF}_f(x)$ for some rational number x. We envision a deterministic finite automaton which reads a purported finite expansion $\mathbf{a} = (a_0, a_1, \ldots a_n)$ and reaches a final state on the last input iff $\mathbf{a}$ is valid.

Definition 3.9. A finite automaton is a 5-tuple $(Q, \Sigma, \delta, q_0, F)$ where Q is a finite set of states, Σ is an (not necessarily finite) input alphabet, $q_0 \in Q$ is the initial state, $F \subseteq Q$ is the set of final states, and δ is the transition function mapping $Q \times \Sigma$ to Q. The transition function δ may be a partial function; i.e., $\delta(q, a)$ may be undefined for some pairs q, a.

We extend δ to a function which maps $Q \times \Sigma^*$ to Q in the obvious fashion.

The reader to whom these definitions are unfamiliar should consult [6].

The acceptance criterion for infinite expansions clearly needs to be different, since in this case there is no "last" partial quotient. We address the case of infinite expansions in Sect. 6.

One minor problem with this model is that the a_i belong to $\mathbb{Z}$, but in defining finite automata we usually insist that our alphabet Σ be finite. We can get around this in one of two ways: first, we could expand the definition of finite automata so that there can be infinitely many transitions (but still only finitely many states). However, such a model is too arbitrary, since allowing infinitely many transitions allows us to accept a set of expansions that is not even recursively enumerable. It suffices to allow only *finitely many* transitions, where each transition must be

- Either a single integer, or
- A set of the form $\{x \in \mathbb{Z} : x \geq \alpha\}$, or
- A set of the form $\{x \in \mathbb{Z} : x \leq \alpha\}$.

As an alternative, we could redefine our strings as numbers encoded in a particular base. That this is equivalent is clear, since the base-k representation of sets like $\{x \in \mathbb{Z} : x \geq \alpha\}$ forms a regular language. So either approach is satisfactory, but for simplicity we choose the first.

Notation. If $A \subseteq \mathbb{R}$ is a set, then by A^{-1} we mean the set of reciprocals $\{x \in \mathbb{R} - \{0\} : x^{-1} \in A\}$. Thus, for example, $[-\frac{1}{2}, \frac{1}{2}]^{-1} = (-\infty, -2] \cup [2, \infty)$ and $[1, \infty)^{-1} = (0, 1]$. If f is a function, then by $f^{-1}[a]$ we mean, as usual, the set $\{x \in \mathbb{R} : f(x) = a\}$. If A is a set, then by $A - a$ we mean the set $\{x : x + a \in A\}$. We will say x is *quadratic* if x is the real root of a quadratic equation with integer coefficients.

Definition 3.10. Let f be a real integer function. Then we say that the finite automaton $A = (Q, \mathbb{Z}, \delta, q_0, F)$ *accepts the outputs* of the algorithm CF_f if $\delta(q_0, a_0 a_1 a_2 \cdots a_n) \in F$ iff there exists $x \in \mathbb{Q}$ such that $\mathrm{CF}_f(x) = (a_0, a_1, \ldots, a_n)$.

The object of this paper is to prove the following theorem:

Theorem 3.11. *Let f be an integer function and suppose $f^{-1}[0]$ is the finite union of intervals. Then there exists a finite automaton accepting the outputs of CF_f iff all the endpoints of the intervals are rational or quadratic.*

In Sect. 4, we will prove one direction of this theorem; in Sect. 5, we prove the other.

Now we give these automata for the three continued fraction algorithms discussed so far: SCF, CCF, and NICF (Figs. 1–3).

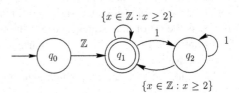

Fig. 1 Automaton for the simple continued fraction algorithm SCF

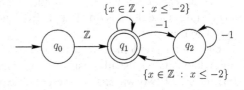

Fig. 2 Automaton for the ceiling algorithm CCF

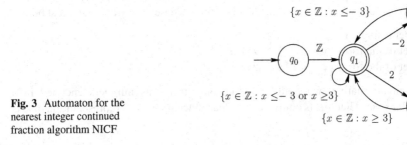

Fig. 3 Automaton for the nearest integer continued fraction algorithm NICF

Remark 12. No simple characterization seems to exist in the case where f is not the finite union of intervals. In Sect. 5, we will give an example of an f that is accepted by a finite automaton, but $f^{-1}[0]$ is not the finite union of intervals.

4 One Direction of the Theorem

We first prove the following theorem:

Theorem 4.12. *Suppose $f^{-1}[0]$ is the finite union of intervals with rational or quadratic endpoints. Consider all possible sets $S_0, S_1, \ldots$, constructed in Theorem 3.5, corresponding to all possible input sequences. Then among these there are only finitely many distinct sets.*

Proof. If S is a finite union of intervals, then $S^{-1} - a$ is also such a union. Hence $T = (S^{-1} - a) \cap f^{-1}[0]$ is such a union. The endpoints e of intervals of T are either those of $f^{-1}[0]$, or $e = 1/d - f(1/d)$, where d is an endpoint of S. Since $f(1/d)$ equals either $\lfloor 1/d \rfloor$ or $\lceil 1/d \rceil$, it will suffice to prove the following:

Lemma 4.14. *Define $s_1 : x \to (1/x) - \lfloor 1/x \rfloor$ and $s_2 : x \to (1/x) - \lceil 1/x \rceil$. Consider the monoid u formed by the maps s_1 and s_2 under composition. Let $u(x)$ be the orbit of x under elements of u.*

Then $u(x)$ is finite iff x is rational or quadratic.

Proof. One direction is easy. Assume $u(x)$ is finite. Then in particular the set

$$x, s_1(x), s_1^{(2)}(x), \ldots$$

is finite. Hence we have $s_1^{(j)}(x) = s_1^{(k)}(x)$ for some $j \neq k$. But it is easily proved by induction that

$$x = [0, a_0, a_1, \ldots, a_{n-1} + s_1^{(n)}(x)]$$

for some sequence of integers $a_0, a_1, \ldots$; hence there exist integers such that

$$x = \frac{a_j + b_j s_1^{(j)}(x)}{c_j + d_j s_1^{(j)}(x)},$$

and similarly

$$x = \frac{a_k + b_k s_1^{(k)}(x)}{c_k + d_k s_1^{(k)}(x)}.$$

Thus we see that $s_1^{(j)}(x)$ is the root of a quadratic equation, and so is either quadratic or rational. Thus x itself is either quadratic or rational.

If x is rational, the other direction follows easily, using an argument exactly like that in the proof of Theorem 3.2. If x is the root of a quadratic equation with integer coefficients, the result follows immediately from an old theorem of Blumer [3, Satz IX, p. 50]; for a more recent proof, see [16, Theorem A, p. 225]. □

This completes the proof of Theorem 4.12. □

We can now prove one direction of Theorem 3.11. Given an integer function f such that $f^{-1}[0]$ is the finite union of intervals with rational or quadratic endpoints, we create a finite automaton A_f as follows: the states of A_f are the distinct sets $S_f(n)$ constructed in Theorem 3.5 for all possible real numbers, together with a start state q_0. For convenience, we rename the states as $q_0, q_1, \ldots, q_t$ for some $t \geq 1$. From our results above, we know that the number of states is finite. We define $\delta(q_0, a_0) = q_1 = f^{-1}[0]$ for all $a_0 \in \mathbb{Z}$ and inductively define

$$\delta(q_i, a) = q_j$$

where $q_j = (q_i^{-1} \cap f^{-1}[a]) - a$, provided this set is nonempty. We say $q_i \in F$ if $0 \in q_i$.

It remains to verify that (a) the automaton accepts CF_f; and (b) the transitions can be characterized finitely, as discussed previously.

Corollary 4.15. $\delta(q_0, a_0 a_1 \cdots a_n) \in F$ iff there exists $x \in \mathbb{Q}$ such that $CF_f(x) = (a_0, a_1, \ldots, a_n)$.

Proof. Assume $\delta(q_0, a_0 a_1 \cdots a_n) \in F$. Then by the definition of the set of final states F, we must have $0 \in \delta(q_0, a_0 a_1 \cdots a_n)$. By Theorem 3.2 the first $n + 1$ outputs of the algorithm CF_f on input $[a_0, a_1, \ldots, a_n]$ are precisely $(a_0, a_1, \ldots, a_n)$. Hence we may take $x = [a_0, a_1, \ldots, a_n]$.

Now assume that there exists $x \in \mathbb{Q}$ such that $CF_f(x) = (a_0, a_1, \ldots, a_n)$. Then from the definition of CF_f, we see that $x_n = a_n$; hence

$$0 = x_n - a_n \in \delta(q_0, a_0 a_1 \cdots a_n)$$

which shows that $\delta(q_0, a_0 a_1 \cdots a_n)$ is a final state. □

The final step is to characterize the transitions. If the transition comes from q_0, then it is labeled $\mathbb{Z}$, which we can write, for example, $\{x : x \geq 0\}$ together with $\{x : x < 0\}$. Otherwise consider a transition of the form $\delta(q_i, a) = q_j$ where

$$q_j = (q_i^{-1} \cap f^{-1}[a]) - a = (q_i^{-1} - a) \cap f^{-1}[0].$$

If q_i is the finite union of intervals, then so is q_i^{-1} and $q_i^{-1} - a$. If q_i^{-1} is bounded, then as a ranges over all integers, there are only finitely many nonempty intersections of $(q_i^{-1} - a)$ with $f^{-1}[0]$. If q_i^{-1} is unbounded, say on the positive axis, then the intersection of $(q_i^{-1} - a)$ with $f^{-1}[0]$ is the same for all sufficiently large a. The same result holds when q_i^{-1} is unbounded on the negative axis. Hence there exists α, β such that the transition on each $x \geq \alpha$ is the same, and the transition on each $x \leq \beta$ is the same.

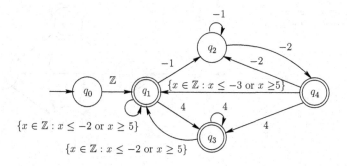

Fig. 4 Automaton corresponding to $f^{-1}[0] = [-\frac{\sqrt{2}}{2}, \frac{2-\sqrt{2}}{2})$

Combining this observation with Lemma 4.14 completes the proof of one direction of Theorem 3.11.   □

We now give some examples of the construction of the finite automaton.

Example 4.16. Let us obtain the description of the outputs for CF_f for $f(x) = \lfloor x + \frac{\sqrt{2}}{2} \rfloor$. We find

$$q_1 = f^{-1}[0] = [-\frac{\sqrt{2}}{2}, \frac{2-\sqrt{2}}{2})$$

$$q_2 = [-\frac{\sqrt{2}}{2}, 1 - \sqrt{2}]$$

$$q_3 = (\sqrt{2} - 2, \frac{2-\sqrt{2}}{2})$$

$$q_4 = [1 - \sqrt{2}, \frac{2-\sqrt{2}}{2}).$$

The full automaton is given below in Fig. 4.

Example 4.17. Our next example corresponds to the integer function defined by $f^{-1}[0] = (-1, -\frac{1}{2}] \cup \{0\} \cup (\frac{1}{2}, 1)$. We have

$$q_1 = f^{-1}[0]$$

$$q_2 = \{0\} \cup (\frac{1}{2}, 1)$$

$$q_3 = (-1, -\frac{1}{2}]$$

$$q_4 = (\frac{1}{2}, 1).$$

This gives the automaton below (Fig. 5).

Fig. 5 Automaton generating
bounded partial quotients

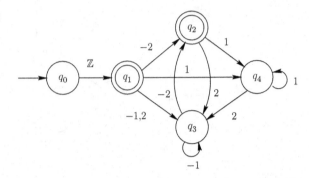

These expansions were introduced by Lehner [11] and further studied by Dajani and Kraaikamp [5]. An interesting feature of this expansion is that the partial quotients all lie in the set $\{-2, -1, 1, 2\}$. For example, the expansion of $\frac{52}{43}$ is $(2, -2, 2, -1, -1, -2, 2, -1, -1, -2)$. One undesirable aspect of these expansions is their slow convergence; the expansion of of $\frac{1}{2n}$ is $(1, \overbrace{-2, 2, -2, 2, \ldots}^{2n-1})$. Another undesirable aspect is that there exist infinite paths through the automaton (such as $(1, -2, 2, -2, 2, -2, 2, \ldots)$) which do not correspond to the expansion of any real x. However, this is essentially the only problematic case, as we will see below in Theorem 6.14.

5 Completing the Proof of Theorem 3.11

Proof. We now wish to show that if $f^{-1}[0]$ consists of the finite union of intervals, but one of those intervals has an endpoint that is not rational or quadratic, then no finite automaton can accept CF_f.

Assume that such an automaton A exists. Then we may assume that each state is in fact reachable from q_0; otherwise this state may be discarded without affecting A. For each state q_j, construct an input sequence $a_0 a_1 \cdots a_i$ such that $\delta(q_0, a_0 a_1 \cdots a_i) = q_j$. Let us label each state q_j with a subset of $\mathbb{Q}$, $L(q_j)$, by the following rule: If $\delta(q_0, a_0 a_1 \cdots a_i) = q_j$, then

$$L(q_j) = \{x \in \mathbb{Q} : \text{pref}_i(CF_f([a_0, a_1, \ldots, a_{i-1}, a_i + x])) = (a_0, a_1, \ldots, a_i)\}.$$

We need to show that this map is indeed well defined, in the sense that different paths from q_0 to q_j give the same labels $L(q_j)$. Assume that

$$\delta(q_0, a_0 a_1 \cdots a_i) = q_j$$

and
$$\delta(q_0, b_0 b_1 \cdots b_k) = q_j,$$

and there exists a rational number p such that

$$p \in S_1 = \{x \in \mathbb{Q} \ : \ \mathrm{CF}_f([a_0, a_1, \ldots, a_{i-1}, a_i + x]) = (a_0, a_1, \ldots, a_i, \ldots)\}$$

but

$$p \notin S_2 = \{x \in \mathbb{Q} \ : \ \mathrm{CF}_f([b_0, b_1, \ldots, b_{k-1}, b_k + x]) = (b_0, b_1, \ldots b_k, \ldots)\}.$$

Write $\mathrm{CF}_f(p) = (0, a_{i+1}, \ldots, a_n)$; by our definition of what it means to accept the output of CF_f, we know that

$$\delta(q_j, a_{i+1} \cdots a_n) = q_r \in F,$$

a final state. Let $y = [b_0, b_1, \ldots, b_k, a_{i+1}, \ldots a_n]$. Then since the automaton is in state q_j upon reading inputs $b_0 b_1 \cdots b_k$, we have

$$\delta(q_0, b_0 b_1 \cdots b_k a_{i+1} \cdots a_n) = q_r.$$

Hence $\mathrm{CF}_f(y) = (b_0, b_1, \ldots, b_k, a_{i+1}, \ldots a_n)$. But then $y = [b_0, b_1, \ldots, b_k + p]$ which shows that indeed $p \in S_2$, a contradiction.

Thus we may assume that sets $L_i = L(q_i)$ are well defined. Let $\bar{A}$ denote the *closure* of the set A in $\mathbb{R}$, and consider the sets $\bar{L}_i$. I claim that since $f^{-1}[0]$ consists of the finite union of intervals, so does each of the sets $\bar{L}_i$; this follows easily from the definition of CF_f. Suppose $\delta(q_i, a) = q_j$; then the endpoints e of intervals of $\bar{L}_j$ are those of $f^{-1}[0]$ or are related to the endpoints E of $\bar{L}_i$ by the equation

$$e = \frac{1}{E} - a.$$

Since $f^{-1}[0]$ contains an endpoint which is not rational or quadratic, so must $\bar{L}_0$. Hence there exists a transition $\delta(q_0, a) = q_i$ such that $\bar{L}_i$ contains an endpoint which is not rational or quadratic. Continuing in this fashion, and remembering that there are only a finite number of states, we eventually return to a state previously visited, which gives one of the two equations

$$e = [0, a_1, \ldots, a_k]$$

or

$$e = [0, a_1, \ldots, a_k + e]$$

which shows that e is rational or quadratic, contrary to assumption.

This completes the proof of Theorem 3.11.   □

Now let us give an example of an f such that $f^{-1}[0]$ is not the finite union of intervals, but nevertheless there is a finite automaton accepting CF_f.

Let $f(x)$ be defined by

$$f(x) = \begin{cases} \lfloor x \rfloor, & \text{if } x \text{ is rational;} \\ \lceil x \rceil, & \text{if } x \text{ is irrational.} \end{cases}$$

Then

$$f^{-1}[0] = \{x \; : \; x \text{ rational}, \; 0 \leq x < 1\} \cup \{x \; : \; x \text{ irrational}, \; -1 < x < 0\}.$$

Clearly $f^{-1}[0]$ cannot be written as the finite union of intervals. Then it is easily verified that the procedure of Sect. 4 generates a finite automaton with four states that accepts CF_f.

It may be of interest to remark that the automata accepting the result of CF_f may be arbitrarily complex. For example, it can be shown that the automaton corresponding to

$$f^{-1}[0] = [-\frac{F_{n-1}}{F_n}, \frac{F_{n-2}}{F_n})$$

has $n + 1$ states. (Here F_n denotes the nth Fibonacci number.)

6 Infinite Expansions

So far we have just addressed the case of finite expansions, the ones arising from rational number. In this section we handle irrational numbers. We say that an infinite sequence $(a_0, a_1, \ldots)$ is valid for an integer function f if it is the output of the algorithm CF_f on some input x.

In the remainder of this section we assume that f is an integer function such that $f^{-1}[0]$ consists of a finite union of intervals with rational or quadratic endpoints. We construct the associated automaton A_f as in Sect. 3. We would expect that infinite paths through A_f correspond in a 1-1 fashion with outputs of CF_f. However, this is not quite true; there can be certain infinite paths that do not correspond to any output of CF_f. By ruling these out, we can get the correspondence we desire.

Theorem 6.13. *If* $CF_f(x) = \mathbf{a} = (a_0, a_1, \ldots)$ *and* $A_f = (Q, \Sigma, \delta, q_0, F)$, *then* $\delta(q_0, a_0 a_1 \cdots a_n)$ *exists for all* $n \geq 0$.

Proof. By induction on n. Clearly this is true for $n = 0$, since there is a transition labeled with each $a \in \mathbb{Z}$ leaving q_0 to q_1. Now assume the claim is true for all $n' < n$; we prove it for n. Now $\operatorname{pref}_n(CF_f(x)) = (a_0, a_1, \ldots, a_n)$ if and only if $\operatorname{pref}_{n-1}(CF_f(x)) = (a_0, \ldots, a_{n-1})$ and $x_n = (x_{n-1} - a_{n-1})^{-1}$ and $a_n = f(x_n)$. Let $q = \delta(q_0, a_0 \ldots a_{n-1})$; then by definition we have a transition out of q labeled a_n if and only if the set $q^{-1} \cap f^{-1}[a_n]$ is nonempty. But $x_n \in q^{-1} \cap f^{-1}[a_n]$, so there is indeed such a transition. $\qquad\square$

Finally, we characterize the infinite paths through A_f that correspond to the expansion associated with some irrational number.

Theorem 6.14. *Let $\mathbf{a} = (a_0, a_1, \ldots)$ be an infinite path through A_f that has no infinite suffix of the form $(2, -2, 2, -2, \ldots)$. Then there is an irrational real number x such that $CF_f(x) = \mathbf{a}$.*

Proof. Take an infinite path through A_f labeled with $(a_0, a_1, \ldots)$. For each finite prefix $(a_0, a_1, \ldots, a_n)$, we can consider the continued fraction $[a_0, \ldots, a_n]$ and the corresponding convergents p_n, q_n. Then $CF_f(p_n/q_n) = (a_0, a_1, \ldots, a_n)$. Furthermore, the p_n/q_n converge to $x = [a_0, a_1, \ldots]$ and $|x - p_n/q_n| < |1/q_n|$.

From our correspondence with semiregular continued fractions given in Sect. 2, the continued fraction $[a_0, a_1, \ldots]$ corresponds to a certain semiregular continued fraction, with pattern of signs in the numerator dependent on the pattern of signs of the a_i, as given in the equivalence between (1) and (2). However, from Theorem 2.1, for a given pattern of signs, an infinite semiregular continued fraction expansion is unique, provided it obeys the rule that $a_i + \varepsilon_{i+1} \geq 2$ infinitely often. We have already seen that $a_i + \varepsilon_{i+1} \geq 1$ for our expansions. The case where $a_i + \varepsilon_{i+1} = 1$ for all but finitely many i corresponds to (in our notation) an expansion that looks like $[\ldots, 2, -2, 2, -2, \ldots]$ with an infinite suffix of $(2, -2)$ repeating. However, by hypothesis, our path A_f has no such suffix. Therefore, it corresponds to a unique real number x. □

7 Variations

Exactly the same result holds for minor variations of our continued fraction algorithm. For example, suppose we have some finite list of integer functions $f_0, f_1, \ldots, f_{n-1}$ and apply them periodically, as follows:

Algorithm PSCF(x); outputs $(a_0, a_1, \ldots)$:
PSCF1. Set $x_0 \leftarrow x$; set $i \leftarrow 0$.
PSCF2. Set $a_i \leftarrow f_{i \bmod n}(x)$.
PSCF3. If $a_i = x_i$ then stop. Otherwise set $x_{i+1} \leftarrow 1/(x_i - a_i)$; set $i \leftarrow i+1$ and
      go to step PSCF2.

Then the analogous version of Theorem 3.11 holds. The only difference is that the automaton needs to keep track of the current value of i, taken modulo n. For example, suppose we let $n = 2$ and $f_0(x) = \lceil x \rceil$ and $f_1(x) = \lfloor x \rfloor$. The resulting algorithm gives what is often called the *reduced simple continued fraction expansion* in the literature. The corresponding automaton is given below (Fig. 6).

Fig. 6 Automaton for the
reduced simple continued
fraction algorithm

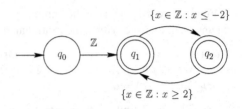

$\{x \in \mathbb{Z} : x \leq -2\}$

$\mathbb{Z}$

q_0 q_1 q_2

$\{x \in \mathbb{Z} : x \geq 2\}$

Because the signs of terms alternate, these continued fractions are often written
in the form

$$a_0 - \cfrac{1}{a_1 - \cfrac{1}{a_2 - \cdots}},$$

where $a_i \in \mathbb{Z}$ and $a_i \geq 2$ for $i \geq 1$.

Another variation is to treat positive and negative numbers differently. For
example, we could define

$$f(x) = \begin{cases} \lfloor x + \frac{1}{2} \rfloor, & \text{if } x \geq 0; \\ \lceil x - \frac{1}{2} \rceil, & \text{if } x < 0. \end{cases}$$

Our results, with small differences, also apply here.

8 Concluding Remarks

Our results apply to, for example, the α-continued fractions of Tanaka and Ito [17],
which correspond to the integer function $f(x) = \lfloor x - \alpha + 1 \rfloor$, where $\frac{1}{2} \leq \alpha \leq 1$ is a
real number.

Several other writers have noted connections between finite automata and
continued fractions. One of the best-known papers is that of Raney, who showed
how to obtain the simple continued fraction for

$$\beta = \frac{a\alpha + b}{c\alpha + d}$$

in terms of the continued fraction for α. See [2, 14] for more details.

Istrail considered the language consisting of all prefixes of the continued fraction
for x, and observed that this language is context-free and non-regular iff x is a
quadratic irrational [9].

Allouche discusses several applications of finite automata to number theory,
including continued fractions [1].

In this paper, we have been concerned with a different approach; namely,
describing the "set of rules" associated with a generalized continued fraction

algorithm. One immediately wonders if similar theorems may be obtained for continued fraction algorithms in $\mathbb{Z}[i]$, such as those discussed by Hurwitz [7] and McDonnell [12].

In [15] the author proved that the McDonnell's complex continued fraction algorithm can be described by a finite automaton with 25 states. The corresponding result for Hurwitz's algorithm is not known.

Acknowledgements I am very grateful to the referee for many suggestions that considerably improved the chapter.

Supported by a grant from NSERC.

References

1. J.-P. Allouche, Automates finis en théorie des nombres. Exposition Math. **5**, 239–266 (1987)
2. W.M. Beynon, A formal account of some elementary continued fraction algorithms. J. Algorithms **4**, 221–240 (1983)
3. F. Blumer, Über die verschiedenen Kettenbruchentwicklungen beliebiger reeller Zahlen und die periodischen Kettenbruchentwicklungen quadratischer Irrationalitäten. Acta Arith. **3**, 3–63 (1939)
4. C. Brezinski, *History of Continued Fractions and Padé Approximants* (Springer, New York, 1991)
5. K. Dajani, C. Kraaikamp, The mother of all continued fractions. Colloq. Math. **84/85**(Pt 1), 109–123 (2000)
6. J.E. Hopcroft, J.D. Ullman, *Introduction to Automata Theory, Languages, and Computation* (Addison-Wesley, Boston, 1979)
7. A. Hurwitz, Über die Entwicklung complexer Grössen in Kettenbrüche. Acta Math. **11**, 187–200 (1888) [= Werke, II, pp. 72–83]
8. A. Hurwitz, Über eine besondere Art der Kettenbruch-Entwicklung reeller Grössen. Acta Math. **12**, 367–405 (1889) [= Werke, II, pp. 84–115]
9. S. Istrail, On formal construction of algebraic numbers of degree two. Rev. Roum. Math. Pures Appl. **22**, 1235–1239 (1977)
10. D.E. Knuth, *The Art of Computer Programming, V. II (Seminumerical Algorithms)*, 2nd edn (Addison-Wesley, Boston, 1981)
11. J. Lehner, Semiregular continued fractions whose partial denominators are 1 or 2, in *The Mathematical Legacy of Wilhelm Magnus: Groups, Geometry, and Special Functions*, ed. by W. Abikoff, J.S. Birman, K. Kuiken (Amer. Math. Soc., New York, 1994), pp. 407–410
12. E.E. McDonnell, Integer functions of complex numbers, with applications. IBM Philadelphia Scientific Center, Technical Report 320-3015, February 1973
13. O. Perron, Die Lehre von den Kettenbrüchen, in *Band I: Elementare Kettenbrüche* (Teubner, 1977)
14. G.N. Raney, On continued fractions and finite automata. Math. Ann. **206**, 265–283 (1973)
15. J.O. Shallit, Integer functions and continued fractions. A.B. Thesis, Princeton University, 1979. Available at http://www.cs.uwaterloo.ca/~shallit/papers.html
16. J. Steinig, On the complete quotients of semi-regular continued fractions for quadratic irrationals. Arch. Math. **43**, 224–228 (1984)
17. S. Tanaka, S. Ito, On a family of continued-fraction transformations and their ergodic properties. Tokyo J. Math. **4**, 153–175 (1981)
18. H. Tietze, Über Kriterien für Konvergenz und Irrationalität unendlichen Kettenbrüche. Math. Annalen **70**, 236–265 (1911)

On Prime Factors of Terms of Linear Recurrence Sequences

C.L. Stewart

In memory of Alf van der Poorten

1 Introduction

Let k be a positive integer, $r_1, \ldots, r_k$ and $u_0, \ldots, u_{k-1}$ be integers and put

$$u_n = r_1 u_{n-1} + \cdots + r_k u_{n-k},$$

for $n = k, k+1, \ldots$. Suppose that r_k is non-zero and that $u_0, \ldots, u_{k-1}$ are not all zero. The sequence $(u_n)_{n=0}^\infty$ is a recurrence sequence of order k. It has a characteristic polynomial $G(z)$ given by

$$G(z) = z^k - r_1 z^{k-1} - \cdots - r_k.$$

Let

$$G(z) = \prod_{i=1}^{t} (z - \alpha_i)^{\ell_i},$$

with $\alpha_1, \ldots, \alpha_t$ distinct. Then, see Theorem C.1 of [34], there exist polynomials $f_1, \ldots, f_t$ of degrees less than $\ell_1, \ldots, \ell_t$, respectively, and with coefficients from $\mathbb{Q}(\alpha_1, \ldots, \alpha_t)$ such that

$$u_n = f_1(n) \alpha_1^n + \cdots + f_t(n) \alpha_t^n, \tag{1}$$

C.L. Stewart (✉)

Department of Pure Mathematics, University of Waterloo, Waterloo, Ontario, Canada N2L 3G1

e-mail: cstewart@uwaterloo.ca

J.M. Borwein et al. (eds.), *Number Theory and Related Fields: In Memory of Alf van der Poorten*, Springer Proceedings in Mathematics & Statistics 43, DOI 10.1007/978-1-4614-6642-0_18, © Springer Science+Business Media New York 2013

for $n = 0, 1, 2, \ldots$. The recurrence sequence $(u_n)_{n=0}^{\infty}$ is said to be degenerate if α_i/α_j is a root of unity for a pair (i, j) with $1 \leq i < j \leq t$ and is said to be non-degenerate otherwise. In 1935 Mahler [20] proved that

$$|u_n| \to \infty \quad \text{as} \quad n \to \infty$$

whenever $(u_n)_{n=0}^{\infty}$ is a non-degenerate linear recurrence sequence. Mahler's proof is not effective in the following sense. Given a positive integer m the proof does not yield a number $C(m)$ which is effectively computable in terms of m, such that $|u_n| > m$ whenever $n > C(m)$. However, Schmidt [31, 32], Allen [1] and Amoroso and Viada [2] have given estimates in terms of t only for the number of times $|u_n|$ assumes a given value when the recurrence sequence is non-degenerate.

For any integer n let $P(n)$ denote the greatest prime factor of n with the convention that $P(0) = P(\pm 1) = 1$. Suppose that in (1) $t > 1$, $f_1, \ldots, f_t$ are polynomials which are not the zero polynomial and that $\alpha_1, \ldots, \alpha_t$ are non-zero. van der Poorten and Schlickewei [25] in 1982 and independently Evertse [12] proved, under the above assumption, that if the sequence $(u_n)_{n=0}^{\infty}$ is non-degenerate then

$$P(u_n) \to \infty \quad \text{as} \quad n \to \infty. \qquad (2)$$

A key feature of the work of van der Poorten and Schlickewei and of Evertse is an appeal to a p-adic version of Schmidt's Subspace Theorem due to Schlickewei [30] and so (2) is also an ineffective result.

We may suppose, without loss of generality, that

$$|\alpha_1| \geq |\alpha_2| \geq \cdots \geq |\alpha_t| > 0.$$

If $|\alpha_1| > |\alpha_2|$ then plainly $|u_n|$ tends to infinity with n. In this case Shparlinski [35] and Stewart [40] independently obtained effective lower bounds for $P(u_n)$ which tend to infinity with n. The sharpest result obtained to date [41] when u_n is the nth term of a non-degenerate linear recurrence as in (1) with $|\alpha_1| > |\alpha_2|$ and $u_n \neq f_1(n)\alpha_1^n$ is that there are positive numbers c_1 and c_2, which are effectively computable in terms of $r_1, \ldots, r_k$ and $u_0, \ldots, u_{k-1}$, such that

$$P(u_n) > c_1 \log n \frac{\log \log n}{\log \log \log n}, \qquad (3)$$

provided that n exceeds c_2. A key tool in the proof of (3) is a lower bound, due to Matveev [23], for linear forms in the logarithms of algebraic numbers.

2 Binary Recurrence Sequences

When the minimal order k of the recurrence is 2 the sequence is known as a binary recurrence sequence. In this case, for $n \geq 0$,

$$u_n = a\alpha^n + b\beta^n, \tag{4}$$

where α and β are the roots of the characteristic polynomial $x^2 - r_1 x - r_2$ and

$$a = \frac{u_1 - u_0\beta}{\alpha - \beta}, \qquad b = \frac{u_0\alpha - u_1}{\alpha - \beta} \tag{5}$$

when $\alpha \neq \beta$. Since the recurrence sequence has order 2, r_2 is non-zero and so $\alpha\beta$ is non-zero. When $(u_n)_{n=0}^{\infty}$ is non-degenerate $\alpha \neq \beta$ and we see that $ab \neq 0$ since the recurrence sequence has minimal order 2. We may assume, without loss of generality that

$$|\alpha| \geq |\beta| > 0.$$

In 1934 Mahler [19] employed a p-adic version of the Thue-Siegel theorem in order to prove that if u_n is the nth term of a non-degenerate binary recurrence sequence then

$$P(u_n) \to \infty \quad \text{as } n \to \infty.$$

Mahler's result is not effective. This defect was remedied by Schinzel [28] in 1967. He refined work of Gelfond on estimates for linear forms in the logarithms of two algebraic numbers in order to prove that if $(u_n)_{n=0}^{\infty}$ is a non-degenerate binary recurrence sequence then there exists a positive number C_0 which is effectively computable in terms of a, b, α and β and positive numbers c_1 and c_2 such that

$$P(u_n) > C_0 n^{c_1} (\log n)^{c_2},$$

where

$$(c_1, c_2) = \begin{cases} (1/84, 7/12) & \text{if } \alpha \text{ and } \beta \text{ are integers} \\ (1/133, 7/19) & \text{otherwise.} \end{cases}$$

In 1982 Stewart [40] used estimates for linear forms in the logarithms of algebraic numbers due to Waldschmidt [44] in the Archimedean setting and due to van der Poorten [24] in the non-Archimedean setting to prove that there is a positive number C_3, which is effectively computable in terms of u_0, u_1, r_1 and r_2, such that for $n > 1$,

$$P(u_n) > C_3 (n/\log n)^{1/(d+1)} \tag{6}$$

where d is the degree of α over the rationals. In 1995 Yu and Hung [46] were able to refine (6) by replacing the term $n/\log n$ by n. We are now able to make a further improvement on (6).

Theorem 1. *Let u_n, as in (4), be the nth term of a non-degenerate binary recurrence sequence with $ab\alpha\beta \neq 0$. There exists a positive number C which is effectively computable in terms of u_0, u_1, r_1 and r_2 such that for $n > C$*

$$P(u_n) > n^{1/2}\exp(\log n/104\log\log n). \tag{7}$$

The proof of Theorem 1 makes use of ideas from [42] which we will discuss in the next section. They were essential in resolving a conjecture made by Erdős in 1965 [11].

It is possible to sharpen (7) for most integers n. In [40] Stewart proved that if $(u_n)_{n=0}^{\infty}$ is a non-degenerate binary recurrence sequence then for all integers n, except perhaps a set of asymptotic density zero,

$$P(u_n) > \varepsilon(n)n\log n,$$

where $\varepsilon(n)$ is any real-valued function for which $\lim_{n\to\infty}\varepsilon(n) = 0$. Furthermore it is possible to strengthen (7) whenever u_n is non-zero and is divisible by a prime p which does not divide u_m for any non-zero u_m with $0 \leq m < n$. In this case Stewart [40] proved that there is a positive number C_4, which is effectively computable in terms of a and b only such that

$$P(u_n) > n - C_4.$$

Luca [17] strengthened (7) when $(u_n)_{n=0}^{\infty}$ is a binary recurrence sequence as in (4) with a/b and α/β multiplicatively dependent. He proved that then there exists a positive number C_5, which is effectively computable in terms of a, b, α and β, such that

$$P(u_n) > n - C_5 \tag{8}$$

for all positive integers n. Schinzel [28] had earlier obtained such a result in the case that α and β are real numbers.

3   Lucas Sequences

Let a and b be integers with $a > b > 0$ and consider the binary recurrence sequence $(a^n - b^n)_{n=0}^{\infty}$. In 1892 Zsigmondy [47], and independently in 1904 Birkhoff and Vandiver [8], proved that for $n > 2$

$$P(a^n - b^n) \geq n + 1. \tag{9}$$

This result had been established by Bang [6] in 1886 for the case when $b = 1$. Schinzel [26] proved in 1962 that if a and b are coprime and ab is a square or twice a square then

$$P(a^n - b^n) \geq 2n + 1$$

provided that (a, b, n) is not $(2, 1, 4)$, $(2, 1, 6)$ or $(2, 1, 12)$.

In 1965 Erdős [11] conjectured that

$$\frac{P(2^n - 1)}{n} \to \infty \quad \text{as } n \to \infty.$$

In 2000 Murty and Wong [22] proved that if ε is a positive real number and a and b are integers with $a > b > 0$ then

$$P(a^n - b^n) > n^{2-\varepsilon},$$

for n sufficiently large in terms of a, b and ε subject to the abc conjecture [43]. A few years later Murata and Pomerance [21] assumed the truth of the generalized Riemann hypothesis and deduced that

$$P(2^n - 1) > n^{4/3} \log\log n$$

for a set of positive integers n of asymptotic density 1.

In 1975 Stewart [36] proved that the Erdős conjecture holds when we restrict n to run over those integers with at most $\kappa \log\log n$ distinct prime factors where κ is any real number less than $1/\log 2$. In 2009 Ford, Luca and Shparlinski [13] proved that the series

$$\sum_{n=1}^{\infty} 1/P(2^n - 1)$$

is convergent. Recently Stewart [42] established the conjecture of Erdős by proving that if a and b are positive integers then

$$P(a^n - b^n) > n \exp(\log n / 104 \log\log n) \tag{10}$$

provided that n is sufficiently large in terms of the number of distinct prime factors of ab.

Suppose that $(u_n)_{n=0}^{\infty}$ is a non-degenerate binary recurrence sequence with $u_0 = 0$ and $u_1 = 1$. Then, recall (4) and (5),

$$u_n = \frac{\alpha^n - \beta^n}{\alpha - \beta} \tag{11}$$

for $n = 0, 1, 2, \ldots$. Lucas [18] undertook an extensive study of the divisibility properties of such numbers in 1878 and we now refer to sequences $(u_n)_{n=0}^{\infty}$ with u_n given by (11) as Lucas sequences. In 1912 Carmichael [9] proved that if α and β are real, $n > 12$ and u_n is the nth term of a Lucas sequence then

$$P(u_n) \geq n - 1. \tag{12}$$

Schinzel [27] established the same estimate in the case when α and β are not real for n sufficiently large in terms of α and β. Both results were proved by showing that u_n possesses a primitive divisor for n sufficiently large. A prime p which divides u_n but does not divide $(\alpha - \beta)^2 u_2 \cdots u_{n-1}$ is known as a primitive divisor of u_n. Let us assume that $\alpha + \beta$ and $\alpha\beta$ are coprime. Then Schinzel [29], in 1974, proved that there is a positive number C_6, which does not depend on α and β, such that u_n has a primitive divisor for n greater than C_6. In [39] Stewart proved that one can take C_6 to be $e^{452} 2^{67}$. Further he showed that one can take C_6 to be 6 with finitely many exceptions and that these exceptions may be found by solving a large but finite collection of Thue equations. Bilu, Hanrot and Voutier [7] were able to determine all exceptions and as a consequence deduce that

$$P(u_n) \geq n - 1,$$

for $n > 30$.

Stewart [38], when α and β are real, and Shorey and Stewart [33], otherwise, extended the work of Stewart [36] to Lucas sequences. Let u_n be the nth term of a non-degenerate Lucas sequence with r_1 and r_2 coprime. Let $\varphi(n)$ denote Euler's function, let $q(n)$ denote the number of square-free divisors of n and let κ denote a positive real number with $\kappa < 1/\log 2$. They proved that if n (> 3) has at most $\kappa \log\log n$ distinct prime factors then

$$P(u_n) > C_7(\varphi(n)\log n)/q(n),$$

where C_7 is a positive number which is effectively computable in terms of α, β and κ only. The proofs depend on estimates for linear forms in the logarithms of algebraic numbers, in the complex case due to Baker [4] and in the p-adic case due to van der Poorten [24].

In [42] Stewart proved that estimate (10) holds with $a^n - b^n$ replaced by u_n where u_n is the nth term of a non-degenerate Lucas sequence. In fact, see [42], the same estimate also applies with $a^n - b^n$ replaced by $\tilde{u}_n$ where $\tilde{u}_n$ denotes the nth term of a non-degenerate Lehmer sequence. (The Lehmer sequences, see [15, 38], are closely related to the Lucas sequences and they possess similar divisibility properties.) For the proofs of these results estimates for linear forms in the logarithms of algebraic numbers again play a central role. In the Archimedean case we apply an estimate of Baker [3] while in the non-Archimedean case we appeal to an estimate of Yu [45].

4   Preliminaries for the Proof of Theorem 1

Let K be a finite extension of $\mathbb{Q}$ and let $\wp$ be a prime ideal in the ring of algebraic integers $\mathcal{O}_K$ of K. Let $\mathcal{O}_\wp$ consist of 0 and the non-zero elements α of K for which $\wp$ has a non-negative exponent in the canonical decomposition of the fractional ideal generated by α into prime ideals. Then let P be the unique prime ideal of $\mathcal{O}_\wp$ and put $\overline{K_{\wp}} = \mathcal{O}_{\wp}/P$. Further for any α in $\mathcal{O}_{\wp}$ we let $\overline{\alpha}$ be the image of α under the residue class map that sends α to $\alpha + P$ in $\overline{K_{\wp}}$.

Let p be an odd prime and let d be an integer coprime with p. The Legendre symbol $\left(\frac{d}{p}\right)$ is 1 if d is a quadratic residue modulo p and is -1 if d is a quadratic non-residue modulo p.

Lemma 1. *Let d be a square-free integer different from 1. Let θ be an algebraic number of degree 2 over $\mathbb{Q}$ in $\mathbb{Q}(\sqrt{d})$, let θ' denote the algebraic conjugate of θ over $\mathbb{Q}$ and let a_0 be the leading coefficient in the minimal polynomial of θ in $\mathbb{Z}[x]$. Suppose that p is a prime which does not divide $2a_0^2\theta\theta'$. Let $\wp$ be a prime ideal of the ring of algebraic integers of $\mathbb{Q}(\sqrt{d})$ lying above p. The order of $\overline{\theta/\theta'}$ in $(\mathbb{Q}(\sqrt{d})_{\wp})^{\times}$ is a divisor of 2 if p divides $a_0^4(\theta^2 - \theta'^2)^2$ and a divisor of $p - \left(\frac{d}{p}\right)$ otherwise.*

Proof. Note that $\gamma = a_0\theta$ is an algebraic integer with algebraic conjugate $\gamma' = a_0\theta'$. Thus $\gamma/\gamma' = \theta/\theta'$ and our result follows from Lemma 2.2 of [42]. $\square$

For any algebraic number γ let $h(\gamma)$ denote the absolute logarithmic height of γ. Thus if $a_0(x - \gamma_1)\cdots(x - \gamma_d)$ in $\mathbb{Z}[x]$ is the minimal polynomial of γ over $\mathbb{Z}$ then

$$h(\gamma) = \frac{1}{d}\left(\log a_0 + \sum_{j=1}^{d}\log\max(1, |\gamma_j|)\right).$$

Let $\alpha_1, \ldots, \alpha_n$ be non-zero algebraic numbers and put $K = \mathbb{Q}(\alpha_1, \ldots, \alpha_n)$ and $d = [K : \mathbb{Q}]$. Let $\wp$ be a prime ideal of the ring $\mathcal{O}_K$ of algebraic integers in K lying above the prime number p. Denote by $e_{\wp}$ the ramification index of $\wp$ and by $f_{\wp}$ the residue class degree of $\wp$. For α in K with $\alpha \neq 0$ let $\text{ord}_{\wp}\alpha$ be the exponent to which $\wp$ divides the principal fractional ideal generated by α in K and put $\text{ord}_{\wp}0 = \infty$. For any positive integer m let $\zeta_m = e^{2\pi i/m}$ and put $\alpha_0 = \zeta_{2^u}$ where ζ_{2^u} is in K and $\zeta_{2^{u+1}}$ is not in K.

Suppose that $\alpha_1, \ldots, \alpha_n$ are multiplicatively independent $\wp$-adic units in K. Let $\overline{\alpha_0}, \overline{\alpha_1}, \ldots, \overline{\alpha_n}$ be the images of $\alpha_0, \alpha_1, \ldots, \alpha_n$ respectively, under the residue class map at $\wp$ from the ring of $\wp$-adic integers in K onto the residue class field $\overline{K_{\wp}}$ at $\wp$. For any set X let $|X|$ denote its cardinality. Let $\langle\overline{\alpha_0}, \overline{\alpha_1}, \ldots, \overline{\alpha_n}\rangle$ be the subgroup of $(\overline{K_{\wp}})^{\times}$ generated by $\overline{\alpha_0}, \ldots, \overline{\alpha_n}$. We define δ by

$$\delta = 1 \quad \text{if} \quad \left[K\left(\alpha_0^{1/2}, \alpha_1^{1/2}, \ldots, \alpha_n^{1/2}\right) : K\right] < 2^{n+1}$$

and

$$\delta = (p^{f_{\wp}} - 1)/|\langle\overline{\alpha_0}, \overline{\alpha_1}, \ldots, \overline{\alpha_n}\rangle|$$

if

$$\left[K\left(\alpha_0^{1/2}, \alpha_1^{1/2}, \ldots, \alpha_n^{1/2}\right) : K\right] = 2^{n+1}.$$

Denote $\log\max(x, e)$ by $\log^* x$.

Lemma 2. *Let p be a prime with $p \geq 5$ and let $\wp$ be an unramified prime ideal of $\mathcal{O}_K$ lying above p. Let $\alpha_1, \ldots, \alpha_n$ be multiplicatively independent $\wp$-adic units. Let $b_1, \ldots, b_n$ be integers, not all zero, and put*

$$B = \max(2, |b_1|, \ldots, |b_n|).$$

Then

$$\operatorname{ord}_\wp(\alpha_1^{b_1} \cdots \alpha_n^{b_n} - 1) < Ch(\alpha_1) \cdots h(\alpha_n) \max(\log B, (n+1)(5.4n + \log d))$$

where

$$C = 376(n+1)^{1/2} \left(7e \frac{p-1}{p-2} \right)^n d^{n+2} \log^* d \log(e^4(n+1)d) \cdot$$

$$\max \left(\frac{p^{f_p}}{\delta} \left(\frac{n}{f_p \log p} \right)^n, e^n f_p \log p \right).$$

Proof. This is Lemma 3.1 of [42] and it follows from the work of Yu [45].   □

The next result we require is proved using class field theory and the Chebotarev Density Theorem.

Lemma 3. *Let d be a square-free integer different from 1 and let p_k denote the kth smallest prime of the form $N(\pi_k) = p_k$ where N denotes the norm from $\mathbb{Q}(\sqrt{d})$ to $\mathbb{Q}$ and π_k is an algebraic integer in $\mathbb{Q}(\sqrt{d})$. Let ε be a positive real number. There is a positive number C, which is effectively computable in terms of ε and d, such that if k exceeds C then*

$$\log p_k < (1 + \varepsilon) \log k.$$

Proof. This is Lemma 2.4 of [42].   □

We shall also require an estimate for the rate of growth of a non-degenerate binary recurrence sequence.

Lemma 4. *Let u_n, as in (4), be the nth term of a non-degenerate binary recurrence sequence. Suppose that $|\alpha| \geq |\beta|$. Then there exist positive numbers C_1 and C_2, which are effectively computable in terms of a and b, such that if n exceeds C_1 then*

$$|u_n| > |\alpha|^{n - C_2 \log n}.$$

Proof. This is Lemma 3.2 of [37]; see also Lemma 5 of [40].   □

Lemma 5. *Let K be a finite extension of $\mathbb{Q}$ and let p be a prime number. Let $\alpha_1, \ldots, \alpha_n$ be non-zero elements of K and let $\alpha_1^{1/p}, \ldots, \alpha_n^{1/p}$ denote fixed pth roots of $\alpha_1, \ldots, \alpha_n$, respectively. Put $K' = K(\alpha_1^{1/p}, \ldots, \alpha_{n-1}^{1/p})$. Then either $K'(\alpha_n^{1/p})$ is an extension of K' of degree p or we have*

$$\alpha_n = \alpha_1^{j_1} \cdots \alpha_{n-1}^{j_{n-1}} \gamma^p$$

for some γ in K and some integers $j_1, \ldots, j_{n-1}$ with $0 \le j_i < p$ for $i = 1, \ldots, n-1$.

Proof. This is Lemma 3 of Baker and Stark [5]. □

Lemma 6. *Let n be a positive integer and let $\alpha_0, \alpha_1, \ldots, \alpha_n$ be multiplicatively dependent non-zero elements of a number field K of degree $d \ge 2$ over $\mathbb{Q}$. Suppose that any n from $\alpha_0, \ldots, \alpha_n$ are multiplicatively independent. Then there are non-zero rational integers $b_0, \ldots, b_n$ with*

$$\alpha_0^{b_0} \cdots \alpha_n^{b_n} = 1$$

and

$$|b_i| \le 58(n!e^n/n^n)d^{n+1}(\log d)h(\alpha_0) \cdots h(\alpha_n)/h(\alpha_i)$$

for $i = 0, \ldots, n$.

Proof. This is Corollary 3.2 of Loher and Masser [16]. They attribute the result to Yu. □

Lemma 7. *Let $(u_n)_{n=0}^{\infty}$ be a non-degenerate binary recurrence sequence as in (4) with $ab\alpha\beta \ne 0$ and a/b and α/β multiplicatively independent. There exists a positive number C which is effectively computable in terms of a, b, α and β such that if p exceeds C then*

$$\mathrm{ord}_p\, u_n < p\exp(-\log p/51.9\log\log p)\log n.$$

Proof. Our proof will be modelled on the proof of Lemma 4.3 in [42]. Let $c_1, c_2, \ldots$ denote positive numbers which are effectively computable in terms of a, b, α and β. Let p be a prime which does not divide $2(\alpha - \beta)^4 ab\alpha\beta$.

Put $K = \mathbb{Q}(\alpha/\beta)$ and

$$\alpha_0 = \begin{cases} i & \text{if } i \in K \\ -1 & \text{otherwise.} \end{cases}$$

Let d be a non-zero square-free integer for which $K = \mathbb{Q}(\sqrt{d})$. Let v be the largest integer for which

$$\alpha/\beta = \alpha_0^j \theta^{2^v} \tag{13}$$

with $0 \le j \le 3$ and θ in K.

Note that v exists since α/β is not a root of unity and thus θ is not a root of unity. Further, by Dobrowolski's theorem $h(\alpha/\beta) > c_1 > 0$ and

$$h(\alpha/\beta) = 2^v h(\theta).$$

Thus v cannot be arbitrarily large. Observe also that by Lemma 5

$$\left[K\left(\alpha_0^{1/2}, \theta^{1/2}\right) : K\right] = 4.$$

Next we choose w maximal so that there exists γ in K with

$$\frac{a}{b} = \alpha_0^{j_0} \theta^{j_1} \gamma^{2^w} \qquad (14)$$

and $0 \le j_0 \le 3$, $0 \le j_1 \le 2^w$. Such a choice is possible as we shall now show. First observe that

$$2^w h(\gamma) \le h\left(\frac{a}{b}\right) + j_1 h(\theta_1)$$

so

$$h(\gamma) \le c_2. \qquad (15)$$

Further we have from (14) that

$$\left(\frac{a}{b}\right)^{-4} \theta^{4j_1} \gamma^{2^{w+2}} = 1. \qquad (16)$$

Next notice that if two of the three numbers a/b, θ and γ are multiplicatively dependent then a/b and θ are multiplicatively dependent; hence, by (13), a/b and α/β are multiplicatively dependent. Therefore we may suppose that any two of the three numbers a/b, θ and γ are multiplicatively independent. Thus, by Lemma 6, there are non-zero integers b_1, b_2, b_3, with

$$\left(\frac{a}{b}\right)^{b_1} \theta^{b_2} \gamma^{b_3} = 1 \qquad (17)$$

and with

$$|b_i| \le c_3 \qquad (18)$$

for $i = 1, 2, 3$. It follows from (16) and (17) that

$$\left(\frac{a}{b}\right)^{b_1 2^{w+2}} \theta^{b_2 2^{w+2}} = \left(\frac{a}{b}\right)^{-4b_3} \theta^{4j_1 b_3}.$$

Since a/b and θ are multiplicatively independent and b_1 is non-zero it follows from (18) that w is at most c_4.

Next we observe that since w is maximal we have

$$\left[K\left(\alpha_0^{1/2}, \theta^{1/2}, \gamma^{1/2}\right) : K\right] = 8 \qquad (19)$$

for otherwise by Lemma 5 there is γ_1 in K and integers j_0 and j_1 with $0 \le j_i < 2$ for $i = 0, 1$ such that

$$\gamma = \alpha_0^{j_0} \theta^{j_1} \gamma_1^2 \qquad (20)$$

and substituting for γ in (14) using (20) we would contradict the maximality of w.

Let $\wp$ be a prime ideal of $\mathcal{O}_K$ lying above the rational prime p. Then since $p \nmid \alpha\beta ab(\alpha - \beta)^4$

$$\operatorname{ord}_p u_n \leq \operatorname{ord}_\wp ((a/b)(\alpha/\beta)^n - 1)$$

$$\leq \operatorname{ord}_\wp ((a/b)^4 (\alpha/\beta)^{4n} - 1).$$

Thus, by (13) and (14),

$$\operatorname{ord}_p u_n \leq \operatorname{ord}_\wp \left(\gamma^{2^{w+2}} \theta^{4j_1 + 2^{v+2}n} - 1 \right). \qquad (21)$$

For any real number x let $[x]$ denote the greatest integer less than or equal to x. Put

$$k = \left[\frac{\log p}{51.8 \log \log p} \right]. \qquad (22)$$

Then, for $p > c_5$, $k > 2$ and

$$\max \left(p \left(\frac{k}{\log p} \right)^k, e^k \log p \right) = p \left(\frac{k}{\log p} \right)^k. \qquad (23)$$

Our proof now splits depending on whether $\mathbb{Q}(\alpha/\beta) = \mathbb{Q}$ or not. Let us first suppose that $\mathbb{Q}(\alpha/\beta) = \mathbb{Q}$ so that α and β are integers. For any positive integer j let p_j denote the $j - 2$th smallest prime which does not divide $2p(\alpha - \beta)^4 ab\alpha\beta$. We put

$$m = 4j_1 + 2^{v+2}n \qquad (24)$$

and

$$\alpha_1 = \theta / p_3 \cdots p_k.$$

Then

$$\gamma^{2^{w+2}} \theta^m = \alpha_1^m \gamma^{2^{w+2}} p_3^m \cdots p_k^m$$

so by (21)

$$\operatorname{ord}_p u_n \leq \operatorname{ord}_p (\alpha_1^m \gamma^{2^{w+2}} p_3^m \cdots p_k^m - 1). \qquad (25)$$

Note that $\alpha_1, \gamma, p_3, \ldots, p_k$ are multiplicatively independent since θ and γ are multiplicatively independent and $p_3, \ldots, p_k$ are primes which do not divide $2p(\alpha - \beta)^4 ab\alpha\beta$. Further since $p_3, \ldots, p_k$ are different from p and p does not divide $2(\alpha - \beta)^4 ab\alpha\beta$ we see that $\alpha_1, \gamma, p_3, \ldots, p_k$ are p-adic units.

We now apply Lemma 2 with $\delta = 1$, $d = 1$, $f_\wp = 1$ and $n = k$ to conclude that

$$\operatorname{ord}_p(\alpha_1^m \gamma^{2^{w+2}} p_3^m \cdots p_k^m - 1) \le c_6(k+1)^3 \left(7e\frac{p-1}{p-2}\right)^k$$

$$\max\left(p\left(\frac{k}{\log p}\right)^k, e^k \log p\right) \log(2^{w+2}m)h(\alpha_1)h(\gamma) \log p_3 \cdots \log p_k. \tag{26}$$

For any non-zero integer n let $\omega(n)$ denote the number of distinct prime factors of n. Put

$$t = \omega(2p(\alpha - \beta)^4 ab\alpha\beta) \tag{27}$$

and let q_i denote the ith prime number. Note that

$$p_k \le q_{k+t}$$

and thus

$$\log p_3 + \cdots + \log p_k \le (k-2)\log q_{k+t}.$$

By the prime number theorem with error term, for $k > c_7$,

$$\log p_3 + \cdots + \log p_k \le 1.001(k-2)\log k. \tag{28}$$

By the arithmetic-geometric mean inequality

$$\log p_3 \cdots \log p_k \le \left(\frac{\log p_3 + \cdots + \log p_k}{k-2}\right)^{k-2}$$

and so, by (28),

$$\log p_3 \cdots \log p_k \le (1.001 \log k)^{k-2}. \tag{29}$$

Since $h(\alpha_1) \le h(\theta) + \log p_3 \cdots p_k$ it follows from (28) that

$$h(\alpha_1) \le c_8 k \log k.$$

Further

$$2^{w+2}m = 2^{w+2}(4j_1 + 2^{v+2}n) < c_9 n$$

and so

$$\log(2^{w+2}m) < c_{10} \log n. \tag{30}$$

Thus, by (23), (25), (26) and (28)–(30),

$$\operatorname{ord}_p u_n < c_{11}k^4 p\left(7e\frac{p-1}{p-2}\frac{1.001k\log k}{\log p}\right)^k \log n.$$

Therefore, by (22), for $p > c_{12}$

$$\mathrm{ord}_p u_n < pe^{-\frac{\log p}{51.9 \log \log p}} \log n. \tag{31}$$

We now suppose that $[\mathbb{Q}(\alpha/\beta) : \mathbb{Q}] = 2$. Let $\pi_3, \ldots, \pi_k$ be elements of $\mathcal{O}_K$ with the property that $N(\pi_i) = p_i$ where N denotes the norm from K to $\mathbb{Q}$ and where p_i is the $(i-2)$th smallest rational prime number of this form which does not divide $2p\alpha\beta ab(\alpha - \beta)^4$. We now put $\theta_i = \pi_i/\pi_i'$ where π_i' denotes the algebraic conjugate of π_i in $\mathbb{Q}(\alpha/\beta)$. Notice that p does not divide $\pi_i\pi_i' = p_i$ and if p does not divide $(\pi_i - \pi_i')^2$ then

$$\left(\frac{(\pi_i - \pi_i')^2}{p}\right) = \left(\frac{d}{p}\right)$$

since $\mathbb{Q}(\alpha/\beta) = \mathbb{Q}(\sqrt{d}) = \mathbb{Q}(\pi_i)$. Thus, by Lemma 1, the order of θ_i in $(\overline{\mathbb{Q}(\alpha/\beta)_\wp})^\times$ is a divisor of 2 if p divides $(\pi_i^2 - \pi_i'^2)^2$ and a divisor of $p - \left(\frac{d}{p}\right)$ otherwise. Since p is odd and p is different from p_i we observe that the order of θ_i in $(\overline{\mathbb{Q}(\alpha/\beta)_\wp})^\times$ is a divisor of $p - \left(\frac{d}{p}\right)$.

Recall (22) and put

$$\alpha_1 = \theta/\theta_3 \cdots \theta_k.$$

Then $\alpha_1^m \theta_3^m \cdots \theta_k^m = \theta^m$ and by (21) and (24) we see that

$$\mathrm{ord}_p u_n \leq \mathrm{ord}_\wp(\alpha_1^m \gamma^{2^{w+2}} \theta_3^m \cdots \theta_k^m - 1). \tag{32}$$

Observe that $\alpha_1, \gamma, \theta_3, \ldots, \theta_k$ are multiplicatively independent since θ and γ are multiplicatively independent and $p_3, \ldots, p_k$ are primes which do not divide $2(\alpha - \beta)^4 ab\alpha\beta$ and the principal prime ideals $[\pi_i]$ for $i = 3, \ldots, k$ do not ramify since $p_i \nmid 2d$. Since $p_3, \ldots, p_k$ are different from p and p does not divide $2(\alpha - \beta)^4 ab\alpha\beta$ we see that $\alpha_1, \gamma, \theta_3, \ldots, \theta_k$ are p-adic units.

Notice that

$$K\left(\alpha_0^{1/2}, \theta^{1/2}, \gamma^{1/2}, \theta_3^{1/2}, \ldots, \theta_k^{1/2}\right) = K\left(\alpha_0^{1/2}, \alpha_1^{1/2}, \gamma^{1/2}, \theta_3^{1/2}, \ldots, \theta_k^{1/2}\right).$$

Further, by (19),

$$\left[K\left(\alpha_0^{1/2}, \theta^{1/2}, \gamma^{1/2}, \theta_3^{1/2}, \ldots \theta_k^{1/2}\right) : K\right] = 2^{k+1}, \tag{33}$$

since otherwise by Lemma 5 there is an integer i with $3 \leq i \leq k$ and integers $j_0, \ldots, j_{i-1}$ with $0 \leq j_b \leq 1$ for $b = 0, \ldots, i-1$ and an element ψ of K for which

$$\theta_i = \alpha_0^{j_0} \theta^{j_1} \gamma^{j_2} \theta_3^{j_3} \cdots \theta_{i-1}^{j_{i-1}} \psi^2. \tag{34}$$

But then the order of the prime ideal $[\pi_i]$ on the left-hand side of (34) is even which is a contradiction. Thus (33) holds.

Since p does not divide the discriminant of K and $[K : \mathbb{Q}] = 2$ either p splits, in which case $f_\wp = 1$ and $\left(\frac{d}{p}\right) = 1$, or p is inert, in which case $f_\wp = 2$ and $\left(\frac{d}{p}\right) = -1$, see [14]. Put

$$\delta = (p^{f_\wp} - 1)/|\langle \overline{\alpha_0}, \overline{\alpha_1}, \overline{\gamma}, \overline{\theta_3}, \ldots, \overline{\theta_k} \rangle|.$$

Observe that if $\left(\frac{d}{p}\right) = 1$ then

$$p^{f_\wp}/\delta \le p. \tag{35}$$

Let us now determine $|\langle \overline{\alpha_0}, \overline{\alpha_1}, \overline{\gamma}, \overline{\theta_3}, \ldots, \overline{\theta_k} \rangle|$ in the case $\left(\frac{d}{p}\right) = -1$. We have shown that the order of $\overline{\theta_i}$ is a divisor of $p+1$ for $i = 3, \ldots, k$. Since α and β are conjugates $N(\alpha/\beta)$, the norm from K to $\mathbb{Q}$ of α/β is 1. Therefore by (13), $N(\theta) = \pm 1$. Similarly a and b are conjugates over $\mathbb{Q}$ so $N(a/b) = 1$ and thus $N(\gamma) = \pm 1$. By Hilbert's Theorem 90, see Theorem 14.35 of [10], $\theta^2 = \rho/\rho'$ where ρ and ρ' are conjugate algebraic integers in K. Similarly, by (13) and (14), $\gamma^2 = \lambda/\lambda'$ where λ and λ' are conjugate algebraic integers in K.

Note that we may suppose that the principal ideals $[\rho]$ and $[\rho']$ have no non-trivial principal ideal divisors in common. Further since p does not divide $2(\alpha - \beta)^2 ab\alpha\beta$ and since $\left(\frac{d}{p}\right) = -1$, $[p]$ is a principal ideal of $\mathcal{O}_K$ and p does not divide $\rho\rho'$. The order of θ^2 in $(\overline{K_\wp})^\times$ is a divisor of $p+1$ by Lemma 1 and thus θ has order a divisor of $2(p+1)$. By the same reasoning as above we find that the order of γ^2 in $(\overline{K_\wp})^\times$ is a divisor of $p+1$ and so, by Lemma 1, γ has order a divisor of $2(p+1)$. Since $\alpha_0^4 = 1$ and, as we have already established, the order of θ_i is a divisor of $p+1$ for $i = 3, \ldots, k$ we see that

$$|\langle \overline{\alpha_0}, \overline{\theta}, \overline{\gamma}, \overline{\theta_3}, \ldots, \overline{\theta_k} \rangle| \le 2(p+1)$$

hence

$$|\langle \overline{\alpha_0}, \overline{\alpha_1}, \overline{\gamma}, \overline{\theta_3}, \ldots, \overline{\theta_k} \rangle| \le 2(p+1).$$

Therefore

$$\delta = (p^2 - 1)/|\langle \overline{\alpha_0}, \overline{\alpha_1}, \overline{\gamma}, \overline{\theta_3}, \ldots, \overline{\theta_k} \rangle| \ge (p-1)/2. \tag{36}$$

We now apply Lemma 2, noting, by (35) and (36), that

$$p^{f_\wp}/\delta \le 2p^2/(p-1).$$

Thus, by (23),

$$\mathrm{ord}_\wp(\alpha_1^m \gamma^{2^{w+2}} \theta_3^m \cdots \theta_k^m - 1) \le c_{12} k^3 \log p \left(7e\frac{p-1}{p-2}\right)^k$$
$$2^k p \left(\frac{k}{\log p}\right)^k (\log m) h(\alpha_1) h(\gamma) h(\theta_3) \cdots h(\theta_k). \tag{37}$$

Observe that $\theta_i = \pi_i/\pi_i'$ and that $p_i(x - \pi_i/\pi_i')(x - \pi_i'/\pi_i) = p_i x^2 - (\pi_i^2 + \pi_i'^2)x + p_i$ is the minimal polynomial of θ_i over the integers since $[\pi_i]$ is unramified. Now either the discriminant of K is negative in which case $|\pi_i| = |\pi_i'|$ or it is positive in which case there is a fundamental unit $\varepsilon > 1$ in $\mathcal{O}_K$. As in [42] we may replace π_i by $\pi_i \varepsilon^u$ for any integer u. Without loss of generality we may suppose that $p_i^{1/2} \le |\pi_i| \le p_i^{1/2}\varepsilon$ and hence that $p_i^{1/2}\varepsilon^{-1} \le |\pi_i'| \le p_i^{1/2}$. Therefore

$$h(\theta_i) \le \frac{1}{2}\log p_i \varepsilon^2 = \frac{1}{2}\log p_i + \log \varepsilon \quad \text{for } d > 0$$

and

$$h(\theta_i) \le \frac{1}{2}\log p_i \quad \text{for } d < 0.$$

Put

$$R = \begin{cases} \log \varepsilon & \text{for } d > 0 \\ 0 & \text{for } d < 0. \end{cases}$$

Then

$$h(\theta_i) \le \frac{1}{2}\log p_i + R$$

for $i = 3, \ldots, k$. We also can ensure that

$$h(\theta_3 \cdots \theta_k) \le \frac{1}{2}\log p_3 \cdots p_k + R$$

and so

$$h(\alpha_1) \le h(\theta) + \frac{1}{2}\log p_3 \cdots p_k + R. \tag{38}$$

Let t be given by (27) and let q_i denote the ith prime number which is representable as the norm of an element of $\mathcal{O}_K$. Note that

$$p_k \le q_{k+t}$$

and so

$$\log p_3 + \cdots + \log p_k \le (k-2)\log q_{k+t}.$$

Therefore by Lemma 3 for $k > c_{13}$

$$(\log p_3 + 2R) + \cdots + (\log p_k + 2R) < (k-2)(1.0005\log k + 2R) < 1.001(k-2)\log k \tag{39}$$

and so, by the arithmetic-geometric mean inequality,

$$(\log p_3 + 2R) \cdots (\log p_k + 2R) < (1.001\log k)^{k-2}.$$

Thus, since p_k is at least k, for $k > c_{14}$,

$$2^{k-2}h(\theta_3)\cdots h(\theta_k) \leq (\log p_3 + 2R)\cdots(\log p_k + 2R) < (1.001\log k)^{k-2}. \qquad (40)$$

By (38) and (39)

$$h(\alpha_1) < c_{15}k\log k$$

and by (13), (14) and (24)

$$m \leq c_{16}n. \qquad (41)$$

Thus, by (32), (38), (40) and (41),

$$\text{ord}_p\, u_n \leq c_{17}k^4 p\log p\left(7e\frac{p-1}{p-2}1.001\frac{k\log k}{\log p}\right)^k\log n.$$

Therefore, by (22), for $p > c_{18}$, we obtain (31) in this case also and our result follows.   □

5   Proof of Theorem 1

Let $K = \mathbb{Q}(\alpha)$ and let $\mathcal{O}_K$ denote the ring of algebraic integers of K. For any θ in $\mathcal{O}_K$ let $[\theta]$ denote the ideal in $\mathcal{O}_K$ generated by θ. We have

$$u_n = r_1 u_{n-1} + r_2 u_{n-2} \quad \text{for } n = 2, 3, \ldots.$$

Let l denote the greatest common divisor of r_1^2 and r_2. Then α^2/l and β^2/l are algebraic integers in K. Further $\frac{r_1^2 + 2r_2}{l}$ and $(r_2/l)^2$ are coprime hence, as in Lemma A.10 of [34], $\left(\left[\frac{\alpha^2}{l}\right], \left[\frac{\beta^2}{l}\right]\right) = ([1])$. We may put

$$v_n = l^{-n}u_{2n} = a\left(\frac{\alpha^2}{l}\right)^n + b\left(\frac{\beta^2}{l}\right)^n$$

and

$$w_n = l^{-n}u_{2n+1} = a\alpha\left(\frac{\alpha^2}{l}\right)^n + b\beta\left(\frac{\beta^2}{l}\right)^n,$$

for $n = 0, 1, 2, \ldots$. Recall $r_2 = \alpha\beta$. For any prime p which does not divide r_2 we have

$$\text{ord}_p(u_{2n}) = \text{ord}_p(v_n) \quad \text{and} \quad \text{ord}_p(u_{2n+1}) = \text{ord}_p(w_n).$$

Further a/b and α/β are multiplicatively independent if and only if a/b and $(\alpha/\beta)^2$ are multiplicatively independent. Similarly a/b and α/β are multiplicatively independent if and only if $a\alpha/b\beta$ and $(\alpha/\beta)^2$ are multiplicatively independent.

Therefore, by considering the non-degenerate binary recurrence sequences $(v_n)_{n=0}^{\infty}$ and $(w_n)_{n=0}^{\infty}$ in place of $(u_n)_{n=0}^{\infty}$, we may assume, without loss of generality, that $([\alpha], [\beta]) = [1]$.

Let $c_1, c_2, \ldots$ denote positive numbers which are effectively computable in terms of a, b, α and β. By the result of Luca given in (8) the theorem follows if a/b and α/β are multiplicatively dependent. We may assume therefore that a/b and α/β are multiplicatively independent. For any integer h and prime p define $|h|_p$ by

$$|h|_p = p^{-\operatorname{ord}_p h}.$$

It follows from the proof of Theorem 1 of [40] that for any prime p and integer $n \geq 2$

$$\log\left(|u_n|_p^{-1}\right) < c_1 p^2 (\log n)^2. \tag{42}$$

By Lemma 7 for $p > c_2$,

$$\log\left(|u_n|_p^{-1}\right) < p \log p \exp(-\log p / 51.9 \log\log p) \log n. \tag{43}$$

By Lemma 4

$$\log|u_n| > c_3 n. \tag{44}$$

Write

$$|u_n| = p_1^{\ell_1} \cdots p_r^{\ell_r} \tag{45}$$

where $p_1, \ldots, p_r$ are distinct primes and $\ell_1, \ldots, \ell_r$ are positive integers. It follows from (42)–(45) that

$$\frac{n}{\log n} < c_4 \sum_{i=1}^{r} p_i \log p_i \exp(-\log p_i / 51.9 \log\log p_i). \tag{46}$$

Put $p_r = P(u_n)$. The right-hand side of inequality (46) is at most

$$r p_r \log p_r \exp(-\log p_r / 51.9 \log\log p_r)$$

and so by the prime number theorem

$$c_5 \frac{n}{\log n} < p_r^2 \exp(-\log p_r / 51.9 \log\log p_r).$$

Therefore

$$P(u_n) = p_r > c_6 n^{1/2} \exp(\log n / 103.99 \log\log n),$$

and our result now follows.

Acknowledgements Research supported in part by the Canada Research Chairs Program and by Grant A3528 from the Natural Sciences and Engineering Research Council of Canada.

References

1. P.B. Allen, On the multiplicity of linear recurrence sequences. J. Number Theory **126**, 212–216 (2007)
2. F. Amoroso, E. Viada, On the zeros of linear recurrence sequences. Acta Arith. **147**, 387–396 (2011)
3. A. Baker, A sharpening of the bounds for linear forms in logarithms. Acta Arith. **21**, 117–129 (1972)
4. A. Baker, The theory of linear forms in logarithms. In *Transcendence Theory: Advances and Applications*, ed. by A. Baker, D.W. Masser (Academic, London, 1977), pp. 1–27
5. A. Baker, H.M. Stark, On a fundamental inequality in number theory. Ann. Math. **94**, 190–199 (1971)
6. A.S. Bang, Taltheoretiske undersøgelser. Tidsskrift for Mat. **4**, 70–78, 130–137 (1886)
7. Y. Bilu, G. Hanrot, P.M. Voutier, Existence of primitive divisors of Lucas and Lehmer numbers. J. reine angew. Math. **539**, 75–122 (2001)
8. G.D. Birkhoff, H.S. Vandiver, On the integral divisors of $a^n - b^n$. Ann. Math. **5**, 173–180 (1904)
9. R.D. Carmichael, On the numerical factors of the arithmetic forms $\alpha^n \pm \beta^n$. Ann. Math. **15**, 30–70 (1913)
10. H. Cohn, *A Classical Invitation to Algebraic Numbers and Class Fields* (Springer, New York, 1978)
11. P. Erdős, Some recent advances and current problems in number theory. In *Lectures on Modern Mathematics*, vol. III, ed. by T.L. Saaty (Wiley, New York, 1965), pp. 196–244
12. J.H. Evertse, On sums of S-units and linear recurrences. Compositio Math. **53**, 225–244 (1984)
13. K. Ford, F. Luca, I. Shparlinski, On the largest prime factor of the Mersenne numbers. Bull. Austr. Math. Soc. **79**, 455–463 (2009)
14. E. Hecke, Lectures on the theory of algebraic numbers, in *Graduate Texts in Mathematics*, vol. 77 (Springer, New York, 1981)
15. D.H. Lehmer, An extended theory of Lucas' functions. Ann. Math. **31**, 419–448 (1930)
16. T. Loher, D. Masser, Uniformly counting points of bounded height. Acta Arith. **111**, 277–297 (2004)
17. F. Luca, Arithmetic properties of members of a binary recurrent sequence. Acta Arith. **109**, 81–107 (2003)
18. E. Lucas, Théorie des fonctions numériques simplement périodiques. Amer. J. Math. **1**, 184–240, 289–321 (1878)
19. K. Mahler, Eine arithmetische Eigenschaft der rekurrierenden Reihen. Mathematica (Leiden) **3**, 153–156 (1934–1935)
20. K. Mahler, Eine arithmetische Eigenschaft der Taylor-Koeffizienten rationaler Funktionen. Proc. Akad. Wetensch. Amsterdam **38**, 50–60 (1935)
21. L. Murata, C. Pomerance, On the largest prime factor of a Mersenne number. In *CRM Proc. Lecture Notes*, vol. 36, ed. by H. Kisilevsky, E.Z. Goren (American Mathematical Society, Providence, 2004), pp. 209–218
22. R. Murty, S. Wong, The *ABC* conjecture and prime divisors of the Lucas and Lehmer sequences. In *Number Theory for the Millennium*, vol. III, ed. by M.A. Bennett, B.C. Berndt, N. Boston, H.G. Diamond, A.J. Hildebrand, W. Philipp, A.K. Peters (Natick, MA, 2002), pp. 43–54
23. E.M. Matveev, An explicit lower bound for a homogeneous rational linear form in the logarithms of algebraic numbers II (Russian). Izv. Ross. Akad. Nauk Ser. Mat. **64**, 125–180 (2000)
24. A.J. van der Poorten, Linear forms in logarithms in the p-adic case. In *Transcendence Theory: Advances and Applications,* ed. by A. Baker, D.W. Masser (Academic, London, 1977), pp. 29–57
25. A.J. van der Poorten, H.P. Schlickewei, The growth conditions for recurrence sequences. Macquarie Math. Reports 82–0041 (1982)

26. A. Schinzel, On primitive prime factors of $a^n - b^n$. Proc. Cambridge Philos. Soc. **58**, 555–562 (1962)
27. A. Schinzel, The intrinsic divisors of Lehmer numbers in the case of negative discriminant. Ark. Mat. **4**, 413–416 (1962)
28. A. Schinzel, On two theorems of Gelfond and some of their applications. Acta Arith. **13**, 177–236 (1967)
29. A. Schinzel, Primitive divisors of the expression $A^n - B^n$ in algebraic number fields. J. reine angew. Math. **268/269**, 27–33 (1974)
30. H.P. Schlickewei, Linearformen mit algebraischen Koeffizienten. Manuscripta Math. **18**, 147–185 (1976)
31. W.M. Schmidt, The zero multiplicity of linear recurrence sequences. Acta Math. **182**, 243–282 (1999)
32. W.M. Schmidt, Zeros of linear recurrence sequences. Publ. Math. Debrecen **56**, 609–630 (2000)
33. T.N. Shorey, C.L. Stewart, On divisors of Fermat, Fibonacci, Lucas and Lehmer numbers II. J. London Math. Soc. **23**, 17–23 (1981)
34. T.N. Shorey, R. Tijdeman, *Exponential Diophantine Equations*. Cambridge Tracts in Mathematics, vol. 87 (Cambridge University Press, Cambridge, 1986)
35. I.E. Shparlinski, Prime divisors of recurrent sequences. Isv. Vyssh. Uchebn. Zaved. Math. **215**, 101–103 (1980)
36. C.L. Stewart, The greatest prime factor of $a^n - b^n$. Acta Arith. **26**, 427–433 (1975)
37. C.L. Stewart, Divisor properties of arithmetical sequences. Ph.D. thesis, Cambridge, 1976
38. C.L. Stewart, On divisors of Fermat, Fibonacci, Lucas and Lehmer numbers. Proc. London Math. Soc. **35**, 425–447 (1977)
39. C.L. Stewart, Primitive divisors of Lucas and Lehmer numbers. In *Transcendence Theory: Advances and Applications*, ed. by A. Baker, D.W. Masser (Academic, London, 1977), pp. 79–92
40. C.L. Stewart, On divisors of terms of linear recurrence sequences. J. Reine Angew. Math. **333**, 12–31 (1982)
41. C.L. Stewart, On the greatest square-free factor of terms of a linear recurrence sequence. In *Diophantine Equations*, ed. by N. Saradha (Narosa Publishing House, New Delhi, 2008), pp. 257–264
42. C.L. Stewart, On divisors of Lucas and Lehmer numbers. Acta Math. (to appear)
43. C.L. Stewart, K. Yu, On the *abc* conjecture II. Duke Math. J. **108**, 169–181 (2001)
44. M. Waldschmidt, A lower bound for linear forms in logarithms. Acta Arith. **37**, 257–283 (1980)
45. K. Yu, P-adic logarithmic forms and a problem of Erdős. Acta Math. (to appear)
46. K. Yu, L.-k. Hung, On binary recurrence sequences. Indag. Mathem., N.S. **6**, 341–354 (1995)
47. K. Zsigmondy, Zur Theorie der Potenzreste. Monatsh. Math. **3**, 265–284 (1892)

Some Notes on Weighted Sum Formulae
for Double Zeta Values

James Wan

Abstract We present a unified approach which gives completely elementary proofs
of three weighted sum formulae for double zeta values. This approach also leads to
new evaluations of sums involving the harmonic numbers, the alternating double
zeta values, and the Mordell–Tornheim double sum. We discuss a heuristic for
finding or dismissing the existence of similar simple sums. We also produce some
new sums from recursions involving the Riemann zeta and the Dirichlet beta
functions.

1 Introduction

Multiple zeta values are a natural generalisation of the Riemann zeta function at
the positive integers; for our present purposes, we shall only consider multiple zeta
values of length 2 (or *double zeta values*), defined for integers $a \geq 2$ and $b \geq 1$ by

$$\zeta(a,b) = \sum_{n=1}^{\infty} \sum_{m=1}^{n-1} \frac{1}{n^a m^b}. \tag{1}$$

It is rather immediate from series manipulations that

$$\zeta(a,b) + \zeta(b,a) = \zeta(a)\zeta(b) - \zeta(a+b), \tag{2}$$

thus we can compute in closed form $\zeta(a,a)$, though it is not a priori obvious that
many other multiple zeta values can be factored into Riemann zeta values. Euler was

J. Wan (✉)
CARMA, The University of Newcastle, Callaghan, NSW, Australia
e-mail: james.wan@newcastle.edu.au

J.M. Borwein et al. (eds.), *Number Theory and Related Fields: In Memory of Alf van der
Poorten*, Springer Proceedings in Mathematics & Statistics 43,
DOI 10.1007/978-1-4614-6642-0_19, © Springer Science+Business Media New York 2013

among the first to study multiple zeta values; indeed, he gave the sum formula
(for $s \geq 3$)

$$\sum_{j=2}^{s-1} \zeta(j, s-j) = \zeta(s). \qquad (3)$$

When $s = 3$, this formula reduces to the celebrated result $\zeta(2,1) = \zeta(3)$, which
has many other proofs [6]. Formula (3) itself may be shown in many ways, one
of which uses partial fractions, telescoping sums and change of summation order,
which we present in Sect. 2. Given the ease with which formula (3) may be derived
or even experimentally observed (see Sect. 4), it is perhaps surprising that a similar
equation, with "weights" 2^j inserted, was only first discovered in 2007 [14]:

$$\sum_{j=2}^{s-1} 2^j \zeta(j, s-j) = (s+1)\zeta(s). \qquad (4)$$

Formula (4) was originally proven in [14] using the closed form expression for
$\zeta(n, 1)$ (which follows from (2) and (3)), together with induction on shuffle
relations—relations arising from iterated integration of generalised polylogarithms
which encapsulate the multiple zeta values. Equation (4) has been generalised to
more sophisticated weights other than 2^j using generating functions and to lengths
greater than 2 (see e.g. [9]).

In conjunction, (2)–(4) can be used to find a closed form for $\zeta(a, b)$ for $a+b \leq 6$.
Indeed, it is a result Euler wrote down and first elucidated in [5] that all $\zeta(a, b)$ with
$a + b$ odd may be expressed in terms of Riemann zeta value (in contrast, $\zeta(5,3)$ is
conjectured not reducible to more fundamental constants).

The third weighted sum we will consider is

$$\sum_{j=2}^{2s-1} (-1)^j \zeta(j, 2s-j) = \frac{1}{2}\zeta(2s). \qquad (5)$$

Given that all known proofs of (4) had their genesis in more advanced areas, one
purpose of this note is to show that (4) and the alternating (5) are not intrinsically
harder than (3) and can be proven in a few short lines. We use the same techniques in
Sect. 3 to give similar identities involving closely related functions. We also observe
that some double zeta values sums are related to recursions (or convolutions)
satisfied by the Riemann zeta function, a connection which we exploit in Sect. 4.
Lastly, we use such recursions and a reflection formula to produce new results for
character sums as defined in [4].

2 Elementary Proofs

In the proofs below, the orders of summation may be interchanged freely, as the
sums involved are absolutely convergent.

Proof of (4). We write the left-hand side of (4) as

$$\sum_{j=2}^{s-1} \sum_{m=1}^{\infty} \sum_{n=1}^{\infty} \frac{2^j}{n^{s-j}(n+m)^j}.$$

We consider the two cases, $m = n$ and $m \neq n$. In the former case the sum immediately yields $(s-2)\zeta(s)$. In the latter case, we do the geometric sum in j first to obtain

$$\sum_{\substack{m,n>0 \\ m \neq n}} \frac{2^s}{(n^2 - m^2)(n+m)^{s-2}} - \frac{4}{(n^2-m^2)n^{s-2}}. \tag{6}$$

The first summand in (6) has antisymmetry in the variables m, n and hence vanishes when summed.

For the second term in (6), we use partial fractions to obtain

$$\sum_{\substack{m>0 \\ m \neq n}} \frac{1}{m^2 - n^2} = \frac{1}{2n} \sum_{\substack{m>0 \\ m \neq n}} \frac{1}{m-n} - \frac{1}{m+n} = \frac{3}{4n^2},$$

as the last sum telescopes (this is easy to see by first summing up to $m = 3n$, then looking at the remaining terms $2n$ at a time).

Therefore, summing over n in the second term of (6) gives $3\zeta(s)$. The result follows. □

Our proof suggests that the base "2" in the weighted sum is rather special as it induces antisymmetry. Another special case is obtained by replacing the 2 by a 1, and the same method proves Euler's result.

Proof of (3). We apply the same procedure as in the previous proof and sum the geometric series first, so the left-hand side becomes

$$\sum_{m,n>0} \frac{1}{m(m+n)n^{s-2}} - \frac{1}{m(m+n)^{s-1}} = \sum_{n>0} \frac{1}{n^{s-1}} \sum_{m>0} \left(\frac{1}{m} - \frac{1}{m+n}\right) - \zeta(s-1,1)$$

$$= \sum_{n=1}^{\infty} \frac{1}{n^{s-1}} \sum_{k=1}^{n} \frac{1}{k} - \zeta(s-1,1)$$

$$= \sum_{n=1}^{\infty} \frac{1}{n^s} + \sum_{n=1}^{\infty} \frac{1}{n^{s-1}} \sum_{k=1}^{n-1} \frac{1}{k} - \zeta(s-1,1)$$

$$= \zeta(s),$$

where we have used partial fractions for the first equality and telescoping for the second. □

Likewise we may easily prove the alternating sum (5):

Proof of (5). We write the left-hand side out in full as above and then perform the geometric sum first to obtain

$$\sum_{m,n>0} \frac{1}{(m+n)(m+2n)n^{2s-2}} - \frac{1}{(m+2n)(m+n)^{2s-1}}.$$

Let $k = m+n$, so we have

$$\sum_{k>n>0} \frac{1}{k(k+n)n^{2s-2}} - \frac{1}{(k+n)k^{2s-1}}.$$

In the first term, use partial fractions and sum over k from $n+1$ to ∞; in the second term, sum over n from 1 to $k-1$. We get

$$\left(\sum_{n>0} \frac{1}{n^{2s-1}} \sum_{k=n+1}^{2n} \frac{1}{k}\right) - \left(\sum_{k>0} \frac{1}{k^{2s-1}} \sum_{n=k+1}^{2k-1} \frac{1}{n}\right).$$

It now remains to observe that if we rename the variables in the second bracket, then the two sums telescope to $\sum_{n>0} 1/(2n^{2s}) = \zeta(2s)/2$. Hence (5) holds.   □

Remark 1 The final sums we shall consider in this section are

$$\sum_{j=1}^{s-1} \zeta(2j, 2s-2j) = \frac{3}{4}\zeta(2s), \qquad \sum_{j=1}^{s-1} \zeta(2j+1, 2s-2j-1) = \frac{1}{4}\zeta(2s). \quad (7)$$

These results were first given in [8] and later proven in a more direct manner in [13] using recursion of the Bernoulli numbers. The difference of the two equations in (7) is (5) and the sum is a case of (3). Therefore, the elementary nature of (7) is revealed since we have elementary proofs of (3) and (5).

If we add the first equation in (7) to itself but reverse the order of summation, then upon applying (2), we produce the identity

$$\sum_{j=1}^{s-1} \zeta(2j)\zeta(2s-2j) = \left(s+\frac{1}{2}\right)\zeta(2s),$$

which is usually derived from the generating function of the Bernoulli numbers B_n, since $2(2n)!\,\zeta(2n) = (-1)^{n+1}(2\pi)^{2n}B_{2n}$.   □

3 New Sums

We shall see in this section that the elementary methods in Sect. 2 can in fact take us a long way.

3.1 Mordell–Tornheim Double Sum

The *Mordell–Tornheim double sum* (sometimes also known as the Mordell–Tornheim–Witten zeta function) is defined as

$$W(r,s,t) = \sum_{n=1}^{\infty} \sum_{m=1}^{\infty} \frac{1}{n^r m^s (n+m)^t}.$$

Note that $W(r,s,0) = \zeta(r)\zeta(s)$ and $W(r,0,t) = W(0,r,t) = \zeta(t,r)$. Due to the simple recursion $W(r,s,t) = W(r-1,s,t+1) + W(r,s-1,t+1)$, when r,s,t are positive integers, W may be expressed in terms of Riemann zeta or double zeta values (see e.g. [10]).

We again emulate the proof of (4) to obtain what seems to be a new sum over W.

Theorem 3.1 *For integers $a \geq 0$ and $s \geq 3$,*

$$\sum_{j=2}^{s-1} W(s-j,a,j) = (-1)^a \zeta(s+a) + (-1)^a \zeta(s+a-1,1) - \zeta(s-1,a+1)$$

$$- \sum_{i=2}^{a+1} (-1)^{i+a} \zeta(i)\zeta(s+a-i) \tag{8}$$

$$= \sum_{i=2}^{s-1} \binom{i+a-2}{a} \zeta(i+a,s-i) + \sum_{i=s-a}^{s-1} \binom{i+a-2}{s-3} \zeta(i+a,s-i).$$

Proof As the Mordell–Tornheim double sum values can be expressed as double zeta values, the second equality follows after some simplification. For the first equality, we sketch the proof based on that of (4). Writing the left-hand side of (8) as a triple sum, we perform the geometric sum first to produce

$$\sum_{m,n>0} \frac{(-1)^a}{m^{a+1} n^{s-2}(m+n)} - \frac{(-1)^a}{m^{a+1}(m+n)^{s-1}}.$$

To the first term we apply the partial fraction decomposition

$$\frac{1}{m^b(m+n)} = \frac{(-1)^b}{n^b(m+n)} + \sum_{i=1}^{b} \frac{(-1)^{b-i}}{m^i n^{b+1-i}}.$$

We recognise the resulting sums as well as the second term above as Riemann zeta and double zeta values. The result follows readily.　　　　□

When $a = 0$ in (8), we recover (3); when $a = 1$, we obtain the very pretty formula

Corollary 2

$$\sum_{j=2}^{s} W(s-j,1,j) = \zeta(2,s-1), \tag{9}$$

which, by the second equality in (8), is equivalent to

$$\sum_{j=2}^{s-1} j\zeta(j,s-j) = 2\zeta(s) + \zeta(2,s-1) - (s-2)\zeta(s-1,1). \tag{10}$$

When $a = 2$ in (8), we have

$$\sum_{j=2}^{s} W(s-j,2,j) = \zeta(s+2) + \zeta(s+1,1) + \zeta(3,s-1) - \zeta(2,s).$$

A counterpart to (9) is the following alternating sum:

$$\sum_{j=2}^{2s} (-1)^j W(2s+1-j,1,j) = \zeta(2s+1,1) + \frac{1}{4}\zeta(2s+2), \tag{11}$$

and the same procedure can be used to prove this and to provide a closed form for the general case, that is, the alternating sum of $W(s-j,a,j)$, though we omit the details.

3.2 Sums Involving the Harmonic Numbers

The nth *harmonic number* is given by $H_n = \sum_{k=1}^{n} \frac{1}{k}$. If we replace $2s$ by $2s+1$ in the proof of (5) (i.e., when the sum of arguments in the double zeta value is odd instead of even), then we obtain

$$\frac{5}{2}\zeta(2s+1) + 2\zeta(2s,1) + \sum_{j=2}^{2s} (-1)^j \zeta(j,2s+1-j) = 2\sum_{n=1}^{\infty} \frac{H_{2n}}{n^{2s}}. \tag{12}$$

Combined with known double zeta values, we can evaluate the right-hand side, giving

$$\sum_{n=1}^{\infty} \frac{H_{2n}}{n^4} = \frac{37}{4}\zeta(5) - \frac{2}{3}\pi^2\zeta(3),$$

etc., in agreement with results obtained via Mellin transform and generating functions in [4] (in whose notation such sums are related to $[2a,1](2s,1)$ – this notation is explained in Sect. 4). Indeed, replacing our right-hand side with results in [4], we have:

$$\sum_{j=2}^{2s} (-1)^j \zeta(j, 2s+1-j)$$

$$= (4^s - s - 2)\zeta(2s+1) - 2\sum_{k=1}^{s-1} (4^{s-k} - 1)\zeta(2k)\zeta(2s+1-2k). \quad (13)$$

Similarly, using weight $\frac{1}{2}$ (instead of 2), we have another new result:

Lemma 3 *For integer $s \geq 3$,*

$$\sum_{j=2}^{s-1} 2^{1-j}\zeta(j, s-j) = (2^{1-s} - 1)\big(\zeta(s-1, 1) - 2\log(2)\zeta(s-1)\big)$$

$$+ (2^{2-s} - 1)\zeta(s) + \sum_{n=0}^{\infty} \frac{H_n}{(2n+1)^{s-1}}. \quad (14)$$

Therefore, we may produce evaluations such as

$$\sum_{n=0}^{\infty} \frac{H_n}{(2n+1)^4} = \frac{372\zeta(5) - 21\pi^2\zeta(3) - 2\pi^4\log(2)}{96},$$

$$\sum_{n=0}^{\infty} \frac{H_n}{(2n+1)^5} = \frac{\pi^6 - 294\zeta(3)^2 - 744\log(2)\zeta(5)}{384}.$$

Indeed, in (14) the harmonic number sum relates to the functions $[2a, 1]$ and $[2a, 2a]$ in [4], and when s is odd, we use their closed forms to simplify (14):

$$\sum_{j=2}^{2s} 2^{1-j}\zeta(j, 2s+1-j) = (s-1+2^{1-2s})\zeta(2s+1)$$

$$- \sum_{k=1}^{s-1} (4^{-k} - 4^{-s})(4^k - 2)\zeta(2k)\zeta(2s+1-2k). \quad (15)$$

On the other hand, if we chose even s in (14), then $[2a, 1]$, $[2a, 2a]$ seem not to simplify in terms of more basic constants, though below we manage to find a closed form for their difference (the proof is more technical than other results in this section). Combined with (14), we have

Theorem 3.4 *For integer $s \geq 2$,*

$$\sum_{n=0}^{\infty} \frac{H_n}{(2n+1)^{2s-1}} = (1 - 4^{-s})(2s-1)\zeta(2s) - (2 - 4^{1-s})\log(2)\zeta(2s-1)$$

$$+ (1 - 2^{-s})^2\zeta(s)^2 - \sum_{k=2}^{s} 2(1 - 2^{-k})(1 - 2^{k-2s})\zeta(k)\zeta(2s-k); \quad (16)$$

$$\sum_{j=2}^{2s-1} 2^{1-j}\zeta(j,2s-j) = \frac{1}{2}(1-2^{1-s})^2\zeta(s)^2 + \frac{1}{2}(2^{3-2s}+2s-3)\zeta(2s)$$

$$-\sum_{k=2}^{s}(2^{k-1}-1)(2^{1-k}-4^{1-s})\zeta(k)\zeta(2s-k). \qquad (17)$$

Proof We only need to prove the first equality as the second follows from (14); to achieve this, we borrow techniques from [4].

Using the fact that the harmonic number sum is $2([2a,1](2s-1,t)-[2a,2a](2s-1,t))$ in the notation of [4], we use the results therein (obtained using Mellin transforms) to write down its integral equivalent:

$$\sum_{n=0}^{\infty}\frac{H_n}{(2n+1)^{2s-1}} = \int_0^1 \frac{\log(x)^{2s-2}\log(1-x^2)}{\Gamma(2s-1)(x^2-1)}\,dx.$$

We denote its generating function by $F(w)$, and after interchanging orders of summation and integration, we obtain

$$F(w) := \sum_{s=2}^{\infty}\left[\int_0^1 \frac{\log(x)^{2s-2}\log(1-x^2)}{\Gamma(2s-1)(x^2-1)}\,dx\right]w^{2s-2}$$

$$= \int_0^1 \frac{x^{-w}(x^w-1)^2\log(1-x^2)}{2(x^2-1)}\,dx$$

$$= -\frac{1}{2}\int_0^1 \frac{d}{dq}\left[x^{-w}(x^w-1)^2(1-x^2)^{q-1}\right]_{q=0}\,dx.$$

Next, we interchange the order of differentiation and integration; the result is a Beta integral which evaluates to

$$F(w) = \frac{1}{4}\frac{d}{dq}\left[\frac{2\Gamma(1/2)\Gamma(q)}{\Gamma(q+1/2)} - \frac{\Gamma((1-w)/2)\Gamma(q)}{\Gamma((1-w)/2+q)} - \frac{\Gamma((1+w)/2)\Gamma(q)}{\Gamma((1+w)/2+q)}\right]_{q=0}$$

$$= \frac{1}{8}\left[8\log(2)^2 - \pi^2 + \pi^2\sec^2\left(\frac{\pi w}{2}\right) - \left[\psi\left(\frac{1-w}{2}\right)+\gamma\right]^2 - \left[\psi\left(\frac{1+w}{2}\right)+\gamma\right]^2\right],$$

where ψ denotes the digamma function and γ is the Euler–Mascheroni constant. The desired equality follows using the series expansions

$$-\psi\left(\frac{1-w}{2}\right) = \gamma + 2\log(2) + \sum_{k=1}^{\infty}(2-2^{-k})\zeta(k+1)w^k,$$

$$\frac{\pi^2}{2}\sec^2\left(\frac{\pi w}{2}\right) = \sum_{k=0}^{\infty}(4-4^{-k})(2k+1)\zeta(2k+2)w^{2k}. \qquad \square$$

Remark 2 Thus Theorem 3.4, together with [4], completes the evaluation of

$$\sum_{n=0}^{\infty} \frac{H_n}{(2n+1)^s}$$

in terms of well-known constants for integer $s \geq 2$. In [1, Theorem 6.5] it is claimed that the said sum may be evaluated in terms of Riemann zeta values alone, but the claim is unsubstantiated by numerical checks and notably the constant $\log(2)$ is missing from the purported evaluation. $\qquad\square$

3.3 Alternating Double Zeta Values

The *alternating* double zeta values $\zeta(a,\overline{b})$ are defined as

$$\zeta(a,\overline{b}) = \sum_{n=1}^{\infty} \sum_{m=1}^{n-1} \frac{1}{n^a} \frac{(-1)^{m-1}}{m^b},$$

with $\zeta(\overline{a},b)$ and $\zeta(\overline{a},\overline{b})$ defined similarly (the bar indicates the position of the -1). In [4], explicit evaluations of $\zeta(s,\overline{1}), \zeta(\overline{2s},1)$ and $\zeta(\overline{2s},\overline{1})$ are given in terms of Riemann zeta values and $\log(2)$; in [5], it is shown that $\zeta(a,\overline{b})$ etc. with $a+b$ odd may be likewise reduced; small examples include (see also [6] for the first one)

$$\zeta(\overline{2},1) = -\frac{\zeta(3)}{8}, \quad \zeta(2,\overline{1}) = \frac{\pi^2 \log(2)}{4} - \zeta(3), \quad \zeta(\overline{2},\overline{1}) = \frac{\pi^2 \log(2)}{4} - \frac{13\zeta(3)}{8}.$$

Again, if we follow closely the proof of (3), we arrive at new summation formulae such as

$$\sum_{j=2}^{s-1} \zeta(j,\overline{s-j}) = (1-2^{1-s})\zeta(s) + \zeta(\overline{s-1},1) + \zeta(\overline{s-1},\overline{1}), \qquad (18)$$

and so on. When s is odd, we simplify the right-hand side using results in [4], thus

$$\sum_{j=2}^{2s} \zeta(j,\overline{2s+1-j}) = 2(1-4^{-s})\log(2)\zeta(2s) - \zeta(2s+1)$$

$$+ \sum_{k=1}^{s-1} (2^{1-2k} - 4^{k-s})\zeta(2k)\zeta(2s+1-2k),$$

$$\sum_{j=2}^{2s} (-1)^j \zeta(j,\overline{2s+1-j}) = 2(1-4^{-s})\log(2)\zeta(2s) + \left((s+1)4^{-s} - \frac{1}{2} - s\right)\zeta(2s+1)$$

$$+ \sum_{k=1}^{s-1} (1 - 4^{k-s})\zeta(2k)\zeta(2s+1-2k).$$

The last two formulae may be added or subtracted to give sums for even or odd j's, for instance,

$$\sum_{j=1}^{s-1} \zeta(2j+1, \overline{2s-2j}) = \left(\frac{2s-1}{4} - \frac{s+1}{2^{2s+1}}\right)\zeta(2s+1)$$

$$- \sum_{k=1}^{s-1}\left(\frac{1}{2} - \frac{1}{4^k}\right)\zeta(2k)\zeta(2s+1-2k). \qquad (19)$$

With perseverance, we may produce a host of similar identities for the three alternating double zeta functions. We only give some examples below; as they have similar proofs, we omit the details.

When we evaluate $\sum_{j=2}^{s-1}(\pm 1)^j\zeta(\overline{j}, s-j)$, we get, for example,

$$\sum_{j=2}^{s-1} \zeta(\overline{j}, s-j) = (1 - 2^{1-s})\left(\zeta(s) + \zeta(s-1, 1) - 2\log(2)\zeta(s-1)\right)$$

$$- \zeta(\overline{s-1}, 1) - \sum_{n=0}^{\infty} \frac{H_n}{(2n+1)^{s-1}},$$

and by applying (14) to the results, we obtain

$$\sum_{j=2}^{2s}\left(2^{1-j}\zeta(j, 2s+1-j) + \zeta(\overline{j}, 2s+1-j)\right) = 4^{-s}\zeta(2s+1) - \zeta(\overline{2s}, 1), \qquad (20)$$

$$2\sum_{j=1}^{s}\zeta(\overline{2j}, 2s+1-2j) = \frac{4^{-s}-1}{2}\zeta(2s+1) - \zeta(\overline{2s}, 1), \qquad (21)$$

where the right-hand side of both equations may be reduced to Riemann zeta values by results in [4].

Likewise, for $\zeta(\overline{j}, \overline{s-j})$, we may deduce

$$\frac{1}{2}\sum_{j=2}^{2s}\zeta(\overline{j}, \overline{2s+1-j}) = (1 - 4^{-s})\log(2)\zeta(2s) - \frac{2s(2^{2s+1}-1)-1}{4^{s+1}}\zeta(2s+1)$$

$$+ \sum_{k=1}^{s-1}(4^k - 1)(4^{-k} - 4^{-s})\zeta(2k)\zeta(2s+1-2k); \qquad (22)$$

$$\sum_{j=1}^{s-1}\zeta(\overline{2j+1}, \overline{2s-2j}) = \left(\frac{1}{2}(4^s - 3s - 2) + 4^{-s}(s+1)\right)\zeta(2s+1)$$

$$- \sum_{k=1}^{s-1} 4^{-(s+k)}(4^s - 4^k)^2\zeta(2k)\zeta(2s+1-2k). \qquad (23)$$

Therefore, for the sums of the three alternating double zeta values, we have succeeded in giving closed forms when s (the sum of the arguments) is odd and the summation index j is odd, even or unrestricted; it is interesting to compare this to the non-alternating case, whose sum is simpler when s is even (see Remark 1). A notable exception is the following formula, whose proof is similar to that of (5):

Theorem 3.5 *For integer $s \geq 2$,*

$$4 \sum_{j=1}^{s-1} \zeta(2j, \overline{2s-2j}) = (4^{1-s} - 1)\zeta(2s). \tag{24}$$

Remark 3 Though it is believed that $\zeta(\overline{s}, 1)$ and $\zeta(\overline{s}, \overline{1})$ cannot be simplified in terms of well-known constants for odd s, their difference can (this situation is analogous to Theorem 3.4 and can be proven using the same method):

$$\zeta(\overline{s}, 1) - \zeta(\overline{s}, \overline{1}) = (1 - 2^{-s})(s\zeta(s+1) - 2\log(2)\zeta(s))$$
$$- \sum_{k=1}^{s-2}(1 - 2^{-k})\zeta(k+1)\zeta(s-k).$$

Moreover, some of the sums involving $\zeta(\overline{s}, 1)$ and $\zeta(\overline{s}, \overline{1})$ are much neater when the summation index j starts from 1 instead of 2, for instance,

$$\sum_{j=1}^{s-1} \zeta(\overline{j}, s-j) = (2^{2-s} - 1)\log(2)\zeta(s-1) - \zeta(\overline{s-1}, 1),$$

$$\sum_{j=1}^{s-1} \zeta(\overline{j}, \overline{s-j}) = \zeta(s-1, \overline{1}) - \log(2)\zeta(s-1),$$

$$2 \sum_{j=1}^{2s} (-1)^j \zeta(\overline{j}, 2s+1-j) = (2 - 4^{1-s})\log(2)\zeta(2s) - (1 - 4^{-s})\zeta(2s+1). \quad \square$$

We wrap up this section with a surprising result, an alternating analog of (4):

Theorem 3.6 *For integer $s \geq 3$,*

$$\sum_{j=2}^{s-1} 2^j \zeta(\overline{j}, s-j) = (3 - 2^{2-s} - s)\zeta(s). \tag{25}$$

Proof The proof is very similar to that of (4): we write the left-hand side as a triple sum and first take care of the $m = n$ case. Then we sum the geometric series to obtain

$$\sum_{\substack{m,n>0 \\ m \neq n}} \frac{(-1)^{m+n}2^s}{(m-n)(m+n)^{s-1}} - \frac{4(-1)^{m+n}}{(m-n)(m+n)n^{s-2}}.$$

The first term vanishes due to antisymmetry, and the second term telescopes due to

$$\sum_{\substack{m>0 \\ m \neq n}} \frac{(-1)^m}{m^2 - n^2} = \frac{2 + (-1)^n}{4n^2}.$$

Now summing over n proves the result. $\square$

With (25) and results in [4], we can evaluate $\zeta(\bar{a}, b)$ etc. with $a + b = 4$, for instance,

$$\zeta(\bar{2}, 2) = \frac{\log(2)^4}{6} - \frac{\log(2)^2 \pi^2}{6} + \frac{7 \log(2) \zeta(3)}{2} - \frac{13 \pi^4}{288} + 4 \mathrm{Li}_4\left(\frac{1}{2}\right),$$

where Li_4 is the polylogarithm of order 4.

4   More Sums from Recursions

In this section we first provide some experimental evidence which suggests that the sums in Sect. 2 (almost) exhaust all "simple" and "nice" sums in some sense. We then use a simple procedure which may be used to produce more weighted sums of greater complexity but of less elegance.

4.1   Experimental Methods

It is a curiosity why (4) had not been observed empirically earlier. As we can express all $\zeta(a, b)$ with $a + b \leq 7$ in terms of the Riemann zeta function, it is a simple matter of experimentation to try all combinations of the form

$$\sum_j (a \cdot b^j + c^s \cdot d^j) \zeta(j, s - j) = f(s) \zeta(s), \tag{26}$$

with j or s being even, odd or any integer (so there are nine possibilities), $a, b, c, d \in \mathbb{Q}$ and $f : \mathbb{N} \to \mathbb{Q}$ is a (reasonable) function to be found.

Now if we assume that $\pi, \zeta(3), \zeta(5), \zeta(7), \dots$ are algebraically independent over $\mathbb{Q}$ (which is widely believed to be true, though proof-wise we are a long way off, for instance, apart from π only $\zeta(3)$ is known to be irrational – see [17]), then we can substitute a few small values of s into (26) and solve for a, b, c, d in that order.

For instance, assuming a formula of the form $\sum_{j=2}^{s-1} a^j \zeta(j, s - j) = f(s) \zeta(s)$ holds, using $s = 5$ forces us to conclude that $a = 1$ or $a = 2$.

Indeed, when we carry out the experiment outlined above, it is revealed that the sums (3)–(5) are essentially the only ones in the form of (26), except for the case

$$\sum_{j=1}^{s-1}(d^j+d^{s-j})\zeta(2j,2s-2j),$$

(note the factor in front of the ζ has to be invariant under $j \mapsto s-j$). Here, the choice of $d=4$ leads to

$$\sum_{j=1}^{s-1}(4^j+4^{s-j})\zeta(2j,2s-2j) = \left(s+\frac{4}{3}+\frac{2}{3}4^{s-1}\right)\zeta(2s),$$

a result which first appeared in [13] and was proven using the generating function of Bernoulli polynomials.

Sums of the form

$$\sum_{j}p(s,j)\zeta(j,s-j) = f(s)\zeta(s),$$

where p is a nonconstant 2-variable polynomial with rational coefficients, can also be subject to experimentation. If the degree of p is restricted to 2, then $j(s-j)\zeta(2j,2s-2j)$ is the only candidate which can give a closed form. Indeed, this sum was essentially considered in [13], using an identity found in [16]:

$$6\sum_{j=2}^{s-2}(2j-1)(2s-2j-1)\zeta(2j)\zeta(2s-2j) = (s-3)(4s^2-1)\zeta(2s).$$

The identity was first due to Ramanujan [3, chapter 15, formula (14.2)] (and hence not original as claimed in [16]). Applying (2), the result can be neatly written as

$$\sum_{j=2}^{s-2}(2j-1)(2s-2j-1)\zeta(2j,2s-2j) = \frac{3}{4}(s-3)\zeta(2s). \tag{27}$$

Searches for "simple" weighted sums of length 3 multiple zeta values, and for q-analogs of (4), have so far proved unsuccessful (except for [7] which contains a generalisation of (5)).

4.2   Recursions of the Zeta Function

We observe that any recursion of the Riemann zeta values—or of Bernoulli numbers—of the form

$$\sum_{j}g(s,j)\zeta(2j)\zeta(2s-2j)$$

for some function g would lead to a sum formula for double zeta values, due to (2). This was the idea behind (27) and was also hinted at in Remark 1. We flesh out the details in some examples below.

One such recursion is formula (14.14) in chapter 15 of [3], which can be written as

$$\sum_{j=1}^{n-1} j(2j+1)(n-j)(2n-2j+1)\zeta(2j+2)\zeta(2n-2j+2)$$

$$= \frac{1}{60}(n+1)(2n+3)(2n+5)(2n^2-5n+12)\zeta(2n+4) - \frac{\pi^4}{15}(2n-1)\zeta(2n). \quad (28)$$

Upon applying (2) to the recursion, we obtain the new sum:

Theorem 4.7 *For integer* $n \geq 4$,

$$\sum_{j=2}^{n-2}(j-1)(2j-1)(n-j-1)(2n-2j-1)\zeta(2j,2n-2j)$$

$$= \frac{3}{8}(n-1)(3n-2)\zeta(2n) - 3(2n-5)\zeta(4)\zeta(2n-4). \quad (29)$$

Next, we use a result from [12], which states

$$\sum_{k=1}^{n-1}\left[1-\binom{2n}{2k}\right]\frac{B_{2k}B_{2n-2k}}{(2k)(2n-2k)} = \frac{H_{2n}}{n}B_{2n}. \quad (30)$$

We apply (2) to the left-hand side to obtain a sum of double zeta values; unfortunately, one term of the sum involves $\sum_{k=1}^{n-1}(2k-1)!(2n-2k-1)!$ which has no nice closed form. On the other hand, a twin result in [11] gives

$$\sum_{k=1}^{n-2}\left[n-\binom{2n}{2k}\right]B_{2k}B_{2n-2k-2} = (n-1)(2n-1)B_{2n-2}. \quad (31)$$

When we apply (2) to it, we end up with a sum involving $\sum_{k=1}^{n-2}(2k)!(2n-2k-2)!$, which again has no nice closed form.

Yet, it is straightforward to show by induction that

$$\sum_{k=0}^{m}\frac{(-1)^k}{\binom{n}{k}} = \frac{(n+1)!+(-1)^m(m+1)!(n-m)!}{(n+2)n!},$$

hence, when $m=n$, the sum vanishes if n is odd and equates to $2(n+1)/(n+2)$ when n is even. In other words, $\sum_{k=0}^{2n}(-1)^k k!(2n-k)!$ has a closed form, and accordingly we subtract the sums obtained from (30) and (31) to produce:

Proposition 8 *For integer $n \geq 2$,*

$$\sum_{k=1}^{n-1} \left\{ \left[1 - \frac{1}{\binom{2n+2}{2k}} \right] \frac{n+1}{k(n+1-k)} \zeta(2k, 2n+2-2k) + \frac{\zeta(2n+2)}{\zeta(2n)} \times \right.$$

$$\left. \left[\frac{2}{(2n+1)\binom{2n}{2k}} - \frac{(2k-n)^2 + (n+1)(n+2)}{(n-k+1)(2n-2k+1)(k+1)(2k+1)} \right] \zeta(2k, 2n-2k) \right\}$$

$$= 3 \left[H_{2n-1} - H_{n-1} - \frac{2n^2+n+1}{2n(2n+1)} \right] \zeta(2n+2) - \frac{2n+3}{2n+1} \zeta(2n, 2). \qquad (32)$$

Our next result uses equation (7.2) recored in [2], whose special case gives

$$B_n^2 + \frac{B_{2n}}{\binom{2n}{n}} = \frac{4nn!}{(2\pi)^{2n}} \sum_{k=0}^{[n/2]} \frac{(2n-2k)!}{(n-k)(n-2k)!} \zeta(2k)\zeta(2n-2k). \qquad (33)$$

Upon applying (2) and much algebra, we arrive at

Proposition 9 *For integer $n \geq 2$,*

$$\sum_{k=1}^{n-1} \frac{(n+|n-2k|)!}{(n+|n-2k|)|n-2k|!} \zeta(2k, 2n-2k)$$

$$= \left[\frac{(1+(-2)^{1-n})(n-1)!}{4} + \frac{3(2n-1)!}{2n!} - \frac{(2n+1)!}{n(n+1)!} {}_2F_1\left(\begin{matrix} 1, 2n+2 \\ n+2 \end{matrix} \middle| -1 \right) \right] \zeta(2n),$$

where ${}_2F_1$ is the Gaussian hypergeometric function.

Proof The only non-trivial step to check here is that the claimed ${}_2F_1$ is produced when we sum the fraction in (33), that is, we wish to prove the claim

$$\sum_{k=0}^{[n/2]} \frac{(-1)^n n}{2(n-k)} \binom{2n-2k}{n-2k} = \frac{1}{(1-x)^{n+1}} - x^m \binom{n+m}{n} {}_2F_1\left(\begin{matrix} 1, n+m+1 \\ m+1 \end{matrix} \middle| x \right) \bigg|_{x=-1, m=n+1}.$$

We observe that for x near the origin, the right-hand side is simply $\sum_{k=0}^{m-1} x^k \binom{n+k}{k}$, as they have the same recursion and initial values in m; hence, when $x = -1$ they also agree by analytic continuation. This sum (in the limit $x = -1$, $m = n+1$), as a function of n, also satisfies the recursion

$$4f(n) - 2f(n-1) = 3(-1)^n \binom{2n}{n}, \quad f(1) = -1,$$

which is the same recursion for the sum on the left-hand side of the claim – as may be checked using Celine's method [15]. Thus equality is established. $\qquad \square$

Remark 4 It is clear that a large number of (uninteresting) identities similar to those recorded in the two propositions may be easily produced. Using [2, (7.2) with $k = n + 1$], for instance, a very similar proof to the above gives

$$\sum_{k=1}^{n-1} \frac{4(n+1+|n-2k|)!}{n!(1+|n-2k|)!} \zeta(2k, 2n-2k) = (1+(-1)^n)(1-n)\zeta(n)^2 + \zeta(2n) \times$$

$$\left\{ n+3+(-1)^n(n-1+2^{-n}) + 2\binom{2n+1}{n} \left[3 - \frac{4(2n+3)}{n+1} \, {}_2F_1\left(\begin{matrix} 1, 2n+4 \\ n+3 \end{matrix} \middle| -1 \right) \right] \right\}.$$

Care must be exercised when consulting the literature, however, as the author found in the course of this work that many recorded recursions of the Bernoulli numbers (or of the even Riemann zeta values) are in fact combinations and reformulations of the formula behind (27) and the basic identity appearing in Remark 1.   □

4.3   The Reflection Formula

Formula (2) is but a special case of a more general *reflection formula*. To state the reflection formula, we will need some notation from [4], which we have tried to avoid until now to keep the exposition elementary.

Let $\chi_p(n)$ denote a 4-periodic function on n; for different p's we tabulate values of χ_p below:

$p \backslash n$	1	2	3	4
1	1	1	1	1
2a	1	0	1	0
2b	1	−1	1	−1
−4	1	0	−1	0

We now define the series L_p by

$$L_p(s) = \sum_{n=1}^{\infty} \frac{\chi_p(n)}{n^s},$$

and $L_{pq}(s)$ means $\sum_{n>0} \chi_p(n)\chi_q(n)/n^s$. Finally, we define *character sums*, which generalise the double zeta values, by

$$[p, q](s, t) = \sum_{n=1}^{\infty} \sum_{m=1}^{n-1} \frac{\chi_p(n)}{n^s} \frac{\chi_q(m)}{m^t}. \qquad (34)$$

In this notation, $\zeta(s, t) = [1, 1](s, t)$, $\zeta(s, \bar{t}) = [1, 2b](s, t)$ etc. We can now state the reflection formula [4, equation (1.7)]:

$$[p,q](s,t) + [q,p](t,s) = L_p(s)L_q(t) - L_{pq}(s+t). \tag{35}$$

Remark 5 With the exception of χ_{2b}, χ_p are examples of Dirichlet characters and L_p are the corresponding Dirichlet series. Indeed,

$$L_1(s) = \zeta(s), \quad L_{2a}(s) = (1 - 2^{-s})\zeta(s) = \lambda(s),$$

$$L_{2b}(s) = (1 - 2^{1-s})\zeta(s) = \eta(s), \quad L_{-4}(s) = \beta(s),$$

where the last three are the Dirichlet lambda, eta and beta functions, respectively.

Moreover, $2(2n)!\beta(2n+1) = (-1)^n(\pi/2)^{2n+1}E_{2n}$ for non-negative integer n, where E_n denotes the nth Euler number. Using generating functions, one may deduce convolution formulae for the Euler numbers, an example of which is

$$\sum_{k=0}^{n-2} \binom{n-2}{k} E_k E_{n-2-k} = 2^n(2^n - 1)\frac{B_n}{n}.$$

Many of our results in the previous sections would look neater had we used $\lambda(s)$ and $\eta(s)$ instead of $\zeta(s)$. □

Using the standard convolution formulae of the Bernoulli and the Euler polynomials, and aided by the reflection formula (35), we can produce the following sums for $[p,q](s,t)$ as we did for $\zeta(s,t)$.

Theorem 4.10 *Using the notation of* (34), *we have, for integer* $n \geq 2$,

$$2\sum_{k=1}^{n-1}[1,2b](2k,2n-2k) + [2b,1](2k,2n-2k) = -4\sum_{k=1}^{n-1}[2b,2b](2k,2n-2k)$$

$$= (1 - 4^{1-n})\zeta(2n); \tag{36}$$

$$4\sum_{k=1}^{n-1}[2a,2a](2k,2n-2k) = -4\sum_{k=0}^{n-1}[-4,-4](2k+1,2n-1-2k)$$

$$= \sum_{k=1}^{n-1}[1,2a](2k,2n-2k) + [2a,1](2k,2n-2k) = (1 - 4^{-n})\zeta(2n); \tag{37}$$

$$\sum_{k=1}^{n}[2a,-4](2k,2n+1-2k) + [-4,2a](2n+1-2k,2k)$$

$$= \sum_{k=1}^{n-1}[2a,2b](2k,2n-2k) + [2b,2a](2n-2k,2k) = 0; \tag{38}$$

$$2 \sum_{k=1}^{n} [1,-4](2k,2n+1-2k) + [-4,1](2n+1-2k,2k)$$

$$= -2 \sum_{k=1}^{n} [2b,-4](2k,2n+1-2k) + [-4,2b](2n+1-2k,2k)$$

$$= \beta(2n+1) + (16^{-n} - 2^{-1-2n})\pi\,\zeta(2n), \tag{39}$$

We note that (36) concerns the alternating double zeta values studied in Sect. 3 (cf. the more elementary Theorem 3.5). As mentioned before, the identities above rest on well-known recursions, for instance, the second equality in (38) is equivalent to the recursion

$$\sum_{k=1}^{n} \beta(2n+1-2k)\lambda(2k) = n\beta(2n+1).$$

Also, the many pairs of equalities within each numbered equation in the theorem are not all coincidental but stem from the identity in [4]:

$$[1,q] + [2b,q] = 2[2a,q],$$

where $q = 1, 2a, 2b$ or -4.

Moreover, one can show the following equations; as character sums are not the main object of our study, we omit the details:

$$\sum_{k=1}^{n} 4^{-k} \big([-4,1](2n+1-2k,2k) + [1,-4](2k,2n+1-2k) \big)$$

$$= \frac{1 + 2^{1-2n}}{6} \beta(2n+1) + (2^{-1-4n} - 4^{-1-n})\pi\,\zeta(2n),$$

$$\sum_{k=1}^{n-1} 4^{k} \big([2a,1](2k,2n-2k) + [1,2a](2n-2k,2k) \big)$$

$$= \frac{(1-4^{-n})(8+4^{n})}{6} \zeta(2n).$$

Since there is an abundance of recursions involving the Bernoulli and the Euler numbers, many more such identities may be produced using the reflection formula.

Acknowledgements The author wishes to thank John Zucker and Wadim Zudilin for illuminating discussions.

References

1. T.M. Apostol, A. Basu, A new method for investigating Euler sums. Ramanujan J. **4**, 397–419 (2000)
2. T. Agoh, K. Dilcher, Convolution identities and lacunary recurrences for Bernoulli numbers. J. Number Theory **124**, 105–122 (2007)
3. B.C. Berndt, *Ramanujan's Notebooks, Part II* (Springer, New York, 1989)
4. J. Boersma, J.M. Borwein, I.J. Zucker, The evaluation of character Euler double sums. Ramanujan J. **15**, 377–405 (2008)
5. D. Borwein, J.M. Borwein, R. Girgensohn, Explicit evaluation of Euler sums. Proc. Edinburgh Math. Soc. **38**, 277–294 (1995)
6. J.M. Borwein, D.M. Bradley, Thirty-two Goldbach variations. Int. J. Number Theory **2**(1), 65–103 (2006)
7. T. Cai, Z. Shen, Some identities for multiple zeta values. J. Number Theory **132**, 314–323 (2012)
8. H. Gangl, M. Kaneko, D. Zagier, Double zeta values and modular forms, in *Automorphic Forms and Zeta Functions. Proc. of Conf. in Memory of Tsuneo Arakawa*, ed. by Böcherer et al. (World Scientific, Hackensack, 2006), pp. 71–106
9. L. Guo, B. Xie, Weighted sum formula for multiple zeta values. J. Number Theory **129**, 2747–2765 (2009)
10. J.G. Huard, K.S. Williams, N.Y. Zhang, On Tornheim's double series. Acta Arith. **75**, 105–117 (1996)
11. Y. Matiyasevich, Identities with Bernoulli numbers (1997). Available at http://logic.pdmi.ras.ru/~yumat/Journal/Bernoulli/bernulli.htm
12. H. Miki, A relation between Bernoulli numbers. J. Number Theory **10**, 297–302 (1978)
13. T. Nakamura, Restricted and weighted sum formulas for double zeta values of even weight. Šiauliai Math. Semin. **4**(12), 151–155 (2009)
14. Y. Ohno, W. Zudilin, Zeta stars. Commun. Number Theory Phys. **2**(2), 324–347 (2008)
15. M. Petkovsek, H. Wilf, D. Zeilberger, $A = B$ (AK Peters, Wellesley, 1996)
16. G. Shimura, *Elementary Dirichlet Series and Modular Forms* (Springer, New York, 2007)
17. A. van der Poorten, A proof that Euler missed... Apéry's proof of the irrationality of $\zeta(3)$. Math. Intell. **1**(4), 195–203 (1979)

Period(d)ness of L-Values

Wadim Zudilin

Abstract In our recent work with Rogers on resolving some of Boyd's conjectures on two-variate Mahler measures, a new analytical machinery was introduced to write the values $L(E,2)$ of L-series of elliptic curves as periods in the sense of Kontsevich and Zagier. Here we outline, in slightly more general settings, the novelty of our method with Rogers and provide two illustrative period evaluations of $L(E,2)$ and $L(E,3)$ for a conductor 32 elliptic curve E.

Key words Modular form • L-series • Period • Arithmetic differential equation • Dedekind's eta function

1 Introduction

A *period* is a complex number whose real and imaginary parts are values of absolutely convergent integrals of rational functions with rational coefficients, over domains in $\mathbb{R}^n$ given by polynomial inequalities with rational coefficients [5]. Without much harm, the three appearances of the adjective "rational" can be replaced by "algebraic". The set of periods $\mathscr{P}$ is countable and admits a ring structure. But what is probably most exciting about the set — it contains a lot of "important" numbers, mathematical constants like π [2] and $\zeta(3)$ [11].

The extended period ring $\hat{\mathscr{P}} := \mathscr{P}[1/\pi] = \mathscr{P}[(2\pi i)^{-1}]$ (rather than the period ring $\mathscr{P}$ itself) contains many natural examples, like values of generalised hypergeometric functions [1] at algebraic points and special L-values. For example, a general theorem [5] due to Beilinson and Deninger–Scholl states that the (non-critical)

W. Zudilin (✉)
School of Mathematical and Physical Sciences, The University of Newcastle,
Callaghan, NSW 2308, Australia
e-mail: wadim.zudilin@newcastle.edu.au

J.M. Borwein et al. (eds.), *Number Theory and Related Fields: In Memory of Alf van der Poorten*, Springer Proceedings in Mathematics & Statistics 43,
DOI 10.1007/978-1-4614-6642-0_20, © Springer Science+Business Media New York 2013

value of the L-series attached to a cusp form $f(\tau)$ of weight k at a positive integer $m \geq k$ (cf. formula (2) below) belongs to $\mathscr{P}$. In spite of the effective nature of the proof of the theorem, computing these L-values as periods remains a difficult problem even for particular examples; it is this odd difficulty which lets us refer to the property of being a period as "period(d)ness". Most such computations are motivated by (conjectural) evaluations of the logarithmic Mahler measures of multivariate polynomials.

With the purpose of establishing such evaluations in the two-variate case, Rogers and the present author [8] have developed a machinery for writing the L-values $L(f,2)$ attached to cusp forms $f(\tau)$ of weight 2 as periods, the machinery which is different from that of Beilinson. In this note, we give an overview of the method of [8, 9] on a particular example of $L(E,2)$ in Sect. 2 and then attempt in Sect. 3 to describe a general algorithm behind the method. In Sect. 4 we present an example of evaluating $L(E,3)$ as a period, a computation we failed to find in the existing literature. Finally, in Sect. 5 we demonstrate that the two particular evaluations discussed can be further reduced to a hypergeometric form; such reduction is not expected to be available for general special L-values and so far is known for very few instances. In the examples of Sects. 2, 4 (and 5), E stands for an elliptic curve of conductor 32. There are at least two reasons for choosing this conductor. First of all, it is not discussed in our joint work [8, 9], and secondly, the modular parametrisations involved are sufficiently classical and remarkably simple.

Throughout the note we keep the notation $q = e^{2\pi i \tau}$ for τ from the upper half-plane $\operatorname{Im} \tau > 0$, so that $|q| < 1$. Our basic constructor of modular forms and functions is Dedekind's eta function

$$\eta(\tau) := q^{1/24} \prod_{m=1}^{\infty} (1 - q^m) = \sum_{n=-\infty}^{\infty} (-1)^n q^{(6n+1)^2/24}$$

with its modular involution

$$\eta(-1/\tau) = \sqrt{-i\tau}\,\eta(\tau). \tag{1}$$

We also set $\eta_k := \eta(k\tau)$ for short.

For functions of τ or $q = e^{2\pi i \tau}$, we use the differential operator

$$\delta := \frac{1}{2\pi i} \frac{\mathrm{d}}{\mathrm{d}\tau} = q \frac{\mathrm{d}}{\mathrm{d}q}$$

and denote by δ^{-1} the corresponding anti-derivative normalised by 0 at $\tau = i\infty$ (or $q = 0$):

$$\delta^{-1} f = \int_0^q f \frac{\mathrm{d}q}{q}.$$

In particular, for a modular form $f(\tau) = \sum_{n=1}^{\infty} a_n q^n$, whose expansion vanishes at infinity, we have

$$L(f,m) = \frac{1}{(m-1)!} \int_0^1 f \log^{m-1} q \, \frac{dq}{q} = \sum_{n=1}^{\infty} \frac{a_n}{n^m} = (\delta^{-m} f)|_{q=1}, \qquad (2)$$

whenever the latter sum makes sense.

The generalised hypergeometric function is defined by the series

$$_{k+1}F_k \left(\begin{matrix} a_0, a_1, \ldots, a_k \\ b_1, \ldots, b_k \end{matrix} \; \middle| \; z \right) = \sum_{n=0}^{\infty} \frac{(a_0)_n (a_1)_n \cdots (a_k)_n}{(b_1)_n \cdots (b_k)_n} \frac{z^n}{n!}$$

in the disc $|z| < 1$; here $(a)_n := \Gamma(a+n)/\Gamma(a) = \prod_{m=0}^{n-1}(a+m)$ denotes Pochhammer's symbol. Details about the analytic continuation and integral representations of the function can be found in Bailey's classical treatise [1]; relevant references are made explicit in Sect. 5.

2 $L(E,2)$

For a conductor 32 elliptic curve E, the L-series is known to coincide with that for the cusp form $f(\tau) := \eta_4^2 \eta_8^2$.

Note the (Lambert series) expansion

$$\frac{\eta_8^4}{\eta_4^2} = \sum_{m \geq 1} \left(\frac{-4}{m} \right) \frac{q^m}{1-q^{2m}} = \sum_{\substack{m,n \geq 1 \\ n \text{ odd}}} \left(\frac{-4}{m} \right) q^{mn} = \sum_{m,n \geq 1} a(m)b(n)q^{mn},$$

$$\text{where} \quad a(m) := \left(\frac{-4}{m} \right), \quad b(n) := n \bmod 2,$$

and $\left(\frac{-4}{m} \right)$ denotes the quadratic residue character modulo 4.

Then

$$f(it) = \frac{\eta_8^4}{\eta_4^2} \frac{\eta_4^4}{\eta_8^2} \bigg|_{\tau=it} = \frac{\eta_8^4}{\eta_4^2} \bigg|_{\tau=it} \cdot \frac{1}{2t} \frac{\eta_8^4}{\eta_4^2} \bigg|_{\tau=i/(32t)}$$

$$= \frac{1}{2t} \sum_{m_1,n_1 \geq 1} a(m_1)b(n_1)e^{-2\pi m_1 n_1 t} \sum_{m_2,n_2 \geq 1} b(m_2)a(n_2)e^{-2\pi m_2 n_2/(32t)}, \qquad (3)$$

where $t > 0$ and the modular involution (1) was used.

Now,

$$L(E,2) = L(f,2) = \int_0^1 f \log q \, \frac{dq}{q} = -4\pi^2 \int_0^\infty f(it)t \, dt$$

$$= -2\pi^2 \int_0^\infty \sum_{m_1,n_1,m_2,n_2 \geq 1} a(m_1)b(n_1)b(m_2)a(n_2)$$

$$\times \exp\left(-2\pi\left(m_1 n_1 t + \frac{m_2 n_2}{32t}\right)\right) dt$$

$$= -2\pi^2 \sum_{m_1,n_1,m_2,n_2 \geq 1} a(m_1)b(n_1)b(m_2)a(n_2)$$

$$\times \int_0^\infty \exp\left(-2\pi\left(m_1 n_1 t + \frac{m_2 n_2}{32t}\right)\right) dt.$$

Here comes the crucial transformation of purely analytical origin: we make the change of variable $t = n_2 u/n_1$. This does not change the form of the integrand but affects the differential, and we obtain

$$L(E,2) = -2\pi^2 \sum_{m_1,n_1,m_2,n_2 \geq 1} \frac{a(m_1)b(n_1)b(m_2)a(n_2)n_2}{n_1}$$

$$\times \int_0^\infty \exp\left(-2\pi\left(m_1 n_2 u + \frac{m_2 n_1}{32u}\right)\right) du$$

$$= -2\pi^2 \int_0^\infty \sum_{m_1,n_2 \geq 1} a(m_1)a(n_2)n_2 e^{-2\pi m_1 n_2 u}$$

$$\times \sum_{m_2,n_1 \geq 1} \frac{b(m_2)b(n_1)}{n_1} e^{-2\pi m_2 n_1/(32u)} du.$$

The first double series in the integrand corresponds to

$$\sum_{m,n \geq 1} a(m)a(n)n q^{mn} = \sum_{m,n \geq 1} \left(\frac{-4}{mn}\right) n q^{mn} = \sum_{n \geq 1} n \left(\frac{-4}{n}\right) \frac{nq^n}{1+q^{2n}} = \frac{\eta_2^4 \eta_8^4}{\eta_4^4},$$

while the second one is

$$\sum_{m,n \geq 1} \frac{b(m)b(n)}{n} q^{mn} = \sum_{m,n \geq 1} \frac{q^{mn}}{n} - \frac{q^{(2m)n}}{n} - \frac{q^{m(2n)}}{2n} + \frac{q^{(2m)(2n)}}{2n}$$

$$= \frac{1}{2} \sum_{m,n \geq 1} \frac{2q^{mn} - 3q^{2mn} + q^{4mn}}{n}$$

$$= -\frac{1}{2} \log \prod_{m \geq 1} \frac{(1-q^m)^2(1-q^{4m})}{(1-q^{2m})^3} = -\frac{1}{2} \log \frac{\eta_1^2 \eta_4}{\eta_2^3};$$

hence,

$$L(E,2) = \pi^2 \int_0^\infty \frac{\eta_2^4 \eta_8^4}{\eta_4^4}\bigg|_{\tau=iu} \cdot \log \frac{\eta_1^2 \eta_4}{\eta_2^3}\bigg|_{\tau=i/(32u)} du.$$

Applying the involution (1) to the eta quotient under the logarithm sign, we obtain

$$L(E,2) = \pi^2 \int_0^\infty \frac{\eta_2^4 \eta_8^4}{\eta_4^4} \log \frac{\sqrt{2}\eta_8 \eta_{32}^2}{\eta_{16}^3}\bigg|_{\tau=iu} du.$$

Now comes the modular magic: choosing a particular modular function $x(\tau) := \eta_2^4 \eta_8^2/\eta_4^6$, which ranges from 0 to 1 when τ ranges from 0 to $i\infty$, one can easily verify that

$$\frac{1}{2\pi i} \frac{x\,dx}{2\sqrt{1-x^4}} = -\frac{\eta_2^4 \eta_8^4}{\eta_4^4} d\tau \quad \text{and} \quad \left(\frac{\sqrt{2}\eta_8 \eta_{32}^2}{\eta_{16}^3}\right)^2 = \frac{1-x}{1+x}.$$

Thus, we arrive at the following result.

Theorem 1. *For an elliptic curve E of conductor 32,*

$$L(E,2) = \frac{\pi}{8} \int_0^1 \frac{x}{\sqrt{1-x^4}} \log \frac{1+x}{1-x} dx = 0.9170506353\ldots.$$

3 General L-Values

To summarise our evaluation of $L(E,2) = L(f,2)$ in Sect. 2, we first split $f(\tau)$ into a product of two Eisenstein series of weight 1 and at the end we arrive at a product of two Eisenstein(-like) series $g_2(\tau)$ and $g_0(\tau)$ of weights 2 and 0, respectively, so that $L(f,2) = c\pi L(g_2 g_0, 1)$ for some rational c. The latter object is doomed to be a period as $g_0(\tau)$ is a logarithm of a modular function, while $2\pi i g_2(\tau) d\tau$ is, up to a modular function multiple, the differential of a modular function, and finally any two modular functions are connected by an algebraic relation over $\overline{\mathbb{Q}}$.

The method can be further formalised to more general settings, and it is this extension which we attempt to outline in this section.

For two *bounded* sequences $a(m)$, $b(n)$, we refer to an expression of the form

$$g_k(\tau) = a + \sum_{m,n \geq 1} a(m)b(n)n^{k-1}q^{mn} \tag{4}$$

as an Eisenstein-like series of weight k, especially in the case when $g_k(\tau)$ is a modular form of certain level, that is, when it transforms sufficiently 'nicely' under $\tau \mapsto -1/(N\tau)$ for some positive integer N. This automatically happens when $g_k(\tau)$

is indeed an Eisenstein series (for example, when $a(m) = 1$ and $b(n)$ is a Dirichlet character modulo N of designated parity, $b(-1) = (-1)^k$), in which case $\hat{g}_k(\tau) :=$ $g_k(-1/(N\tau))(\sqrt{-N}\tau)^{-k}$ is again an Eisenstein series. It is worth mentioning that the above notion makes perfect sense in the case $k \le 0$ as well. Indeed, modular units, or weak modular forms of weight 0 that are the logarithms of modular functions are examples of Eisenstein-like series $g_0(\tau)$. Also, for $k \le 0$ examples are given by Eichler integrals, the $(1-k)$th τ-antiderivatives of holomorphic Eisenstein series of weight $2-k$, a consequence of the famous lemma of Hecke [12, Section 5].

Suppose we are interested in the L-value $L(f, k_0)$ of a cusp form $f(\tau)$ of weight $k = k_1 + k_2$ which can be represented as a product (in general, as a linear combination of several products) of two Eisenstein(-like) series $g_{k_1}(\tau)$ and $\hat{g}_{k_2}(\tau)$, where the first one vanishes at infinity ($a = g_{k_1}(i\infty) = 0$ in (4)) and the second one vanishes at zero ($\hat{g}_{k_2}(i0) = 0$). (The vanishing happens because the product is a cusp form!) In reality, we need the series $g_{k_2}(\tau) := \hat{g}_{k_2}(-1/(N\tau))(\sqrt{-N}\tau)^{-k_2}$ to be Eisenstein like:

$$g_{k_1}(\tau) = \sum_{m,n \ge 1} a_1(m)b_1(n)n^{k_1-1}q^{mn} \quad \text{and} \quad g_{k_2}(\tau) = \sum_{m,n \ge 1} a_2(m)b_2(n)n^{k_2-1}q^{mn}.$$

We have

$$L(f, k_0) = L(g_{k_1}\hat{g}_{k_2}, k_0) = \frac{1}{(k_0-1)!} \int_0^1 g_{k_1}\hat{g}_{k_2} \log^{k_0-1} q \, \frac{dq}{q}$$

$$= \frac{(-1)^{k_0-1}(2\pi)^{k_0}}{(k_0-1)!} \int_0^\infty g_{k_1}(it)\hat{g}_{k_2}(it)t^{k_0-1} \, dt$$

$$= \frac{(-1)^{k_0-1}(2\pi)^{k_0}}{(k_0-1)!N^{k_2/2}} \int_0^\infty g_{k_1}(it)g_{k_2}(i/(Nt))t^{k_0-k_2-1} \, dt$$

$$= \frac{(-1)^{k_0-1}(2\pi)^{k_0}}{(k_0-1)!N^{k_2/2}} \int_0^\infty \sum_{m_1,n_1 \ge 1} a_1(m_1)b_1(n_1)n_1^{k_1-1}e^{-2\pi m_1 n_1 t}$$

$$\times \sum_{m_2,n_2 \ge 1} a_2(m_2)b_2(n_2)n_2^{k_2-1}e^{-2\pi m_2 n_2/(Nt)}t^{k_0-k_2-1} dt$$

$$= \frac{(-1)^{k_0-1}(2\pi)^{k_0}}{(k_0-1)!N^{k_2/2}} \sum_{m_1,n_1,m_2,n_2 \ge 1} a_1(m_1)b_1(n_1)a_2(m_2)b_2(n_2)n_1^{k_1-1}n_2^{k_2-1}$$

$$\times \int_0^\infty \exp\left(-2\pi\left(m_1 n_1 t + \frac{m_2 n_2}{Nt}\right)\right)t^{k_0-k_2-1}dt;$$

the interchange of integration and summation is legitimate because of the exponential decay of the integrand at the endpoints. After performing the change of variable $t = n_2 u/n_1$ and interchanging summation and integration back again, we obtain

$$L(f,k_0) = \frac{(-1)^{k_0-1}(2\pi)^{k_0}}{(k_0-1)!N^{k_2/2}} \sum_{m_1,n_1,m_2,n_2 \geq 1} a_1(m_1)b_1(n_1)a_2(m_2)b_2(n_2)n_1^{k_1+k_2-k_0-1}n_2^{k_0-1}$$

$$\times \int_0^\infty \exp\left(-2\pi\left(m_1n_2u + \frac{m_2n_1}{Nu}\right)\right)u^{k_0-k_2-1}du$$

$$= \frac{(-1)^{k_0-1}(2\pi)^{k_0}}{(k_0-1)!N^{k_2/2}} \int_0^\infty \sum_{m_1,n_2 \geq 1} a_1(m_1)b_2(n_2)n_2^{k_0-1}e^{-2\pi m_1 n_2 u}$$

$$\times \sum_{m_2,n_1 \geq 1} a_2(m_2)b_1(n_1)n_1^{k_1+k_2-k_0-1}e^{-2\pi m_2 n_1/(Nu)}u^{k_0-k_2-1}du$$

$$= \frac{(-1)^{k_0-1}(2\pi)^{k_0}}{(k_0-1)!N^{k_2/2}} \int_0^\infty g_{k_0}(iu)g_{k_1+k_2-k_0}(i/(Nu))u^{k_0-k_2-1}du.$$

Assuming a modular transformation of the Eisenstein-like series $g_{k_1+k_2-k_0}(\tau)$ under $\tau \mapsto -1/(N\tau)$, we can realise the resulting integral as $c\pi^{k_0-k_1}L(g_{k_0}\hat{g}_{k_1+k_2-k_0}, k_1)$, where c is algebraic (plus some extra terms when $g_{k_1+k_2-k_0}(\tau)$ is an Eichler integral). Alternatively, if $g_{k_0}(\tau)$ transforms under the involution, we perform the transformation and switch to the variable $v = 1/(Nu)$ to arrive at $c\pi^{k_0-k_2}L(\hat{g}_{k_0}g_{k_1+k_2-k_0}, k_2)$. In both cases we obtain an identity which relates the starting L-value $L(f,k_0)$ to a different "L-value" of a modular-like object of the same weight.

The case $k_1 = k_2 = 1$ and $k_0 = 2$, discussed in [8, 9] and in Sect. 2 above, allows one to reduce the L-values to periods. As we will see in Sect. 4, the period(d)ness can be achieved in a more general situation, based on the fact that Eichler integrals are related to solutions of inhomogeneous linear differential equations.

4 L(E, 3)

To manipulate with $L(E,3)$ for a conductor 32 elliptic curve, we use again $L(E,3) = L(f,3)$ with $f(\tau) = \eta_4^2\eta_8^2$ and write the decomposition in (3) as

$$f(it) = \frac{1}{2t} \sum_{m_1,n_1 \geq 1} b(m_1)a(n_1)e^{-2\pi m_1 n_1 t} \sum_{m_2,n_2 \geq 1} b(m_2)a(n_2)e^{-2\pi m_2 n_2/(32t)}.$$

Then

$$L(E,3) = L(f,3) = \frac{1}{2}\int_0^1 f \log^2 q \frac{dq}{q} = 4\pi^3 \int_0^\infty f(it)t^2 dt$$

$$= 2\pi^3 \int_0^\infty \sum_{m_1,n_1,m_2,n_2 \geq 1} b(m_1)a(n_1)b(m_2)a(n_2)$$

$$\times \exp\left(-2\pi\left(m_1n_1t + \frac{m_2n_2}{32t}\right)\right)t\,dt$$

$$= 2\pi^3 \sum_{m_1,n_1,m_2,n_2 \geq 1} b(m_1)a(n_1)b(m_2)a(n_2)$$

$$\times \int_0^\infty \exp\left(-2\pi\left(m_1 n_1 t + \frac{m_2 n_2}{32t}\right)\right) t\, dt$$

(here we perform the change of variable $t = n_2 u / n_1$)

$$= 2\pi^3 \sum_{m_1,n_1,m_2,n_2 \geq 1} \frac{b(m_1)a(n_1)b(m_2)a(n_2)n_2^2}{n_1^2}$$

$$\times \int_0^\infty \exp\left(-2\pi\left(m_1 n_2 u + \frac{m_2 n_1}{32u}\right)\right) u\, du$$

$$= 2\pi^3 \int_0^\infty \sum_{m_1,n_2 \geq 1} b(m_1)a(n_2)n_2^2 e^{-2\pi m_1 n_2 u}$$

$$\times \sum_{m_2,n_1 \geq 1} \frac{b(m_2)a(n_1)}{n_1^2} e^{-2\pi m_2 n_1 /(32u)} u\, du.$$

Furthermore,

$$\sum_{m,n \geq 1} b(m)a(n)n^2 q^{mn} = \sum_{\substack{m,n \geq 1 \\ m\ \text{odd}}} \left(\frac{-4}{n}\right) n^2 q^{mn} = \frac{\eta_2^8 \eta_8^4}{\eta_4^6},$$

$$\sum_{m,n \geq 1} b(m)a(n)m^2 q^{mn} = \sum_{\substack{m,n \geq 1 \\ m\ \text{odd}}} \left(\frac{-4}{n}\right) m^2 q^{mn} = \frac{\eta_4^{18}}{\eta_2^8 \eta_8^4},$$

so that

$$r(\tau) := \sum_{m,n \geq 1} \frac{b(m)a(n)}{n^2} q^{mn} = \delta^{-2}\left(\frac{\eta_4^{18}}{\eta_2^8 \eta_8^4}\right). \tag{5}$$

Continuing the previous computation,

$$L(E,3) = 2\pi^3 \int_0^\infty \frac{\eta_2^8 \eta_8^4}{\eta_4^6}\bigg|_{\tau=iu} \cdot r(i/(32u))\, u\, du$$

(we apply the involution to the eta quotient)

$$= \frac{\pi^3}{8} \int_0^\infty \frac{\eta_4^4 \eta_{16}^8}{\eta_8^6} r(\tau)\bigg|_{\tau=i/(32u)} \frac{du}{u^2}$$

(we change the variable $u = 1/(32v)$)

$$= 4\pi^3 \int_0^\infty \frac{\eta_4^4 \eta_{16}^8}{\eta_8^6} r(\tau) \bigg|_{\tau=iv} dv.$$

This is so far the end of the algorithm we have discussed in Sect. 3. In order to show that the resulting integral is a period, we require to do one step more. As in Sect. 2 we make a modular parametrisation; this time we take the modular function $x(\tau) := 4\eta_2^4 \eta_8^8 / \eta_4^{12}$ which ranges from 0 to 1 when τ goes from $i\infty$ to 0. Then

$$\delta x = \frac{4\eta_2^{12} \eta_8^8}{\eta_4^{16}}, \quad (1-x^2)^{1/4} = \frac{\eta_2^4 \eta_8^2}{\eta_4^6}, \quad s(x) := \frac{(1-\sqrt{1-x^2})^2}{x(1-x^2)^{3/4}} = \frac{16\eta_4^{10} \eta_{16}^8}{\eta_2^8 \eta_8^{10}}.$$

Furthermore, the substitution $z = x^2(\tau)$ into the hypergeometric function

$$F(z) := {}_2F_1\left(\begin{matrix} \frac{1}{2}, \frac{1}{2} \\ 1 \end{matrix} \bigg| z\right) = \frac{2}{\pi} \int_0^1 \frac{dy}{\sqrt{(1-y^2)(1-zy^2)}}$$

results in the modular form

$$\varphi(\tau) := F(x^2) = \sum_{n=0}^\infty \binom{2n}{n}^2 \left(\frac{x}{4}\right)^{2n} = \frac{\eta_4^{10}}{\eta_2^4 \eta_8^4}$$

of weight 1. Because $F(z)$ (along with $F(1-z)$) satisfy the hypergeometric differential equation

$$z(1-z)\frac{d^2 F}{dz^2} + (1-2z)\frac{dF}{dz} - \frac{1}{4}F = 0,$$

it is not hard to write down the corresponding linear second-order differential operator

$$\mathscr{L} := x(1-x^2)\frac{d^2}{dx^2} + (1-3x^2)\frac{d}{dx} - x$$

(in terms of x) such that $\mathscr{L}\varphi = 0$.

With this notation in mind, we obtain

$$L(E, 3) = \pi^3 \int_0^\infty \frac{\eta_4^{10} \eta_{16}^8}{\eta_2^8 \eta_8^{10}} \varphi(\tau) r(\tau) \delta x \bigg|_{\tau=iv} dv$$

$$= \frac{\pi^3}{16} \int_0^\infty s(x(\tau)) \varphi(\tau) r(\tau) \delta x \bigg|_{\tau=iv} dv, \qquad (6)$$

and at this point we make an observation that the function $h(\tau) := 4\varphi(\tau)r(\tau)$ solves the inhomogeneous differential equation

$$\mathscr{L}h = \frac{1}{1-x^2} \qquad \left(\text{which is nothing but } [10,13] \quad \frac{\delta^2 r}{\delta x \cdot \varphi} = \frac{\eta_4^{24}}{4\eta_2^{16}\eta_8^8}\right),$$

so that it can be written as an integral using the method of variation of parameters:

$$h = \frac{\pi}{2}\left(F(x^2)\int \frac{F(1-x^2)}{1-x^2}\,dx - F(1-x^2)\int \frac{F(x^2)}{1-x^2}\,dx\right)$$

$$= \frac{\pi x}{2}\int_0^1 \frac{F(x^2)F(1-x^2w^2) - F(1-x^2)F(x^2w^2)}{1-x^2w^2}\,dw.$$

This implies that

$$L(E,3) = \frac{\pi^2}{128}\int_0^1 s(x)\,h(x)\,dx,$$

an expression which can be clearly transformed into a (complicated) real integral.

The recipe of expressing Eisenstein series of negative weight via solutions of inhomogeneous linear differential equations is standard [13] and applicable in any situation similar to the one considered above. The Eisenstein series (5) of weight -1 however possesses a different treatment because of a special formula due to Ramanujan [4, eq. (2·2)]:

$$r(\tau) = \sum_{\substack{m,n\geq 1\\ m\ \mathrm{odd}}} \left(\frac{-4}{n}\right)\frac{q^{mn}}{n^2} = \frac{\tilde{x}G(-\tilde{x}^2)}{4F(-\tilde{x}^2)},$$

where $\tilde{x}(\tau) := 4\eta_8^4/\eta_2^4$, $F(-\tilde{x}^2) = \eta_2^4/\eta_4^2$ and

$$G(z) := {}_3F_2\!\left(\begin{matrix}1,1,1\\ \frac{3}{2},\frac{3}{2}\end{matrix}\,\bigg|\,z\right) = \frac{1}{4}\int_0^1\int_0^1 \frac{(1-x_1)^{-1/2}(1-x_2)^{-1/2}}{1-zx_1x_2}\,dx_1\,dx_2.$$

The latter integral also gives the analytic continuation of the hypergeometric ${}_3F_2$-series to the domain $\mathrm{Re}\,z < 1$ (see [6, Lemma 2]); the change $y = (1-x_1)^{1/2}$, $w = (1-x_2)^{1/2}$ translates the integral into the form

$$G(z) = \int_0^1\int_0^1 \frac{dy\,dw}{1-z(1-y^2)(1-w^2)}.$$

Rolling back to the modular function $x(\tau) = 4\eta_2^4\eta_8^8/\eta_4^{12}$ and noting that $\tilde{x} = x/\sqrt{1-x^2}$, we may now write (8) as

$$L(E,3) = \frac{\pi^3}{64} \int_0^\infty \frac{s(x(\tau))x(\tau)}{1-x(\tau)^2} G\left(-\frac{x(\tau)^2}{1-x(\tau)^2}\right) \delta x \Big|_{\tau=iv} dv.$$

After performing the modular substitution $x = x(\tau)$, we finally arrive at

Theorem 2. *For an elliptic curve E of conductor 32,*

$$L(E,3) = \frac{\pi^2}{128} \int_0^1 \frac{(1-\sqrt{1-x^2})^2}{(1-x^2)^{3/4}} dx \int_0^1 \int_0^1 \frac{dy\, dw}{1-x^2(1-(1-y^2)(1-w^2))}$$

$$= 0.9826801478\ldots.$$

5 Hypergeometric Evaluations

A remarkable feature of the integrals in Theorems 1 and 2 is the possibility to reduce them further to hypergeometric functions [3].

Theorem 3. *For an elliptic curve E of conductor 32,*

$$L(E,2) = \frac{\pi^{1/2}\Gamma(\frac{1}{4})^2}{96\sqrt{2}} \, {}_3F_2\left(\begin{matrix} 1,1,\frac{1}{2} \\ \frac{7}{4},\frac{3}{2} \end{matrix}\Big| 1\right) + \frac{\pi^{1/2}\Gamma(\frac{3}{4})^2}{8\sqrt{2}} \, {}_3F_2\left(\begin{matrix} 1,1,\frac{1}{2} \\ \frac{5}{4},\frac{3}{2} \end{matrix}\Big| 1\right), \qquad (7)$$

$$L(E,3) = \frac{\pi^{3/2}\Gamma(\frac{1}{4})^2}{768\sqrt{2}} \, {}_4F_3\left(\begin{matrix} 1,1,1,\frac{1}{2} \\ \frac{7}{4},\frac{3}{2},\frac{3}{2} \end{matrix}\Big| 1\right) + \frac{\pi^{3/2}\Gamma(\frac{3}{4})^2}{32\sqrt{2}} \, {}_4F_3\left(\begin{matrix} 1,1,1,\frac{1}{2} \\ \frac{5}{4},\frac{3}{2},\frac{3}{2} \end{matrix}\Big| 1\right)$$

$$+ \frac{\pi^{3/2}\Gamma(\frac{1}{4})^2}{256\sqrt{2}} \, {}_4F_3\left(\begin{matrix} 1,1,1,\frac{1}{2} \\ \frac{3}{4},\frac{3}{2},\frac{3}{2} \end{matrix}\Big| 1\right). \qquad (8)$$

Proof. In the integral representation for $L(E,2)$ in Theorem 1, write

$$\log\frac{1+x}{1-x} = \frac{2}{3}x^3 \, {}_2F_1\left(\begin{matrix} \frac{3}{4},1 \\ \frac{7}{4} \end{matrix}\Big| x^4\right) + 2x\, {}_2F_1\left(\begin{matrix} \frac{1}{4},1 \\ \frac{5}{4} \end{matrix}\Big| x^4\right)$$

and change the variable $x^4 = x_0$ to get

$$L(E,2) = \frac{\pi}{48} \int_0^1 \left(x_0^{1/4} \, {}_2F_1\left(\begin{matrix} \frac{3}{4},1 \\ \frac{7}{4} \end{matrix}\Big| x_0\right) + 3x_0^{-1/4} \, {}_2F_1\left(\begin{matrix} \frac{1}{4},1 \\ \frac{5}{4} \end{matrix}\Big| x_0\right)\right)(1-x_0)^{-1/2}dx_0$$

$$= \frac{\pi^{3/2}}{48} \frac{\Gamma(\frac{5}{4})}{\Gamma(\frac{7}{4})} \, {}_3F_2\left(\begin{matrix} \frac{3}{4},\frac{5}{4},1 \\ \frac{7}{4},\frac{7}{4} \end{matrix}\Big| 1\right) + \frac{\pi^{3/2}}{16} \frac{\Gamma(\frac{3}{4})}{\Gamma(\frac{5}{4})} \, {}_3F_2\left(\begin{matrix} \frac{1}{4},\frac{3}{4},1 \\ \frac{5}{4},\frac{5}{4} \end{matrix}\Big| 1\right);$$

the representation (7) now follows from application of Thomae's transformation [1, Eq. 3.2.(1)] to the both ${}_3F_2$ series.

In the integral of Theorem 2, let $x_0 = x^2$, $x_1 = 1 - y^2$ and $x_2 = 1 - w^2$:

$$L(E,3) = \frac{\pi^2}{1024} \int_0^1 x_0^{-1/2} \left((1-x_0)^{-3/4} - 2(1-x_0)^{-1/4} + (1-x_0)^{1/4} \right) dx_0$$

$$\times \int_0^1 \int_0^1 \frac{(1-x_1)^{-1/2}(1-x_2)^{-1/2}}{1 - x_0(1 - x_1 x_2)} \, dx_1 \, dx_2.$$

First consider the integral (see [1, Eqs. 1.5.(1) and 1.4.(1)])

$$\int_0^1 \frac{x_0^{-1/2}(1-x_0)^{a-1}}{1 - x_0 z} \, dx_0 = \frac{\Gamma(\frac{1}{2})\Gamma(a)}{\Gamma(a+\frac{1}{2})} \, {}_2F_1 \left(\begin{matrix} 1, \frac{1}{2} \\ a+\frac{1}{2} \end{matrix} \middle| z \right)$$

$$= \frac{\Gamma(\frac{1}{2})\Gamma(a-1)}{\Gamma(a-\frac{1}{2})} \, {}_2F_1 \left(\begin{matrix} 1, \frac{1}{2} \\ 2-a \end{matrix} \middle| 1-z \right)$$

$$+ \Gamma(a)\Gamma(1-a)(1-z)^{a-1} {}_2F_1 \left(\begin{matrix} a-\frac{1}{2}, a \\ a \end{matrix} \middle| 1-z \right)$$

$$= \frac{\sqrt{\pi}\,\Gamma(a-1)}{\Gamma(a-\frac{1}{2})} \, {}_2F_1 \left(\begin{matrix} 1, \frac{1}{2} \\ 2-a \end{matrix} \middle| 1-z \right) + \frac{\pi}{\sin \pi a} \frac{(1-z)^{a-1}}{z^{a-1/2}}$$

for $z = 1 - x_1 x_2$. Secondly,

$$\int_0^1 \int_0^1 (1-x_1)^{-1/2}(1-x_2)^{-1/2} {}_2F_1 \left(\begin{matrix} 1, \frac{1}{2} \\ 2-a \end{matrix} \middle| x_1 x_2 \right) dx_1 \, dx_2$$

$$= 4 \, {}_4F_3 \left(\begin{matrix} 1, 1, 1, \frac{1}{2} \\ 2-a, \frac{3}{2}, \frac{3}{2} \end{matrix} \middle| 1 \right)$$

and

$$\int_0^1 \int_0^1 \frac{x_1^{a-1}(1-x_1)^{-1/2} x_2^{a-1}(1-x_2)^{-1/2}}{(1-x_1 x_2)^{a-1/2}} \, dx_1 \, dx_2$$

$$= \left(\frac{\Gamma(\frac{1}{2})\Gamma(a)}{\Gamma(a+\frac{1}{2})} \right)^2 {}_3F_2 \left(\begin{matrix} a, a, a-\frac{1}{2} \\ a+\frac{1}{2}, a+\frac{1}{2} \end{matrix} \middle| 1 \right)$$

$$= 2\pi^{1/2}\Gamma(a)\Gamma(\tfrac{3}{2} - a) \cdot {}_3F_2 \left(\begin{matrix} \frac{1}{2}, \frac{1}{2}, \frac{3}{2} - a \\ 1, \frac{3}{2} \end{matrix} \middle| 1 \right),$$

where Thomae's transformation [1, Eq. 3.2.(1)] is used on the last step.

The computation above means that if we take

$$I_1(a) := \frac{\pi^{5/2}\Gamma(a-1)}{256\Gamma(a-\frac{1}{2})} \, {}_4F_3\left(\begin{matrix} 1,1,1,\frac{1}{2} \\ 2-a,\frac{3}{2},\frac{3}{2} \end{matrix} \,\middle|\, 1\right),$$

$$I_2(a) := \frac{\pi^{7/2}\Gamma(a)\Gamma(\frac{3}{2}-a)}{512\sin\pi a} \, {}_3F_2\left(\begin{matrix} \frac{1}{2},\frac{1}{2},\frac{3}{2}-a \\ 1,\frac{3}{2} \end{matrix} \,\middle|\, 1\right),$$

then

$$L(E,3) = \left(I_1(\tfrac{1}{4}) - 2I_1(\tfrac{3}{4}) + I_1(\tfrac{5}{4})\right) + \left(I_2(\tfrac{1}{4}) - 2I_2(\tfrac{3}{4}) + I_2(\tfrac{5}{4})\right).$$

For $I_2(\tfrac{3}{4})$, the Watson–Whipple summation [1, Eq. 3.3.(1)] results in

$${}_3F_2\left(\begin{matrix} \frac{1}{2},\frac{1}{2},\frac{3}{4} \\ 1,\frac{3}{2} \end{matrix} \,\middle|\, 1\right) = \frac{\Gamma(\frac{1}{2})\Gamma(\frac{5}{4})}{\Gamma(\frac{3}{4})},$$

so that $I_2(\tfrac{3}{4}) = \pi^5/1024$. Furthermore,

$$\begin{aligned}
I_2(\tfrac{1}{4}) + I_2(\tfrac{5}{4}) &= \frac{\pi^{7/2}\Gamma(\frac{1}{4})\Gamma(\frac{5}{4})}{256\sqrt{2}}\left({}_3F_2\left(\begin{matrix} \frac{1}{2},\frac{1}{2},\frac{5}{4} \\ 1,\frac{3}{2} \end{matrix} \,\middle|\, 1\right) - {}_3F_2\left(\begin{matrix} \frac{1}{2},\frac{1}{2},\frac{1}{4} \\ 1,\frac{3}{2} \end{matrix} \,\middle|\, 1\right)\right) \\
&= \frac{\pi^{7/2}\Gamma(\frac{1}{4})^2}{1024\sqrt{2}} \sum_{n=0}^{\infty} \frac{(\frac{1}{2})_n^2 \cdot ((\frac{5}{4})_n - (\frac{1}{4})_n)}{(1)_n(\frac{3}{2})_n n!} \\
&= \frac{\pi^{7/2}\Gamma(\frac{1}{4})^2}{1024\sqrt{2}} \sum_{n=1}^{\infty} \frac{(\frac{1}{2})_n^2(\frac{5}{4})_{n-1}}{(1)_n(\frac{3}{2})_n (n-1)!} \\
&= \frac{\pi^{7/2}\Gamma(\frac{1}{4})^2}{1024\sqrt{2}\cdot 6} \, {}_3F_2\left(\begin{matrix} \frac{3}{2},\frac{3}{2},\frac{5}{4} \\ 2,\frac{5}{2} \end{matrix} \,\middle|\, 1\right) \\
&= \frac{\pi^{7/2}\Gamma(\frac{1}{4})^2}{1024\sqrt{2}\cdot 6} \, \frac{\Gamma(\frac{1}{2})\Gamma(\frac{1}{4})\Gamma(\frac{7}{4})}{\Gamma(\frac{5}{4})^2} = \frac{\pi^5}{512},
\end{aligned}$$

where again the Watson–Whipple summation was applied.

To summarise, $L(E,3) = I_1(\tfrac{1}{4}) - 2I_1(\tfrac{3}{4}) + I_1(\tfrac{5}{4})$, which is exactly Equation (8). $\square$

Theorem 3 produces amazingly similar hypergeometric forms of $L(E,2)$ and $L(E,3)$. In the notation

$$F_k(a) := \frac{\pi^{k-1/2}\Gamma(a)}{2^{3k-1}\Gamma(a+\frac{1}{2})} \, {}_{k+1}F_k\left(\begin{matrix} \overbrace{1,\ldots,1}^{k\text{ times}},\frac{1}{2} \\ a+\frac{1}{2},\underbrace{\frac{3}{2},\ldots,\frac{3}{2}}_{k-1\text{ times}} \end{matrix} \,\middle|\, 1\right),$$

relations (7) and (8) can be alternatively written as

$$L(E,2) = F_2\left(\tfrac{5}{4}\right) + F_2\left(\tfrac{3}{4}\right) \quad \text{and} \quad L(E,3) = F_3\left(\tfrac{5}{4}\right) + 2F_3\left(\tfrac{3}{4}\right) + F_3\left(\tfrac{1}{4}\right). \qquad (9)$$

In view of the known formula

$$L(E,1) = \frac{\pi^{-1/2}\Gamma(\tfrac{1}{4})^2}{8\sqrt{2}} = \frac{\pi^{-1/2}\Gamma(\tfrac{1}{4})^2}{24\sqrt{2}} \; {}_3F_2\left(\begin{matrix} 1, \tfrac{1}{2} \\ \tfrac{7}{4} \end{matrix} \,\middle|\, 1\right) = 2F_1\left(\tfrac{5}{4}\right),$$

we can conclude that, for $k = 1$, 2 or 3, the L-value $L(E,k)$ can be written as a (simple) $\mathbb{Q}$-linear combination of $F_k\left(\tfrac{7}{4} - \tfrac{m}{2}\right)$ for $m = 1, \ldots, k$. However, this pattern does not seem to work for $k > 3$.

The formulae (9) in turn can be transformed into the period representations of $L(E,2)$ and $L(E,3)$ which differ from the ones in Theorems 2 and 3:

$$L(E,2) = \frac{\pi}{16} \int_0^1 \frac{1 + \sqrt{1 - x^2}}{(1 - x^2)^{1/4}} \, dx \int_0^1 \frac{dy}{1 - x^2(1 - y^2)},$$

$$L(E,3) = \frac{\pi^2}{128} \int_0^1 \frac{(1 + \sqrt{1 - x^2})^2}{(1 - x^2)^{3/4}} \, dx \int_0^1 \int_0^1 \frac{dy\,dw}{1 - x^2(1 - y^2)(1 - w^2)}.$$

An interesting problem is identifying the (linear combination of the) hypergeometric series involved in the right-hand side of (8) with a linear combinations of 3-variable Mahler measures. There are related results in [7], although not written hypergeometrically enough.

A crucial ingredient in deducing the integral representation in Theorem 2 is the hypergeometric form of an Eisenstein series of negative weight given in [4]. Are there other results of this type?

Acknowledgements I am thankful to Anton Mellit, Mat Rogers, Evgeny Shinder, Masha Vlasenko and James Wan for fruitful conversations on the subject and to Don Zagier for his encouragement to isolate the L-series transformation part from [8,9]. I thank the anonymous referee for his careful reading of an earlier version and for his suggesting some useful corrections.

The principal part of this work was done during my visit in the Max Planck Institute for Mathematics (Bonn). I would like to thank the staff of the institute for hospitality and enjoyable working conditions.

This work is supported by the Max Planck Institute for Mathematics (Bonn, Germany) and the Australian Research Council (grant DP110104419).

References

1. W.N. Bailey, *Generalized Hypergeometric Series*. Cambridge Tracts in Math., vol 32 (Cambridge Univ. Press, Cambridge, 1935); 2nd reprinted edn. (Stechert-Hafner, New York/London, 1964)
2. J.L. Berggren, J.M. Borwein, P. Borwein, *Pi: A Source Book*, 3rd edn. (Springer, New York, 2004)

3. J.M. Borwein, R.E. Crandall, Closed forms: what they are and why we care. Notices Am. Math. Soc. **60**, 50–65 (2013)
4. W. Duke, Some entries in Ramanujan's notebooks. Math. Proc. Camb. Phil. Soc. **144**, 255–266 (2008)
5. M. Kontsevich, D. Zagier, Periods, in *Mathematics Unlimited—2001 and Beyond* (Springer, Berlin, 2001), pp. 771–808
6. Yu.V. Nesterenko, Integral identities and constructions of approximations to zeta-values. J. Theor. Nombres Bordeaux **15**, 535–550 (2003)
7. M.D. Rogers, A study of inverse trigonometric integrals associated with three-variable Mahler measures, and some related identities. J. Number Theory **121**, 265–304 (2006)
8. M. Rogers, W. Zudilin, From L-series of elliptic curves to Mahler measures. Compos. Math. **148**, 385–414 (2012)
9. M. Rogers, W. Zudilin, On the Mahler measure of $1 + X + 1/X + Y + 1/Y$. Intern. Math. Research Notices, 22 pp. (2013, in press, doi:10.1093/imrn/rns285)
10. E. Shinder, M. Vlasenko, Linear Mahler measures and double L-values of modular forms (2012), 22 pp. Preprint at http://arxiv.org/abs/1206.1454
11. A. van der Poorten, A proof that Euler missed... Apéry's proof of the irrationality of $\zeta(3)$. Math. Intell. **1**, 195–203 (1978/1979)
12. A. Weil, Remarks on Hecke's lemma and its use, in *Algebraic Number Theory. Kyoto Internat. Sympos., Res. Inst. Math. Sci., Univ. Kyoto*, Kyoto 1976 (Japan Soc. Promotion Sci., Tokyo, 1977), pp. 267–274
13. Y. Yang, Apéry limits and special values of L-functions. J. Math. Anal. Appl. **343**, 492–513 (2008)

Printed in the United States
By Bookmasters